THE GEOMETRY OF MODULI SPACES
OF SHEAVES
Second Edition

Now back in print, this highly regarded book has been updated to reflect recent advances in the theory of semistable coherent sheaves and their moduli spaces, which include moduli spaces in positive characteristic, moduli spaces of principal bundles and of complexes, Hilbert schemes of points on surfaces, derived categories of coherent sheaves, and moduli spaces of sheaves on Calabi-Yau threefolds. The authors review changes in the field since the publication of the original edition in 1997 and point the reader towards further literature. References have been brought up to date and errors removed.

Developed from the authors' lectures, this book is ideal as a text for graduate students as well as a valuable resource for any mathematician with a background in algebraic geometry who wants to learn more about Grothendieck's approach.

'This book fills a great need: it is almost the only place the foundations of the moduli theory of sheaves on algebraic varieties appears in any kind of expository form. The material is of basic importance to many further developments: Donaldson-Thomas theory, mirror symmetry, and the study of derived categories.'
Rahul Pandharipande, Princeton University

'This is a wonderful book; it's about time it was available again. It is the definitive reference for the important topics of vector bundles, coherent sheaves, moduli spaces and geometric invariant theory; perfect as both an introduction to these subjects for beginners, and as a reference book for experts. Thorough but concise, well written and accurate, it is already a minor modern classic. The new edition brings the presentation up to date with discussions of more recent developments in the area.'
Richard Thomas, Imperial College London

THE GEOMETRY OF MODULI SPACES
OF SHEAVES

Second Edition

DANIEL HUYBRECHTS AND MANFRED LEHN

CAMBRIDGE
UNIVERSITY PRESS

University Printing House, Cambridge CB2 8BS, United Kingdom

Cambridge University Press is part of the University of Cambridge.

It furthers the University's mission by disseminating knowledge in the pursuit of education, learning and research at the highest international levels of excellence.

www.cambridge.org
Information on this title: www.cambridge.org/9780521134200

First edition published in 1997 by Vieweg
Second Edition 2010
Reprinted 2011

A catalogue record for this publication is available from the British Library

Library of Congress Cataloguing in Publication data

Huybrechts, Daniel.
The geometry of moduli spaces of sheaves / Daniel Huybrechts, Manfred Lehn. – 2nd ed.
p.   cm. – (Cambridge mathematical library)
Includes bibliographical references and index.
ISBN 978-0-521-13420-0 (Pbk.)
1. Sheaf theory.   2. Moduli theory.   3. Surfaces, Algebraic.
I. Lehn, Manfred.   II. Title.   III. Series.
QA612.36.H89 2010
514′.224–dc22

2010001103

ISBN 978-0-521-13420-0 Paperback

Contents

Preface to the second edition

The first edition of this book has been out of print for some years now. But as the book still seems a useful source for the main techniques, results and open problems in this area and as it has been appreciated by newcomers wanting to learn the material from scratch, there have been frequent requests for a new edition.

Since the first appearance of the book, the field has branched out in various directions: moduli spaces in positive characteristic, moduli spaces of principal bundles and of complexes, Hilbert schemes of points on surfaces, derived categories of coherent sheaves, moduli spaces of sheaves on Calabi-Yau threefolds, and many, many others. In the new appendices we make comments on some of these interesting new research directions without aiming at completeness nor going into the technical details. The main text has been left unchanged as far as possible. We used however the opportunity to correct a number of mistakes known to us and tried to improve the presentation at certain places.

We are aware of a number of excellent recent textbooks that are closely related to the topics treated here. Friedman's book [322] is a good point to start for anybody who wants to learn about vector bundles on surfaces. It combines an introduction to the theory of algebraic surfaces with a study of vector bundles. Moduli spaces of vector bundles on curves are constructed and discussed in Le Potier's book [367]. The monograph of Schmitt [429] gives a very detailed account of GIT phenomena encountered in the construction of moduli spaces of sheaves. Two further new textbooks on GIT by Dolgachev and Mukai are highly recommended, [312] and [394].

Many people have indicated inaccuracies and mistakes in the first edition and have made valuable suggestions on how to improve the text. We would like to thank in particular Holger Brenner, Alastair King, Adrian Langer, and Alexander Schmitt.

We wish to thank Cambridge University Press for convincing us that a new edition was a good idea and for letting us decide to what extent we wanted to modify and augment the previous version.

Preface to the first edition

The topic of this book is the theory of semistable coherent sheaves on a smooth algebraic surface and of moduli spaces of such sheaves. The content ranges from the definition of a semistable sheaf and its basic properties over the construction of moduli spaces to the birational geometry of these moduli spaces. The book is intended for readers with some background in Algebraic Geometry, as for example provided by Hartshorne's textbook [98].

There are at least three good reasons to study moduli spaces of sheaves on surfaces. Firstly, they provide examples of higher dimensional algebraic varieties with a rich and interesting geometry. In fact, in some regions in the classification of higher dimensional varieties the only known examples are moduli spaces of sheaves on a surface. The study of moduli spaces therefore sheds light on some aspects of higher dimensional algebraic geometry. Secondly, moduli spaces are varieties naturally attached to any surface. The understanding of their properties gives answers to problems concerning the geometry of the surface, e.g. Chow group, linear systems, etc. From the mid-eighties till the mid-nineties most of the work on moduli spaces of sheaves on a surface was motivated by Donaldson's ground breaking results on the relation between certain intersection numbers on the moduli spaces and the differentiable structure of the four-manifold underlying the surface. Although the interest in this relation has subsided since the introduction of the extremely powerful Seiberg-Witten invariants in 1994, Donaldson's results linger as a third major motivation in the background; they throw a bridge from algebraic geometry to gauge theory and differential geometry.

Part I of this book gives an introduction to the general theory of semistable sheaves on varieties of arbitrary dimension. We tried to keep this part to a large extent self-contained. In Part II, which deals almost exclusively with sheaves on algebraic surfaces, we occasionally sketch or even omit proofs. This area of research is still developing and we feel that some of the results are not yet in their final form.

Some topics are only touched upon. Many interesting results are missing, e.g. the Fourier-Mukai transformation, Picard groups of moduli spaces, bundles on the projective plane (or more generally on projective spaces, see [230]), computation of Donaldson polynomials on algebraic surfaces, gauge theoretical aspects of moduli spaces (see the book of Friedman and Morgan [71]). We also wish to draw the readers' attention to the forthcoming book of R. Friedman [69].

Usually, we give references and sometimes historical remarks in the Comments at the end of each chapter. If not stated otherwise, all results should be attributed to others. We apologize for omissions and inaccuracies that we may have incorporated in presenting their work.

These notes grew out of lectures delivered by the authors at a summer school at Humboldt-Universität zu Berlin in September 1995. Every lecture was centered around one topic. In writing up these notes we tried to maintain this structure. By adding the necessary background to the orally presented material, some chapters have grown out of size and the global structure of the book has become rather non-linear. This has two effects. It should be possible to read some chapters of Part II without going through all the general theory presented in Part I. On the other hand, some results had to be referred to before they were actually introduced.

We wish to thank H. Kurke for the invitation to Berlin and I. Quandt for the organization of the summer school. We are grateful to F. Hirzebruch for his encouragement to publish these notes in the MPI-subseries of the Aspects of Mathematics. We also owe many thanks to S. Bauer and the SFB 343 at Bielefeld, who supported the preparation of the manuscript.

Many people have read portions and preliminary versions of the text. We are grateful for their comments and criticism. In particular, we express our gratitude to: S. Bauer, V. Brinzanescu, R. Brussee, H. Esnault, L. Göttsche, G. Hein, L. Hille, S. Kleiman, A. King, J. Klein, J. Li, S. Müller-Stach, and K. O'Grady.

While working on these notes the first author was supported by the Max-Planck-Institut für Mathematik (Bonn), the Institute for Advanced Study (Princeton), the Institut des Hautes Etudes Scientifiques (Bures-sur-Yvette), the Universität Essen and by a grant from the DFG. The second author was supported by the SFB 343 'Diskrete Strukturen in der Mathematik' at the Universität Bielefeld.

Bielefeld, December 1996   Daniel Huybrechts, Manfred Lehn

Introduction

It is one of the deep problems in algebraic geometry to determine which cohomology classes on a projective variety can be realized as Chern classes of vector bundles. In low dimensions the answer is known. On a curve X any class $c_1 \in H^2(X, \mathbb{Z})$ can be realized as the first Chern class of a vector bundle of prescribed rank r. In dimension two the existence of bundles is settled by Schwarzenberger's result, which says that for given cohomology classes $c_1 \in H^2(X, \mathbb{Z}) \cap H^{1,1}(X)$ and $c_2 \in H^4(X, \mathbb{Z}) \cong \mathbb{Z}$ on a complex surface X there exists a vector bundle of prescribed rank ≥ 2 with first and second Chern class c_1 and c_2, respectively.

The next step in the classification of bundles aims at a deeper understanding of the set of all bundles with fixed rank and Chern classes. This naturally leads to the concept of moduli spaces.

The case $r = 1$ is a model for the theory. By means of the exponential sequence, the set $\mathrm{Pic}^{c_1}(X)$ of all line bundles with fixed first Chern class c_1 can be identified, although not canonically for $c_1 \neq 0$, with the abelian variety $H^1(X, \mathcal{O}_X)/H^1(X, \mathbb{Z})$. Furthermore, over the product $\mathrm{Pic}^{c_1}(X) \times X$ there exists a 'universal line bundle' with the property that its restriction to $[L] \times X$ is isomorphic to the line bundle L on X. The following features are noteworthy here: Firstly, the set of all line bundles with fixed Chern class carries a natural scheme structure, such that there exists a universal line bundle over the product with X. This is roughly what is called a moduli space. Secondly, if c_1 is in the Neron-Severi group $H^2(X, \mathbb{Z}) \cap H^{1,1}(X)$, the moduli space is a nonempty projective scheme. Thirdly, the moduli space is irreducible and smooth. And, last but not least, the moduli space has a distinguished geometric structure: it is an abelian variety. This book is devoted to the analogous questions for bundles of rank greater than one. Although none of these features generalizes literally to the higher rank situation, they serve as a guideline for the investigation of the intricate structures encountered there.

For $r > 1$ one has to restrict oneself to semistable bundles in order to get a separated finite type scheme structure for the moduli space. Pursuing the natural desire to work with complete spaces, one compactifies moduli spaces of bundles by adding semistable non-locally free sheaves. The existence of semistable sheaves on a surface, i.e. the non-emptiness of the moduli spaces, can be ensured for large c_2 while r and c_1 are fixed. Under the same assumptions, the moduli spaces turn out to be irreducible. Moduli spaces of sheaves of rank ≥ 2 on a surface are not smooth, unless we consider sheaves with special invariants on special surfaces. Nevertheless, something is known about the type of singularities they can attain. Concerning the geometry of moduli spaces of sheaves of higher rank, there are two guiding principles for the investigation. Firstly, the geometric structure of sheaves of rank $r > 1$ reveals itself only for large second Chern number c_2 while c_1 stays fixed. In other words, only high dimensional moduli spaces display the properties one expects them to have. Secondly, contrary to the case $r = 1$, $c_2 = 0$, where $\mathrm{Pic}^{c_1}(X)$ is always an abelian variety no matter whether X is ruled, abelian, or of general type, moduli spaces of sheaves of higher rank are expected to inherit geometric properties from the underlying surface. In particular, the position of the surface in the Enriques classification is of uttermost importance for the geometry of the moduli spaces of sheaves on it. Much can be said about the geometry, but at least as much has yet to be explored. The variety of geometric structures exposed by moduli spaces, which in general are far from being 'just' abelian, makes the subject highly attractive to algebraic geometers.

Let us now briefly describe the contents of each single chapter of this book. We start out in Chapter 1 by providing the basic concepts in the field. Stability, as it was first introduced for bundles on curves by Mumford and later generalized to sheaves on higher dimensional varieties by Takemoto, Gieseker, Maruyama, and Simpson, is the topic of Section 1.2. This notion is natural from an algebraic as well as from a gauge theoretical point of view, for there is a deep relation between stability of bundles and existence of Hermite-Einstein metrics. This relation, known as the Kobayashi-Hitchin correspondence, was established by the work of Narasimhan-Seshadri, Donaldson and Uhlenbeck-Yau. We will elaborate on the algebraic aspects of stability, but refer to Kobayashi's book [127] for the analytic side (see also [157]). Vector bundles are best understood on the projective line where they always split into the direct sum of line bundles due to a result usually attributed to Grothendieck (1.3.1). In the general situation, this splitting is replaced by the Harder-Narasimhan filtration, a filtration with semistable factors (Section 1.3). If the sheaf is already semistable, then the Jordan-Hölder filtration filters it further, so that the factors become stable. Following Seshadri, the associated graded object is used to define S-equivalence of semistable sheaves (Section 1.5). Stability in higher dimensions can be introduced in various ways, all generalizing Mumford's original concept. In Section 1.6 we provide a framework to compare the different possibilities. The Mumford-Castelnuovo regularity and Kleiman's boundedness results, which are stated without proof in Section 1.7, are fundamental for the construction of the moduli space. They are needed to ensure that the set of semistable sheaves is small enough to be parametrized by a scheme of finite type. Another important ingredient is Grothendieck's Lemma (1.7.9) on the boundedness of subsheaves.

Moduli spaces are not just sets of objects; they can be endowed with a scheme structure. The notion of families of sheaves gives a precise meaning to the intuition of what this structure should be. Chapter 2 is devoted to some aspects related to families of sheaves. In Section 2.1 we first construct the flattening stratification for any sheaf and then consider flat families of sheaves and some of their properties. The Grothendieck Quot-scheme, one of the fundamental objects in modern algebraic geometry, together with its local description will be discussed in Section 2.2 and Appendix 2.A. In this context we also recall the notion of corepresentable functors which will be important for the definition of moduli spaces as well. As a consequence of the existence of the Quot-scheme, a relative version of the Harder-Narasimhan filtration is constructed. This and the openness of stability, due to Maruyama, will be presented in Section 2.3. In Appendix 2.A we introduce flag-schemes, a generalization of the Quot-scheme, and sketch some aspects of the deformation theory of sheaves, quotient sheaves and, more general, flags of sheaves. In Appendix 2.B we present a result of Langton showing that the moduli space of semistable sheaves is, a priori, complete.

Chapter 3 establishes the boundedness of the set of semistable sheaves. The main tool here is a result known as the Grauert-Mülich Theorem. Barth, Spindler, Maruyama, Hirschowitz, Forster, and Schneider have contributed to it in its present form. A complete proof is given in Section 3.1. At first sight this result looks rather technical, but it turns out to be powerful in controlling the behaviour of stability under basic operations like tensor products or restrictions to hypersurfaces. We explain results of Gieseker, Maruyama and Takemoto related to tensor products and pull-backs under finite morphisms in Section 3.2. In the proof of boundedness (Section 3.3), we essentially follow arguments of Simpson and Le Potier. The theory would not be complete without mentioning the famous Bogomolov Inequality. We reproduce its by now standard proof in Section 3.4 and give an alternative one later (Section 7.3). The Appendix to Chapter 3 uses the aforementioned boundedness results to prove a technical proposition due to O'Grady which comes in handy in Chapter 9.

The actual construction of the moduli space takes up all of Chapter 4. The first construction, due to Gieseker and Maruyama, differs from the one found by Simpson some ten years later in the choice of a projective embedding of the Quot-scheme. We present Simpson's approach (Sections 4.3 and 4.4) as well as a sketch of the original construction (Appendix 4.A). Both will be needed later. We hope that Section 4.2, where we recall some results concerning group actions and quotients, makes the construction accessible even for the reader not familiar with the full machinery of Geometric Invariant Theory. In Section 4.5 deformation theory is used to obtain an infinitesimal description of the moduli space, including bounds for its dimension and a formula for the expected dimension in the surface case. In particular, we prove the smoothness of the Hilbert scheme of zero-dimensional subschemes of a smooth projective surface, which is originally due to Fogarty. In contrast to the rank one case, a universal sheaf on the product of the moduli space and the variety does not always exist. Conditions for the existence of a (quasi)-universal family are discussed in Section 4.6. In Appendix 4.B moduli spaces of sheaves with an additional structure, e.g. a global section, are discussed. As an application we construct a 'quasi-universal family' over a projective birational model of the moduli space of semistable sheaves. This will be useful for later arguments. The dependence of stability on the fixed ample line bundle on the variety was neglected for many years. Only in connection with the Donaldson invariants was its significance recognized. Friedman and Qin studied the question from various angles and revealed interesting phenomena. We only touch upon this in Appendix 4.C, where it is shown that for two fixed polarizations on a surface the corresponding moduli spaces are birational for large second Chern number. Other aspects concerning fibred surfaces will be discussed in Section 5.3.

From Chapter 5 on we mainly focus on sheaves on surfaces. Chapter 5 deals with the existence of stable bundles on surfaces. The main techniques are Serre's construction (Section 5.1) and Maruyama's elementary transformations (Section 5.2). With these techniques at hand, one produces stable bundles with prescribed invariants like rank, determinant, Chern classes, Albanese classes, etc. Sometimes, on special surfaces, the same methods can in fact be used to describe the geometry of the moduli spaces. Bundles on elliptic surfaces were quite intensively studied by Friedman. Only a faint shadow of his results can be found in Sections 5.3, where we treat fibred surfaces in some generality and two examples for K3 surface.

We continue to consider special surfaces in Chapter 6. Mukai's beautiful results concerning two-dimensional moduli spaces on K3 surfaces are presented in Section 6.1. Some of the results, due to Beauville, Göttsche-Huybrechts, O'Grady, concerning higher dimensional moduli spaces will be mentioned in Section 6.2. In the course of this chapter we occasionally make use of the irreducibility of the Quot-scheme of all zero-dimensional quotients of a locally free sheaf on a surface. This is a result originally due to Li and Li-Gieseker. We present a short algebraic proof due to Ellingsrud and Lehn in Appendix 6.A.

As a sequel to the Grauert-Mülich theorem we discuss other restriction theorems in Chapter 7. Flenner's theorem (Section 7.1) is an essential improvement of the former and allows one to predict the μ-semistability of the restriction of a μ-semistable sheaf to hyperplane sections. The techniques of Mehta-Ramanathan (Section 7.2) are completely different and allow one to treat the μ-stable case as well. Bogomolov exploited his inequality to prove the rather surprising result that the restriction of a μ-stable vector bundle on a surface to any curve of high degree is stable (Section 7.3).

In Chapter 8 we strive for an understanding of line bundles on moduli spaces. Line bundles of geometric significance can be constructed using the technique of determinant bundles (Section 8.1). Unfortunately, Li's description of the full Picard group is beyond the scope of these notes, for it uses gauge theory in an essential way. We only state a special case of his result (8.1.6) which can be formulated in our framework. Section 8.2 is devoted to the study of a particular ample line bundle on the moduli space and a comparison between the algebraic and the analytic (Donaldson-Uhlenbeck) compactification of the moduli space of stable bundles. We build upon work of Le Potier and Li. As a result we construct algebraically a moduli space of μ-semistable sheaves on a surface. By work of Li and Huybrechts, the canonical class of the moduli space can be determined for a large class of surfaces (Section 8.3).

Chapter 9 is almost entirely a presentation of O'Grady's work on the irreducibility and generic smoothness of moduli spaces. Similar results were obtained by Gieseker and Li. Their techniques are completely different and are based on a detailed study of bundles on ruled surfaces. The main result roughly says that for large second Chern number the moduli space of semistable sheaves is irreducible and the bad locus of sheaves, which are not μ-stable or which correspond to singular points in the moduli space, has arbitrary high codimension.

In Chapter 10 we show how one constructs holomorphic one- and two-forms on the moduli space starting with such forms on the surface. This reflects rather nicely the general philosophy that moduli spaces inherit properties from the underlying surface. We provide the necessary background like Atiyah class, trace map, cup product, Kodaira-Spencer map, etc., in Section 10.1. In Section 10.2 we describe the tangent bundle of the smooth part of the moduli space in terms of a universal family. In fact, this result has been used already in earlier chapters. The actual construction of the forms is given in Section 10.3 where we also prove their closedness. The most famous result concerning forms on the moduli space is Mukai's theorem on the existence of a non-degenerate symplectic structure on the moduli space of stable sheaves on K3 surfaces (Section 10.4). O'Grady pursued this question for surfaces of general type.

Chapter 11 combines the results of Chapter 8 and 10 and shows that moduli spaces of semistable sheaves on surfaces of general type are of general type as well. We start with a proof of this result for the case of rank one sheaves, i.e. the Hilbert scheme. Our presentation of the higher rank case deviates slightly from Li's original proof. Other results on the birational type of moduli spaces are listed in Section 11.2. We conclude this chapter with two rather general examples where the birational type of moduli spaces of sheaves on (certain) K3 surfaces can be determined.

Part I

General Theory

1

Preliminaries

This chapter provides the basic definitions of the theory. After introducing pure sheaves and their homological aspects we discuss the notion of reduced Hilbert polynomials in terms of which the stability condition is formulated. Harder-Narasimhan and Jordan-Hölder filtrations are defined in Section 1.3 and 1.5, respectively. Their formal aspects are discussed in Section 1.6. In Section 1.7 we recall the notion of bounded families and the Mumford-Castelnuovo regularity. The results of this section will be applied later (cf. 3.3) to show the boundedness of the family of semistable sheaves. This chapter is slightly technical at times. The reader may just skim through the basic definitions at first reading and come back to the more technical parts whenever needed.

1.1 Some Homological Algebra

Let X be a Noetherian scheme. By $\mathrm{Coh}(X)$ we denote the category of coherent sheaves on X. For $E \in \mathrm{Ob}(\mathrm{Coh}(X))$, i.e. a coherent sheaf on X, one defines:

Definition 1.1.1 — *The support of E is the closed set* $\mathrm{Supp}(E) = \{x \in X | E_x \neq 0\}$. *Its dimension is called the dimension of the sheaf E and is denoted by* $\dim(E)$.

The annihilator ideal sheaf of E, i.e. the kernel of $\mathcal{O}_X \to \mathcal{E}nd(E)$, defines a subscheme structure on $\mathrm{Supp}(E)$.

Definition 1.1.2 — *E is pure of dimension d if $\dim(F) = d$ for all non-trivial coherent subsheaves $F \subset E$.*

Equivalently, E is pure if and only if all associated points of E (cf. [172] p. 49) have the same dimension.

3

Example 1.1.3 — The structure sheaf $\mathcal{O}_Y$ of a closed subscheme $Y \subset X$ is of dimension $\dim(Y)$. It is pure if Y has no components of dimension less than $\dim(Y)$ and no embedded points.

Definition 1.1.4 — *The torsion filtration of a coherent sheaf E is the unique filtration*

$$0 \subset T_0(E) \subset \ldots \subset T_d(E) = E,$$

where $d = \dim(E)$ and $T_i(E)$ is the maximal subsheaf of E of dimension $\leq i$.

The existence of the torsion filtration is due to the fact that the sum of two subsheaves $F, G \subset E$ of dimension $\leq i$ has also dimension $\leq i$. Note that by definition $T_i(E)/T_{i-1}(E)$ is zero or pure of dimension i. In particular, E is a pure sheaf of dimension d if and only if $T_{d-1}(E) = 0$.

Recall that a coherent sheaf E on an integral scheme X is torsion free if for each $x \in X$ and $s \in \mathcal{O}_{X,x} \setminus \{0\}$ multiplication by s is an injective homomorphism $E_x \to E_x$. Using the torsion filtration, this is equivalent to $T(E) := T_{\dim(X)-1}(E) = 0$. Thus, the property of a d-dimensional sheaf E to be pure is a generalization of the property to be torsion free.

Definition 1.1.5 — *The saturation of a subsheaf $F \subset E$ is the minimal subsheaf F' containing F such that E/F' is pure of dimension $d = \dim(E)$ or zero.*

Clearly, the saturation of F is the kernel of the surjection

$$E \to E/F \to (E/F)/T_{d-1}(E/F).$$

Next, we briefly recall the notions of *depth* and *homological dimension*. Let M be a module over a local ring A. Recall that an element a in the maximal ideal $\mathfrak{m}$ of A is called M-regular, if the multiplication by a defines an injective homomorphism $M \to M$. A sequence $a_1, \ldots, a_\ell \in \mathfrak{m}$ is an M-regular sequence if a_i is $M/(a_1, \ldots, a_{i-1})M$-regular for all i. The maximal length of an M-regular sequence is called the depth of M. On the other hand the homological dimension, denoted by $\mathrm{dh}(M)$, is defined as the minimal length of a projective resolution of M. If A is a regular ring, these two notions are related by the Auslander-Buchsbaum formula:

$$\mathrm{dh}(M) + \mathrm{depth}(M) = \dim(A) \tag{1.1}$$

For a coherent sheaf E on X one defines $\mathrm{dh}(E) = \max\{\mathrm{dh}(E_x) \mid x \in X\}$. If X is not regular, the homological dimension of E might be infinite. For regular X it is bounded by $\dim(X)$ and $\mathrm{dh}(E) \leq \dim(X) - 1$ for a torsion free sheaf. Both statements follow from (1.1). Also note that for a regular closed point $x \in X$, one has $\mathrm{dh}(k(x)) = \dim(X)$ and for a short exact sequence $0 \to E \to F \to G \to 0$ with F locally free one has $\mathrm{dh}(E) = \max\{0, \mathrm{dh}(G) - 1\}$.

In the sequel we discuss some more homological algebra. In particular, we will study the restriction of pure (torsion free, reflexive, ...) sheaves to hypersurfaces. The reader interested in vector bundles or sheaves on surfaces exclusively might want to skip the next part and to go directly to 1.1.16 or even to the next section. For the sake of completeness and in order to avoid many ad hoc arguments later on we explain this part in broader generality.

Let X be a smooth projective variety of dimension n over a field k. Consider a coherent sheaf E of dimension d. The codimension of E is by definition $c := n - d$. The following generalizes Serre's conditions S_k ($k \geq 0$):

$$S_{k,c} : \text{depth}(E_x) \geq \min\{k, \dim(\mathcal{O}_{X,x}) - c\} \text{ for all } x \in \text{Supp}(E).$$

The condition $S_{0,c}$ is vacuous. Condition $S_{1,c}$ is equivalent to the purity of E. Indeed, $S_{1,c}$ is equivalent to the following: if $x \in \text{Supp}(E)$ with $\dim(\mathcal{O}_{X,x}) > c$, then depth $(E_x) \geq 1$. But $\text{depth}(E_x) \geq 1$ if and only if $k(x) = \mathcal{O}_{X,x}/m_x$ does not embed into E_x, i.e. x is not an associated point of E. Hence E satisfies $S_{1,c}$ if and only if E is pure. Note, for $c = 0$ the condition $S_{1,c}$ implies that the set of singular points $\{x \in X | \text{dh}(E_x) \neq 0\}$ has codimension ≥ 2. More generally, if $\text{Supp}(E)$ is normal, then $S_{1,c}$ implies that E is locally free on an open subset of $\text{Supp}(E)$ whose complement in $\text{Supp}(E)$ has at least codimension two.

The conditions $S_{k,c}$ can conveniently be expressed in terms of the dimension of certain local Ext-sheaves.

Proposition 1.1.6 — *Let E be a coherent sheaf of dimension d and codimension $c := n - d$ on a smooth projective variety X.*
i) The sheaves $\mathcal{E}xt_X^q(E, \omega_X)$ are supported on $\text{Supp}(E)$ and $\mathcal{E}xt_X^q(E, \omega_X) = 0$ for all $q < c$. Moreover, $\text{codim}(\mathcal{E}xt_X^q(E, \omega_X)) \geq q$ for $q \geq c$.
ii) E satisfies the condition $S_{k,c}$ if and only if $\text{codim}(\mathcal{E}xt_X^q(E, \omega_X)) \geq q + k$ for all $q > c$.

Proof. The first statement in *i)* is trivial. For the second, one takes m large enough such that $H^0(X, \mathcal{E}xt_X^q(E, \omega_X) \otimes \mathcal{O}(m)) = H^0(X, \mathcal{E}xt_X^q(E, \omega_X(m))) \cong \text{Ext}^q(E, \omega_X(m))$ and uses Serre duality $\text{Ext}^q(E, \omega_X(m)) \cong H^{n-q}(X, E(-m))^\vee$ to prove $\mathcal{E}xt_X^q(E, \omega_X) = 0$ for $n - q > d$. For *ii)* we apply (1.1) and the fact that for a finite module M over a regular ring A one has $\text{dh}(M) = \max\{q | \text{Ext}_A^q(M, A) \neq 0\}$. Then

$$\text{depth}(E_x) \geq \min\{k, \dim \mathcal{O}_{X,x} - c\}$$
$$\Leftrightarrow \max\{\dim \mathcal{O}_{X,x} - k, c\} \geq \text{dh}(E_x) = \max\{q | \text{Ext}^q(E_x, \mathcal{O}_{X,x}) \neq 0\}$$
$$\Leftrightarrow \text{Ext}^q(E_x, \mathcal{O}_{X,x}) = 0 \quad \forall q > \max\{\dim \mathcal{O}_{X,x} - k, c\}$$
$$\Leftrightarrow \text{For all } q > c \text{ and } x \in X \text{ the following holds:}$$
$$\text{Ext}^q(E_x, \mathcal{O}_{X,x}) = \mathcal{E}xt_X^q(E, \omega_X)_x \neq 0 \Rightarrow \dim \mathcal{O}_{X,x} \geq q + k.$$

$\square$

For a sheaf E of dimension n, the dual $\mathcal{H}om(E, \mathcal{O}_X)$ is a non-trivial torsion free sheaf. If the dimension of E is less than n, then, with this definition, the dual is always trivial. Thus a modification for sheaves of smaller dimension is in order.

Definition 1.1.7 — *Let E be a coherent sheaf of dimension d and let $c = n - d$ be its codimension. The dual sheaf is defined as $E^D = \mathcal{E}xt^c_X(E, \omega_X)$.*

If $c = 0$, then E^D differs from the usual definition by the twist with the line bundle ω_X, i.e. $E^D \cong E^{\vee} \otimes \omega_X$. The definition of the dual in this form has the advantage of being independent of the ambient space. Namely, if X and Y are smooth, $i : X \subset Y$ is a closed embedding and E is a sheaf on X, then $(i_*E)^D \cong i_*(E^D)$. In particular, this property can be used to define the dual of a sheaf even if the ambient space is not smooth.

Lemma 1.1.8 — *There is a spectral sequence*

$$E^{pq}_2 = \mathcal{E}xt^p_X(\mathcal{E}xt^{-q}_X(E, \omega_X), \omega_X) \Rightarrow E.$$

In particular, there is a natural homomorphism $\theta_E : E \to E^{c,-c}_2 = E^{DD}$.

Proof. The existence of the spectral sequence is standard: take a locally free resolution $L_\bullet \to E$ and an injective resolution $\omega_X \to I^\bullet$ and compare the two possible filtrations of the total complex associated to the double complex $\mathcal{H}om(\mathcal{H}om(L_\bullet, \omega_X), I^\bullet)$. Note that one has $\mathrm{codim}(\mathcal{E}xt^q_X(E, \omega_X)) \geq q$ and therefore $E^{pq}_2 = 0$ if $p < -q$. Hence the only non-vanishing E_2-terms lie within the triangle cut out by the conditions $p + q \geq 0$, $p \leq \dim(X)$ and $q \leq -c$. Moreover, $E^{c,-c}_\infty \subset E^{c,-c}_2$ and thus $\theta_E : E \to E^{c,-c}_\infty \subset E^{c,-c}_2 = E^{DD}$ is naturally defined.   □

The spectral sequence also shows that $E^{p,-p}_2 = \mathcal{E}xt^p_X(\mathcal{E}xt^p_X(E, \omega_X), \omega_X)$ is pure of codimension p or trivial. Indeed, one first shows that $\mathcal{E}xt^c_X(E, \omega_X)$ is pure of codimension c. Then the assertion for $\mathcal{E}xt^c_X(\mathcal{E}xt^c_X(E, \omega_X), \omega_X)$ follows directly. In fact, we show that $\mathcal{E}xt^c_X(E, \omega_X)$ even satisfies $S_{2,c}$: Since $\mathrm{codim}(E^{pq}_2) \geq p$ and $E^{p,-c}_\infty = 0$ for $p > c$, the exact sequences

$$0 \to E^{p,-c}_{r+1} \to E^{p,-c}_r \to E^{p+r,-c-r+1}_r$$

show

$$\dim(E^{p,-c}_2) \leq \max\{\dim(E^{p,-c}_3), \dim(E^{p+2,-c-2+1}_2)\}$$
$$\vdots$$
$$\leq \max_{r \geq 2}\{\dim(E^{p+r,-c-r+1}_r)\}.$$

Hence $\mathrm{codim}(E^{p,-c}_2) \geq p + 2$ for $p > c$.

Definition 1.1.9 — *A coherent sheaf E of codimension c is called reflexive if θ_E is an isomorphism. E^{DD} is called the reflexive hull of E.*

We summarize the results:

Proposition 1.1.10 — *Let E be a coherent sheaf of codimension c on a smooth projective variety X. Then the following conditions are equivalent:*
1) *E is pure*
2) $\operatorname{codim}(\mathcal{E}xt^q(E, \omega_X)) \geq q + 1$ *for all $q > c$*
3) *E satisfies $S_{1,c}$*
4) *θ_E is injective.*
Similarly, the following conditions are equivalent:
1') *E is reflexive, i.e. θ_E is an isomorphism*
2') *E is the dual of a coherent sheaf of codimension c*
3') $\operatorname{codim}(\mathcal{E}xt^q(E, \omega_X)) \geq q + 2$ *for all $q > c$*
4') *E satisfies $S_{2,c}$.*

Proof. i) 1) $\Leftrightarrow$ 2) $\Leftrightarrow$ 3) have been shown above. If θ_E is injective, then E is a subsheaf of the pure sheaf $\mathcal{E}xt^c_X(\mathcal{E}xt^c_X(E, \omega_X), \omega_X)$. Hence E is pure as well. If E is pure, then $\operatorname{codim}(\mathcal{E}xt^q_X(E, \omega_X)) \geq q + 1$ for $q > c$. Hence $\mathcal{E}xt^p_X(\mathcal{E}xt^q_X(E, \omega_X), \omega_X) = 0$ for $p < q + 1$. In particular, $E_2^{q, -q} = 0$ for $q > c$ and, therefore, θ_E is injective.

ii) 3') $\Leftrightarrow$ 4') follows from 1.1.6. Also 1') $\Rightarrow$ 2') is obvious. Now assume that condition 3') holds true, i.e. that we have $\operatorname{codim}(\mathcal{E}xt^q_X(E, \omega_X)) \geq q + 2$ for $q > c$. Then $\mathcal{E}xt^p_X(\mathcal{E}xt^q_X(E, \omega_X), \omega_X) = 0$ for $p < q + 2$. Hence $E_2^{p, -q} = 0$ for $p < q + 2 > c + 2$. This shows $E_2^{c, -c} = E_\infty^{c, -c}$ and $E_2^{q, -q} = 0$ for $q \neq c$. Hence θ_E is an isomorphism, i.e. 1') holds. It remains to show 2') $\Rightarrow$ 3'), but this was explained after the proof of the previous lemma. $\qquad\square$

Note that the proposition justifies the term reflexive hull for E^{DD}. A familiar example of a reflexive sheaf is the following: if $Y \subset X$ is a proper normal projective subvariety of X, then $\mathcal{O}_Y$ is a reflexive sheaf of dimension $\dim(Y)$ on X. Indeed, Serre's condition S_2 is equivalent to $S_{2,c}$ where $c = \operatorname{codim}(Y)$

The interpretation of homological properties of a coherent sheaf E in terms of local Ext-sheaves enables us to control whether the restriction $E|_H$ to a hypersurface H shares these properties. Roughly, the properties discussed above are preserved under restriction to hypersurfaces which are regular with respect to the sheaf. Both concepts generalize naturally to sheaves as follows:

Definition 1.1.11 — *Let X be a Noetherian scheme, let E be a coherent sheaf on X and let L be a line bundle on X. A section $s \in H^0(X, L)$ is called E-regular if and only if*

$E \otimes L^{\vee} \xrightarrow{\cdot s} E$ is injective. A sequence $s_1, \ldots, s_\ell \in H^0(X, L)$ is called E-regular if s_i is $E/(s_1, \ldots, s_{i-1})(E \otimes L^{\vee})$-regular for all $i = 1, \ldots, \ell$.

Obviously, $s \in H^0(X, L)$ is E-regular if and only if its zero set $H \in |L|$ contains none of the associated points of E. We also say that the divisor $H \in |L|$ is E-regular if the corresponding section $s \in H^0(X, L)$ is E-regular. The existence of regular sections is ensured by

Lemma 1.1.12 — *Assume X is a projective scheme defined over an infinite field k. Let E be a coherent sheaf and let L be a globally generated line bundle on X. Then the E-regular divisors in the linear system $|L|$ form a dense open subscheme.*

Proof. Let $x_1, \ldots, x_N$ denote the associated points of E, and let $\mathcal{I}_{X_i}$ be the ideal sheaves of the reduced closed subschemes $X_i = \overline{\{x_i\}}$. Then $H \in |L|$ contains x_i if and only if H is contained in the linear subspace $P_i = |\mathcal{I}_{X_i} \otimes L| \subset |L|$. Since L is globally generated, $h^0(X, \mathcal{I}_{X_i} \otimes L) < h^0(X, L)$, so that the linear subspaces P_i are proper subspaces in $|L|$ and their complement is open and dense. $\square$

Lemma 1.1.13 — *Let X be a smooth projective variety and $H \in |L|$.*
i) If E is a coherent sheaf of codimension c satisfying $S_{k,c}$ for some integer $k \geq 1$ and H is E-regular, then $E|_H$, considered as a sheaf on X, satisfies $S_{k-1,c+1}$.
ii) If in addition H is $\mathcal{E}xt_X^q(E, \omega_X)$-regular for all $q \geq 0$, then $\mathcal{E}xt_X^{q+1}(E|_H, \omega_X) \cong \mathcal{E}xt_X^q(E, \omega_X) \otimes L|_H$. In particular, if E satisfies $S_{k,c}$, then $E|_H$ satisfies $S_{k,c+1}$.

Proof. By assumption we have an exact sequence $0 \to E \otimes L^{\vee} \to E \to E|_H \to 0$. The associated long exact sequence

$$\ldots \to \mathcal{E}xt_X^{q-1}(E \otimes L^{\vee}, \omega_X) \to \mathcal{E}xt_X^q(E|_H, \omega_X) \to \mathcal{E}xt_X^q(E, \omega_X) \ldots$$

gives

$$\mathrm{codim}(\mathcal{E}xt_X^q(E|_H, \omega_X)) \geq \min\{\mathrm{codim}(\mathcal{E}xt_X^{q-1}(E \otimes L^{\vee}, \omega_X)), \mathrm{codim}(\mathcal{E}xt_X^q(E, \omega_X))\}.$$

The second regularity assumption implies that the above complex of $\mathcal{E}xt$-groups splits up into short exact sequences

$$0 \to \mathcal{E}xt_X^q(E, \omega_X) \to \mathcal{E}xt_X^q(E, \omega_X) \otimes L \to \mathcal{E}xt_X^{q+1}(E \otimes \mathcal{O}_H, \omega_X) \to 0.$$

This gives the second assertion. $\square$

Corollary 1.1.14 — *Let X be a smooth projective variety and $H \in |L|$.*
i) If E is a reflexive sheaf of codimension c and H is E-regular then $E|_H$ is pure of codimension $c + 1$.

ii) If E is pure (reflexive) and H is E-regular and $\mathcal{E}xt_X^q(E, \omega_X)$-regular for all $q \geq 0$ then $E|_H$ is pure (reflexive) of codimension $c + 1$. $\qquad\square$

Corollary 1.1.15 — *Let X be a normal closed subscheme in $\mathbb{P}^N$ and k an infinite field. Then there is a dense open subset U of hyperplanes $H \in |\mathcal{O}(1)|$ such that H intersects X properly and such that $X \cap H$ is again normal.*

Proof. One must show that $X \cap H$ is regular in codimension one and satisfies property S_2. By assumption $\mathcal{O}_X$ is a reflexive sheaf on $\mathbb{P}^N$. Hence Corollary 1.1.14 implies that $\mathcal{O}_{X \cap H}$ is reflexive again for all H in a dense open subset of $|\mathcal{O}(1)|$. Let $X' \subset X$ be the set of singular points of X. Then $\operatorname{codim}_X(X') \geq 2$. If H intersects X' properly, then $\operatorname{codim}_{X \cap H}(X' \cap H) \geq 2$, too. Hence it is enough to show that a general hyperplane H intersects the regular part X_{reg} of X transversely, but this is the content of the Bertini Theorem. $\qquad\square$

Example 1.1.16 — For later use we bring the results down to earth and specify them in the case of projective curves and surfaces.

First, let X be a smooth curve. Then a coherent sheaf E might be zero or one-dimensional. If $\dim(E) = 0$, then $\operatorname{Supp}(E)$ is a finite collection of points. In general, $E = T(E) \oplus E/T(E)$, where $E/T(E)$ is locally free. Indeed, a sheaf on a smooth curve is torsion free if and only if it is locally free.

If X is a smooth surface, then a sheaf E of dimension two is reflexive if and only if it is locally free. Any torsion free sheaf E embeds into its reflexive hull $E^{\vee\vee}$ such that $E^{\vee\vee}/E$ has dimension zero. In particular, a torsion free sheaf of rank one is of the form $\mathcal{I}_Z \otimes M$, where M is a line bundle and $\mathcal{I}_Z$ is the ideal sheaf of a codimension two subscheme. Note that for a torsion free sheaf E on a surface $\operatorname{dh}(E) \leq 1$. The support of $E^{\vee\vee}/E$ is called the set of *singular points* of the torsion free sheaf E. We will also use the fact that if a locally free sheaf F is a subsheaf of a torsion free sheaf E, then $T_0(E/F) = 0$. The restriction results are quite elementary on a surface: if E is of dimension two and reflexive, i.e. locally free, then the restriction to any curve is locally free. If E is purely two-dimensional, i.e. torsion free, then the restriction to any curve avoiding the finitely many singular points of E is locally free.

1.1.17 Determinant bundles — Recall the definition of the *determinant* of a coherent sheaf. If E is locally free of rank s, then $\det(E)$ is by definition the line bundle $\Lambda^s(E)$. More generally, let E be a coherent sheaf that admits a finite locally free resolution

$$0 \to E_n \to E_{n-1} \to \ldots \to E_0 \to E \to 0.$$

Define $\det(E) = \bigotimes \det(E_i)^{(-1)^i}$. The definition does not depend on the resolution. If X is a smooth variety, every coherent sheaf admits a finite locally free resolution. See exc. III 6.8 and 6.9 in [98] for the non-projective case. If $\dim(E) \leq \dim(X) - 2$, then $\det(E) \cong \mathcal{O}_X$.

1.2 Semistable Sheaves

Let X be a projective scheme over a field k. Recall that the Euler characteristic of a coherent sheaf E is $\chi(E) := \sum (-1)^i h^i(X, E)$, where $h^i(X, E) = \dim_k H^i(X, E)$. If we fix an ample line bundle $\mathcal{O}(1)$ on X, then the *Hilbert polynomial* $P(E)$ is given by

$$m \mapsto \chi(E \otimes \mathcal{O}(m)).$$

Lemma 1.2.1 — *Let E be a coherent sheaf of dimension d and let $H_1, \ldots, H_d \in |\mathcal{O}(1)|$ be an E-regular sequence. Then*

$$P(E, m) = \chi(E \otimes \mathcal{O}(m)) = \sum_{i=0}^{d} \chi\left(E|_{\cap_{j \leq i} H_j}\right) \binom{m + i - 1}{i}.$$

Proof. We proceed by induction. If $d = 0$ the assertion is trivial. Assume that $d > 0$ and that the assertion of the lemma has been proved for all sheaves of dimension $< d$. Let $H = H_1$ and consider the short exact sequence

$$0 \to E(m - 1) \to E(m) \to E(m)|_H \to 0$$

Then by the induction hypothesis

$$\chi(E(m)) - \chi(E(m - 1)) = \chi(E(m)|_H) = \sum_{i=0}^{d-1} \chi\left(E|_{\cap_{j \leq i+1} H_j}\right) \binom{m + i - 1}{i}.$$

This means that if $f(m)$ denotes the difference of $\chi(E(m))$ and the term on the right hand side in the lemma, then $f(m) - f(m - 1) = 0$. But clearly $f(0) = 0$, so that f vanishes identically. $\square$

In particular, $P(E)$ can be uniquely written in the form

$$P(E, m) = \sum_{i=0}^{\dim(E)} \alpha_i(E) \frac{m^i}{i!}$$

with rational coefficients $\alpha_i(E)$ ($i = 0, \ldots, \dim(E)$). Furthermore, if $E \neq 0$ the leading coefficient $\alpha_{\dim(E)}(E)$, called the *multiplicity*, is always positive. Note that $\alpha_{\dim(X)}(\mathcal{O}_X)$ is the degree of X with respect to $\mathcal{O}(1)$.

Definition 1.2.2 — *If E is a coherent sheaf of dimension $d = \dim(X)$, then*

$$\text{rk}(E) := \frac{\alpha_d(E)}{\alpha_d(\mathcal{O}_X)}$$

is called the rank of E.

On an integral scheme X of dimension d there exists for any d-dimensional sheaf E an open dense subset $U \subset X$ such that $E|_U$ is locally free. Then $\text{rk}(E)$ is the rank of the vector bundle $E|_U$. In general, $\text{rk}(E)$ need not be integral, and if X is reducible it might even depend on the polarization.

Definition 1.2.3 — *The reduced Hilbert polynomial $p(E)$ of a coherent sheaf E of dimension d is defined by*

$$p(E, m) := \frac{P(E, m)}{\alpha_d(E)}$$

Recall that there is a natural ordering of polynomials given by the lexicographic order of their coefficients. Explicitly, $f \leq g$ if and only if $f(m) \leq g(m)$ for $m \gg 0$. Analogously, $f < g$ if and only if $f(m) < g(m)$ for $m \gg 0$. We are now prepared for the definition of stability.

Definition 1.2.4 — *A coherent sheaf E of dimension d is semistable if E is pure and for any proper subsheaf $F \subset E$ one has $p(F) \leq p(E)$. E is called stable if E is semistable and the inequality is strict, i.e. $p(F) < p(E)$ for any proper subsheaf $F \subset E$.*

We want to emphasize that the notion of stability depends on the fixed ample line bundle on X. However, replacing $\mathcal{O}(1)$ by $\mathcal{O}(m)$ has no effect. We come back to this problem in 4.C.

Notation 1.2.5 — In order to avoid case considerations for stable and semistable sheaves we will occasionally employ the following short-hand notation: if in a statement the word "(semi)stable" appears together with relation signs "($\leq$)" or "($<$)", the statement encodes in fact two assertions: one about semistable sheaves and relation signs "$\leq$" and "$<$", respectively, and one about stable sheaves and relation signs "$<$" and "$\leq$", respectively. For example, we could say that E is *(semi)stable if and only if it is pure and $p(F) (\leq) p(E)$ for every proper subsheaf $F \subset E$.*

An alternative definition of stability would have been the following: a coherent sheaf E of dimension d is (semi)stable if $\alpha_d(E) \cdot P(F) (\leq) \alpha_d(F) \cdot P(E)$ for all proper subsheaves $F \subset E$. This is obviously the same definition except that it does not require explicitly that E is pure. But applying the inequality to $F = T_{d-1}(E)$ and using $\alpha_d(T_{d-1}(E)) = 0$ we get $P(T_{d-1}(E)) \leq 0$. This immediately implies $T_{d-1}(E) = 0$, i.e. E is pure.

Proposition 1.2.6 — *Let E be a coherent sheaf of dimension d and assume E is pure. Then the following conditions are equivalent:*
i) E is (semi)stable.
ii) For all proper saturated subsheaves $F \subset E$ one has $p(F)(\leq)p(E)$.
iii) For all proper quotient sheaves $E \to G$ with $\alpha_d(G) > 0$ one has $p(E)(\leq)p(G)$.
iv) For all proper purely d-dimensional quotient sheaves $E \to G$ one has $p(E)(\leq)p(G)$.

Proof. The implications $i) \Rightarrow ii)$ and $iii) \Rightarrow iv)$ are obvious. Consider an exact sequence

$$0 \to F \to E \to G \to 0.$$

Using $\alpha_d(E) = \alpha_d(F) + \alpha_d(G)$ and $P(E) = P(F) + P(G)$, we get $\alpha_d(F) \cdot (p(F) - p(E)) = \alpha_d(G) \cdot (p(E) - p(G))$. Since G is pure and d-dimensional if and only if F is saturated, this yields $i) \Rightarrow iii)$ and $ii) \Leftrightarrow iv)$. Finally, $ii) \Rightarrow i)$ follows from $\alpha_d(F) = \alpha_d(F')$ and $P(F) \leq P(F')$, where F' is the saturation of F in E. $\qquad\square$

Proposition 1.2.7 — *Let F and G be semistable purely d-dimensional coherent sheaves. If $p(F) > p(G)$, then $\mathrm{Hom}(F, G) = 0$. If $p(F) = p(G)$ and $f : F \to G$ is non-trivial then f is injective if F is stable and surjective if G is stable. If $p(F) = p(G)$ and $\alpha_d(F) = \alpha_d(G)$ then any non-trivial homomorphism $f : F \to G$ is an isomorphism provided F or G is stable.*

Proof. Let $f : F \to G$ be a non-trivial homomorphism of semistable sheaves with $p(F) \geq p(G)$. Let E be the image of f. Then $p(F) \leq p(E) \leq p(G)$. This contradicts immediately the assumption $p(F) > p(G)$. If $p(F) = p(G)$ it contradicts the assumption that F is stable unless $F \to E$ is an isomorphism, and the assumption that G is stable unless $E \to G$ is an isomorphism. If F and G have the same Hilbert polynomial $\alpha_d(F) \cdot p(F) = \alpha_d(G) \cdot p(G)$, then any homomorphism $f : F \to G$ is an isomorphism if and only if f is injective or surjective. $\qquad\square$

Corollary 1.2.8 — *If E is a stable sheaf, then $\mathrm{End}(E)$ is a finite dimensional division algebra over k. In particular, if k is algebraically closed, then $k \cong \mathrm{End}(E)$, i.e. E is a simple sheaf.*

Proof. If E is stable then according to the proposition any endomorphism of E is either 0 or invertible. The last statement follows from the general fact that any finite dimensional division algebra D over an algebraically closed field is trivial: any element $x \in D \setminus k$ would generate a finite dimensional and hence algebraic commutative field extension of k in D. $\qquad\square$

The converse of the assertion in the corollary is not true: if E is simple, i.e. $\text{End}(E) \cong k$, then E need not be stable. An example will be given in §1.2.10.

Definition 1.2.9 — *A coherent sheaf E is geometrically stable if for any base field extension $X_K = X \times_k \text{Spec}(K) \to X$ the pull-back $E \otimes_k K$ is stable.*

A stable sheaf need not be geometrically stable. An example will be given in 1.3.9. But note that a stable sheaf on a variety over an algebraically closed field is also geometrically stable (cf. 1.5.11). The corresponding notion of geometrically semistable sheaves does not differ from the ordinary semistability due to the uniqueness of the Harder-Narasimhan filtration (cf. 1.3.7).

Historically, the notion of stability for coherent sheaves first appeared in the context of vector bundles on curves [191]: let X be a smooth projective curve over an algebraically closed field k, and let E be a locally free sheaf of rank r. The Riemann-Roch Theorem for curves says

$$\chi(E) = \deg(E) + r(1 - g),$$

where g is the genus of X. Accordingly, the Hilbert polynomial is

$$P(E, m) = r \deg(X)m + \deg(E) + r(1 - g) = (\deg(X)m + \mu(E) + (1 - g)) \cdot r,$$

where $\mu(E) := \deg(E)/r$ is called the *slope* of E. Then E is said to be (semi)stable, if for all subsheaves $F \subset E$ with $0 < \text{rk}(F) < \text{rk}(E)$ one has $\mu(F)(\leq)\mu(E)$. Note that this is equivalent to our stability condition $p(F)(\leq)p(E)$.

Example 1.2.10 — Examples of stable or semistable bundles are easily available: any line bundle is stable. Furthermore, if $0 \to L_0 \to F \to L_1 \to 0$ is a non-trivial extension with line bundles L_0 and L_1 of degree 0 and 1, respectively, then F is stable: since the degree is additive, we have $\deg(F) = 1$ and $\mu(F) = 1/2$. Let $M \subset F$ be an arbitrary subsheaf. If $\text{rk}(M) = 2$, then F/M is a sheaf of dimension zero of length, say $\ell > 0$, and $\mu(M) = \mu(F) - \ell/2 < \mu(F)$. If $\text{rk}(M) = 1$ consider the composition $M \to L_1$. This is either zero or injective. In the first case $M \subset L_0$ and therefore $\mu(M) \leq \mu(L_0) = 0 < 1/2$. In the second case $M \subset L_1$ and therefore $\mu(M) \leq \mu(L_1) = 1$. If $\mu(M) = 1$, then necessarily $M = L_1$ and M would provide a splitting of the extension in contrast to the assumption. Hence again $\mu(M) \leq 0 < 1/2$. On the other hand, a direct sum $L_0 \oplus L_1$ of line bundles of different degree is not even semistable. By a similar technique, one can also construct semistable bundles which are not stable, but simple: let X be a projective curve of genus $g \geq 2$ over an algebraically closed field k and let E_1 and E_2 be two non-isomorphic stable vector bundles of rank r_1 and r_2, respectively, with $\mu(E_1) = \mu(E_2)$. Then $\text{Hom}(E_2, E_1) = 0$ by Proposition 1.2.7. Hence the dimension of $\text{Ext}^1(E_2, E_1)$ can be computed using the Riemann-Roch formula:

$$\dim(\mathrm{Ext}^1(E_2, E_1)) = -\chi(E_2^{\vee} \otimes E_1) = r_1 \cdot r_2 \cdot (g - 1).$$

Therefore, there are non-trivial extensions $0 \to E_1 \to E \to E_2 \to 0$. Of course, E is semistable, but not stable. We show that E is simple: Suppose $\phi : E \to E$ is a non-trivial endomorphism. Then the composition $E_1 \to E \xrightarrow{\phi} E \to E_2$ must vanish, hence $\phi(E_1) \subset E_1$. Since E_1 is simple, $\phi|_{E_1} = \lambda \cdot \mathrm{id}_{E_1}$ for some scalar $\lambda \in k$. Consider $\psi = \phi - \lambda \cdot \mathrm{id}_E$. Then $\psi : E \to E$ is trivial when restricted to E_1 and hence factorizes through a homomorphism $\psi' : E_2 \to E$. If the composition $\psi' : E_2 \to E \to E_2$ were non-zero, it would be an isomorphism and hence a multiple of the identity and would provide a splitting of the sequence defining E. Hence ψ' factorizes through some homomorphism $E_2 \to E_1$. But since $\mathrm{Hom}(E_2, E_1) = 0$, one concludes $\psi = 0$.

If we pass from sheaves on curves to higher dimensional sheaves the notion of stability can be generalized in different ways. One, using the reduced Hilbert polynomial, was presented above. This version of stability is sometimes called Gieseker-stability. Another possible generalization uses the slope of a sheaf. The resulting stability condition is called Mumford-Takemoto-stability or μ-stability. Compared with the notion of Gieseker-stability μ-stability behaves better with respect to standard operations like tensor products, restrictions to hypersurfaces, pull-backs, etc., which are important technical tools. We want to give the definition of μ-stability in the case of a sheaf of dimension $d = \dim(X)$. For a completely general treatment compare Section 1.6

Definition 1.2.11 — *Let E be a coherent sheaf of dimension $d = \dim(X)$. The degree of E is defined by*

$$\deg(E) := \alpha_{d-1}(E) - \mathrm{rk}(E) \cdot \alpha_{d-1}(\mathcal{O}_X)$$

and its slope by

$$\mu(E) := \frac{\deg(E)}{\mathrm{rk}(E)}.$$

On a smooth projective variety the Hirzebruch-Riemann-Roch formula shows $\deg(E) = c_1(E).H^{d-1}$, where H is the ample divisor. In particular, $\deg(E) = \deg(\det(E))$. If we want to emphasize the dependence on the ample divisor H we write $\deg_H(E)$ and $\mu_H(E)$. Obviously, $\deg_{nH}(E) = n^{d-1} \deg_H(E)$ and $\mu_{nH}(E) = n^{d-1}\mu_H(E)$.

Definition 1.2.12 — *A coherent sheaf E of dimension $d = \dim(X)$ is μ-(semi)stable if $T_{d-2}(E) = T_{d-1}(E)$ and $\mu(F)(\leq)\mu(E)$ for all subsheaves $F \subset E$ with $0 < \mathrm{rk}(F) < \mathrm{rk}(E)$.*

The condition on the torsion filtration just says that any torsion subsheaf of E has codimension at least two. Observe, that a coherent sheaf of dimension $\dim(E) = \dim(X)$ is μ-(semi)stable if and only if $\mathrm{rk}(E) \cdot \deg(F)(\leq)\mathrm{rk}(F) \cdot \deg(E)$ for all subsheaves $F \subset E$ with $\mathrm{rk}(F) < \mathrm{rk}(E)$ (compare the arguments after 1.2.4). One easily proves

Lemma 1.2.13 — *If E is a pure coherent sheaf of dimension $d = \dim(X)$, then one has the following chain of implications*

$$E \text{ is } \mu\text{-stable} \Rightarrow E \text{ is stable} \Rightarrow E \text{ is semistable} \Rightarrow E \text{ is } \mu\text{-semistable.}$$

$\square$

For later use, we also formulate the following easy observation.

Lemma 1.2.14 — *Let X be integral. If a coherent sheaf E of dimension $d = \dim(X)$ is μ-semistable and $\mathrm{rk}(E)$ and $\deg(E)$ are coprime, then E is μ-stable.*

Proof. If E is not μ-stable, then there exists a subsheaf $F \subset E$ with $0 < \mathrm{rk}(F) < \mathrm{rk}(E)$ and $\deg(F) \cdot \mathrm{rk}(E) = \deg(E) \cdot \mathrm{rk}(F)$. This clearly contradicts the assumption g.c.d.$(\mathrm{rk}(E), \deg(E)) = 1$. $\square$

1.3 The Harder-Narasimhan Filtration

Before we state the general theorem, let us consider the special situation of vector bundles on $\mathbb{P}^1$ over a field k.

Theorem 1.3.1 — *Let E be a vector bundle of rank r on $\mathbb{P}^1$. There is a uniquely determined decreasing sequence of integers $a_1 \geq a_2 \geq \ldots \geq a_r$ such that $E \cong \mathcal{O}(a_1) \oplus \ldots \oplus \mathcal{O}(a_r)$.*

Proof. The theorem is clear for $r = 1$. Assume that the theorem holds for all vector bundles of rank $< r$ and that E is a vector bundle of rank r. Then there is a line bundle $\mathcal{O}(a) \subset E$ such that the quotient is again a vector bundle: simply take the saturation of any rank 1 subsheaf of E. Let a_1 be maximal with this property, and let $\bigoplus_{i=2}^{r} \mathcal{O}(a_i)$ be a decomposition of the quotient $E/\mathcal{O}(a_1)$. Consider the twisted extension:

$$0 \to \mathcal{O}(-1) \to E(-1-a_1) \to \bigoplus_{i=2}^{r} \mathcal{O}(a_i - a_1 - 1) \to 0.$$

Any section of $E(-1-a_1)$ would induce a non-trivial homomorphism $\mathcal{O}(1+a_1) \to E$, contradicting the maximality of a_1. Hence, $H^0(E(-1-a_1)) = 0$. Since $H^1(\mathcal{O}(-1)) = 0$

we have also $H^0(\mathcal{O}(a_i - 1 - a_1)) = 0$ for all i. This implies $a_i < a_1 + 1$, so that $a_1 \geq a_2 \geq \ldots \geq a_r$. It remains to show that the sequence splits. But clearly

$$\mathrm{Ext}^1(\bigoplus_{i\geq 2}\mathcal{O}(a_i), \mathcal{O}(a_1))^\vee \cong \bigoplus_{i\geq 2}\mathrm{Hom}(\mathcal{O}(a_1), \mathcal{O}(a_i - 2)) = 0,$$

since $a_1 \geq a_i > a_i - 2$. We can rephrase the existence part of the theorem as follows:

There is an isomorphism

$$E \cong \bigoplus_{a\in\mathbb{Z}} V_a \otimes_k \mathcal{O}(a)$$

for finite dimensional vector spaces V_a, almost all of which vanish. To prove uniqueness amounts to showing that E determines the dimensions $\dim(V_a)$.

We define a filtration of E in the following way: for every integer b let

$$H^0(\mathbb{P}^1, E(b)) \otimes \mathcal{O}(-b) \longrightarrow E$$

denote the canonical evaluation map and E_b its image. Since $E(b)$ has no global sections for very negative b and is globally generated for very large b, we get a finite increasing filtration

$$\ldots \subset E_{-2} \subset E_{-1} \subset E_0 \subset E_1 \subset \ldots$$

Moreover, it is clear that, if $E \cong \bigoplus_a V_a \otimes_k \mathcal{O}(a)$, then $E_b \cong \bigoplus_{a\geq -b} V_a \otimes_k \mathcal{O}(a)$. This shows: $\dim(V_a) = \mathrm{rk}(E_{-a}/E_{-a-1})$.   □

In fact, we proved more than the theorem required, namely the existence of a certain unique split filtration, though the splitting homomorphisms are not unique. In general, we still have a filtration for a given coherent sheaf with similar properties as above but which is non-split.

The following definition and theorem give a first justification for the notion of a semistable sheaf: we can think of semistable sheaves as building blocks for arbitrary pure dimensional sheaves. Let X be a projective scheme with a fixed ample line bundle.

Definition 1.3.2 — *Let E be a non-trivial pure sheaf of dimension d. A Harder-Narasimhan filtration for E is an increasing filtration*

$$0 = \mathrm{HN}_0(E) \subset \mathrm{HN}_1(E) \ldots \subset \mathrm{HN}_\ell(E) = E,$$

such that the factors $\mathrm{gr}_i^{\mathrm{HN}} = \mathrm{HN}_i(E)/\mathrm{HN}_{i-1}(E)$ for $i = 1, \ldots, \ell$, are semistable sheaves of dimension d with reduced Hilbert polynomials p_i satisfying

$$p_{\max}(E) := p_1 > \ldots > p_\ell =: p_{\min}(E).$$

Obviously, E is semistable if and only if E is pure and $p_{\max}(E) = p_{\min}(E)$. A priori, the definition of the maximal and minimal p of a sheaf E depends on the filtration. We will see in the next theorem, that the Harder-Narasimhan filtration is uniquely determined, so that there is no ambiguity in the notation. For the following lemma, however, we fix Harder-Narasimhan filtrations for both sheaves:

Lemma 1.3.3 — *If F and G are pure sheaves of dimension d with $p_{\min}(F) > p_{\max}(G)$, then $\mathrm{Hom}(F, G) = 0$.*

Proof. Suppose $\psi : F \to G$ is non-trivial. Let $i > 0$ be minimal with $\psi(\mathrm{HN}_i(F)) \neq 0$ and let $j > 0$ be minimal with $\psi(\mathrm{HN}_i(F)) \subset \mathrm{HN}_j(G)$. Then there is a non-trivial homomorphism $\bar{\psi} : gr_i^{\mathrm{HN}}(F) \to gr_j^{\mathrm{HN}}(G)$. By assumption $p(gr_i^{\mathrm{HN}}(F)) \geq p_{\min}(F) > p_{\max}(G) \geq p(gr_j^{\mathrm{HN}}(G))$. This contradicts Proposition 1.2.7. $\qquad\square$

Theorem 1.3.4 — *Every pure sheaf E has a unique Harder-Narasimhan filtration.*

We will prove the theorem in a number of steps:

Lemma 1.3.5 — *Let E be a purely d-dimensional sheaf. Then there is a subsheaf $F \subset E$ such that for all subsheaves $G \subset E$ one has $p(F) \geq p(G)$, and in case of equality $F \supset G$. Moreover, F is uniquely determined and semistable.*

Definition 1.3.6 — *F is called the maximal destabilizing subsheaf of E.*

Proof. Clearly, the last two assertions follow directly from the first.

Let us define an order relation on the set of non-trivial subsheaves of E by $F_1 \leq F_2$ if and only if $F_1 \subset F_2$ and $p(F_1) \leq p(F_2)$. Since any ascending chain of subsheaves terminates, we have for every subsheaf $F \subset E$ a subsheaf $F \subset F' \subset E$ which is maximal with respect to $\leq$. Let $F \subset E$ be $\leq$-maximal with minimal multiplicity $\alpha_d(F)$ among all maximal subsheaves. We claim that F has the asserted properties.

Suppose there exists $G \subset E$ with $p(G) \geq p(F)$. First, we show that we can assume $G \subset F$ by replacing G by $G \cap F$. Indeed, if $G \not\subset F$, then F is a proper subsheaf of $F + G$ and hence $p(F) > p(F + G)$. Using the exact sequence

$$0 \to F \cap G \to F \oplus G \to F + G \to 0$$

one finds $P(F) + P(G) = P(F \oplus G) = P(F \cap G) + P(F + G)$ and $\alpha_d(F) + \alpha_d(G) = \alpha_d(F \oplus G) = \alpha_d(F \cap G) + \alpha_d(F + G)$. Hence, $\alpha_d(F \cap G)(p(G) - p(F \cap G)) = \alpha_d(F + G)(p(F + G) - p(F)) + (\alpha_d(G) - \alpha_d(F \cap G))(p(F) - p(G))$. Together with the two inequalities $p(F) \leq p(G)$ and $p(F) > p(F + G)$ this shows $p(F) \leq p(G) < p(F \cap G)$. Next, fix $G \subset F$ with $p(G) > p(F)$ which is maximal in F with respect to $\leq$. Then let

G' contain G and be $\leq$-maximal in E. In particular, $p(F) < p(G) \leq p(G')$. By the maximality of G' and F we know $G' \not\subset F$, since otherwise $\alpha_d(G') < \alpha_d(F)$ contradicting the minimality of $\alpha_d(F)$. Hence, F is a proper subsheaf of $F + G'$. Therefore, $p(F) > p(F + G')$. As before the inequalities $p(F) < p(G')$ and $p(F) \geq p(F + G')$ imply $p(F \cap G') > p(G') \geq p(G)$. Since $G \subset F \cap G' \subset F$, this contradicts the assumption on G. $\square$

The lemma allows to prove the existence part of the theorem: let E be a pure sheaf of dimension d and let E_1 be the maximal destabilizing subsheaf. By induction we can assume that E/E_1 has a Harder-Narasimhan filtration $0 = G_0 \subset G_1 \subset \ldots \subset G_{\ell-1} = E/E_1$. If $E_{i+1} \subset E$ denotes the pre-image of G_i, all that is left is to show that $p(E_1) > p(E_2/E_1)$. But if this were false, we would have $p(E_2) \geq p(E_1)$ contradicting the maximality of E_1.

For the uniqueness part assume that $E_\bullet$ and $E'_\bullet$ are two Harder-Narasimhan filtrations. Without loss of generality $p(E'_1) \geq p(E_1)$. Let j be minimal with $E'_1 \subset E_j$. Then the composition $E'_1 \to E_j \to E_j/E_{j-1}$ is a non-trivial homomorphism of semistable sheaves. This implies $p(E_j/E_{j-1}) \geq p(E'_1) \geq p(E_1) \geq p(E_j/E_{j-1})$ by Proposition 1.2.7. Hence, equality holds everywhere, implying $j = 1$ so that $E'_1 \subset E_1$. But then $p(E'_1) \leq p(E_1)$ because of the semistability of E_1, and one can repeat the argument with the rôles of $E'_\bullet$ and $E_\bullet$ reversed. This shows: $E'_1 = E_1$. By induction we can assume that uniqueness holds for the Harder-Narasimhan filtrations of E/E_1. This shows $E'_i/E_1 = E_i/E_1$ and finishes the proof of the uniqueness part of the theorem. $\square$

Theorem 1.3.7 — *Let E be a pure sheaf of dimension d and let K be a field extension of k. Then*

$$\mathrm{HN}_\bullet(E \otimes_k K) = \mathrm{HN}_\bullet(E) \otimes_k K,$$

i.e. the Harder-Narasimhan filtration is stable under base field extension.

Proof. If $F \subset E$ is a destabilizing subsheaf then so is $F \otimes K \subset E \otimes K$. Hence if $E \otimes K$ is semistable, then E is also semistable. It therefore suffices to prove: there exists a filtration $E_\bullet$ of E such that $\mathrm{HN}_i(E \otimes K) = E_i \otimes K$. The sheaves $\mathrm{HN}_i(E \otimes K)$ are finitely presented and hence defined over some field L, $k \subset L \subset K$, which is finitely generated over k. Filtering L by appropriate subfields we can reduce to the case that $K = k(x)$ for some single element $x \in K$ and that either

1. K/k is purely transcendental or separable, or
2. K/k is purely inseparable.

If in the first case the extension is separable but not normal, pass to a normal hull. Then we can assume that in the first case k is the fixed field under the action of $G = \mathrm{Aut}_k(K)$. In general any submodule $N_K \subset E \otimes K$ is of the form $N_K = N_k \otimes K$ for some submodule $N_k \subset E$ if and only if N_K is invariant under the induced action of G on N_K. This applies to all members of the Harder-Narasimhan filtration: For any $g \in G$, $g(\text{HN}_\bullet(E \otimes K))$ is again an HN-filtration, and hence coincides with $\text{HN}_\bullet(E \otimes K)$.

In the second case, the algebra $A = \text{Der}_k(K)$ acts on $E \otimes K$, and $N_K \subset E \otimes K$ can be written as $N_K = N_k \otimes K$ for some $N_k \subset E$ if and only if $\delta(N_K) \subset N_K$ for all $\delta \in A$ (Jacobson descent). Let $F = \text{HN}_i(E \otimes K)$ and consider the composition

$$\psi : F \longrightarrow E \otimes K \xrightarrow{\delta} E \otimes K \longrightarrow (E \otimes K)/F.$$

Though δ certainly is not K-linear, the composition ψ is:

$$\psi(f \cdot \lambda) = \psi(f) \cdot \lambda + f \cdot \delta(\lambda) = \psi(f) \cdot \lambda \quad \text{mod } F.$$

Lemma 1.3.3 imlies $\psi = 0$. This means $\delta(F) \subset F$, we are done. $\qquad\square$

A special case of the theorem is the following:

Corollary 1.3.8 — *If E is a semistable sheaf and K is a field extension of k, then $E \otimes_k K$ is semistable as well.* $\qquad\square$

Example 1.3.9 — Here we provide an example of a stable sheaf which is not geometrically stable. Let $X = \text{Proj}(\mathbb{R}[x_0, x_1, x_2]/(x_0^2 + x_1^2 + x_2^2))$ and let $\mathbb{H}$ be the skew field of real quaternions, i.e. the real algebra with generators I, J and K and relations $I \cdot J = K = -J \cdot I$ and $I^2 = J^2 = K^2 = -1$. Define a homomorphism

$$\varphi : \mathbb{H} \otimes_{\mathbb{R}} \mathcal{O}_X(-1) \longrightarrow \mathbb{H} \otimes_{\mathbb{R}} \mathcal{O}_X$$

of $\mathbb{H} \otimes_{\mathbb{R}} \mathcal{O}_X$-*left* bimodules as *right*-multiplication by the element $I \otimes x_0 + J \otimes x_1 + K \otimes x_2$. The $\mathbb{H} \otimes_{\mathbb{R}} \mathcal{O}_X$-structure inherited by $F := \text{coker}(\varphi)$ induces an $\mathbb{R}$-algebra homomorphism $\mathbb{H} \to \text{End}_X(F)$, which is injective as $\mathbb{H}$ is a skew field. Complexifying, we get identifications

$$i : \mathbb{P}_{\mathbb{C}}^1 = \text{Proj}(\mathbb{C}[u, v]) \cong X \times \text{Spec}(\mathbb{C}), \quad i^* \mathcal{O}_X(1) \cong \mathcal{O}_{\mathbb{P}_{\mathbb{C}}^1}(2)$$

via

$$x_0 = \frac{1}{2}(u^2 - v^2), \quad x_1 = uv, \quad x_2 = \frac{i}{2}(u^2 + v^2)$$

and $\mathbb{H} \otimes_{\mathbb{R}} \mathbb{C} \cong M_2(\mathbb{C})$ with

$$I = \begin{pmatrix} 0 & -1 \\ 1 & 0 \end{pmatrix}, \quad J = \begin{pmatrix} i & 0 \\ 0 & -i \end{pmatrix}, \quad K = \begin{pmatrix} 0 & i \\ i & 0 \end{pmatrix}.$$

With respect to these identifications, $\varphi_{\mathbb{C}}$ is right multiplication by

$$\begin{pmatrix} uvi & -u^2 \\ -v^2 & -uvi \end{pmatrix} = \begin{pmatrix} u \\ vi \end{pmatrix} \cdot \begin{pmatrix} vi & -u \end{pmatrix},$$

so that $\varphi_{\mathbb{C}}$ factors as follows

$$M_2(\mathbb{C}) \otimes \mathcal{O}_{\mathbb{P}^1_{\mathbb{C}}}(-2) \xrightarrow{\begin{pmatrix} u \\ vi \end{pmatrix}} \mathbb{C}^2 \otimes \mathcal{O}_{\mathbb{P}^1_{\mathbb{C}}}(-1) \xrightarrow{\cdot \begin{pmatrix} vi & -u \end{pmatrix}} M_2(\mathbb{C}) \otimes \mathcal{O}_{\mathbb{P}^1_{\mathbb{C}}}.$$

From this we get $i^* F_{\mathbb{C}} \cong \mathbb{C}^2 \otimes \mathcal{O}_{\mathbb{P}^1_{\mathbb{C}}}(1)$. Since $i^* F_{\mathbb{C}}$ is locally free and semistable by 1.3.8, the same holds for F, but obviously F is not geometrically stable. Moreover, by the flat base extension theorem, $\dim_{\mathbb{R}} \operatorname{End}_X(F) = \dim_{\mathbb{C}} \operatorname{End}_{\mathbb{P}^1_{\mathbb{C}}}(\mathcal{O}_{\mathbb{P}^1_{\mathbb{C}}}(1)^2) = 4$ which implies $\mathbb{H} \cong \operatorname{End}_X(F)$. We claim that F is stable. For otherwise there would exist a short exact sequence $0 \to \mathcal{L} \to F \to \mathcal{L}' \to 0$ with line bundles $\mathcal{L}$ and $\mathcal{L}'$ of the same degree. Comparison with the complexified situation implies that $F \cong \mathcal{L} \oplus \mathcal{L}' \cong \mathcal{L}^{\oplus 2}$ which leads to the contradiction $\operatorname{End}_X(F) \cong M_2(\mathbb{R}) \ncong \mathbb{H}$. $\qquad\square$

1.4 An Example

Here we want to show that the cotangent bundle of the projective space is stable and at the same time supply ourselves with some detailed information which will be needed later in the proof of Flenner's Restriction Theorem 7.1.1. At one point in the proof we will use the existence and the uniqueness of the Harder-Narasimhan filtration.

Let k be algebraically closed and of characteristic 0. Let $n \geq 2$ be an integer and V a k-vector space of dimension $n+1$. We want to study a sequence of vector bundles on $\mathbb{P}(V)$ related to the cotangent bundle $\Omega = \Omega_{\mathbb{P}(V)}$. It is well known that the cotangent bundle is given by the Euler sequence

$$0 \to \Omega(1) \to V \otimes \mathcal{O}_{\mathbb{P}} \to \mathcal{O}_{\mathbb{P}}(1) \to 0 \tag{1.2}$$

Here the homomorphism $\delta : V \otimes \mathcal{O}_X \to \mathcal{O}_{\mathbb{P}}(1)$ is the evaluation map for the global sections of $\mathcal{O}_{\mathbb{P}}(1)$. Symmetrizing sequence (1.2) we get exact sequences

$$0 \longrightarrow S^d(\Omega(1)) \longrightarrow S^d V \otimes \mathcal{O}_{\mathbb{P}} \xrightarrow{\delta_d} S^{d-1} V \otimes \mathcal{O}_{\mathbb{P}}(1) \longrightarrow 0, \tag{1.3}$$

where the map δ_d at a closed point corresponding to a hyperplane $W \subset V$ is given by

$$(v_1 \vee \ldots \vee v_d) \otimes 1 \mapsto \sum_{i=1}^{d} (v_1 \ldots \vee \hat{v}_i \vee \ldots \vee v_d) \otimes (v_i \bmod W).$$

The assumption that the characteristic of k be zero is necessary for the surjectivity of δ_d.

More general, we consider the epimorphisms

$$\delta_d^i := \delta_{d-i+1}(i-1) \circ \dots \circ \delta_d : S^d V \otimes \mathcal{O} \longrightarrow S^{d-i} V \otimes \mathcal{O}(i),$$

where $\delta_{d-i+1}(i-1)$ is short for $\delta_{d-i+1} \otimes \mathrm{id}_{\mathcal{O}(i-1)}$, and we agree that $\delta_d^0 = \mathrm{id}$ and $\delta_d^i = 0$ for $i > d$. In particular, $\delta_d^1 = \delta_d$. The subbundles $\mathcal{K}_d^i := \ker(\delta_d^i)$, $i = 0, \dots, d+1$, form a filtration

$$0 = \mathcal{K}_d^0 \subset \mathcal{K}_d^1 \subset \dots \subset \mathcal{K}_d^d \subset \mathcal{K}_d^{d+1} = S^d V \otimes \mathcal{O}, \tag{1.4}$$

with factors of the same nature:

Lemma 1.4.1 — *For $0 < i < j \le d+1$ there are natural short exact sequences*

$$0 \to \mathcal{K}_d^i \to \mathcal{K}_d^j \to \mathcal{K}_{d-i}^{j-i}(i) \to 0.$$

If $j = i + 1$, the sequence is non-split.

Proof. The first claim follows from the identity $\delta_d^j = \delta_{d-i}^{j-i}(i) \circ \delta_d^i$. In particular, for $i = j - 1$ one gets:

$$\mathcal{K}_d^{i+1}/\mathcal{K}_d^i = \mathcal{K}_{d-i}^1(i) = S^{d-i}(\Omega(1)) \otimes \mathcal{O}(i).$$

If the corresponding short exact sequence were split, there would be a non-trivial homomorphism

$$S^{d-i}(\Omega(1)) \longrightarrow \mathcal{K}_d^{i+1}(-i) \longrightarrow S^d(V) \otimes \mathcal{O}(-i).$$

On the other hand, applying $\mathrm{Hom}(.\, , S^d(V) \otimes \mathcal{O}(-i))$ to the short exact sequence (1.3) (with d replaced by $d - i$), one gets the exact sequence

$$\mathrm{Hom}(S^{d-i}V \otimes \mathcal{O}, S^d V \otimes \mathcal{O}(-i)) \quad \to \quad \mathrm{Hom}(S^{d-i}(\Omega(1)), S^d V \otimes \mathcal{O}(-i)) \to$$
$$\to \quad \mathrm{Ext}^1(S^{d-i-1}V \otimes \mathcal{O}(1), S^d V \otimes \mathcal{O}(-i)),$$

where the exterior terms vanish, and hence the one in the middle as well. $\qquad\square$

Lemma 1.4.2 — *The slopes of the sheaves $\mathcal{K}_d^i$ satisfy the following relations:*

$$i) \quad \mu(S^d(\Omega(1))) = -\frac{d}{n}$$

$$ii) \quad \mu(\mathcal{K}_d^1) < \mu(\mathcal{K}_d^2) < \dots < \mu(\mathcal{K}_d^d) < 0.$$

Proof. From the exact sequence (1.3) we deduce:

$$\mu(S^d(\Omega(1))) = -\frac{\dim S^{d-1}V}{\dim S^d V - \dim S^{d-1}V} = -\frac{\binom{n+d-1}{n}}{\binom{n+d}{n} - \binom{n+d-1}{n}} = -\frac{d}{n}.$$

Therefore, the slopes

$$\mu(\mathcal{K}_d^{i+1}/\mathcal{K}_d^i) = \mu(S^{d-i}(\Omega(1)) \otimes \mathcal{O}(i)) = i - \frac{d-i}{n} = i\left(1 + \frac{1}{n}\right) - \frac{d}{n}$$

are strictly increasing with i. Since the last term of the sequence is $\mu(\mathcal{K}_d^{d+1}) = \mu(S^d V \otimes \mathcal{O}) = 0$, the lemma is proved. $\qquad\square$

The group $\mathrm{SL}(V)$ acts naturally on $\mathbb{P}(V)$. The sheaves $\mathcal{O}(\ell)$, $S^d V \otimes \mathcal{O}$ and Ω also carry a natural $\mathrm{SL}(V)$-action with respect to which the homomorphisms δ_d^i are equivariant.

Lemma 1.4.3 — *The vector bundles $S^d(\Omega(1))$ have no proper invariant subsheaves.*

Proof. Any invariant subsheaf G must necessarily be a subbundle, since $\mathrm{SL}(V)$ acts transitively on $\mathbb{P}(V)$. Let $x \in \mathbb{P}(V)$ be a closed point corresponding to a hyperplane $W \subset V$. The isotropy subgroup $\mathrm{SL}(V)_x$ acts via the canonical surjection $\mathrm{SL}(V)_x \to \mathrm{GL}(W)$ on the fibre $S^d(\Omega(1))(x) = S^d W$. For any invariant subbundle G the fibre $G(x) \subset S^d W$ is an $\mathrm{GL}(W)$-subrepresentation. But $S^d W$ is an irreducible representation, so that $G(x) = 0$ or $= S^d W$, which means $G = 0$ or $G = S^d(\Omega(1))$. $\qquad\square$

Lemma 1.4.4 — *The bundles $\mathcal{K}_d^i$ are the only invariant subsheaves of $S^d V \otimes \mathcal{O}$.*

Proof. We proceed by induction on d. The case $d = 0$ is trivial. Hence, assume that $d > 0$ and that the assertion is true for all $d' < d$. Let $G \subset S^d V \otimes \mathcal{O}$ be a proper invariant subbundle. Then $G_i := G \cap \mathcal{K}_d^i \subset \mathcal{K}_d^i$ and $\overline{G}_i = G_i/G_{i-1} \subset S^{d+1-i}(\Omega(1)) \otimes \mathcal{O}(i-1)$ are also invariant subbundles. Let i be minimal with $G_i \neq 0$. Then $G_i \cong \overline{G}_i = S^{d+1-i}(\Omega(1)) \otimes \mathcal{O}(i-1)$ because of 1.4.3. But this isomorphism provides a splitting of the exact sequence

$$0 \to \mathcal{K}_d^{i-1} \to \mathcal{K}_d^i \to S^{d+1-i}(\Omega(1)) \otimes \mathcal{O}(i-1) \to 0.$$

According to Lemma 1.4.1 this is impossible unless $i = 1$. Since $\mathcal{K}_d^1 \cong S^d(\Omega(1))$ is irreducible, $G_1 = \mathcal{K}_d^1$. Therefore, let $\nu \geq 1$ be the maximal index such that $G_\nu = \mathcal{K}_d^\nu$. If $G = G_\nu$ we are done. If not, $G' := G/G_\nu$ is a proper invariant subbundle of $S^d V \otimes \mathcal{O}/\mathcal{K}_d^\nu = S^{d-\nu} V \otimes \mathcal{O}(\nu)$. By the induction hypothesis

$$\overline{G}_{\nu+1} = G_{\nu+1}/G_\nu \cong G' \cap \mathcal{K}_{d-\nu}^1(\nu) = \mathcal{K}_{d-\nu}^1(\nu) = \mathcal{K}_d^{\nu+1}/\mathcal{K}_d^\nu,$$

so that $G_{\nu+1} = \mathcal{K}_d^{\nu+1}$ contradicting the maximality of ν. $\qquad\square$

Lemma 1.4.5 — *The vector bundles $\mathcal{K}_d^i$ are semistable. Moreover, $\Omega(1) = \mathcal{K}_1^1$ is μ-stable, hence stable.*

Proof. The Harder-Narasimhan filtration of $\mathcal{K}_d^i$ is invariant under the action of $\mathrm{SL}(V)$ because of its uniqueness. By the previous lemma, all subsheaves of the Harder-Narasimhan filtration also appear in the filtration (1.4). But according to Lemma 1.4.2 one has $\mu(\mathcal{K}_d^j) < \mu(\mathcal{K}_d^i)$ for all $j < i$. Hence, none of these bundles can have a bigger reduced Hilbert polynomial than $\mathcal{K}_d^i$, i.e. $\mathcal{K}_d^i$ is semistable. The last assertion follows from $\mu(\Omega(1)) = -1/n$, since μ-semistability implies μ-stability whenever degree and rank are coprime (1.2.14). $\square$

1.5 Jordan-Hölder Filtration and S-Equivalence

Just as the Harder-Narasimhan filtration splits every sheaf in semistable factors the Jordan-Hölder filtration splits a semistable sheaf in its stable components. More precisely,

Definition 1.5.1 — *Let E be a semistable sheaf of dimension d. A Jordan-Hölder filtration of E is a filtration*

$$0 = E_0 \subset E_1 \subset \ldots \subset E_\ell = E,$$

such that the factors $gr_i(E) = E_i/E_{i-1}$ are stable with reduced Hilbert polynomial $p(E)$.

Note that the sheaves E_i, $i > 0$, are also semistable with Hilbert polynomial $p(E)$. Taking the direct sum of two line bundles of the same degree one immediately finds that a Jordan-Hölder filtration need not be unique.

Proposition 1.5.2 — *Jordan-Hölder filtrations always exist. Up to isomorphism, the sheaf $gr(E) := \bigoplus_i gr_i(E)$ does not depend on the choice of the Jordan-Hölder filtration.*

Proof. Any filtration of E by semistable sheaves with reduced Hilbert polynomial $p(E)$ has a maximal refinement, whose factors are necessarily stable. Now, suppose that $E_\bullet$ and $E'_\bullet$ are two Jordan-Hölder filtrations of length ℓ and ℓ', respectively, and assume that the uniqueness of $gr(F)$ has been proved for all F with $\alpha_d(F) < \alpha_d(E)$, where d is the dimension of E and α_d is the multiplicity. Let i be minimal with $E_1 \subset E'_i$. Then the composite map $E_1 \to E'_i \to E'_i/E'_{i-1}$ is non-trivial and therefore an isomorphism, for both E_1 and E'_i/E'_{i-1} are stable and $p(E_1) = p(E'_i/E'_{i-1})$. Hence $E'_i \cong E'_{i-1} \oplus E_1$, so that there is a short exact sequence

$$0 \to E'_{i-1} \to E/E_1 \to E/E'_i \to 0.$$

The sheaf $F = E/E_1$ inherits two Jordan-Hölder filtrations: firstly, let $F_j = E_{j+1}/E_1$ for $j = 0, \ldots, \ell - 1$. And secondly, let $F'_j = E'_j$ for $j = 0, \ldots, i - 1$ and let F'_j be the preimage of E'_{j+1}/E'_i for $j = i, \ldots, \ell' - 1$. The induction hypothesis applied to F gives

$\ell = \ell'$ and

$$\bigoplus_{j \neq 1} E_j / E_{j-1} \cong \bigoplus_{j \neq i} E'_j / E'_{j-1}.$$

Since $E_1 \cong E'_i / E'_{i-1}$, we are done. □

Definition 1.5.3 — *Two semistable sheaves E_1 and E_2 with the same reduced Hilbert polynomial are called S-equivalent if $gr(E_1) \cong gr(E_2)$.*

The importance of this definition will become clear in Section 4. Roughly, the moduli space of semistable sheaves parametrizes only S-equivalence classes of semistable sheaves.

We conclude this section by introducing the concepts of polystable sheaves and of the socle and the extended socle of a semistable sheaf.

Definition 1.5.4 — *A semistable sheaf E is called polystable if E is the direct sum of stable sheaves.*

As we saw above, every S-equivalence class of semistable sheaves contains exactly one polystable sheaf up to isomorphism. Thus, the moduli space of semistable sheaves in fact parametrizes polystable sheaves.

Lemma 1.5.5 — *Every semistable sheaf E contains a unique non-trivial maximal polystable subsheaf of the same reduced Hilbert polynomial. This sheaf is called the socle of E.*

Proof. Any semistable sheaf E admits a Jordan-Hölder filtration. Thus there always exists a non-trivial stable subsheaf with Hilbert polynomial $p(E)$. If there were two maximal polystable subsheaves, then, similarly to the proof of 1.5.2, one inductively proves that every direct summand of the first also appears in the second. □

Definition 1.5.6 — *The extended socle of a semistable sheaf E is the maximal subsheaf $F \subset E$ with $p(F) = p(E)$ and such that all direct summands of $gr(F)$ are direct summands of the socle.*

Lemma 1.5.7 — *Let F be the extended socle of a semistable sheaf E. Then there are no non-trivial homomorphisms form F to E/F, i.e. $\mathrm{Hom}(F, E/F) = 0$.*

Proof. If $\overline{G} \subset E/F$ is the image of a non-trivial homomorphism $F \to E/F$ and G denotes its pre-image in E, then G contains F properly and the direct summands of $gr(G)$ and $gr(F)$ coincide. This contradicts the maximality of the extended socle. □

Example 1.5.8 — Let X be a curve and let $0 \to L_1 \to E \to L_2 \to 0$ be a non-trivial extension of two line bundles of the same degree. The socle of E is L_1. The extended socle of E is E itself if $L_1 \cong L_2$ and it is L_1 otherwise.

Lemma 1.5.9 — *The socle and the extended socle of a semistable sheaf E are invariant under automorphisms of X and E. Moreover, if E is simple, semistable, and equals its extended socle, then E is stable.*

Proof. The first assertion is clear. Suppose that E is not stable. If E equals its socle F', then E is not simple. Suppose $F' \neq E$. Since the last factor of a Jordan-Hölder filtration of E/F' is isomorphic to a submodule in F' there is a non-trivial homomorphism $E/F' \to F'$, inducing a non-trivial nilpotent endomorphism of E. $\square$

Lemma 1.5.10 — *If E is a simple sheaf, then E is stable if and only if E is geometrically stable.*

Proof. Assume E is simple and stable but not geometrically stable. Let K be a field extension of k. According to the previous lemma, the extended socle E' of $K \otimes E$ is a proper submodule. The extended socle is invariant under all automorphisms of K/k and satisfies the condition $\operatorname{Hom}(E', K \otimes E/E') = 0$. Thus E' is already defined over k. (Compare the arguments in the proof of Theorem 1.3.7.) $\square$

Combined with 1.2.8 this lemma shows:

Corollary 1.5.11 — *If k is algebraically closed and E is a stable sheaf, then E is also gemetrically stable.* $\square$

Remark 1.5.12 — Consider the full subcategory $\mathcal{C}(p)$ of $\operatorname{Coh}(X)$ consisting of all semistable sheaves E with reduced Hilbert polynomial p. Then $\mathcal{C}(p)$ is an abelian category in which all objects are Noetherian and Artinian. All definitions and statements made in this section are just specializations of corresponding definitions and statements within this more general framework. Our stable and polystable sheaves are the simple and semisimple objects in $\mathcal{C}(p)$. Be aware of the very different meanings that the word "simple" assumes in these contexts.

1.6 μ-Semistability

We have encountered already two different stability concepts; using the Hilbert polynomial and the slope, respectively. In fact there are others. We present an approach which allows one to deal with the different stability definitions in a uniform manner. In particular, for μ-stability it takes care of things happening in codimension two which do not effect the stability condition. As it turns out, almost everything we have said about Harder-Narasimhan and Jordan-Hölder filtrations remains valid in the more general framework.

Let us first introduce the appropriate categories.

Definition 1.6.1 — $\mathrm{Coh}_d(X)$ *is the full subcategory of* $\mathrm{Coh}(X)$ *whose objects are sheaves of dimension* $\leq d$.

For two integers $0 \leq d' \leq d \leq \dim(X)$ the category $\mathrm{Coh}_{d'}(X)$ is a full subcategory of $\mathrm{Coh}_d(X)$. In fact, $\mathrm{Coh}_{d'}(X)$ is a Serre subcategory, i.e. it is closed with respect to subobjects, quotients objects and extensions. Therefore, we can form the quotient category.

Definition 1.6.2 — $\mathrm{Coh}_{d,d'}(X)$ *is the quotient category* $\mathrm{Coh}_d(X)/\mathrm{Coh}_{d'-1}(X)$.

Recall that $\mathrm{Coh}_{d,d'}(X)$ has the same objects as $\mathrm{Coh}_d(X)$. A morphism $f : F \to G$ in $\mathrm{Coh}_{d,d'}(X)$ is an equivalence class of diagrams $F \xleftarrow{s} G' \longrightarrow G$ of morphisms in $\mathrm{Coh}_d(X)$ such that $\ker(s)$ and $\mathrm{coker}(s)$ are at most $(d'-1)$-dimensional. G and F are isomorphic in $\mathrm{Coh}_{d,d'}(X)$ if they are isomorphic in dimension d'. Moreover, we say that $E \in \mathrm{Ob}(\mathrm{Coh}_{d,d'}(X))$ is *pure*, if $T_{d-1}(E) \cong 0$ in $\mathrm{Coh}_{d,d'}(X)$, i.e. $T_{d-1}(E) = T_{d'-1}(E)$, and that $F \subset E$ is *saturated*, if E/F is pure in $\mathrm{Coh}_{d,d'}(X)$.

Similarly, if we let $\mathbb{Q}[T]_d = \{P \in \mathbb{Q}[T] | \deg(P) \leq d\}$, then $\mathbb{Q}[T]_{d'-1} \subset \mathbb{Q}[T]_d$ is a linear subspace, and the quotient space $\mathbb{Q}[T]_{d,d'} = \mathbb{Q}[T]_d/\mathbb{Q}[T]_{d'-1}$ inherits a natural ordering. There is a well defined map

$$P_{d,d'} : \mathrm{Coh}_{d,d'}(X) \to \mathbb{Q}[T]_{d,d'},$$

given by taking the residue class of the Hilbert polynomial. For if E and F are d-dimensional sheaves which are isomorphic as objects in $\mathrm{Coh}_{d,d'}(X)$ then $P(E,m) = P(F,m)$ modulo terms of degree $< d'$. In particular, $P_{d,d'}(E) = 0$ if and only if $E \cong 0$ in $\mathrm{Coh}_{d,d'}(X)$. The reduced Hilbert polynomials $p_{d,d'}$ are defined analogously.

We can now introduce a notion of stability in the categories $\mathrm{Coh}_{d,d'}(X)$ which generalizes the notion given in Section 1.2:

Definition 1.6.3 — $E \in \mathrm{Ob}(\mathrm{Coh}_{d,d'}(X))$ *is (semi)stable, if and only if E is pure in* $\mathrm{Coh}_{d,d'}(X)$ *and if for all proper non-trivial subsheaves F one has* $p_{d,d'}(F) (\leq) p_{d,d'}(E)$.

Lemma 1.2.13 immediately generalizes to the following

Lemma 1.6.4 — *If E is a pure sheaf of dimension d and $j < i$, then one has:*

$$E \text{ is stable in } \mathrm{Coh}_{d,i}(X) \quad \Rightarrow \quad E \text{ is stable in } \mathrm{Coh}_{d,j}(X)$$

$$\Downarrow$$

$$E \text{ is semistable in } \mathrm{Coh}_{d,i}(X) \quad \Leftarrow \quad E \text{ is semistable in } \mathrm{Coh}_{d,j}(X)$$

Example 1.6.5 — By definition $\mathrm{Coh}_{d,0}(X) = \mathrm{Coh}_d(X)$ and $P_{d,0} = P$. In the case $d' = d - 1$ one has

$$P_{d,d-1}(E) = \alpha_d(E)\frac{T^d}{d!} + \alpha_{d-1}(E)\frac{T^{d-1}}{(d-1)!}$$

in $\mathbb{Q}[T]_{d,d-1}$ and hence

$$p_{d,d-1}(E) = \frac{T^d}{d!} + (\alpha_{d-1}(E)/\alpha_d(E))\frac{T^{d-1}}{(d-1)!}.$$

Hence, for $d = \dim(X)$ and a sheaf E of dimension d the (semi)stability in the category $\mathrm{Coh}_{d,d-1}(X)$ is equivalent to the μ-(semi)stability in the sense of 1.2.12.

The verification of the following meta-theorem is left to the reader.

Theorem 1.6.6 — *All the statements of the previous sections remain true for the categories* $\mathrm{Coh}_{d,d'}(X)$ *if appropriately adopted. The proofs carry over literally.* □

Two results, however, shall be mentioned explicitly.

Theorem 1.6.7 — *i) If E is a sheaf of dimension d and pure as an object of the category* $\mathrm{Coh}_{d,d'}(X)$, *then there exists a unique filtration in* $\mathrm{Coh}_{d,d'}(X)$ *(the Harder-Narasimhan filtration)*

$$0 = E_0 \subset E_1 \subset \ldots \subset E_\ell = E$$

such that the factors E_i/E_{i-1} are semistable in $\mathrm{Coh}_{d,d'}(X)$ *and their reduced Hilbert polynomials satisfy $p_{d,d'}(E_1) > \ldots > p_{d,d'}(E/E_{\ell-1})$.*
ii) If $E \in \mathrm{Ob}(\mathrm{Coh}_{d,d'}(X))$ is semistable, then there exists a filtration in $\mathrm{Coh}_{d,d'}(X)$ *(the Jordan-Hölder filtration)*

$$0 = E_0 \subset E_1 \subset \ldots \subset E_\ell = E$$

such that the factors $E_i/E_{i-1} \in \mathrm{Ob}(\mathrm{Coh}_{d,d'}(X))$ and $p_{d,d'}(E_i/E_{i-1}) = p_{d,d'}(E)$. The graded sheaf $\mathrm{gr}^{JH}(E)$ of the filtration is uniquely determined as an object in $\mathrm{Coh}_{d,d'}(X)$.

Note that for a pure sheaf the Harder-Narasimhan filtration with respect to ordinary stability is a refinement of the Harder-Narasimhan filtration in $\mathrm{Coh}_{d,d'}(X)$, whereas the Jordan-Hölder filtration in $\mathrm{Coh}_{d,d'}(X)$ is a refinement of the standard Jordan-Hölder filtration provided the sheaf E is semistable.

Example 1.6.5 suggests to extend the definition of μ-stability to sheaves of dimension less than $\dim(X)$. We first introduce a modified slope which comes in handy at various places later on.

Definition 1.6.8 — *Let E be a coherent sheaf of dimension d. Then $\alpha_{d-1}(E)/\alpha_d(E)$ is denoted by $\hat{\mu}(E)$. For a polynomial $P = \sum_{i=0}^{d} \alpha_i \frac{m^i}{i!}$ of degree d we write $\hat{\mu}(P) := \alpha_{d-1}/\alpha_d$.*

When working with Hilbert polynomials $\hat{\mu}$ is the more natural slope, but for historical reasons $\mu(E) = \deg(E)/\operatorname{rk}(E)$ for a sheaf of dimension $\dim(X)$ will be used whenever possible. Note that for $d = \dim(X)$ the usual slope $\mu(E)$ differs from $\hat{\mu}(E)$ by the constant factor $\alpha_d(\mathcal{O}_X)$ and the constant term $\alpha_{d-1}(\mathcal{O}_X)$. More precisely, $\mu(E) = \alpha_d(\mathcal{O}_X) \cdot \hat{\mu}(E) - \alpha_{d-1}(\mathcal{O}_X)$.

Definition and Corollary 1.6.9 — *A coherent sheaf E of dimension d is called μ-(semi)-stable if it is (semi)stable as an object in $\operatorname{Coh}_{d,d-1}(X)$. Then, E is μ-(semi)stable if and only if $T_{d-1}(E) = T_{d-2}(E)$ and $\hat{\mu}(F)(\leq)\hat{\mu}(E)$ for all $0 \subsetneq F \subsetneq E$ in $\operatorname{Coh}_{d,d-1}(X)$.*

If $d = \dim(X)$ the Harder-Narasimhan and Jordan-Hölder filtration of a torsion free sheaf considered as an object in $\operatorname{Coh}_{d,d-1}(X)$ are also called μ-Harder-Narasimhan and μ-Jordan-Hölder filtration, respectively. In this case, if E is torsion free and we require that in the Harder-Narasimhan filtration all factors are torsion free, then the filtration is unique in $\operatorname{Coh}(X)$. The maximal and minimal slope of the factors of the μ-Harder-Narasimhan filtration of a torsion free sheaf E will be denoted $\mu_{\max}(E)$ respectively $\mu_{\min}(E)$. On the other hand, for a torsion free μ-semistable sheaf the graded sheaf $gr^{JH}(E)$ is uniquely defined only in codimension one. Since two reflexive sheaves which are isomorphic in codimension one are isomorphic, we have

Corollary 1.6.10 — *If E is a μ-semistable torsion free sheaf of dimension $d = \dim(X)$, then the reflexive hull $gr^{JH}(E)^{\sim}$ of the graded sheaf is independent of the choice of the Jordan-Hölder filtration.* $\square$

The concept of polystability also naturally generalizes to objects in $\operatorname{Coh}_{d,d'}(X)$: a sheaf $E \in \operatorname{Ob}(\operatorname{Coh}_{d,d'}(X))$ is polystable if $E \cong \oplus E_i$ in $\operatorname{Coh}_{d,d'}(X)$, where the sheaves E_i are stable in $\operatorname{Coh}_{d,d'}(X)$ and $p_{d,d'}(E_i) = p_{d,d'}(E)$. Again, for $d' = d - 1$ such a sheaf E is called μ-polystable. Since a saturated sheaf of a locally free sheaf is reflexive and a direct summand of a locally free sheaf is locally free, one has

Corollary 1.6.11 — *A locally free sheaf E on X is polystable in $\operatorname{Coh}_{d,d-1}(X)$ if and only if $E \cong \oplus E_i$ in $\operatorname{Coh}(X)$, where the sheaves E_i are μ-stable locally free sheaves with*

$\mu(E_i) = \mu(E)$. *In this case any saturated non-trivial subsheaf $F \subset E$ with $\mu(F) = \mu(E)$ is a direct summand of E.* □

There are a number of useful formulae for the slope of which we indicate a few. They are easy to prove and left to the reader as an exercise. The reader will also observe that these formulae, which we state for simplicity for $\mu(E)$ with E torsion free, can be generalized to $\hat{\mu}$, $p_{d,d'}$, and p accordingly. So for torsion free sheaves E and F on a normal variety X one has:

- $\mu(E(a)) = \mu(E) + a \cdot \alpha_d(\mathcal{O}_X) = \mu(E) + a \deg(X)$ and analogously for $\mu_{\min}$ and $\mu_{\max}$.
- $\mu_{\min}(E \oplus F) = \min\{\mu_{\min}(E), \mu_{\min}(F)\}$ and analogously for $\mu_{\max}$.
- $\mu_{\min}(F) \geq \mu_{\min}(F)$ for $E \twoheadrightarrow F$ and $\mu_{\max}(E) \leq \mu_{\max}(F)$ for $E \hookrightarrow F$.
- $\mu(E \otimes F) = \mu(E) + \mu(F)$ and $\mu_{\min}(E \otimes F) = \mu_{\min}(E) + \mu_{\min}(F)$ and similarly for $\mu_{\max}$. (In the last equality only $\leq$ is elementary. The other inequality needs the semi-stability of tensor products which is Theorem 3.1.4.)

1.7 Boundedness I

In order to construct moduli spaces one first has to ensure that the set of sheaves one wants to parametrize is not too big. In fact, this is one of the two reasons why one restricts attention to semistable sheaves. As we eventually will show in Section 3.3 the family of semistable sheaves is bounded, i.e. it is reasonably small. This problem is rather intriguing. Here, we give the basic definitions, discuss some fundamental results and prove the boundedness of semistable sheaves on a smooth projective curve.

Let X be a projective scheme over a field k and let $\mathcal{O}(1)$ be a very ample line bundle.

Definition 1.7.1 — *Let m be an integer. A coherent sheaf F is said to be m-regular, if*

$$H^i(X, F(m - i)) = 0 \quad \text{for all } i > 0.$$

For the proof of the next lemma we refer the reader to [192] or [124].

Lemma 1.7.2 — *If F is m-regular, then the following holds:*
i) F is m'-regular for all integers $m' \geq m$. ii) $F(m)$ is globally generated.
iii) The natural homomorphisms $H^0(X, F(m)) \otimes H^0(X, \mathcal{O}(n)) \to H^0(X, F(m+n))$ are surjective for $n \geq 0$.

Because of Serre's vanishing theorem, for any sheaf F there is an integer m such that F is m-regular. And because of *i)* the following definition makes sense:

Definition 1.7.3 — *The Mumford-Castelnuovo regularity of a coherent sheaf F is the number* $\mathrm{reg}(F) = \inf\{m \in \mathbb{Z}|F \text{ is } m\text{-regular}\}$.

The regularity is $\mathrm{reg}(F) = -\infty$ if and only if F is 0-dimensional.
The following important proposition allows to estimate the regularity of a sheaf F in terms of its Hilbert polynomial and the number of global sections of the restriction of F to a sequence of iterated hyperplane sections. For the proof we again refer to [124].

Proposition 1.7.4 — *There are universal polynomials $P_i \in \mathbb{Q}[T_0, \ldots, T_i]$ such that the following holds: Let F be a coherent sheaf of dimension $\leq d$ and let $H_1, \ldots, H_d$ be an F-regular sequence of hyperplane sections. If $\chi(F|_{\cap_{j\leq i}H_j}) = a_i$ and $h^0(F|_{\cap_{j\leq i}H_j}) \leq b_i$ then*

$$\mathrm{reg}(F) \leq P_d(a_0 - b_0, a_1 - b_1, \ldots, a_d - b_d).$$

$\square$

Definition 1.7.5 — *A family of isomorphism classes of coherent sheaves on X is bounded if there is a k-scheme S of finite type and a coherent $\mathcal{O}_{S \times X}$-sheaf F such that the given family is contained in the set $\{F|_{\mathrm{Spec}(k(s)) \times X}|s \text{ a closed point in } S\}$.*

Note that later we use the word family of sheaves in a different setting (cf. Chapter 2.) Here it still has its set-theoretical meaning.

Lemma 1.7.6 — *The following properties of a family of sheaves $\{F_\iota\}_{\iota \in I}$ are equivalent:*
i) The family is bounded.
ii) The set of Hilbert polynomials $\{P(F_\iota)\}_{\iota \in I}$ is finite and there is a uniform bound $\mathrm{reg}(F_\iota) \leq \rho$ for all $\iota \in I$.
iii) The set of Hilbert polynomials $\{P(F_\iota)\}_{\iota \in I}$ is finite and there is a coherent sheaf F such that all F_ι admit surjective homomorphisms $F \to F_\iota$.

$\square$

As an example consider the family of locally free sheaves on $\mathbb{P}^1$ with Hilbert polynomial $P(m) = 2m + 2$, that is, bundles of rank 2 and degree 0. We know that any such sheaf is isomorphic to $F_a := \mathcal{O}(a) \oplus \mathcal{O}(-a)$ for some $a \geq 0$. And it is clear that $\mathrm{reg}(F_a) = a$. In particular, this family cannot be bounded, since the regularity can get arbitrarily large. The lemma already suffices to prove the boundedness of semistable sheaves on curves:

Corollary 1.7.7 — *The family of semistable sheaves with fixed Hilbert polynomial P on a smooth projective curve is bounded.*

Proof. The family of zero-dimensional sheaves with fixed Hilbert polynomial, i.e. of fixed length, is certainly bounded. Any integer can be taken as a uniform regularity. For one-dimensional semistable sheaves one applies Serre duality

$$H^1(X, E(m-1)) = \operatorname{Hom}(E, \omega_X(1-m))^\vee.$$

The latter space vanishes due to the semistability of E if

$$m > \frac{2g(X) - 2 - d/r}{\deg(\mathcal{O}(1))} + 1,$$

where d and r are given by $P = r(\deg(\mathcal{O}(1)) \cdot m + 1 - g) + d$. $\qquad\square$

Combining Lemma 1.7.6 with Proposition 1.7.4 we get the following crucial boundedness criterion:

Theorem 1.7.8 (Kleiman Criterion) — *Let $\{F_\iota\}$ be a family of coherent sheaves on X with the same Hilbert polynomial P. Then this family is bounded if and only if there are constants C_i, $i = 0, \ldots, d = \deg(P)$ such that for every F_ι there exists an F_ι-regular sequence of hyperplane sections $H_1, \ldots, H_d$, such that $h^0(F|_{\cap_{j \le i} H_j}) \le C_i$.* $\qquad\square$

Next, we prove a useful boundedness result for quotient sheaves of a given sheaf.

Lemma 1.7.9 (Grothendieck) — *Let P be a polynomial and ρ an integer. Then there is a constant C depending only on P and ρ such that the following holds: if X is a projective k-scheme with a very ample line bundle $\mathcal{O}(1)$, E is a d-dimensional sheaf with Hilbert polynomial P and Mumford-Castelnuovo regularity $\operatorname{reg}(E) \le \rho$ and if F is a purely d-dimensional quotient sheaf of E then $\hat{\mu}(F) \ge C$. Moreover, the family of purely d-dimensional quotients F with $\hat{\mu}(F)$ bounded from above is bounded.*

Proof. We can assume that X is a projective space: choose an embedding $j : X \to \mathbb{P}^N$ and replace E by j_*E. Then we can choose a linear subspace L in $\mathbb{P}^N$ of dimension $N - d - 1$ disjoint from $\operatorname{Supp}(E)$. The linear projection $\pi : \mathbb{P}^N - L \to \mathbb{P}^d$ induces a finite map $\pi : \operatorname{Supp}(E) \to \mathbb{P}^d$ with $\pi^*(\mathcal{O}_{\mathbb{P}^d}(1)) = \mathcal{O}_{\operatorname{Supp}(E)}(1)$. If G is a coherent sheaf on $\operatorname{Supp}(E)$, then $G' = \pi_*G$ is also coherent, and if G is purely d-dimensional, then the same is true for G', which in this case is the same as saying that G' is torsion free. Moreover, using the projection formula, we see that G and G' have the same Hilbert polynomial, regularity and $\hat{\mu}$. But this implies, that we can safely replace E by E' and hence assume that E is a coherent sheaf of dimension d on $\mathbb{P}^d$. The assumption on the regularity allows to write down a surjective homomorphism

$$G := V \otimes \mathcal{O}_{\mathbb{P}^d}(-\rho) \longrightarrow E,$$

where V is a vector space of dimension $P(\rho)$. Note that the bundle G depends on P and ρ only. Any quotient of E is a quotient of G as well, and we may therefore replace E by G. Let $q : G \to F$ be a surjective homomorphism onto a torsion free coherent sheaf of rank $0 < s \le \mathrm{rk}(G) = P(\rho)$. Then q induces a generically surjective homomorphism

$$\Lambda^s q : \Lambda^s G = \Lambda^s V \otimes \mathcal{O}_{\mathbb{P}^d}(-s\rho) \longrightarrow \det(F) \cong \mathcal{O}_{\mathbb{P}^d}(\deg(F)).$$

This shows that $\deg(F) \ge -s\rho$, and hence $\hat{\mu}(F) \ge \rho + \alpha_{d-1}(\mathcal{O}_{\mathbb{P}^d})$ is uniformly bounded. This proves the first part of the theorem. Now fix C'. In order to prove the second assertion it is enough to show that the family of pure quotient sheaves F of rank $0 < s \le \mathrm{rk}(G) = P(\rho)$ and with $\ell := \deg(F) = s \cdot (C' - \alpha_{d-1}(\mathcal{O}_{\mathbb{P}^d}))$ is bounded. For a given quotient $q : G \to F$ with $\deg(F) = \ell$ and $\mathrm{rk}(F) = s$ consider the induced homomorphism

$$\psi : G \otimes \Lambda^{s-1} G \xrightarrow{\ \wedge\ } \Lambda^s G \xrightarrow{\ \det(q)\ } \mathcal{O}(\ell)$$

and the adjoint homomorphism

$$\hat{\psi} : G \to \mathcal{O}(\ell) \otimes \Lambda^{s-1} G^{\vee}.$$

Let $U \subset \mathbb{P}^d$ denote the dense open subscheme where F is locally free. Then $\ker(\hat{\psi})|_U = \ker(q)|_U$. Since the quotients of G corresponding to these two subsheaves of G are torsion free and since they coincide on a dense open subscheme of $\mathbb{P}^d$, we must have $\ker(\hat{\psi}) = \ker(q)$ everywhere, i.e. $F \cong \mathrm{im}(\hat{\psi})$. Now, the family of such image sheaves certainly is bounded.   □

Remark 1.7.10 — Note that in particular the set of Hilbert polynomials of pure quotients with fixed $\hat{\mu}(F)$ is finite.

Comments:
— The presentation of the homological algebra in Section 1.1 is inspired by Le Potier's article [147]. The reader may also consult the books of Okonek, Schneider, Spindler [212] and of Kobayashi [127]. For the details concerning the definition of the determinant 1.1.17 of a coherent sheaf see the article of Knudson and Mumford [126].
— The concept of stable vector bundles on curves goes back to Mumford [191] and was later generalized by Takemoto [244] to μ-stable vector bundles on higher dimensional varieties. The notion of stability using the Hilbert polynomial appears first in Gieseker's paper [77] for sheaves on surfaces and in Maruyama's paper [162] for sheaves on varieties of arbitrary dimension. Later Simpson introduced pure sheaves and their stability ([240], also [145]). This led him to consider the multiplicity $\hat{\mu}$ of a coherent sheaf instead of the slope μ.
— In modern language Theorem 1.3.1 was proved by Grothendieck in [92].
— The Harder-Narasimhan filtration, as the name suggest, was introduced by Harder and Narasimhan in [95]. For generalizations see articles by Maruyama or Shatz [164], [238]. In particular, 1.3.4 in the general form was proved in [238]. Another important notion is the notion of the Harder-Narasimhan polygon which can also be found in Shatz' paper [238].

— The example in Section 1.4 is due to Flenner [63]. For other results concerning bundles on projective spaces see [212].

— S-equivalence was again first defined for bundles on curves by Seshadri [235]. There, two S-equivalent sheaves are called strongly equivalent.

— Langton defined the socle and the extended socle in [135]. For another reference see the paper of Mehta and Ramanathan [176].

— Definitions 1.7.1, 1.7.3 and Lemma 1.7.2 can be found in Mumford's book [192]. Proposition 1.7.4 is proved in [192] for the special case of ideal sheaves and in general in Kleiman's exposé in [124]. Lemmas 1.7.6 and 1.7.9 are taken from [93].

2

Families of Sheaves

In the first chapter we proved some elementary properties of coherent sheaves related to semistability. The main topic of this chapter is the question how these properties vary in algebraic families. A major technical tool in the investigations here is Grothendieck's Quot-scheme. We give a complete existence proof in Section 2.2 and discuss its infinitesimal structure. As an application of this construction we show that the property of being semistable is open in flat families and that for flat families the Harder-Narasimhan filtrations of the members of the family form again flat families, at least generically. In the appendix the notion of the Quot-scheme is slightly generalized to Flag-schemes. We sketch some parts of deformation theory of sheaves and derive important dimension estimates for Flag-schemes that will be used in Chapter 4 to get similar a priori estimates for the dimension of the moduli space of semistable sheaves. In the second appendix to this chapter we prove a theorem due to Langton, which roughly says that the moduli functor of semistable sheaves is proper (cf. Chapter 4 and Section 8.2).

2.1 Flat Families and Determinants

Let $f : X \to S$ be a morphism of finite type of Noetherian schemes. If $g : T \to S$ is an S-scheme we will use the notation X_T for the fibre product $T \times_S X$, and $g_X : X_T \to X$ and $f_T : X_T \to T$ for the natural projections. For $s \in S$ the fibre $f^{-1}(s) = \mathrm{Spec}(k(s)) \times_S X$ is denoted X_s. Similarly, if F is a coherent $\mathcal{O}_X$-module, we write $F_T := g_X^* F$ and $F_s = F|_{X_s}$. Often, we will think of F as a collection of sheaves F_s parametrized by $s \in S$. The requirement that the sheaves F_s and their properties should vary 'continuously' is made precise by the following definition:

Definition 2.1.1 — *A flat family of coherent sheaves on the fibres of f is a coherent $\mathcal{O}_X$-module F which is flat over S.*

Recall that this means that for each point $x \in X$ the stalk F_x is flat over the local ring $\mathcal{O}_{S,f(x)}$. If F is S-flat, then F_T is T-flat for any base change $T \to S$. If $0 \to F' \to F \to F'' \to 0$ is a short exact sequence of coherent $\mathcal{O}_X$-sheaves and if F'' is S-flat then F' is S-flat if and only if F is S-flat. If $X \cong S$ then F is S-flat if and only if F is locally free.

A special case that will occur frequently in these notes is the following: k is a field, S and Y are k-schemes and $X = S \times_k Y$. In this situation the natural projections will almost always be denoted by $p : X \to S$ and $q : X \to Y$, and we will say 'sheaves on Y' rather than 'sheaves on the fibres of p'.

Assume from now on that $f : X \to S$ is a projective morphism and that $\mathcal{O}_X(1)$ is an f-ample line bundle on X, i.e. the restriction of $\mathcal{O}_X(1)$ to any fibre X_s is ample. Let F be a coherent $\mathcal{O}_X$-module. Consider the following assertions:

1. F is S-flat
2. For all sufficiently large m the sheaves $f_*(F(m))$ are locally free.
3. The Hilbert polynomial $P(F_s)$ is locally constant as a function of $s \in S$.

Proposition 2.1.2 — *There are implications $1 \Leftrightarrow 2 \Rightarrow 3$. If S is reduced then also $3 \Rightarrow 1$.*

Proof. Thm. III 9.9 in [98] □

This provides an important flatness criterion. If S is not reduced, it is easy to write down counterexamples to the implication $3 \Rightarrow 1$. However, in the non-reduced case the following criteria are often helpful:

Lemma 2.1.3 — *Let $S_0 \subset S$ be a closed subscheme defined by a nilpotent ideal sheaf $\mathcal{I} \subset \mathcal{O}_S$. Then F is S-flat if and only if F_{S_0} is S_0-flat and the natural multiplication map $\mathcal{I} \otimes_{\mathcal{O}_S} F \to \mathcal{I}F$ is an isomorphism.* □

Lemma 2.1.4 — *Let $0 \to F' \to F \to F'' \to 0$ be a short exact sequence of $\mathcal{O}_X$-modules. If F is S-flat, then F'' is S-flat if and only if for each $s \in S$ the homomorphism $F'_s \to F_s$ is injective.* □

For proofs see Thm. 49 and its Cor. in [172].
The following theorem of Mumford turns out to be extremely useful as it allows us to 'flatten' any coherent sheaf by splitting up the base scheme in an appropriate way.

Theorem 2.1.5 — *Let $f : X \to S$ be a projective morphism of Noetherian schemes, let $\mathcal{O}(1)$ be an invertible sheaf on X which is very ample relative S, and let F be a coherent $\mathcal{O}_X$-module. Then the set $\mathcal{P} = \{P(F_s) | s \in S\}$ of Hilbert polynomials of the fibres of F is* *finite. Moreover, there are finitely many locally closed subschemes $S_P \subset S$, indexed by the polynomials $P \in \mathcal{P}$, with the following properties:*

1. *The natural morphism $j : \coprod_P S_P \to S$ is a bijection.*
2. *If $g : S' \to S$ is a morphism of Noetherian schemes, then $g_X^* F$ is flat over S' if and only if g factorizes through j.*

Such a decomposition is called a *flattening stratification* of S for F. It is certainly unique. We begin with a weaker version of this due to Grothendieck:

Lemma 2.1.6 — *Under the assumptions of the theorem there exist finitely many pairwise disjoint locally closed subschemes S_i of S which cover S such that F_{S_i} is flat over S_i.*

Proof. It suffices to show that there is an open subset $U \subset S$ such that F is flat over U_{red}. Moreover, the problem is local in X and S. One may therefore assume that $S = \text{Spec}(A)$ for some Noetherian integral domain A with quotient field K, that $X = \text{Spec}(B)$ for some finitely generated A-algebra, that $A \to B$ is injective, and that $F = M^\sim$ for some finite B-module M. This module M has a finite filtration by B-submodules with factors of the form $M_i \cong B/p_i$ for prime ideals $p_i \subset B$. It suffices to consider these factors separately, so that we may further reduce to the case that $M = B$ is integral and $A \to B$ injective. By Noether's normalization lemma there are elements $b_1, \ldots, b_n \in B$ such that $K \otimes B$ is a finite module over the polynomial ring $K[b_1, \ldots, b_n]$. 'Clearing denominators' we can find an element $f \in A$ such that $M' := B_f$ is still a finite module over $B' := A_f[b_1, \ldots, b_n]$. Replace M, B and A by M', B' and A' and apply the same procedure again. By induction over the dimension of B we may finally reduce the problem to the case that $M = B$ and B is a polynomial ring over A, in which case flatness is obvious. $\square$

Proof of the theorem. Let $S_0 = \coprod_i S_i$ be a decomposition of S as in the lemma and let $i_0 : S_0 \to S$ be the natural morphism. Then $i_{0,X}^* F$ is flat, and since the Hilbert polynomial of a flat family is locally constant as a function on the base, we conclude that the set $\mathcal{P}$ defined in Theorem 2.1.5 is indeed finite.

For any $m \geq 0$ let $\Gamma_*(F) := \bigoplus_{m \geq 0} \Gamma_m(F) := \bigoplus_{m \geq 0} f_* F(m)$. Recall that there is a functor $\sim$ which converts $\mathbb{Z}$-graded $\mathcal{O}_S$-modules into $\mathcal{O}_X$-modules and is inverse to the functor Γ (cf. [98] II.5.). Thus there is a natural isomorphism $\Gamma_*(F)^\sim \cong F$, and if $g : S' \to S$ is any morphism of Noetherian schemes, then $(g^* \Gamma_*(F))^\sim \cong g_X^* F$. Moreover, there is an integer $m(g)$, depending on g, such that for all $m \geq m(g)$ we have $\Gamma_m(g_X^* F) \cong g^* \Gamma_m(F)$ (cf. [98], exc. II 5.9). We apply this to the case $g = i_0$ and conclude that there is an integer m_0 such that for all $m \geq m_0$ we have

- $H^i(F_s(m)) = 0$ for all $i > 0$ and for all $s \in S$.
- $H^0(F_s(m)) = \Gamma_m(i_{0,X}^* F)(s) = (i_0^* \Gamma_m(F))(s) = \Gamma_m(F)(s)$ for all $s \in S$.

By Proposition 2.1.2, we see that $g_X^* F$ is flat if and only if $g^* \Gamma_m(F)$ is locally free for all sufficiently large m. Fixing m for a moment, we claim that there are finitely many locally closed subschemes $S_{m,r}$ such that

1. $j_m : \coprod_r S_{m,r} \to S$ is a bijection,
2. $\Gamma_m(F)|_{S_{m,r}}$ is locally free of rank r and
3. $g : S' \to S$ factors through j_m if and only if $g^* \Gamma_m(F)$ is locally free.

Set-theoretically, this decomposition is given by $S_{m,r} = \{s \in S | \dim_{k(s)} \Gamma_m(F)(s) = r\}$. We must endow the sets $S_{m,r}$ with appropriate scheme structures. Because of the universal property of these sets, this can be done locally: let s be a point in $S_{m,r}$. Then there is an open neighbourhood U of s in S such that $\Gamma_m(F)|_U$ admits a presentation

$$\mathcal{O}_U^{r'} \xrightarrow{A} \mathcal{O}_U^r \to \Gamma_m(F)|_U \to 0.$$

Let $S_{m,r} \cap U$ be the closed subscheme in U which is defined by the ideal generated by the entries of the $r \times r'$-matrix A and check that it has the required properties.

Now suppose that $g : S' \to S$ is a morphism such that $g_X^* F$ is S'-flat with Hilbert polynomial P. According to what was said before, g must factor through the locally closed subscheme $S_{m,P(m)}$ for all $m \geq m_0$. We therefore consider the *sets*

$$S_P := \{s \in S | P(F_s) = P\} = \bigcap_{m \geq m_0} S_{m,P(m)} \quad \text{for all } P \in \mathcal{P}. \tag{2.1}$$

By 2.1.6 and the *first* description of S_P, we know that S_P is the finite union of locally closed subsets. But then it is evident from the *second* description and the fact that S is Noetherian, that the intersection on the right hand side in (2.1) is in fact finite, even when considered as an intersection of subschemes. Let S_P be endowed with this subscheme structure and check that the collection S_P, $P \in \mathcal{P}$, thus defined has the properties postulated in the theorem. □

Lemma 2.1.7 — *Let F be a coherent $\mathcal{O}_X$-module, $x \in X$ a point and $s = f(x)$. Assume that F_x is flat over $\mathcal{O}_{S,s}$. Then F_x is free if and only if the restriction $(F_s)_x$ is free.*

Proof. The 'only if' direction is trivial. For the 'if' direction let r be the $k(x)$-dimension of $F(x) = F_x/\mathfrak{m}_x F_x$. Then there is a short exact sequence $0 \to K \to \mathcal{O}_{X,x}^r \to F_x \to 0$, and F_x is free if $K = 0$. Let $\mathfrak{m}_s$ denote the maximal ideal of the local ring $\mathcal{O}_{S,s}$. Since F_x is $\mathcal{O}_{S,s}$-flat, $K/\mathfrak{m}_s K$ is the kernel of the isomorphism $\mathcal{O}_{X_s,x}^r \to (F_s)_x$. By Nakayama's Lemma $K = 0$. □

Lemma 2.1.8 — *Let F be a flat family of coherent sheaves. Then the set*

$$\{s \in S | F_s \text{ is a locally free sheaf}\}$$

is an open subset of S.

Proof. The set $A = \{x \in X | F_x \text{ is not locally free at } x\}$ is closed in X, and the set defined in the lemma is the complement of $f(A)$. Since f is projective, $f(A)$ is closed. $\square$

Definition 2.1.9 — *Let P be a property of coherent sheaves on Noetherian schemes. P is said to be an open property, if for any projective morphism $f : X \to S$ of Noetherian schemes and any flat family F of sheaves on the fibres of f the set of points $s \in S$ such that F_s has P is an open subset in S. F is said to be a family of sheaves with P, if for all $s \in S$ the sheaf F_s has P.*

Examples of open properties are: being locally free (as we just saw), of pure dimension, semistable, geometrically stable (as will be proved in Section 2.3).

Proposition 2.1.10 — *Let k be a field, S a k-scheme of finite type and $f : X \to S$ a smooth projective morphism of relative dimension n. If F is a flat family of coherent sheaves on the fibres of f then there is a locally free resolution*

$$0 \to F_n \to F_{n-1} \to \ldots \to F_0 \to F$$

such that $R^n f_* F_\nu$ is locally free for $\nu = 0, \ldots, n$, $R^i f_* F_\nu = 0$ for $i \neq n$ and $\nu = 0, \ldots, n$. Moreover, in this case the higher direct image sheaves $R^\bullet f_* F$ can be computed as the homology of the complex $R^n f_* F_\bullet$: Namely, $R^{n-i} f_* F = h_i(R^n f_* F_\bullet)$.

Proof. Let $\mathcal{O}_X(1)$ be an f-very ample line bundle on X. Since the fibres of f are smooth, it follows from Serre duality and the Base Change Theorem for cohomology that there is an integer m_0 such that for all $m \geq m_0$ the $\mathcal{O}_S$-module $R^n f_* \mathcal{O}_X(-m)$ is locally free and $R^i f_* \mathcal{O}_X(-m)$ vanishes for all $i \neq n$. Define S-flat sheaves K_ν, G_ν for $\nu = 0, 1, \ldots$ inductively as follows: Let $K_0 := F$, and assume that K_ν has been constructed for some $\nu \geq 0$. For sufficiently large $m \gg m_0$ all fibres $(K_\nu)_s$, $s \in S$, are m-regular. Hence $f_* K_\nu(m)$ is locally free and there is a natural surjection $G_\nu := f^*(f_* K_\nu(m))(-m) \to K_\nu$. Then G_ν is locally free and $R^i f_* G_\nu = f_* K_\nu(m) \otimes R^i f_* \mathcal{O}_X(-m)$ by the projection formula. In particular, $R^n f_* G_\nu$ is locally free and the other direct image sheaves vanish. Finally, let $K_{\nu+1}$ be the kernel of the map $G_\nu \to K_\nu$. This procedure yields an (infinite) locally free resolution $G_\bullet \to F$. Since all sheaves involved are flat, it follows that $(G_\bullet)_s$ is a locally free resolution of F_s for all $s \in S$. In particular, $(K_n)_s$ is isomorphic to the kernel of $(G_{n-1})_s \to (G_{n-2})_s$, and as any coherent sheaf on the fibres of f has homological dimension $\leq n$, $(K_n)_s$ is locally free. According to Lemma 2.1.7 the sheaf K_n is itself locally free. Hence we can truncate the resolution $G_\bullet \to F$ at the n-th step and define $F_n = K_n$ and $F_\nu = G_\nu$ for $\nu = 0, \dots, n-1$. To prove the last statement split the resolution $F_\bullet \to F$ into short exact sequences and apply the functors $R^\bullet f_*$. □

Recall the notion of Grothendieck's groups $K^0(X)$ and $K_0(X)$ for a Noetherian scheme X: these are the abelian groups generated by locally free and coherent $\mathcal{O}_X$-modules, respectively, with relations $[F'] - [F] + [F'']$ for any short exact sequence $0 \to F' \to F \to F'' \to 0$. Moreover, the tensor product turns $K^0(X)$ into a commutative ring with $1 = [\mathcal{O}_X]$ and gives $K_0(X)$ a module structure over $K^0(X)$. A projective morphism $f : X \to S$ induces a homomorphism $f_! : K_0(X) \to K_0(S)$ defined by $f_![F] := \sum_{\nu \geq 0} (-1)^\nu [R^\nu f_* F]$.

Corollary 2.1.11 — *Under the hypotheses of Proposition 2.1.10: if F is an S-flat family of coherent sheaves on the fibres of f, then $[F] \in K^0(X)$ and $f_![F] \in K^0(S)$.*

Proof. $[F] = \sum_i (-1)^i [F_i]$ and $f_![F] = \sum_i (-1)^i [R^i f_* F] = \sum_i (-1)^{n-i} [R^n f_* F_i]$. □

Since the determinant is multiplicative in short exact sequences, it defines a homomorphism det $: K^0(X) \to \mathrm{Pic}(X)$ for any Noetherian scheme X (1.1.17). Applying this homomorphism to the elements $[F] \in K^0(X)$ and $f_![F] \in K^0(S)$ in the corollary, we get well defined line bundles

$$\det(F) := \det([F]) \in \mathrm{Pic}(X) \text{ and } \det(Rf_*F) := \det(f_![F]) \in \mathrm{Pic}(S).$$

More explicitly, if $F_\bullet \to F$ is a finite locally free resolution of F as in Proposition 2.1.10, then $\det(F) = \bigotimes_\nu \det(F_\nu)^{(-1)^\nu}$. This construction commutes with base change. For example, there is a natural isomorphism

$$\det(Rf_*F)(s) = \bigotimes_i \det(H^i(F_s))^{(-1)^i}$$

for each $s \in S$.

We conclude this section with a standard construction of a flat family that will be used frequently in the course of these notes.

Example 2.1.12 — Let F_1 and F_2 be coherent $\mathcal{O}_X$-modules on a projective k-scheme X and let $E = \mathrm{Ext}_X^1(F_2, F_1)$. Since elements $\xi \in E$ correspond to extensions

$$0 \to F_1 \to F_\xi \to F_2 \to 0,$$

the space $S = \mathbb{P}(E^\vee)$ parametrizes all non-split extensions of F_2 by F_1 up to scalars. Moreover, there exists a *universal extension*

$$0 \to q^* F_1 \otimes p^* \mathcal{O}_S(1) \to \mathcal{F} \to q^* F_2 \to 0$$

on the product $S \times X$ (with projections p and q to S and X, respectively), such that for each rational point $[\xi] \in S$, the fibre $\mathcal{F}_\xi$ is isomorphic to F_ξ. Indeed, the identity id_E gives a canonical extension class in $E^\vee \otimes_k E = \mathrm{Ext}_X^1(F_2, E^\vee \otimes_k F_1)$. Let π denote the canonical homomorphism $E^\vee \otimes \mathcal{O}_S \to \mathcal{O}_S(1)$ and consider the class $\pi_*(\mathrm{id}_E)$, i.e. the extension defined by the push-out diagram

$$
\begin{array}{ccccccccc}
0 & \longrightarrow & p^*\mathcal{O}_S(1) \otimes q^*F_1 & \longrightarrow & \mathcal{F} & \longrightarrow & q^*F_2 & \longrightarrow & 0 \\
& & \uparrow \pi \otimes 1 & & \uparrow & & \| & & \\
0 & \longrightarrow & E^\vee \otimes q^*F_1 & \longrightarrow & q^*\mathcal{G} & \longrightarrow & q^*F_2 & \longrightarrow & 0,
\end{array}
$$

where the extension in the bottom row is given by id_E. Note that $\mathcal{F}$ is S-flat for the obvious reason that q^*F_1 and q^*F_2 are S-flat.

2.2 Grothendieck's Quot-Scheme

The Quot-scheme is an important technical tool in many branches of algebraic geometry. In the same way as the Grassmann variety $\mathrm{Grass}_k(V, r)$ parametrizes r-dimensional quotient spaces of the k-vector space V, the Quot-scheme $\mathrm{Quot}_X(F, P)$ parametrizes quotient sheaves of the $\mathcal{O}_X$-module F with Hilbert polynomial P. Recall the notion of a representable functor:

Let $\mathcal{C}$ be a category, $\mathcal{C}^o$ the opposite category, i.e. the category with the same objects and reversed arrows, and let $\mathcal{C}'$ be the functor category whose objects are the functors $\mathcal{C}^o \to (Sets)$ and whose morphisms are the natural transformations between functors. The Yoneda Lemma states that the functor $\mathcal{C} \to \mathcal{C}'$ which associates to $x \in \mathrm{Ob}(\mathcal{C})$ the functor $\underline{x} : y \longmapsto \mathrm{Mor}_\mathcal{C}(y, x)$ embeds $\mathcal{C}$ as a full subcategory into $\mathcal{C}'$. A functor in $\mathcal{C}'$ of the form $\underline{x}$ is said to be *represented* by the object x.

Definition 2.2.1 — *A functor $\mathcal{F} \in \mathrm{Ob}(\mathcal{C}')$ is corepresented by $F \in \mathrm{Ob}(\mathcal{C})$ if there is a $\mathcal{C}'$–morphism $\alpha : \mathcal{F} \to \underline{F}$ such that any morphism $\alpha' : \mathcal{F} \to \underline{F'}$ factors through a unique morphism $\beta : \underline{F} \to \underline{F'}$; $\mathcal{F}$ is universally corepresented by $\alpha : \mathcal{F} \to \underline{F}$, if for any morphism $\phi : \underline{T} \to \underline{F}$, the fibre product $\mathcal{T} = \underline{T} \times_{\underline{F}} \mathcal{F}$ is corepresented by T. And $\mathcal{F}$ is represented by F if $\alpha : \mathcal{F} \to \underline{F}$ is a $\mathcal{C}'$–isomorphism.*

If F represents $\mathcal{F}$ then it also universally corepresents $\mathcal{F}$; and if F corepresents $\mathcal{F}$ then it is unique up to a unique isomorphism. This follows directly from the definition. We can rephrase these definitions by saying that F represents $\mathcal{F}$ if $\mathrm{Mor}_\mathcal{C}(y, F) = \mathrm{Mor}_{\mathcal{C}'}(\underline{y}, \mathcal{F})$ for all $y \in \mathrm{Ob}(\mathcal{C})$, and F corepresents $\mathcal{F}$ if $\mathrm{Mor}_\mathcal{C}(F, y) = \mathrm{Mor}_{\mathcal{C}'}(\mathcal{F}, \underline{y})$ for all $y \in \mathrm{Ob}(\mathcal{C})$.

Example 2.2.2 — We sketch the construction of the Grassmann variety. Let k be a field, let V be a finite dimensional vector space and let r be an integer, $0 \le r \le \dim(V)$. Let

$\underline{\text{Grass}}(V, r) : (Sch/k)^\circ \to (Sets)$ be the functor which associates to any k-scheme S of finite type the set of all subsheaves $K \subset \mathcal{O}_S \otimes_k V$ with locally free quotient $F = \mathcal{O}_S \otimes_k V/K$ of constant rank r.

For each r-dimensional linear subspace $W \subset V$ we may consider the subfunctor $\mathcal{G}_W \subset \underline{\text{Grass}}(V, r)$ which for a k-scheme S consists of those locally free quotients $\varphi : \mathcal{O}_S \otimes V \to F$ such that the composition $\mathcal{O}_S \otimes W \to \mathcal{O}_S \otimes V \to F$ is an isomorphism. In this case, the inverse of this isomorphism leads to a homomorphism $g : \mathcal{O}_S \otimes V \to \mathcal{O}_S \otimes W$ which splits the inclusion of W in V. From this one concludes that $\mathcal{G}_W$ is represented by the affine subspace $G_W \subset \text{Hom}(V, W) = \text{Spec} \, S^*\text{Hom}(V, W)^\vee$ corresponding to homomorphisms that split the inclusion map $W \to V$. Now for any element $[\varphi : \mathcal{O}_S \otimes V \to F] \in \underline{\text{Grass}}(V, r)(S)$ there is a maximal open subset $S_W \subset S$ such that $[\varphi|_{S_W}]$ lies in the subset $\mathcal{G}_W(S_W) \subset \underline{\text{Grass}}(V, r)(S_W)$. Moreover, if W runs through the set of all r-dimensional subspaces of V, then the corresponding S_W form an open cover of S. Apply this to the universal families parametrized by G_W and $G_{W'}$ for two subspaces $W, W' \subset V$: because of the universal property of $G_{W'}$ there is a canonical morphism $\alpha_{W, W'} : G_{W, W'} \to G_{W'}$. One checks that $\alpha_{W, W'}$ is an isomorphism onto the open subset $G_{W', W}$ and that for three subspaces the cocycle condition $\alpha_{W', W''} \circ \alpha_{W, W'} = \alpha_{W, W''}$ is satisfied. Hence we can glue the spaces G_W to produce a scheme $\text{Grass}(V, r) =: G$. Then G represents the functor $\underline{\text{Grass}}(V, r)$. Using the valuative criterion, one checks that G is proper. The Plücker embedding

$$\underline{\text{Grass}}(V, r) \to \mathbb{P}(\Lambda^r V), \qquad [\mathcal{O}_S \otimes V \to F] \mapsto [\mathcal{O}_S \otimes \Lambda^r V \to \det(F)]$$

exhibits G as a projective scheme. The local description shows that G is a smooth irreducible variety. $\qquad\square$

Example 2.2.3 — The previous example can be generalized to the case where V is replaced by a coherent sheaf $\mathcal{V}$ on a k-scheme S of finite type. By definition, a *quotient module* of $\mathcal{V}$ is an equivalence class of epimorphisms $q : \mathcal{V} \to F$ of coherent $\mathcal{O}_S$-sheaves, where two epimorphisms $q_i : \mathcal{V} \to F_i$, $i = 1, 2$, are equivalent, if $\ker(q_1) = \ker(q_2)$, or, equivalently, if there is an isomorphism $\Phi : F_1 \to F_2$ with $q_2 = \Phi \circ q_1$. Here and in the following, quotient modules are used rather than submodules because the tensor product is a right exact functor, so that surjectivity of a homomorphism of coherent sheaves is preserved under base change, whereas injectivity is not. Let $\underline{\text{Grass}}_S(\mathcal{V}, r) : (Sch/S)^\circ \to (Sets)$ be the functor which associates to $(T \to S) \in \text{Ob}((Sch/S))$ the set of all locally free quotient modules $q : \mathcal{V}_T = \mathcal{O}_T \otimes_{\mathcal{O}_S} \mathcal{V} \to F$ of rank r. *Then $\underline{\text{Grass}}_S(\mathcal{V}, r)$ is represented by a projective S-scheme $\pi : \text{Grass}_S(\mathcal{V}, r) \to S$.* We reduce the proof of this assertion to the case of the ordinary Grassmann variety of the previous example. First observe, that because of the uniqueness of Grass, if it exists, the problem is local in S, so that one can assume that $S = \mathrm{Spec}(A)$ and $\mathcal{V} = M^{\sim}$ for some finitely generated A-module M. Now let $A^{n'} \xrightarrow{a} A^n \xrightarrow{b} M$ be a finite presentation. Any quotient module $\mathcal{V}_T \to F$ by composition with b gives a quotient $\mathcal{O}_T^n \to F$. Thus b induces an injection

$$b^{\sharp} : \underline{\mathrm{Grass}}_S(\mathcal{V}, r) \to \underline{\mathrm{Grass}}_S(\mathcal{O}_S^n, r) \cong S \times \underline{\mathrm{Grass}}_k(k^n, r).$$

Clearly, the functor on the right hand side is represented by $S \times_k \mathrm{Grass}(k^n, r)$. We must show that $\underline{\mathrm{Grass}}_S(\mathcal{V}, r)$ is represented by a closed subscheme of $\mathrm{Grass}_S(\mathcal{O}_S^n, r)$. This follows from the more general statement: if $q : \mathcal{O}_T^n \to F$ is a locally free quotient module of rank r, then there is closed subscheme $T_0 \subset T$ such that any $g : T' \to T$ factors through T_0 if and only if $g^*(q \circ a_T) = 0$. Again this claim is local in T, and by shrinking T we may assume that $F \cong \mathcal{O}_T^r$. Then $q \circ a_T$ is given by an $r \times n'$-matrix B with values in $\mathcal{O}_T$, and $g^*(q \circ a_T)$ vanishes if and only if g factors through the closed subscheme corresponding to the ideal which is generated by the entries of B. $\qquad\square$

We now turn to the Quot-scheme itself: let k, S and $\mathcal{C} = (Sch/S)$ be as in the second example. Let $f : X \to S$ be a projective morphism and $\mathcal{O}_X(1)$ an f-ample line bundle on X. Let $\mathcal{H}$ be a coherent $\mathcal{O}_X$-module and $P \in \mathbb{Q}[z]$ a polynomial. We define a functor

$$\mathcal{Q} := \underline{\mathrm{Quot}}_{X/S} : (Sch/S)^o \longrightarrow (Sets)$$

as follows: if $T \to S$ is an object in $\mathcal{C}$, let $\mathcal{Q}(T)$ be the set of all T-flat coherent quotient sheaves $\mathcal{H}_T = \mathcal{O}_T \otimes \mathcal{H} \to F$ with Hilbert polynomial P. And if $g : T' \to T$ is an S-morphism, let $\mathcal{Q}(g) : \mathcal{Q}(T) \to \mathcal{Q}(T')$ be the map that sends $\mathcal{H}_T \to F$ to $\mathcal{H}_{T'} \to g_X^* F$. Thus $\mathcal{H}$ here plays the rôle of $\mathcal{V}$ for the Grassmann scheme in the second example above.

Theorem 2.2.4 — *The functor* $\underline{\mathrm{Quot}}_{X/S}(\mathcal{H}, P)$ *is represented by a projective S-scheme* $\pi : \mathrm{Quot}_{X/S}(\mathcal{H}, P) \to S$.

Proof. Step 1. Assume that $S = \mathrm{Spec}(k)$ and that $X = \mathbb{P}_k^N$. It follows from 1.7.6 that there is an integer m such that the following holds: If $[\rho : \mathcal{H}_T \to F] \in \mathcal{Q}(T)$ is any quotient and if $K = \ker(\rho)$ is the corresponding kernel, then for all $t \in T$ the sheaves K_t, $\mathcal{H}_t$ and F_t are m-regular. Applying the functor $f_{T*}(\,.\, \otimes \mathcal{O}(m))$ one gets a short exact sequence

$$0 \to f_{T*}K(m) \to \mathcal{O}_T \otimes H^0(\mathcal{H}(m)) \to f_{T*}F(m) \to 0$$

of locally free sheaves, and all the higher direct image sheaves vanish. Moreover, for any $m' \geq m$ there is an exact sequence

$$f_{T*}K(m) \otimes H^0(\mathcal{O}(m' - m)) \longrightarrow \mathcal{O}_T \otimes H^0(\mathcal{H}(m')) \longrightarrow f_{T*}F(m') \longrightarrow 0 \,,$$

where the first map is given by multiplication of global sections. Thus $f_{T*}K(m)$ completely determines the graded module $\bigoplus_{m' \geq m} f_{T*}F(m')$ which in turn determines F.

This argument shows that sending $[\mathcal{H}_T \to F]$ to $\mathcal{O}_T \otimes_k H^0(\mathcal{H}(m)) \to H^0(F(m))$ gives an injective morphism of functors

$$\underline{\text{Quot}}_{X/k}(\mathcal{H}, P) \longrightarrow \underline{\text{Grass}}_k(H^0(\mathcal{H}(m)), P(m)).$$

Thus we must identify those morphisms $T \to G := \text{Grass}_k(H^0(\mathcal{H}(m)), P(m))$ which are contained in the subset $\mathcal{Q}(T) \subset \underline{G}(T)$. Let

$$0 \to \mathcal{A} \to \mathcal{O}_G \otimes_k H^0(\mathcal{H}(m)) \to \mathcal{B} \to 0$$

be the tautological exact sequence on G. Consider the graded algebra

$$\mathcal{S} = \bigoplus_{\nu \geq 0} H^0(\mathbb{P}_k^N, \mathcal{O}(\nu))$$

and the graded $\mathcal{S}$-module $\Gamma_* \mathcal{H} = \bigoplus_{\nu \geq 0} H^0(\mathbb{P}_k^N, \mathcal{H}(\nu))$. The subbundle $\mathcal{A}$ generates a submodule $\mathcal{A} \cdot \mathcal{S} \subset \mathcal{O}_G \otimes_k \Gamma_* \mathcal{H}$. Let F be the $\mathcal{O}_{\mathbb{P}_G^N}$-module corresponding to the graded $\mathcal{O}_G$-module $\mathcal{O}_G \otimes_k \Gamma_* \mathcal{H} / \mathcal{A} \cdot \mathcal{S}$. Now it is straightforward to check that $\mathcal{Q}$ is represented by the locally closed subscheme $G_P \subset G$ which is the component of the flattening stratification for F corresponding to the Hilbert polynomial P (see Theorem 2.1.5).

It remains to show that Q is projective. Since we already know that Q is quasi-projective, it suffices to show that Q is proper. The valuation criterion requires that if R is a discrete valuation ring with quotient field L and if a commutative diagram

$$\begin{array}{ccc} \text{Spec}(L) & \longrightarrow & Q \\ \downarrow & & \downarrow \\ \text{Spec}(R) & \longrightarrow & \text{Spec}(k) \end{array}$$

is given, then there should exist a morphism $q_R : \text{Spec}(R) \to Q$ such that the whole diagram commutes. The diagram encodes the following data: there is a coherent sheaf F on X_L with $P(F) = P$ and a short exact sequence

$$0 \to K \to \mathcal{H} \otimes_k L \to F \to 0.$$

Certainly, there are coherent subsheaves $K'_R \subset \mathcal{H} \otimes_k R$ which restrict to K over the generic point of $\text{Spec}(R)$. Let K_R be maximal among all these subsheaves, and put $F_R = \mathcal{H} \otimes_k R/K_R$. The maximality of K_R implies that multiplication with the uniformizing parameter induces an injective map $F_R \to F_R$, which means that F_R is R-flat. The classifying map for F_R is the required q_R.

Step 2. Let S and X be arbitrary. Choosing a closed immersion $i : X \to \mathbb{P}_S^N$ and replacing $\mathcal{H}$ by $i_* \mathcal{H}$ we may reduce to the case $X = \mathbb{P}_S^N$. By Serre's theorem there exist presentations

$$\mathcal{O}_{\mathbb{P}_S^N}(-m'')^{n''} \longrightarrow \mathcal{O}_{\mathbb{P}_S^N}(-m')^{n'} \longrightarrow \mathcal{H} \longrightarrow 0.$$

As in Example 2.2.3 any quotient of $\mathcal{H}$ can be considered as a quotient of $\mathcal{O}_{\mathbb{P}^N_T}(-m')^{n'}$. Conversely, a quotient F of $\mathcal{O}_{\mathbb{P}^N_T}(-m')^{n'}$ factors through $\mathcal{H}$, if and only if the composite homomorphism $\mathcal{O}_{\mathbb{P}^N_T}(-m'')^{n''} \to \mathcal{O}_{\mathbb{P}^N_S}(-m')^{n'} \to F$ vanishes. The latter is equivalent to the vanishing of the homomorphism $\mathcal{O}_T \otimes H^0(\mathcal{O}_{\mathbb{P}^N_k}(\ell - m'')) \to f_{T*}F(\ell)$ for some sufficiently large integer ℓ. Hence by the same argument as in Example 2.2.3 the functor $\underline{\mathrm{Quot}}_{\mathbb{P}^N_S/S}(\mathcal{H}, P)$ is represented by a closed subscheme in $\mathrm{Quot}_{\mathbb{P}^N_S/S}(\mathcal{O}(-m')^{n'}, P) = S \times_k \mathrm{Quot}_{\mathbb{P}^N_k/k}(\mathcal{O}(-m')^{n'}, P)$. $\quad\square$

Since $Q := \mathrm{Quot}_{X/S}(\mathcal{H}, P)$ represents the functor $\mathcal{Q} := \underline{\mathrm{Quot}}_{X/S}(\mathcal{H}, P)$, we have $\mathrm{Mor}_{(Sch/S)}(Y, Q) = \mathcal{Q}(Y)$ for any S-scheme Y. Inserting Q for Y we see that the identity map on Q corresponds to a *universal* or *tautological quotient*

$$[\tilde{\rho} : \mathcal{H}_Q \longrightarrow \widetilde{F}] \in \mathcal{Q}(Q).$$

Any quotient $[\rho : \mathcal{H}_T \to F] \in \mathcal{Q}(T)$ is equivalent to the pull-back of $\tilde{\rho}$ under a uniquely determined S-morphism $\phi_\rho : T \to Q$, the classifying map associated to ρ.

In the case $X = S$ the polynomial P reduces to a number and $\mathrm{Quot}_{X/S}(\mathcal{H}, P)$ simply is $\mathrm{Grass}_S(\mathcal{H}, P)$. If $S = \mathrm{Spec}(k)$ and $\mathcal{H} = \mathcal{O}_X$, then quotients of $\mathcal{H}$ correspond to closed subschemes of X. In this context the Quot-scheme is usually called the *Hilbert scheme* of closed subschemes of X of given Hilbert polynomial P and is denoted by $\mathrm{Hilb}^P(X) = \mathrm{Quot}_X(\mathcal{O}_X, P)$. In particular, if $P = \ell$ is a number, the Hilbert scheme $\mathrm{Hilb}^\ell(X)$ parametrizes zero-dimensional subschemes of length ℓ in X.

Proposition 2.2.5 — *Let $\mathcal{H}$ be a coherent $\mathcal{O}_X$-module. Let $\tilde{\rho} : \mathcal{O}_{\mathrm{Quot}} \otimes \mathcal{H} \to \widetilde{F}$ be the universal quotient module parametrized by $\mathrm{Quot}_{X/S}(\mathcal{H}, P)$. Then for sufficiently large ℓ the line bundles $L_\ell = \det(f_{\mathrm{Quot}*}(\widetilde{F} \otimes \mathcal{O}_X(\ell)))$ are S-very ample.*

Proof. The arguments of the first step in the proof of the theorem show that for sufficiently large ℓ there is a closed immersion

$$\iota_\ell : \mathrm{Quot}_{X/S}(\mathcal{H}, P) \longrightarrow \mathrm{Grass}_S(f_*\mathcal{H}(\ell), P(\ell)).$$

Recall the Plücker embedding of the Grassmannian: If $\mathcal{V}$ is a vector bundle on S and if $\mathrm{pr}_S^*\mathcal{V} \to \mathcal{W}$ denotes the tautological quotient on $\mathrm{Grass}(\mathcal{V}, r)$, then the r-th exterior power $\mathrm{pr}_S^* : \Lambda^r\mathcal{V} \longrightarrow \det\mathcal{W}$ induces a closed immersion $\mathrm{Grass}_S(\mathcal{V}, r) \longrightarrow \mathbb{P}(\Lambda^r\mathcal{V})$ of S-schemes, and $\det\mathcal{W}$ is the pull-back of the tautological line bundle $\mathcal{O}_{\mathbb{P}(\Lambda^r\mathcal{V})}(1)$ on $\mathbb{P}(\Lambda^r\mathcal{V})$. Combining the Plücker embedding with the Grothendieck embedding ι_ℓ, we see that the line bundles $L_\ell = \det(f_*\widetilde{F}(\ell))$ are very ample relative to S. $\quad\square$

In general, L_ℓ depends non-linearly on ℓ as we will see later (cf. 8.1.3).
We now turn to the study of some infinitesimal properties of the Quot-scheme. Recall that
the Zariski tangent space of a k-scheme Y at a point y is defined as

$$T_y Y = \mathrm{Hom}_{k(y)}(\mathfrak{m}_y/\mathfrak{m}_y^2, k(y)),$$

where $\mathfrak{m}_y$ is the maximal ideal of $\mathcal{O}_{Y,y}$. Moreover, there is a natural bijection between
tangent vectors at y and morphisms $\tau : \mathrm{Spec}(k(y)[\varepsilon]) \longrightarrow Y$ with set-theoretic image y.
If Y represents a functor, then one expects such morphisms τ to admit an interpretation
in terms of intrinsic properties of the object represented by y. We follow this idea in the
case of the scheme $Q = \mathrm{Quot}_{X/S}(\mathcal{H}, P)$, where $X \to S$ is a projective morphism of
k-schemes, $\mathcal{O}_X(1)$ a line bundle on X, ample relative to S, and $\mathcal{H}$ an S-flat coherent
$\mathcal{O}_X$-module.

Let $(Artin/k)$ denote the category of Artinian local k-algebras with residue field k.
Let $\sigma : A' \to A$ be a surjective morphism in $(Artin/k)$ and suppose that there is a
commutative diagram

$$
\begin{array}{ccc}
\mathrm{Spec}(A) & \xrightarrow{q} & Q \\
\sigma \downarrow & & \downarrow \pi \\
\mathrm{Spec}(A') & \xrightarrow{\psi} & S.
\end{array}
$$

The images of the closed point of $\mathrm{Spec}(A)$ are k-rational points $q_0 \in Q$ and $s \in S$, and q
corresponds to a short exact sequence $0 \to K \to \mathcal{H}_A \to F \to 0$ of coherent sheaves on
$X_A = \mathrm{Spec}(A) \times_S X$ with $\mathcal{H}_A = A \otimes_{\mathcal{O}_S} \mathcal{H}$.

We ask whether the morphism q can be extended to a morphism $q' : \mathrm{Spec}(A') \to Q$
such that $q = q' \circ \sigma$ and $\psi = \pi \circ q'$, and if the answer is yes, how many different extensions
are there? The kernel I of σ is annihilated by some power of $\mathfrak{m}_{A'}$. We can filter I by the
ideals $\mathfrak{m}_{A'}^\nu I$ and in this way break up the extension problem in several smaller ones which
satisfy the additional property that $\mathfrak{m}_{A'} I = 0$. Assume that $0 \to I \to A' \to A \to 0$ is an
extension of this form.

Suppose that an extension q' exists. It corresponds to an exact sequence

$$0 \to K' \to \mathcal{H}_{A'} \to F' \to 0$$

on $X_{A'}$. That q' extends q means that over $X_A \subset X_{A'}$ the quotient $A \otimes_{A'} F'$ is equivalent
to F. Let $F_0 = A/\mathfrak{m}_A \otimes_A F$ etc. Then there is a commutative diagram whose columns

and rows are exact because of the flatness of $\mathcal{H}_{A'}$ and F':

$$
\begin{array}{ccccccccc}
 & & 0 & & 0 & & 0 & & \\
 & & \downarrow & & \downarrow & & \downarrow & & \\
0 & \to & I \otimes_k K_0 & \xrightarrow{1 \otimes i_0} & I \otimes_k \mathcal{H} & \xrightarrow{1 \otimes q_0} & I \otimes_k F_0 & \to & 0 \\
 & & \downarrow & & j \downarrow & & \downarrow & & \\
0 & \to & K' & \xrightarrow{i'} & \mathcal{H}_{A'} & \xrightarrow{q'} & F' & \to & 0 \\
 & & \downarrow & & \sigma \downarrow & & \downarrow & & \\
0 & \to & K & \xrightarrow{i} & \mathcal{H}_A & \xrightarrow{q} & F & \to & 0 \\
 & & \downarrow & & \downarrow & & \downarrow & & \\
 & & 0 & & 0 & & 0 & &
\end{array}
$$

In the first row we have used the isomorphisms $I \otimes_{A'} F' \cong I \otimes_k F_0$ etc. We can recover F' as the cokernel of the homomorphism $\hat{\imath} : K \to \mathcal{H}_{A'}/(1 \otimes i_0)(I \otimes_k K_0)$ induced by i'. Conversely, any $\mathcal{O}_{X_{A'}}$-homomorphism $\hat{\imath}$ which gives i when composed with σ defines an A'-flat extension F' of F (flatness follows from 2.1.3). Thus the existence of F' is equivalent to the existence of $\hat{\imath}$ as above which in turn is equivalent to the splitting of the extension

$$0 \to I \otimes_k F_0 \to B \to K \to 0, \tag{2.2}$$

where B is the middle homology of the complex

$$0 \longrightarrow I \otimes K_0 \xrightarrow{j \cdot (1 \otimes i_0)} \mathcal{H}_{A'} \xrightarrow{q \cdot \sigma} F \longrightarrow 0.$$

Check that though B a priori is an $\mathcal{O}_{X_{A'}}$-module it is in fact annihilated by I, so that B can be considered as an $\mathcal{O}_{X_A}$-module. The extension class

$$\mathfrak{o}(\sigma, q, \psi) \in \mathrm{Ext}^1_{X_A}(K, I \otimes_k F_0)$$

defined by (2.2) is the *obstruction* to extend q to q'. Since K is a A-flat and $I \otimes_k F_0$ is annihilated by $\mathfrak{m}_A$, there is a natural isomorphism

$$\mathrm{Ext}^1_{X_A}(K, I \otimes_k F_0) \cong \mathrm{Ext}^1_{X_s}(K_0, F_0) \otimes_k I.$$

Lemma 2.2.6 — *An extension q' of q exists if and only if $\mathfrak{o}(\sigma, q, \psi)$ vanishes. If this is the case, the possible extensions are given by an affine space with linear transformation group* $\mathrm{Hom}_{X_s}(K_0, F_0) \otimes_k I$.

Proof. The first statement follows from the discussion above. For the second note that, given one splitting $\hat{\imath}$, any other differs by a homomorphism $K \to I \otimes_k F_0$. As before the flatness of K implies that these are elements in $\mathrm{Hom}_{X_s}(K_0, F_0) \otimes_k I$. $\qquad\square$

Proposition 2.2.7 — *Let $f : X \to S$ be a projective morphism of k-schemes of finite type and $\mathcal{O}_X(1)$ an f-ample line bundle on X. Let $\mathcal{H}$ be an S-flat coherent $\mathcal{O}_X$-module, P a polynomial and $\pi : Q = \mathrm{Quot}_{X/S}(\mathcal{H}, P) \to S$ the associated relative Quot-scheme. Let $s \in S$ and $q_0 \in \pi^{-1}(s)$ be k-rational points corresponding to a quotient $\mathcal{H}_s \to F$ with kernel K. Then there is a short exact sequence*

$$0 \longrightarrow \mathrm{Hom}_{X_s}(K, F) \longrightarrow T_{q_0}Q \xrightarrow{T\pi} T_s S \xrightarrow{o} \mathrm{Ext}^1_{X_s}(K, F)$$

Proof. This is just a specialization of the lemma to the case $A = k$, $A' = k[\varepsilon]$. □

Proposition 2.2.8 — *Let X be a projective scheme over k and $\mathcal{H}$ a coherent sheaf on X. Let $[q : \mathcal{H} \to F] \in \mathrm{Quot}(\mathcal{H}, P)$ be a k-rational point and $K = \ker(q)$. Then*

$$\hom(K, F) \geq \dim_{[q]} \mathrm{Quot}(\mathcal{H}, P) \geq \hom(K, F) - \mathrm{ext}^1(K, F).$$

If equality holds at the second place, $\mathrm{Quot}(\mathcal{H}, P)$ is a local complete intersection near $[q]$. If $\mathrm{ext}^1(K, F) = 0$, then $\mathrm{Quot}(\mathcal{H}, P)$ is smooth at $[q]$.

The proof will be given in the appendix to this chapter, see 2.A.13.

Corollary 2.2.9 — *Let F' and F'' be coherent sheaves on a smooth projective curve C of positive ranks r' and r'' and slopes μ' and μ'', respectively. Let $0 \to F' \to F \to F'' \to 0$ be an extension that represents the point $s \in \Sigma := \mathrm{Quot}_C(F, P(F''))$. Then*

$$\dim_s \Sigma \geq \hom(F', F'') - \mathrm{ext}^1(F', F'') =: \chi(F', F'') = r'r''(\mu'' - \mu' + 1 - g).$$

□

Corollary 2.2.10 — *Let V be a k-vector space, $0 < r < \dim V$, and let*

$$0 \longrightarrow \mathcal{A} \longrightarrow V \otimes \mathcal{O}_{\mathrm{Grass}} \longrightarrow \mathcal{B} \longrightarrow 0$$

be the tautological exact sequence on $\mathrm{Grass}(V, r)$. Then the tangent bundle of the smooth variety $\mathrm{Grass}(V, r)$ is given by

$$\mathcal{T}_{\mathrm{Grass}} \cong \mathcal{H}om(\mathcal{A}, \mathcal{B}) .$$

Proof. Let $G = \mathrm{Grass}(V, r)$. Consider the composite homomorphism

$$\Phi : p_1^* \mathcal{A} \longrightarrow V \otimes \mathcal{O}_{G \times G} \longrightarrow p_2^* \mathcal{B}$$

on the product $G \times G$ and its adjoint homomorphism $\widehat{\Phi} : \mathcal{H}om(p_2^*\mathcal{B}, p_1^*\mathcal{A}) \to \mathcal{O}_{G \times G}$. Since Φ clearly vanishes precisely along the diagonal, the image of $\widehat{\Phi}$ is the ideal sheaf of the diagonal. Restricting this homomorphism to the diagonal, we get a surjection

$\mathcal{H}om(\mathcal{B}, \mathcal{A}) \to \Omega_G$ of locally free sheaves of the same rank, which must therefore be an isomorphism.   □

2.3 The Relative Harder-Narasimhan Filtration

In this section we give two applications to the existence of relative Quot-schemes: we prove the openness of (semi)stability in flat families and extend the Harder-Narasimhan filtration, which was constructed in Section 1.3 for a coherent sheaf, to flat families.

Proposition 2.3.1 — *The following properties of coherent sheaves are open in flat families: being simple, of pure dimension, semistable, or geometrically stable.*

Proof. Let $f : X \to S$ be a projective morphism of Noetherian schemes and let $\mathcal{O}_X(1)$ be an f-very ample invertible sheaf on X. Let F be a flat family of d-dimensional sheaves with Hilbert polynomial P on the fibres of f. For each $s \in S$, a sheaf F_s is simple if $\hom_{k(s)}(F_s, F_s) = 1$. Thus openness here is an immediate consequence of the semicontinuity properties for relative Ext-sheaves ([19], Satz 3(i)). The three remaining properties of being of pure dimension P_1, semistable P_2, or geometrically stable P_3 have similar characteristics: they can be described by the absence of certain pure dimensional quotient sheaves. Consider the following sets of polynomials:

$$
\begin{aligned}
A \quad &= \quad \{P''| \deg(P'') = d, \hat{\mu}(P'') \le \hat{\mu}(P) \text{ and there is a geometric point } s \in S \\
&\qquad \text{and a surjection } F_s \to F'' \text{ onto a pure sheaf with } P(F'') = P''\} \\
A_1 \quad &= \quad \{P'' \in A| \deg(P - P'') \le d - 1\}, \\
A_2 \quad &= \quad \{P'' \in A|p'' < p\}, \qquad A_3 \quad = \quad \{P'' \in A|p'' \le p \text{ and } P'' < P\},
\end{aligned}
$$

where as usual, p'' is the reduced polynomial associated to P'' etc. By the Grothendieck Lemma 1.7.9 the set A is finite. For each polynomial $P'' \in A$ we consider the relative Hilbert scheme $\pi : Q(P'') = \text{Quot}_{X/S}(F, P'') \to S$. Since π is projective, the image $\pi(Q(P'')) =: S(P'')$ is a closed subset of S. We see that F_s has property P_i if and only if s is not contained in the finite – and hence closed – union $\bigcup_{P'' \in A_i} S(P'') \subset S$.   □

Theorem 2.3.2 — *Let S be an integral k-scheme of finite type, let $f : X \to S$ be a projective morphism and let $\mathcal{O}_X(1)$ be an f-ample invertible sheaf on X. Let F be a flat family of d-dimensional coherent sheaves on the fibres of f. There is a projective birational morphism $g : T \to S$ of integral k-schemes and a filtration*

$$
0 = \text{HN}_0(F) \subset \text{HN}_1(F) \subset \ldots \subset \text{HN}_\ell(F) = F_T
$$

such that the following holds:

1. *The factors* $\mathrm{HN}_i(F)/\mathrm{HN}_{i-1}(F)$ *are T-flat for all $i = 1, \ldots, \ell$, and*
2. *there is a dense open subscheme $U \subset T$ such that $\mathrm{HN}_\bullet(F)_t = g_X^* \mathrm{HN}_\bullet(F_{g(t)})$ for all $t \in U$.*

Moreover, $(g, \mathrm{HN}_\bullet(F))$ is universal in the sense that if $g' : T' \to S$ is any dominant morphism of integral schemes and if $\mathcal{F}'_\bullet$ is a filtration of $F_{T'}$ satisfying these two properties, then there is an S-morphism $h : T' \to T$ with $\mathcal{F}'_\bullet = h_X^* \mathrm{HN}_\bullet(F)$.

This filtration is called the *relative Harder-Narasimhan filtration* of F.

Proof. It suffices to construct an integral scheme T and a projective birational morphism $g : T \to S$ such that $F_T = g_X^* F$ admits a T-flat quotient F'' which fibrewise gives the minimal destabilizing quotient of F_t for all t in a dense open subscheme of T and such that T is universal in the sense of the theorem. For in that case the kernel F' of the epimorphism $F_T \to F''$ is S-flat and we could iterate the argument with (S, F) replaced by (T, F'). This would result in a finite sequence of morphisms

$$T_\ell \to T_{\ell-1} \to \ldots \to T_1 = T \to S,$$

and the composition of theses morphisms would have the required properties.

As in the proof of the proposition consider the finite set A_4 of polynomials $P'' \in A$ such that $p'' \leq p$. Then S is the (set-theoretic) union of the closed subsets $S(P'')$, $P'' \in A$. Define a total ordering on A_4 as follows: $P_1 \preceq P_2$ if and only if $p_1 \leq p_2$ and $P_1 \geq P_2$ in case $p_1 = p_2$. Since S is irreducible, there is a polynomial P'' with $S(P'') = S$. Let P_- be minimal among all polynomials with this property with respect to $\prec$. Thus

$$\bigcup_{P'' \in A_4, P'' \prec P_-} S(P'')$$

is a proper closed subscheme of S. Let V be its open complement. Consider the morphism $\pi : Q(P_-) \to S$. By definition of P_-, π is surjective. For any point $s \in S$ the fibre of π parametrizes possible quotients of F_s with Hilbert polynomial P_-. If $s \in V$ then any such quotient is minimally destabilizing, by construction of V. By Theorem 1.3.4 the minimal destabilizing quotient is unique and by Theorem 1.3.7 it is defined over the residue field $k(s)$. This implies that $\pi : U := \pi^{-1}(V) \to V$ is bijective, and for each $t \in U$ and $s = \pi(t)$ one has $k(s) \cong k(t)$. Moreover, according to Proposition 2.2.7 the Zariski tangent space to the fibre of π at t is given by $\mathrm{Hom}_{X_s}(F'_s, F''_s)$, where $0 \to F'_t \to F_t \to F''_t \to 0$ is the short exact sequence corresponding to t. But by construction $\mathrm{Hom}(F'_t, F''_t)$ must vanish according to Lemma 1.3.3. This proves that the relative tangent sheaf $\Omega_{U/V}$ is zero, i.e. $\pi : U \to V$ is unramified and bijective. Since V is integral, $\pi : U \to V$ is an isomorphism. Now let T be the closure of U in $Q(P_-)$ with its reduced subscheme structure. Then $U \subset T$ is an open subscheme, and $f = \pi|_T : T \to S$ is a projective birational morphism. To see that T is universal, suppose that T' is an integral scheme with a dominant morphism

$g' : T' \rightarrow S$ and a quotient morphism $F_{T'} \rightarrow G$ such that $P(G_t) = P_-$ for general $t \in T'$. By the universal property of $Q(P_-)$, g' factors through a morphism $h : T' \rightarrow Q(P_-)$ with $g' = \pi \circ g'$. The image of T' is a reduced irreducible subscheme of $Q(P_-)$ and contains an open subscheme of U, since $h|_{(g')^{-1}(V)} = \pi^{-1} \circ g'|_{(g')^{-1}(V)}$. Thus h factors through T. $\square$

Remark 2.3.3 — If under the hypotheses of the theorem the family F is not flat over S or if F_s is d-dimensional only for points s in some open subset of S, one can always find an open subset $S' \subset S$ such that the conditions of the theorem are satisfied for $F_{S'}$. Making S' even smaller if necessary, we can assume that the relative Harder-Narasimhan filtration $HN_\bullet(F_{S'})$ is defined over S'. This filtration can easily be extended to a filtration of F over S by coherent subsheaves (cf. Exc. II 5.15 in [98]), which, however, can no longer be expected to be S-flat or to induce the (absolute) Harder-Narasimhan filtration on all fibres. This filtration satisfies some (weaker) universal property. Nevertheless, we will occasionally use the relative Harder-Narasimhan filtration in this form in order to simplify the notations.

Appendix to Chapter 2

2.A Flag-Schemes and Deformation Theory

2.A.1 Flag-schemes — These are natural generalizations of the Quot-schemes. Let $f :$ $X \to S$ be a projective morphism of Noetherian schemes, $\mathcal{O}_X(1)$ an f-ample line bundle on X, and let $\mathcal{H}$ be an S-flat coherent sheaf on X with Hilbert polynomial P. Fix polynomials $P_1, \ldots, P_\ell$ with $P = \sum P_i$. Let

$$\underline{\mathrm{Drap}}_{X/S}(\mathcal{H}, P_\bullet) : (Sch/S)^o \to (Sets)$$

be the functor which associates to $T \to S$ the set of all filtrations

$$0 = F_0 \mathcal{H}_T \subset F_1 \mathcal{H}_T \subset \ldots \subset F_\ell \mathcal{H}_T = \mathcal{H}_T := \mathcal{O}_T \otimes \mathcal{H}$$

such that the factors $gr_i^F \mathcal{H}_T$ are T-flat and have (fibrewise) the Hilbert polynomial P_i for $i = 1, \ldots, \ell$. Clearly, if $\ell = 1$ then $\underline{\mathrm{Drap}}_{X/S}(\mathcal{H}, P_1) = \underline{\mathrm{Quot}}_{X/S}(\mathcal{H}, P_1)$. In general, $\underline{\mathrm{Drap}}_{X/S}(\mathcal{H}, P_\bullet)$ is represented by a projective S-scheme $\mathrm{Drap}_{X/S}(\mathcal{H}, P_\bullet)$ which can be constructed inductively as follows: let $S_\ell = S$, $X_\ell = X$ and $\mathcal{H}_\ell = \mathcal{H}$. Let $0 < i \leq \ell$, and suppose that S_i, X_i and $\mathcal{H}_i \in \mathrm{Ob}(\mathrm{Coh}(X_i))$ have already been constructed. Let $S_{i-1} := \mathrm{Quot}_{X_i/S_i}(\mathcal{H}_i, P_i)$, $X_{i-1} := S_{i-1} \times_{S_i} X_i$ and let $\mathcal{H}_{i-1}$ be the kernel of the tautological surjection parametrized by S_i. Then $\mathrm{Drap}_{X/S}(\mathcal{H}, P_\bullet) = S_0$.

2.A.2 Ext-groups revisited — If $\mathcal{H}$ is a coherent sheaf together with a flag of subsheaves, we can consider the subgroup $\mathrm{Hom}_-(\mathcal{H}, \mathcal{H})$ of those endomorphisms of $\mathcal{H}$ which preserve the given flag. In analogy to ordinary Ext-groups one is led to the definition of corresponding higher Ext_--groups, which play a rôle in the deformation theory of the flag-schemes: let k be a field and X a k-scheme of finite type. Let $K^\bullet$ and $L^\bullet$ be complexes of $\mathcal{O}_X$-modules which are bounded below. Let $\mathrm{Hom}(K^\bullet, L^\bullet)^\bullet$ be the complex with homogeneous components $\mathrm{Hom}(K^\bullet, L^\bullet)^q = \prod_i \mathrm{Hom}(K^i, L^{i+q})$ and boundary operator $(d^q(f))^i = d^{q+i} \circ f^i + (-1)^q f^{i+1} \circ d^i$. A finite filtration of $K^\bullet$ is a filtration by subcomplexes $F_p K^\bullet$ such that only finitely many of the factor complexes $gr_p K^\bullet = F_p K^\bullet / F_{p-1} K^\bullet$ are nonzero. If $K^\bullet$ and $L^\bullet$ are endowed with finite filtrations then $\mathrm{Hom}(K^\bullet, L^\bullet)^\bullet$ inherits a filtration as well: let

$$F_p \mathrm{Hom}(K^\bullet, L^\bullet)^\bullet = \{f | f(F_j K^\bullet) \subset F_{j+p} L^\bullet \text{ for all } j\}.$$

Let $\mathrm{Hom}_-(K^\bullet, L^\bullet)^\bullet = F_0 \mathrm{Hom}(K^\bullet, L^\bullet)^\bullet$ and

$$\mathrm{Hom}_+(K^\bullet, L^\bullet)^\bullet = \mathrm{Hom}(K^\bullet, L^\bullet)^\bullet / F_0 \mathrm{Hom}(K^\bullet, L^\bullet)^\bullet.$$

A filtered injective resolution of the filtered complex $L^\bullet$ consists of a finitely filtered complex $I^\bullet$ of injective $\mathcal{O}_X$-modules and a filtration preserving augmentation homomorphism $\varepsilon : L^\bullet \to I^\bullet$ such that all factor complexes $gr_p^F I^\bullet$ consist of injective modules and ε induces quasi-isomorphisms $gr_p^F L^\bullet \to gr_p^F I^\bullet$. Such resolutions always exist.

Definition and Theorem 2.A.3 — *Let* $\text{Ext}_-^i(K^\bullet, L^\bullet)$ *and* $\text{Ext}_+^i(K^\bullet, L^\bullet)$ *be the cohomology groups of the complexes* $\text{Hom}_-(K^\bullet, I^\bullet)^\bullet$ *and* $\text{Hom}_+(K^\bullet, I^\bullet)^\bullet$, *respectively. These are up to isomorphism independent of the choice of the resolution.*   □

From the short exact sequence of complexes

$$0 \to \text{Hom}_-(K^\bullet, I^\bullet)^\bullet \to \text{Hom}(K^\bullet, I^\bullet)^\bullet \to \text{Hom}_+(K^\bullet, I^\bullet)^\bullet \to 0$$

one gets a long exact sequence of Ext-groups:

$$\ldots \to \text{Ext}_-^q(K^\bullet, L^\bullet) \to \text{Ext}^q(K^\bullet, L^\bullet) \to \text{Ext}_+^q(K^\bullet, L^\bullet) \to \text{Ext}_-^{q+1}(K^\bullet, L^\bullet) \to \ldots$$

Theorem 2.A.4 — *There are spectral sequences*

$$\text{Ext}_+^{p+q}(K^\bullet, L^\bullet) \quad \Leftarrow \quad E_1^{pq} = \begin{cases} \prod_i \text{Ext}^{p+q}(gr_i K^\bullet, gr_{i-p} L^\bullet) & p < 0 \\ 0 & p \geq 0 \end{cases}$$

$$\text{Ext}_-^{p+q}(K^\bullet, L^\bullet) \quad \Leftarrow \quad E_1^{pq} = \begin{cases} 0 & p < 0 \\ \prod_i \text{Ext}^{p+q}(gr_i K^\bullet, gr_{i-p} L^\bullet) & p \geq 0 \end{cases}.$$

Proof. Use the natural induced filtrations on $\text{Hom}_\pm(K^\bullet, I^\bullet)^\bullet$.   □

2.A.5 Deformation theory — This is a *very short* sketch of *some* aspects of deformation theory which is by no means intended to provide a systematic treatment of the theory. Not all assertions will be justified by explicit computations.

Let $(\Lambda, \mathfrak{m}_\Lambda)$ be a complete Noetherian local ring with residue field k, and let $(Artin/\Lambda)$ be the category of local Artinian Λ-algebras with residue field k. We want to study covariant functors $\mathcal{D} : (Artin/\Lambda) \to (Sets)$ with the property that $\mathcal{D}(k)$ consists of a single element. Suppose we are given a surjective homomorphism $\sigma : A' \to A$ in $(Artin/\Lambda)$. What is the image of the induced map $\mathcal{D}(\sigma) : \mathcal{D}(A') \to \mathcal{D}(A)$, and what can be said about the fibres? We can always factor σ through the rings $A'/\mathfrak{a}^\nu$, $\mathfrak{a} = \ker(\sigma)$, and in this way reduce ourselves to the study of those maps σ which satisfy the additional hypothesis $\mathfrak{m}_{A'}\mathfrak{a} = 0$, $\mathfrak{m}_{A'}$ denoting the maximal ideal of A'. We will refer to such maps as *small extensions*, deviating slightly from the use of this notion by Schlessinger [231]. The functor $\mathcal{D}$ is said to have an *obstruction theory* with values in a (finite dimensional) k-vector space U, if the following hold. (1) For each small extension $A' \to A$ with kernel $\mathfrak{a}$, there is a map (of sets) $\mathfrak{o} : \mathcal{D}(A) \to U \otimes \mathfrak{a}$ such that the sequence $\mathcal{D}(A') \to \mathcal{D}(A) \to U \otimes \mathfrak{a}$ is exact. (2) If $A' \to A$ and $B' \to B$ are small extensions with kernels $\mathfrak{a}$ and $\mathfrak{b}$, respectively, and if $\varphi : A' \to B'$ is a morphism with $\varphi(\mathfrak{a}) \subset \mathfrak{b}$, then the diagram

$$\begin{array}{ccc} \mathcal{D}(A) & \overset{\mathfrak{o}}{\to} & U \otimes \mathfrak{a} \\ \downarrow & & \downarrow \\ \mathcal{D}(B) & \overset{\mathfrak{o}}{\to} & U \otimes \mathfrak{b} \end{array}$$

commutes.

There are essentially two types of examples that concern us here: The problem of deforming a sheaf, and that of deforming a subsheaf, or more generally, a flag of subsheaves within a given sheaf.

2.A.6 Sheaves — Let $\Lambda = k$ be an algebraically closed field, let X be a projective variety over k, and let F be a coherent $\mathcal{O}_X$-module which is simple, i.e. $\mathrm{End}(F) \cong k$. If $A \in \mathrm{Ob}(Artin/k)$, let $\mathcal{D}_F(A)$ be the set of all equivalence classes of pairs (F_A, φ) where F_A is a flat family of coherent sheaves on X parametrized by $\mathrm{Spec}(A)$ and $\varphi : F_A \otimes_A k \to F$ is an isomorphism of $\mathcal{O}_X$-modules. (F_A, φ) and (F'_A, φ') are equivalent if and only if there is an isomorphism $\Phi : F'_A \to F_A$ with $\varphi \circ \Phi = \varphi'$.

Let $(I^\bullet, d^\bullet)$ be a complex of injective $\mathcal{O}_X$-modules and $\varepsilon : F \to I^\bullet$ a quasi-isomorphism. The following assertions can be checked easily with the usual diligence and patience necessary in homological algebra: the cohomology of the complex $\mathrm{Hom}(I^\bullet, I^\bullet)^\bullet$ computes $\mathrm{Ext}(F, F)$. Let $A \in \mathrm{Ob}(Artin/k)$ and suppose we are given a collection of maps $d_A \in \mathrm{Hom}(A \otimes I^\bullet, A \otimes I^\bullet)^1$ which restrict to d over the residue field of A. If $d_A^2 = 0$ then $(A \otimes I^\bullet, d_A^\bullet)$ is in fact an exact (!) complex except in degree 0, and $F_A := H^0(A \otimes I^\bullet, d_A^\bullet)$ is an A-flat extension of F over A, i.e. an element in $\mathcal{D}_F(A)$ (use induction on the length of A and the local flatness criterion 2.1.3). Conversely, any element in $\mathcal{D}_F(A)$ can be represented this way. Suppose that such a boundary map d_A with $d_A^2 = 0$ is given, defining an element $F_A \in \mathcal{D}_F(A)$. Let $\sigma : A' \to A$ be a small extension with kernel $\mathfrak{a}$. Choose a lift $d_{A'}$ of d_A. Since $d_A^2 = 0$, the square $d_{A'}^2$ factors through a homomorphism $\phi : I^\bullet \to I^{\bullet+2} \otimes_k \mathfrak{a}$. This homomorphism is a 2-cocycle, i.e. $d(\phi) = d\phi - \phi d = 0$, and its cohomology class $\mathfrak{o}(F_A, \sigma) := [\phi] \in \mathrm{Ext}_X^2(F, F) \otimes_k \mathfrak{a}$ is independent of the choice of the extension $d_{A'}$. If $d_{A'}^2 = 0$ then $\mathfrak{o}(F_A, \sigma) = 0$, and conversely, if $\mathfrak{o}(F_A, \sigma) = 0$ then $\phi = d(\xi)$ is the boundary of some homomorphism $\xi : I^\bullet \to I^{\bullet+1} \otimes_k \mathfrak{a}$, and $d'_{A'} := d_{A'} - \xi$ satisfies $(d'_{A'})^2 = 0$. Moreover, if $d_{A'}$ and $d'_{A'}$ are two boundary maps extending d_A then they differ by a 1-cocycle ξ, and they are equivalent, if this cocycle is a coboundary (it is at this place that we need the assumption that F be simple). We summarize: the fibres of the map $\mathcal{D}_F(\sigma) : \mathcal{D}_F(A') \to \mathcal{D}_F(A)$ are affine spaces with structure group

$\mathrm{Ext}^1(F, F) \otimes_k \mathfrak{a}$, and the image of $\mathcal{D}_F(\sigma)$ is the preimage of 0 under the obstruction map $\mathfrak{o}(\sigma) : \mathcal{D}_F(A) \to \mathrm{Ext}^2(F, F) \otimes_k \mathfrak{a}$.

2.A.7 Flags of subsheaves — We turn to the description of the deformation obstructions for points in Drap. We will proceed in a similar way as sketched in the previous paragraph. This yields independent proofs of the results of Section 2.2 and a bit more. Again we leave out the details.

Let $(X, \mathcal{O}_X(1))$ be a polarized projective k-scheme, and let $\mathcal{G}$ be a flat family of coherent sheaves on X parametrized by a Noetherian k-scheme S. Let $s \in S$ be a closed point and Λ the completion of the local ring $\mathcal{O}_{S,s}$. A Λ-algebra structure of an Artinian k-algebra A corresponds to a morphism $\mathrm{Spec}(A) \to S$ that maps the maximal ideal of A to s. Let G_A be the corresponding deformation of the fibre G of $\mathcal{G}$ at the point s.

Suppose we are given a flag of submodules $0 = G_0 \subset G_1 \subset \ldots \subset G_\ell = G$ in the coherent $\mathcal{O}_X$-module G. Define a functor $\mathcal{D} = \mathcal{D}_{G_\bullet} : (Artin/\Lambda) \to (Sets)$ as follows: Let $\mathcal{D}(A)$ be the set of all filtrations $0 \subset G_{1,A} \subset \ldots \subset G_{\ell,A} = G_A$ with A-flat factors, whose restriction to $k = A/\mathfrak{m}_A$ equals the given filtration $0 \subset G_1 \subset \ldots \subset G_\ell = G$.

There is an injective resolution $G \to I^\bullet$ of the following special form: in each degree n the module I^n decomposes into a direct sum $\bigoplus_{p=1}^\ell I_p^n$, such that the boundary map $d = (d_{ij})$, $d_{ij} : I_j^n \to I_i^{n+1}$, has upper triangular form, i.e. $d_{ij} = 0$ for $i > j$, and the subcomplexes $I_{\leq p}^\bullet = \bigoplus_{i \leq p} I_i^n$ are injective resolutions for the subsheaves $G_p \subset G$. (To get such a resolution, first choose injective resolutions $gr_p G \to I_p^\bullet$ and then choose appropriate homomorphisms d_{ij}, $i < j$.)

Let $A \in \mathrm{Ob}(Artin/\Lambda)$. The associated deformation G_A of G can be described by an element $d_A \in \mathrm{Hom}(A \otimes I^\bullet, A \otimes I^\bullet)^1$ with $d_A^2 = 0$. A deformation of the flag $G_\bullet$ over A is given by an endomorphism of the form $b_A = 1 + \beta_A \in \mathrm{Hom}(A \otimes I^\bullet, A \otimes I^\bullet)^0$, where β_A is a strictly lower triangular matrix with entries $\beta_{A,ij} : A \otimes I_j^\bullet \to \mathfrak{m}_A \otimes I_i^\bullet$; in particular, b_A is invertible. Moreover, b_A is subject to the condition that the boundary map $\tilde{d}_A := b_A^{-1} d_A b_A$ is filtration preserving, i.e. upper triangular. (To see this, observe that a deformation of the flag is given by (1) deformations of the boundary maps of the complexes $I_{\leq p}^\bullet$ and (2) deformations of the inclusion maps $I_{\leq p}^\bullet \to I_{\leq p+1}^\bullet$. Since we are free to change these by deformations of the identity map of the complex $I^\bullet$, we can in fact assume that the latter are given by a matrix b_A as above. Clearly, the boundary maps of the subcomplexes $I_{\leq p}^\bullet$ then are already determined by the requirement that they commute with the inclusion maps.)

Suppose now that $0 \to \mathfrak{a} \to A' \to A \to 0$ is a small extension in $(Artin/\Lambda)$. Let $d_{A'}$ be a homorphism that yields $G_{A'}$, $d_A = d_{A'} \otimes_{A'} A$ and assume that β_A defines an A-flat extension $G_{\bullet,A}$ of the filtration $G_\bullet$. Choose an arbitrary (strictly lower triangular) extension $\beta_{A'}$ of β_A and let $b_{A'} = 1 + \beta_{A'}$. Let ϕ denote the strictly lower triangular part of $\tilde{d}_{A'} := b_{A'}^{-1} d_{A'} b_{A'}$. Since the strictly lower triangular part of $\tilde{d}_A$ vanishes by the assumptions, ϕ defines an element in $\mathrm{Hom}_+(I^\bullet, I^\bullet)^1 \otimes \mathfrak{a}$. As before, ϕ is in fact a 1-cocycle, and its cohomology class is independent of the choice of $\beta_{A'}$. Let this class be denoted by $\mathfrak{o}(G_{A,\bullet}, \sigma) \in \mathrm{Ext}^1_+(G, G) \otimes \mathfrak{a}$. An extension $G_{\bullet, A'}$ of the filtration $G_{\bullet, A}$ exists if and only if this obstruction class vanishes. Moreover, if the obstruction vanishes then any two admissible choices of $\beta_{A'}$ differ by a cocyle in $\mathrm{Hom}_+(I^\bullet, I^\bullet)^0 \otimes \mathfrak{a}$ and are isomorphic if and only if this cocycle is a coboundary. This means that the fibre of $\mathcal{D}(A') \to \mathcal{D}(A)$ over $G_{\bullet, A}$ is an affine space with structure group $\mathrm{Ext}^0_+(G, G)$.

2.A.8 Comparison of the obstructions — As before, let X be a projective variety, and let $\rho : \mathcal{H} \to F$ be an epimorphism of coherent sheaves with kernel K, such that F is simple. In the last two paragraphs we defined obstruction classes for the deformation of F as an 'individual' sheaf and of F as a quotient of $\mathcal{H}$, or what amounts to the same, of K as a submodule of $\mathcal{H}$. We want to show next that these obstructions are related as follows: Let $\sigma : A' \to A$ be a small extension in $(Artin/k)$ with kernel $\mathfrak{a}$ and let

$$0 \to K_A \to A \otimes_k \mathcal{H} \to F_A \to 0$$

be an extension of the quotient $\mathcal{H} \to F$ over A. Let $\delta : \mathrm{Ext}^1(K, F) \to \mathrm{Ext}^2(F, F)$ be the boundary map associated to $0 \to K \to \mathcal{H} \to F \to 0$. Then

$$(\delta \otimes \mathrm{id}_\mathfrak{a})(\mathfrak{o}(K_A \subset A \otimes \mathcal{H}, \sigma)) = -\mathfrak{o}(F_A, \sigma) \in \mathrm{Ext}^2(F, F) \otimes \mathfrak{a}.$$

(Note that $\mathrm{Ext}^\bullet_+(\mathcal{H}, \mathcal{H}) = \mathrm{Ext}^\bullet(K, F)$.) Following the recipe above, we choose resolutions $K \to (I^\bullet_K, d_K)$ and $F \to (I^\bullet_F, d_F)$, and a homomorphism $\gamma : I^\bullet_F \to I^{\bullet+1}_K$ such that the complex

$$\left(I^\bullet_K \oplus I^\bullet_F, d := \begin{pmatrix} d_K & \gamma \\ 0 & d_F \end{pmatrix} \right)$$

is an injective resolution of $\mathcal{H}$. Note that $d^2 = 0$ implies $d_K \gamma + \gamma d_F = 0$, which means that γ is a 1-cocycle. Its class is precisely the extension class in $\mathrm{Ext}^1(F, K)$ corresponding to $\mathcal{H}$. A deformation of the inclusion over an Artinian algebra A' is given by a matrix

$$b' = \begin{pmatrix} 1 & 0 \\ \beta' & 1 \end{pmatrix}$$

subject to the condition that

$$\tilde{d} = b'^{-1} \circ d \circ b' = \begin{pmatrix} d_K + \gamma \beta' & \gamma \\ d_F \beta' - \beta' d_K - \beta' \gamma \beta' & d_F - \beta' \gamma \end{pmatrix}$$

be upper triangular. Let $\psi = (d_F\beta' - \beta'd_K) - \beta'\gamma\beta'$. Moreover, the induced deformation of F is described by the lower right entry $d_F - \beta'\gamma$. Let $\psi' = (d_F - \beta'\gamma)^2$. Then

$$
\begin{aligned}
\psi' &= d_F^2 - d_F\beta'\gamma - \beta'\gamma d_F + \beta'\gamma\beta'\gamma \\
&= d_F^2 - \beta'(\gamma d_F + d_K\gamma) + (\beta'd_K - d_F\beta' + \beta'\gamma\beta')\gamma.
\end{aligned}
$$

The first two terms in the last line vanish (γ is a cocycle!). Thus $\psi' = -\psi\gamma$. Assuming that the deformation exists over A means that ψ and, therefore, ψ' vanish when restricted to A and thus induce the obstruction classes in $\operatorname{Hom}(I_K^\bullet, I_F^\bullet)^1 \otimes \mathfrak{a}$ and $\operatorname{Hom}(I_F^\bullet, I_F^\bullet)^2 \otimes \mathfrak{a}$, respectively. Note that multiplication by γ gives the boundary map δ. Hence indeed,

$$
\mathfrak{o}(K_A \subset A \otimes \mathcal{H}, \sigma) = -(\delta \otimes \operatorname{id})([\mathfrak{o}(F_A), \sigma]).
$$

Moreover, if the obstruction vanishes for a given β', then any other choice is of the form $\beta' + \xi$ for a cocyle ξ. Note that the boundary operator of $F_{A'}$ then changes by $-\xi\gamma$. Thus the natural map of the fibre of $\mathcal{D}_{KC\mathcal{H}}(A') \to \mathcal{D}_{KC\mathcal{H}}(A)$ over $[K_A \subset \mathcal{H} \otimes A]$ into the fibre of $\mathcal{D}_F(A') \to \mathcal{D}_F(A)$ over $[F_A]$ is compatible with the boundary map δ between the structure groups of these affine spaces.

2.A.9 Dimension estimates — Let $(\Lambda, \mathfrak{m}_\Lambda)$ be a complete Noetherian local k-algebra with residue field k, and let $(R, \mathfrak{m}_R)$ be a complete local Λ-algebra with residue field k. Let $\underline{R}$ denote the functor $\operatorname{Hom}_{\Lambda-alg}(R, \,.\,) : (Artin/\Lambda) \to (Sets)$, and let $\mathcal{D} : (Artin/\Lambda) \to (Sets)$ be a covariant functor as in the previous sections. Though R is not in the category $(Artin/\Lambda)$, the quotients $R/\mathfrak{m}_R^i$ are. Any element $\xi \in \lim_i \mathcal{D}(R/\mathfrak{m}_R^i)$ defines a natural transformation $\underline{\xi} : \underline{R} \to \mathcal{D}$. The *pro-couple* (R, ξ) is said to *pro-represent* $\mathcal{D}$, if $\underline{\xi}$ is an isomorphism of functors.

For example, let $\mathcal{G}$ be an S-flat family of sheaves on X, let $p : Y = \operatorname{Drap}(\mathcal{G}, P_\bullet) \to S$ be a relative flag scheme, and let $y \in Y$ be a closed point that corresponds to a flag $G_\bullet \subset G = \mathcal{G}_s$, $s = p(y) \in S$. Then the functor $\mathcal{D}$ as defined in 2.A.7 is pro-represented by the pro-couple $(\widehat{\mathcal{O}}_{Y,y}, \widehat{\mathcal{G}}_\bullet)$, where $\widehat{\mathcal{O}}_{Y,y}$ is the completion of the local ring $\mathcal{O}_{Y,y}$ at its maximal ideal, and $\widehat{\mathcal{G}}_\bullet$ is the limit of the projective system of flags obtained from restricting the tautological flag on $Y \times X$ to the infinitesimal neighbourhoods $\operatorname{Spec}(\mathcal{O}_{Y,y}/\mathfrak{m}_y^i) \times X$ of $\{y\} \times X$. In particular, Y and $\mathcal{D}$ have the same tangent spaces. The results of Section 2.A.7 say:

Theorem 2.A.10 — *There is an exact sequence*

$$
0 \longrightarrow \operatorname{Ext}_+^0(G, G) \longrightarrow T_yY \xrightarrow{T\pi} T_sS \xrightarrow{\mathfrak{o}} \operatorname{Ext}_+^1(G, G).
$$

$\square$

If $\mathcal{D}$ is pro-represented by (R, ξ) then the deformation theory for $\mathcal{D}$ provides information about the number of generators and relations for R:

Proposition 2.A.11 — *Suppose that $\mathcal{D}$ is pro-represented by a couple (R, ξ) and has an obstruction theory with values in an r-dimensional vector space U. Let $d = \dim(\mathfrak{m}_R/\mathfrak{m}_R^2)$ be the embedding dimension of R. Then*

$$d \geq \dim(R) \geq d - r.$$

Moreover, if $\dim(R) = d - r$, then R is a local complete intersection, and if $r = 0$, then R is isomorphic to a ring of formal power series in d variables.

Proof. Choose representatives $t_1, \ldots, t_d \in \mathfrak{m}_R$ of a k-basis of $\mathfrak{m}_R/\mathfrak{m}_R^2$. Then $R \cong k[[t_1, \ldots, t_d]]/J$ for some ideal J. It suffices to show that J is generated by at most r elements: all statements of the proposition follow immediately from this. Let $\mathfrak{n} := (t_1, \ldots, t_d)$ be the maximal ideal in $k[[t_1, \ldots, t_d]]$. According to the Artin-Rees Lemma one has an inclusion $J \cap \mathfrak{n}^N \subset J\mathfrak{n}$ for sufficiently large N. Consider the small extension $0 \to \mathfrak{a} \to A' \to A \to 0$ with $A = R/\mathfrak{m}_R^N = k[[t_1, \ldots, t_d]]/(J + \mathfrak{n}^N)$, $A' = k[[t_1, \ldots, t_d]]/(\mathfrak{n}J + \mathfrak{n}^N)$ and $\mathfrak{a} = (J + \mathfrak{n}^N)/(\mathfrak{n}J + \mathfrak{n}^N) = J/\mathfrak{n}J$. The natural surjection $R \to A$ defines an element $\xi_A \in \mathcal{D}(A)$, and the obstruction to extend ξ_A to an element $\xi_{A'} \in \mathcal{D}(A')$ is given by an element

$$\mathfrak{o}' = \sum_{\alpha=1}^{r} \psi_\alpha \otimes \bar{f}_\alpha \in U \otimes \mathfrak{a},$$

where $\{\psi_\alpha\}$ is a basis of U and $f_1, \ldots, f_r$ are elements in J. Consider the algebra $A'' = A'/(f_1, \ldots, f_r)$. The obstruction $\mathfrak{o}''$ to extend ξ_A to A'' is the image of $\mathfrak{o}'$ under the map $U \otimes \mathfrak{a} \to U \otimes \mathfrak{a}/(\bar{f}_1, \ldots, \bar{f}_r)$ and therefore vanishes. The existence of an extension $\xi_{A''}$ corresponds to a lift of the natural ring homomorphism $q : R \to A$ to a ring homomorphism $q'' : R \to A''$. And picking pre-images for the generators $t_1, \ldots, t_d$ we can also lift the composite homomorphism $k[[t_1, \ldots, t_d]] \to R \to A''$ to a homomorphism $\Phi : k[[t_1, \ldots, t_d]] \to k[[t_1, \ldots, t_d]]$ such that the following diagram commutes:

$$\begin{array}{ccccc} k[[t_1, \ldots, t_d]] & \longrightarrow & R & \stackrel{q}{\longrightarrow} & A \\ \Phi \downarrow & & q'' \downarrow & & \| \\ k[[t_1, \ldots, t_d]] & \longrightarrow & A'' & \longrightarrow & A. \end{array}$$

In this diagram all horizontal arrows are natural quotient homomorphisms. Φ is an isomorphism, since it induces the identity on $\mathfrak{n}/\mathfrak{n}^2$. For any $x \in k[[t_1, \ldots, t_d]]$ one has $\Phi^{-1}(x) = x \bmod (J + \mathfrak{n}^N)$, which implies $J \subset \Phi(J) + \mathfrak{n}^N$. By construction of Φ, one has $\Phi(J) \subset \mathfrak{n}J + (f_1, \ldots, f_d) + \mathfrak{n}^N$. Combining these two inclusions one gets

$$J \subset \mathfrak{n}J + (f_1, \ldots, f_d) + \mathfrak{n}^N \subset J + \mathfrak{n}^N.$$

Recall the inclusion $J \cap \mathfrak{n}^N \subset \mathfrak{n}J$ we started with. From

$$\left(\mathfrak{n}J + \mathfrak{n}^N + (f_1, \ldots, f_d) \right)/\mathfrak{n}^N \cong J + \mathfrak{n}^N/\mathfrak{n}^N \cong J/J \cap \mathfrak{n}^N \twoheadrightarrow J/\mathfrak{n}J$$

one deduces $J = \mathfrak{n}J + (f_1, \ldots, f_d)$. Nakayama's Lemma therefore implies that J is generated by $f_1, \ldots, f_r$. $\square$

Note that if R is the completion of a local k-algebra $\mathcal{O}$ of finite type, then the statements of the proposition will hold for $\mathcal{O}$ as well, i.e. $d \geq \dim \mathcal{O} \geq d - r$. If $\dim \mathcal{O} = d - r$, then $\mathcal{O}$ is a local complete intersection, and if $r = 0$, then $\mathcal{O}$ is a regular ring: clearly, $\dim(R) = \dim(\widehat{\mathcal{O}}) = \dim(\mathcal{O})$; $\mathcal{O}$ is regular if and only if its completion is isomorphic to a ring of power series. Finally, write $\mathcal{O} = k[x_1, \ldots, x_\ell]_\mathfrak{m}/I$ for some ideal I. Then $R = k[[x_1, \ldots, x_\ell]]/\widehat{I}$, and $I/\mathfrak{m}I = \widehat{I}/\widehat{\mathfrak{m}}\widehat{I}$. Hence by Nakayama's Lemma, if $\widehat{I}$ is generated by, say, s elements, then I is also generated by s elements. This shows that $\mathcal{O}$ is a local complete intersection, if this is true for R.

We can apply the previous proposition and the observation above to flag-schemes and conclude:

Proposition 2.A.12 — *Let G be a coherent sheaf on a projective scheme X. Let y be a closed point in $\mathrm{Drap}_{X/k}(G, P_\bullet)$, defining a filtration of G. Then*

$$\mathrm{ext}^0_+(G, G) \geq \dim_y \mathrm{Drap}_{X/k}(G, P_\bullet) \geq \mathrm{ext}^0_+(G, G) - \mathrm{ext}^1_+(G, G).$$

If equality holds at the second place, then $\mathrm{Drap}_{X/k}(G, P_\bullet)$ is a local complete intersection near y. If $\mathrm{Ext}^1_+(G, G) = 0$, then $\mathrm{Drap}_{X/k}(G, P_\bullet)$ is smooth at y.

Proof. This follows at once from Subsection 2.A.7, Proposition 2.A.11 and the remark following it. $\square$

Note that these estimates can be sharpened if one can show in special cases that all deformation obstructions are contained in a proper linear subspace of $\mathrm{Ext}^1_+(G, G)$.

Remark 2.A.13 — Note that Proposition 2.A.12 contains 2.2.8 as the special case $\ell = 2$: if the filtration of G is given by a single subsheaf K, then $\mathrm{Ext}^i_+(G, G) = \mathrm{Ext}^i(K, G/K)$.

2.B A Result of Langton

Let X be a smooth projective variety over an algebraically closed field k. Let $(R, \mathfrak{m} = (\pi))$ be a discrete valuation ring with residue field k and quotient field K. We write $X_K = X \times \mathrm{Spec}(K)$ etc.

Theorem 2.B.1 — *Let F be an R-flat family of d-dimensional coherent sheaves on X such that $F_K = F \otimes_R K$ is a semistable sheaf in $\mathrm{Coh}_{d,d'}(X_K)$ for some $d' < d$. Then there is a subsheaf $E \subset F$ such that $E_K = F_K$ and such that E_k is semistable in $\mathrm{Coh}_{d,d'}(X)$.*

Proof. It suffices to show the following claim: *If $d > \delta \geq d'$ and if in addition to the assumptions of the theorem F_k is semistable in $\mathrm{Coh}_{d,\delta+1}$ then there is a sheaf $E \subset F$ such that $E_K = F_K$ and such that E_k is semistable in $\mathrm{Coh}_{d,\delta}$.* Clearly, the theorem follows from this by descending induction on δ, beginning with the empty case $\delta = d - 1$.

Suppose the claim were false. Then we can define a descending sequence of sheaves $F = F^0 \supset F^1 \supset F^2 \ldots$ with $F_K = F_K^n$ and F_k^n not semistable in $\mathrm{Coh}_{d,\delta}(X)$ as follows: Suppose F^n has already been defined. Let B^n be the saturated subsheaf in F_k^n which represents the maximal destabilizer of F_k^n in $\mathrm{Coh}_{d,\delta}(X)$. Let $G^n = F_k^n/B^n$ and let F^{n+1} be the kernel of the composite homomorphism $F^n \to F_k^n \to G^n$. As a submodule of an R-flat sheaf, F^{n+1} is R-flat again. There are two exact sequences

$$0 \to B^n \to F_k^n \to G^n \to 0 \quad \text{and} \quad 0 \to G^n \to F_k^{n+1} \to B^n \to 0. \tag{2.3}$$

(To get the second one use the inclusions $\pi F^n \subset F^{n+1} \subset F^n$.) If $C^n := G^n \cap B^{n+1}$ is nonzero, then

$$p(C^n) \leq p_{\max}(G^n) < p(F_k^n) \leq p(B^{n+1}) \mod \mathbb{Q}[T]_{\delta-1}.$$

Thus, in any case B^{n+1}/C^n is isomorphic to a nonzero submodule of B^n and $p_{d,\delta}(B^{n+1}) \leq p_{d,\delta}(B^{n+1}/C^n) \leq p_{d,\delta}(B^n)$ with equality if and only if $C^n = 0$. Since F_k is semistable in $\mathrm{Coh}_{d,\delta+1}$ it follows that $p_{d,\delta+1}(B^n) = p_{d,\delta+1}(F_k) = p_{d,\delta+1}(G^n)$ for all n. In particular, $p_{d,\delta}(B^n) - p_{d,\delta}(F_k) = \beta_n \cdot T^\delta \mod \mathbb{Q}[T]_{\delta-1}$ for a rational number β_n. As β_n is a descending sequence of strictly positive numbers in a lattice $\frac{1}{r!}\mathbb{Z} \subset \mathbb{Q}$, it must become stationary. We may assume without loss of generality that β_n is constant for all n. In this case we must have $G^n \cap B^{n+1} = 0$ for all n. In particular, there are injective homomorphisms $B^{n+1} \subset B^n$ and $G^n \subset G^{n+1}$. Hence there is an integer n_0 such that for all $n \geq n_0$ we have $P(B^n) \equiv P(B^{n+1}) \equiv \ldots \mod \mathbb{Q}[T]_{\delta-1}$ and $P(G^n) \equiv P(G^{n+1}) \equiv \ldots \mod \mathbb{Q}[T]_{\delta-1}$. (Again we may and do assume that $n_0 = 0$.) Now $G^0 \subset G^1 \subset \ldots$ is an increasing sequence of purely d-dimensional sheaves which are isomorphic in dimensions $\geq d - 1$. In particular, their reflexive hulls $(G^n)^{DD}$ are all isomorphic. Therefore, we may consider the G^n as a sequence of subsheaves in some fixed coherent sheaf. As an immediate consequence all injections must eventually become isomorphisms. Again we may assume that $G^n \cong G^{n+1}$ for all $n \geq 0$. This implies: the short exact sequences (2.3) split, and we have $B^n = B$, $G^n = G$ and $F_k^n = B \oplus G$ for all n. Define $Q^n = F/F^n$, $n \geq 0$. Then $Q_k^n \cong G$ and there are short exact sequences $0 \to G \to Q^{n+1} \to Q^n \to 0$ for all n. It follows from the local flatness criterion 2.1.3 that Q^n is an R/π^n-flat quotient of $F/\pi^n F$ for each n. Hence the image of the proper morphism $\sigma : \mathrm{Quot}_{X_R/R}(F, P(G)) \to \mathrm{Spec}(R)$ contains the closed subscheme $\mathrm{Spec}(R/\pi^n)$ for all n. But this is only possible if σ is surjective, so that $F_{K'}$

must also admit a (destabilizing!) quotient with Hilbert polynomial $P(G)$ for some field extension $K' \supset K$. This contradicts the assumption on F_K.   □

Exercise 2.B.2 — Use the same technique to show: if F is an R-flat family of d-dimensional sheaves such that F_K is pure, then there is a subsheaf $E \subset F$ such that $E_K = F_K$ and E_k is pure. Moreover, there is a homomorphism $F_k \to E_k$ which is generically isomorphic.

Comments:
— For a discussion of flatness see the textbooks of Matsumura [172], Atiyah and Macdonald [8] or Grothendieck's EGA [94]. The existence of a flattening stratification in the strong form 2.1.5 is due to Mumford [192]. The weaker form 2.1.6 is due to Grothendieck, cf. [172] 22.A Lemma 1. A detailed discussion of determinant bundles can be found in the paper of Knudson and Mumford [126].

— The notion of a scheme corepresenting a functor is due to Simpson [240]. Quot-schemes were introduced by Grothendieck in his paper [93]. There he also discusses deformations of quotients. Other presentations of the material are in Altman and Kleiman [3], Kollár [128] or Viehweg [260].

— Openness of semistability and torsion freeness is shown in Maruyama's paper [161]. Relative Harder-Narasimhan filtrations are constructed by Maruyama in [164].

— Proposition 2.A.11 is based on Prop. 3 in Mori's article [182] with an additional argument from Li [149] Sect. 1. Flag-schemes and their infinitesimal structure are discussed by Drezet and Le Potier [51].

— The presentation in Appendix 2.A is modelled on a similar discussion of the deformation of modules over an algebra by Laudal [137]. Deformations of sheaves are treated in Artamkin's papers [5] and [7] and by Mukai [187, 189]. For an intensive study of deformations see the forthcoming book of Friedman [69]. For more recent results on deformations and obstructions see the articles of Ran [227] and Kawamata [120, 121].

— The theorem of Appendix 2.B is the main result of Langton's paper [135]. In fact, the original version deals with μ-semistable sheaves. The analogous assertion for semistable sheaves was formulated by Maruyama (Thm. 5.7 in [163]). We have stated and proved it in a more general form which covers both cases.

2.C Further comments (second edition)

I. Construction of the Quot-scheme.

We recommend Nitsure's more detailed text on the construction of the Quot-scheme contained in [321].

In [408] Olsson and Starr generalize Grothendieck's existence result to $X \to S$ with X a separated locally finitely presented Deligne-Mumford stack over an algebraic space S. The resulting Quot-scheme for any locally finitely presented quasi-coherent sheaf $\mathcal{H}$ on X exists. However it is not a scheme any longer but only a separated and locally finitely presented algebraic space over S. This was later further generalized in [409].

There are also results in the literature that show that some of the assumptions cannot be relaxed. E.g. in [377] one finds examples of non-separated schemes for which the Quot-functor cannot be represented by a scheme or an algebraic space.

A very interesting approach to the Quot-scheme was initiated by Ciocan-Fontanine and Kapranov in [296]. Following a general program that replaces classical moduli spaces by their derived versions, the Quot-scheme is constructed as a dg-scheme whose degree 0 truncation is the classical Quot-scheme. The classical Quot-scheme can be very singular while the new *derived Quot-scheme* is always smooth in an appropriate sense. The paper also contains an introduction to the general theory of dg-schemes. The special case of derived Hilbert schemes was further studied in [297].

II. The cotangent sheaf of the Quot-scheme.

Proposition 2.2.7 for the special case of a variety X over a field k states that the Zariski tangent space of the Quot-scheme $Q := \mathrm{Quot}(\mathcal{H}, P)$ at a closed point $q_0 \in Q$ corresponding to the quotient $\mathcal{H} \twoheadrightarrow F$ with kernel K is naturally isomorphic as a k-vector space to $\mathrm{Hom}(K, F)$, i.e. $T_{q_0} Q \cong \mathrm{Hom}(K, F)$. One may wonder whether these descriptions of the Zariski tangent spaces in the closed points of Q glue to an isomorphism of the tangent sheaf T_X with the relative Hom-sheaf $\mathcal{H}om_p(\widetilde{K}, \widetilde{F})$, where $\mathcal{H}_Q \twoheadrightarrow \widetilde{F}$ is the universal quotient on $Q \times X$ with kernel $\widetilde{K}$ and $p : Q \times X \to Q$ is the projection. This is true as long as the dimension of the Zariski spaces $T_{q_0} Q$ stays constant or, more precisely, for smooth Q, but not in general.

As it turns out, however, such a global description can be given for the cotangent sheaf. If X is a smooth variety of dimension d with canonical bundle ω_X, the isomorphism $T_{q_0} Q \cong \mathrm{Hom}(K, F)$ can be dualized to describe the fibre $\Omega_Q(q_0)$ of the cotangent sheaf Ω_Q at the point q_0 as $\mathrm{Ext}^d(F, K \otimes \omega_X)$ and this description glues. More generally, the following result is proved in [361].

Let $X \to S$ be a smooth projective morphism of Noetherian schemes of relative dimension d, with a relative very ample invertible sheaf $\mathcal{O}(1)$ and with the relative dualizing sheaf $\omega_{X/S}$. For a coherent sheaf $\mathcal{H}$ on X and a numerical polynomial P let $Q :=$ $\mathrm{Quot}_{X/S}(\mathcal{H}, P)$. By $0 \to \widetilde{K} \to \mathcal{H}_Q \to \widetilde{F} \to 0$ we denote the universal quotient on $Q \times_S X$ with its kernel.

There exists a natural isomorphism of coherent sheaves on Q

$$\mathcal{E}xt_p^d(\widetilde{F}, \widetilde{K} \otimes q^*\omega_{X/S}) \cong \Omega_{Q/S},$$

where p and q denote the two projections from $Q \times_S X$ to Q respectively X.

This result can be used to deduce a description of the cotangent sheaf of the moduli space of stable sheaves, constructed in Chapter 4.

Let X be a smooth projective variety of dimension d over an algebraically closed field k of characteristic 0. If M^s is a moduli space of stable sheaves on X and $[E] \in M^s$ is a closed point represented by a stable sheaf E, then according to Corollary 4.5.2 the tangent space $T_{[E]}M$ to M in $[E]$ is canonically isomorphic to $\mathrm{Ext}^1(E, E)$. A global description, of how these spaces behave under variation of the point $[E]$, is provided by Thm. 4.1 in [361]. The result is this:

Assume that $\mathcal{E}$ is a universal family on $M^s \times X$. Then there is an isomorphism $\Phi_M :$ $\mathcal{E}xt_p^{d-1}(\mathcal{E}, \mathcal{E} \otimes q^*\omega_X) \to \Omega_{M^s}$.

A universal family always exists at least on the principal bundle $\pi : R^s \to M^s$ in the notation of Section 4.3. The centre of $\mathrm{GL}(V)$ acts trivially on $\mathcal{E}xt_p^{d-1}(\mathcal{E}, \mathcal{E} \otimes q^*\omega_X)$ so that this sheaf descends to M^s, even if a universal family on M^s does not exist.

III. Explicit calculations of Quot-schemes.

In general not much can be said about the geometry of the Quot-scheme beyond its mere existence. Even in simple situations like $\mathcal{H}$ being a locally free sheaf E on a smooth curve X the Quot-scheme is often mysterious. E.g. in [416] one finds upper bounds for the dimension of $\mathrm{Quot}(E, kn + d + k(1 - g))$ in terms of the degree and rank of E and the minimal degree of quotients of rank k of E. In the same paper one finds conditions which ensure that for large d the Quot-scheme is irreducible. See also Section 6.A where we prove that the Quot-scheme of zero-dimensional quotients of a locally free sheaf on a smooth surface is irreducible. All these results are intimately related to the geometry of the curve and very little can be said about Quot-schemes of positive dimensional quotients of locally free sheaves on higher-dimensional varieties.

3

The Grauert-Mülich Theorem

One of the key problems one has to face in the construction of a moduli space for semi-stable sheaves is the boundedness of the family of semistable sheaves with given Hilbert polynomial. In fact, this boundedness is easily obtained for semistable sheaves on a curve, as we have seen before (1.7.7). On the other hand, the Kleiman Criterion for boundedness (Theorem 1.7.8) suggests to restrict semistable sheaves to appropriate hyperplane sections and to proceed by induction on the dimension. In order to follow this idea we would like the restriction $F|_H$ of a semistable sheaf F to be semistable again. There are three obstacles:

- The right stability notion that is well-behaved under restriction to hyperplane sections is μ-semistability. There is no general restriction theorem for semistability.
- In general, the restriction $F|_H$ will have good properties only for a *general* element H in a given ample linear system.
- Even for a general hyperplane section the restriction might well fail to be μ-semistable. But this failure can be numerically controlled.

The Grauert-Mülich Theorem gives a first positive answer to the problem. In its original form, it can be stated as follows:

Theorem 3.0.1 — *Let E be a μ-semistable locally free sheaf of rank r on the complex projective space $\mathbb{P}^n_{\mathbb{C}}$. If L is a general line in $\mathbb{P}^n$ and $E|_L \cong \mathcal{O}_L(b_1) \oplus \ldots \oplus \mathcal{O}_L(b_r)$ with integers $b_1 \geq b_2 \geq \ldots \geq b_r$, then*

$$0 \leq b_i - b_{i+1} \leq 1$$

for all $i = 1, \ldots, r - 1$.

In Section 3.1 we will prove a more general version of this theorem that suffices to establish the boundedness of the family of semistable sheaves in any dimension. This will be done in Section 3.3. As a further application of the Grauert-Mülich theorem we will show in Section 3.2 that the tensor product of two μ-semistable sheaves is again μ-semistable. The chapter ends with a proof of the famous Bogomolov inequality. For all these results it is essential that the characteristic of the base field be zero. It is not known, whether the family of semistable sheaves is bounded, if the characteristic of the base field is positive. The discussion of restriction theorems for semistable sheaves will be resumed in greater detail in Chapter 7.

3.1 Statement and Proof

Let k be an algebraically closed field of characteristic zero, and let X be a normal projective variety over k of dimension $n \geq 2$ endowed with a very ample invertible sheaf $\mathcal{O}_X(1)$. For $a > 0$ let $V_a = H^0(X, \mathcal{O}_X(a))$, and let $\Pi_a := \mathbb{P}(V_a^{\vee}) = |\mathcal{O}_X(a)|$ denote the linear system of hypersurfaces of degree a. Let $Z_a := \{(D, x) \in \Pi_a \times X | x \in D\}$ be the incidence variety with its natural projections

$$
\begin{array}{ccc}
Z_a & \xrightarrow{q} & X \\
{\scriptstyle p}\downarrow & & \\
\Pi_a & &
\end{array}
$$

Scheme-theoretically Z_a can be described as follows: let $\mathcal{K}$ be the kernel of the evaluation map $V_a \otimes \mathcal{O}_X \to \mathcal{O}_X(a)$. Then there is a natural closed immersion $Z_a = \mathbb{P}(\mathcal{K}^{\vee}) \to \mathbb{P}(V_a^{\vee}) \times X$. In particular, q is the projection morphism of a projective bundle, and the relative tangent bundle is given by the Euler sequence

$$0 \to \mathcal{O}_{Z_a} \to q^*\mathcal{K} \otimes p^*\mathcal{O}(1) \to \mathcal{T}_{Z_a/X} \to 0.$$

We slightly generalize this setting: let $(a_1, \dots, a_\ell)$ be a fixed finite sequence of positive integers, $0 < \ell < n$. Let $\Pi := \Pi_{a_1} \times \dots \times \Pi_{a_\ell}$ with projections $pr_i : \Pi \to \Pi_{a_i}$, and let $Z = Z_{a_1} \times_X \dots \times_X Z_{a_\ell}$ with natural morphisms

$$
\begin{array}{ccc}
Z & \xrightarrow{q} & X \\
{\scriptstyle p}\downarrow & & \\
\Pi & &
\end{array}
$$

as above and projections $pr_i : Z \to Z_{a_i}$. Then q is a locally trivial bundle map with fibres isomorphic to products of projective spaces. The relative tangent bundle is given by

$$\mathcal{T}_{Z/X} = pr_1^*\mathcal{T}_{Z_{a_1}/X} \oplus \dots \oplus pr_\ell^*\mathcal{T}_{Z_{a_\ell}/X}.$$

If s is a closed point in Π parametrizing an ℓ-tuple of divisors $D_1, \dots, D_\ell$, then the fibre $Z_s = p^{-1}(s)$ is identified by q with the scheme-theoretic intersection $D_1 \cap \dots \cap D_\ell \subset X$. Next, let E be a torsion free coherent sheaf on X and let $F := q^*E$.

Lemma 3.1.1 — *i) There is a nonempty open subset $S' \subset \Pi$ such that the morphism $p_{S'} : Z_{S'} \to S'$ is flat and such that for all $s \in S'$ the fibre Z_s is a normal irreducible complete intersection of codimension ℓ in X.*
ii) There is a nonempty open subset $S \subset S'$ such that the family $F_S = q^* E|_{Z_S}$ is flat over S and such that for all $s \in S$ the fibre $F_s \cong E|_{Z_s}$ is torsion free.

Proof. Part *i)* follows from Bertini's Theorem and 1.1.15. For *ii)* restrict first to the dense open subset of S' over which F is flat (cf. 2.1.5). Then let S be the intersection with the dense open subset of points $s = (s_1, \ldots, s_\ell) \in \Pi$ which form regular sequences for E and for all $\mathcal{E}xt^i(E, \omega_X)$, $i \geq 0$. Then $E|_{Z_s}$ is torsion free for all $s \in S$ by 1.1.13. (To reduce to the case of a smooth variety, view the torsion free sheaf E on X as a pure sheaf on the ambient projective space.) $\square$

Now we apply Theorem 2.3.2 to the family F_S and conclude that there exists a relative Harder-Narasimhan filtration

$$0 = F_0 \subset F_1 \subset \ldots \subset F_j = F_S$$

such that all factors F_i/F_{i-1} are flat over some nonempty open subset $S_0 \subset S$ and such that for all $s \in S_0$ the fibres $(F_\bullet)_s$ form the Harder-Narasimhan filtration of $F_s \cong E|_{Z_s}$. Without loss of generality we can assume that $S_0 = S$. Since S is connected, there are rational numbers $\mu_1 > \ldots > \mu_j$ with $\mu_i = \mu((F_i/F_{i-1})_s)$ for all $s \in S$. We define

$$\delta\mu = \max\{\mu_i - \mu_{i+1} | i = 1, \ldots, j - 1\}$$

if $j > 1$ and $\delta\mu = 0$ else. Then $\delta\mu = \delta\mu(E|_{Z_s})$ for a general point $s \in \Pi$, and $\delta\mu$ vanishes if and only if $E|_{Z_s}$ is μ-semistable for general s. Using these notations we can state the general form of the Grauert-Mülich Theorem:

Theorem 3.1.2 — *Let E be a μ-semistable torsion free sheaf. Then there is a nonempty open subset $S \subset \Pi$ such that for all $s \in S$ the following inequality holds:*

$$0 \leq \delta\mu(E|_{Z_s}) \leq \max\{a_i\} \cdot \Pi a_i \cdot \deg(X).$$

Proof. If $\delta\mu = 0$, there is nothing to prove. Assume that $\delta\mu$ is positive, and let i be an index such that $\delta\mu = \mu_i - \mu_{i+1}$. Let $F' = F_i$ and $F'' = F/F'$, so that for all $s \in S$ the sheaves F'_s and F''_s are torsion free, and $\mu_{\min}(F'_s) = \mu_i$, $\mu_{\max}(F''_s) = \mu_{i+1}$. Let Z_0 be the maximal open subset of Z_S such that $F|_{Z_0}$ and $F''|_{Z_0}$ are locally free, say of rank r and r''. The surjection $F|_{Z_0} \to F''|_{Z_0}$ defines an X-morphism $\varphi : Z_0 \to \operatorname{Grass}_X(E, r'')$. We are interested in the relative differential $D\varphi : T_{Z/X}|_{Z_0} \to \varphi^* T_{\operatorname{Grass}(E,r'')/X}$ of the map φ. Recall that the relative tangent sheaf of a Grassmann variety can be expressed in terms of the tautological subsheaf $\mathcal{A}$ and the tautological quotient sheaf $\mathcal{B}$ (cf. 2.2.10):

$$\varphi^* \mathcal{T}_{\mathrm{Grass}(E,r'')/X} = \varphi^* \mathcal{H}om(\mathcal{A}, \mathcal{B}) = \mathcal{H}om(\varphi^* \mathcal{A}, \varphi^* \mathcal{B}) = \mathcal{H}om(F', F'')|_{Z_0}.$$

Thus we can consider $D\varphi$ as the adjoint of a homomorphism

$$\Phi : (F' \otimes \mathcal{T}_{Z/X})|_{Z_0} \longrightarrow F''|_{Z_0}.$$

Suppose Φ_s were zero for a general point $s \in S$. This would lead to a contradiction: first, making S smaller if necessary, this supposition would imply that Φ is zero. Let X_0 be the image of Z_0 in X. Since $q : Z \to X$ is a bundle, X_0 is open. In fact, since for any point $s \in S$ the complement of $Z_{0,s}$ in Z_s has codimension ≥ 2 in Z_s, we also have $\mathrm{codim}(X - X_0, X) \geq 2$. Moreover, $E|_{X_0}$ is locally free. Thus we are in the following situation:

$$Z_0 \xrightarrow{\ \varphi\ } \mathrm{Grass}(E|_{X_0}, r'')$$
$$q_0 \searrow \qquad \swarrow$$
$$X_0$$

where q_0 is a smooth map with connected fibres. If $\Phi = 0$, then φ is constant on the fibres of q_0 and hence factors through a morphism $\rho : X_0 \to \mathrm{Grass}(E|_{X_0}, r'')$ (Here we make use of the assumption that the characteristic of the base field is zero). But such a map ρ corresponds to a locally free quotient $E|_{X_0} \to E''$ of rank r'' with the property that $E''|_{Z_s \cap X_0}$ is isomorphic to $F''_s|_{Z_s \cap Z_0}$ for general s. Since by assumption F''_s is a destabilizing quotient of F_s, any extension of E'' as a quotient of E is destabilizing. This contradicts the assumption that E is μ-semistable.

We can rephrase the fact that Φ_s is nonzero for general $s \in S$ as follows: let $\mathcal{C}$ be the quotient category $\mathrm{Coh}_{n-\ell, n-\ell-1}(Z_s)$ as defined in Section 1.6. Then Φ_s is a non-trivial element in $\mathrm{Hom}_{\mathcal{C}}(F'_s \otimes \mathcal{T}_{Z/X}|_{Z_s}, F''_s)$. The appropriately modified version of 1.3.3 says that in this case the following inequality must necessarily hold:

$$\mu_{\min}(F'_s \otimes \mathcal{T}_{Z/X}|_{Z_s}) \leq \mu_{\max}(F''_s). \tag{3.1}$$

The Koszul complex associated to the evaluation map $e : V_a \otimes \mathcal{O}_X \to \mathcal{O}_X(a)$ provides us with a surjection $\Lambda^2 V_a \otimes \mathcal{O}_X(-a) \to \ker(e) = \mathcal{K}$ and hence a surjection

$$\Lambda^2 V_a \otimes q^* \mathcal{O}_X(-a) \otimes p^* \mathcal{O}(1) \to q^* \mathcal{K} \otimes p^* \mathcal{O}(1) \to \mathcal{T}_{Z_a/X}.$$

Using $\mathcal{T}_{Z/X} = \bigoplus pr_i^* \mathcal{T}_{Z_{a_i}/X}$ this yields a surjection

$$\left(\bigoplus_i \Lambda^2 V_{a_i} \otimes_k \mathcal{O}_X(-a_i) \right)\Big|_{Z_s} \to \mathcal{T}_{Z/X}|_{Z_s}.$$

From this we get the estimate

$$
\mu_{\min}(T_{Z/X}|_{Z_s} \otimes F'_s) \geq \mu_{\min}(\bigoplus_i \Lambda^2 V_{a_i} \otimes_k \mathcal{O}_X(-a_i) \otimes F'|_{Z_s})
$$
$$
= \min_i\{\mu_{\min}(\mathcal{O}_{Z_s}(-a_i) \otimes F'_s)\}
$$
$$
= \mu_{\min}(F'_s) - \max\{a_i\} \cdot \deg(Z_s)
$$

Combining this with inequality (3.1) one gets

$$
\delta\mu = \mu_i - \mu_{i+1} = \mu_{\min}(F'_s) - \mu_{\max}(F''_s)
$$
$$
\leq \max\{a_i\} \cdot \deg(Z_s) = \max\{a_i\} \cdot \Pi a_i \cdot \deg(X).
$$

□

Note that if $a_1 = \ldots = a_\ell = 1$, then the bound for $\delta\mu$ is just $\deg(X)$.

Remark 3.1.3 — In the proof above we used an argument involving the relative Grassmann scheme to show the following: if $\operatorname{Hom}(T_{Z/X} \otimes F', F'') = 0$, then there is subsheaf E', namely the kernel of $E|_{X_0} \to E''$, such that $q^*E' = F'|_{Z_0}$. This fact can also be interpreted as follows: Consider the k-linear map $\nabla : q^*E \to \Omega_{Z/X} \otimes q^*E$ given by $\nabla(s \otimes e) = ds \otimes e$, where $s \otimes e \in \mathcal{O}_Z \otimes_{q^{-1}\mathcal{O}_X} q^{-1}E = q^*E$. This is an integrable relative connection in q^*E, i.e. $\nabla(s \cdot e) = s \cdot \nabla(e) + ds \otimes e$ where e is a local section in q^*E and s a local section in $\mathcal{O}_Z$. Since $\operatorname{Hom}(F', \Omega_{Z/X} \otimes F'') = \operatorname{Hom}(T_{Z/X} \otimes F', F'') = 0$, the connection ∇ preserves F', i.e. ∇ induces a relative integrable connection $\nabla' : F' \to F' \otimes \Omega_{Z/X}$. Then $E' := F' \cap q^{-1}E$ defines a coherent subsheaf of E. That we indeed have $q^*E' = F'$ is a local problem, which can be solved by either going to the completion or using the analytic category ([41],[32]), where Deligne has proved an equivalence between coherent sheaves with relative integrable connections and relative local systems.

The last step of the proof of the theorem indicates that there is space for improvement. Indeed, if the inequality for $\mu_{\min}(T_{Z/X}|_{Z_s} \otimes F'|_{Z_s})$ can be sharpened then we automatically get a better bound for $\delta\mu(E|_{Z_s})$. In order to relate $\delta\mu$ and $\mu_{\min}(T_{Z/X}|_{Z_s})$ we need the following important theorem:

Theorem 3.1.4 — *Let X be a normal projective variety over an algebraically closed field of characteristic zero. If F_1 and F_2 are μ-semistable sheaves then $F_1 \otimes F_2$ is μ-semistable, too.*

Remember that even if F_1 and F_2 are torsion free, their tensor product might have torsion, though only in codimension 2. Thus under the assumption of the theorem F_1, F_2 and $F_1 \otimes F_2$ are locally free in codimension 1.

The proof of Theorem 3.1.4 will be given in the next section (3.2.8). It uses the coarse version of the Grauert-Mülich Theorem proved above. If the μ-semistability of the tensor product is granted one can prove a refined version of 3.1.2.

Theorem 3.1.5 — *Let E be μ-semistable. Then there is a nonempty open subset $S \subset \Pi$ such that for all $s \in S$ the following inequality holds:*

$$0 \leq \delta\mu(E|_{Z_s}) \leq -\mu_{\min}(T_{Z/X}|_{Z_s}).$$

Proof. Indeed, it is an immediate consequence of Theorem 3.1.4 that $\mu_{\min}(F_1 \otimes F_2) = \mu_{\min}(F_1) + \mu_{\min}(F_2)$. (Take the tensor product of the Harder-Narasimhan filtrations of F_1 and F_2 and use 3.1.4.) In particular,

$$\mu_{\min}(T_{Z/X}|_{Z_s} \otimes F'|_{Z_s}) = \mu_{\min}(T_{Z/X}|_{Z_s}) + \mu_{\min}(F'|_{Z_s}).$$

Hence, $-\mu_{\min}(T_{Z/X}|_{Z_s}) \geq \delta\mu(E|_{Z_s})$ follows from 3.1. $\qquad\square$

Therefore any further analysis of the minimal slope of the relative tangent bundle is likely to improve the crude bound of the Grauert-Mülich Theorem. This analysis was carried out by Flenner and led to an effective restriction theorem: if the degrees of the hyperplane sections are large enough then $E|_{Z_s}$ is semistable for a general complete intersection Z_s. This result will be discussed in Section 7.1.

Corollary 3.1.6 — *Let X be a normal projective variety of dimension n and let $\mathcal{O}_X(1)$ be a very ample line bundle. Let F be a μ-semistable coherent $\mathcal{O}_X$-module of rank r. Let Y be the intersection of $s < n$ general hyperplanes in the linear system $|\mathcal{O}_X(1)|$. Then*

$$\mu_{\min}(F|_Y) \geq \mu(F) - \deg(X) \cdot \frac{r-1}{2} \text{ and } \mu_{\max}(F|_Y) \leq \mu(F) + \deg(X) \cdot \frac{r-1}{2}.$$

Proof. We may assume that F is torsion free. If $F|_Y$ is μ-semistable there is nothing to prove. Let $\mu_1, \ldots, \mu_j$ and $r_1, \ldots, r_j$ be the slopes and ranks of the factors of the μ-Harder-Narasimhan filtration of $F|_Y$. By Theorem 3.1.2 one has $0 \leq \mu_i - \mu_{i+1} \leq \deg(X)$, and summing up terms from 1 to i: $\mu_i \geq \mu_1 - (i-1)\deg(X)$. This gives

$$\mu(F) = \sum_{i=1}^{j} \frac{r_i}{r}\mu_i \geq \mu_1 - \sum_{i=1}^{j}(i-1)\frac{r_i}{r}\deg(X)$$

$$\geq \mu_1 - \frac{\deg(X)}{r}\sum_{i=1}^{r}(i-1) = \mu_{\max}(F|_Y) - \deg(X)\frac{r-1}{2},$$

and similarly for $\mu_{\min}(F|_Y)$. $\qquad\square$

3.2 Finite Coverings and Tensor Products

In this section we will use the Grauert-Mülich Theorem to prove that the tensor product of μ-semistable sheaves is again μ-semistable. This in turn allows one to improve the Grauert-Mülich Theorem, as has been shown in the previous section, and paves the way to Flenner's Restriction Theorem. The question how μ-semistable sheaves behave under pull-back for finite covering maps enters naturally into the arguments. Conversely, some boundedness problems for pure sheaves can be treated by converting pure sheaves into torsion free ones via an appropriate push-forward.

At various steps in the discussion one needs that the characteristic of the base field is 0, though some of the arguments work in greater generality. We therefore continue to assume that k is an algebraically closed field of characteristic 0.

Let $f : Y \to X$ be a finite morphism of degree d of normal projective varieties over k of dimension n. Let $\mathcal{O}_X(1)$ be an ample invertible sheaf. Then $\mathcal{O}_Y(1) = f^*\mathcal{O}_X(1)$ is ample as well. The functor f_* on coherent sheaves is exact and the higher direct image sheaves vanish. Therefore $H^i(Y, F(m)) = H^i(X, f_*(F(m))) = H^i(X, (f_*F)(m))$ and in particular $P(F) = P(f_*F)$. The sheaf of algebras $\mathcal{A} := f_*\mathcal{O}_Y$ is a torsion free coherent $\mathcal{O}_X$-module of rank d, and f_* gives an equivalence between the category of coherent sheaves on Y and the category of coherent sheaves on X with an $\mathcal{A}$-module structure. Moreover, f_* preserves the dimension of sheaves and purity. On the other hand, since X is normal and $\mathcal{A}$ is torsion free, $\mathcal{A}$ is locally free in codimension 1, which means that f is flat in codimension 1. Thus f^* is exact modulo sheaves of dimension $\leq n - 2$. Moreover, if $F \in \mathrm{Ob}(\mathrm{Coh}(X))$ has no torsion in dimension $n - 1$, then the same is true for f^*F. It is therefore appropriate to work in the categories $\mathrm{Coh}_{n,n-1}$ as we will do throughout this section.

We need to relate rank and slope of F and f^*F:

Lemma 3.2.1 — *Let F be a coherent $\mathcal{O}_X$-module of dimension n with no torsion in codimension 1. Then* $\mathrm{rk}(f^*F) = \mathrm{rk}(F)$ *and* $\mu(f^*F) = d \cdot \mu(F)$. *Let G be a coherent $\mathcal{O}_Y$-module with no torsion in codimension 1. Then* $\mathrm{rk}(f_*G) = d \cdot \mathrm{rk}(G)$ *and* $\mu(G) = d \cdot (\mu(f_*G) - \mu(\mathcal{A}))$.

Proof. For the last assertion note the following identities: $\hat{\mu}(G) = \hat{\mu}(f_*G)$, in particular $\hat{\mu}(\mathcal{O}_Y) = \hat{\mu}(\mathcal{A})$. Moreover, μ and $\hat{\mu}$ are related by $\mu(\mathcal{A}) = \deg(X) \cdot (\hat{\mu}(\mathcal{A}) - \hat{\mu}(\mathcal{O}_X))$, $\mu(G) = \deg(Y) \cdot (\hat{\mu}(G) - \hat{\mu}(\mathcal{O}_Y))$ and $\mu(f_*G) = \deg(X) \cdot (\hat{\mu}(f_*G) - \hat{\mu}(\mathcal{O}_X))$ (See the remark after Definition 1.6.8). Finally, $\deg(Y) = d \cdot \deg(X)$. The assertion follows from this. □

Lemma 3.2.2 — *Let F be an n-dimensional coherent $\mathcal{O}_X$-module. Then F is μ-semistable if and only if f^*F is μ-semistable.*

Proof. Certainly, F has no torsion in codimension 1 if and only if the same is true for f^*F. If $F' \subset F$ is a submodule with $\mu(F') > \mu(F)$ then $\mu(f^*F') > \mu(f^*F)$ by the previous lemma. This shows the 'if'-direction. For the converse, let K be a splitting field of the function field $K(Y)$ over $K(X)$ and let Z be the normalization of Y in K. This gives finite morphisms $Z \to Y \to X$ and, because of the first part of the proof, it is enough to consider the morphism $Z \to X$ instead of $Y \to X$. In other words we may assume that $K(Y) \supset K(X)$ is a Galois extension with Galois group G. Suppose now that F is a torsion free sheaf on X such that f^*F is not μ-semistable, and let $F'_Y \subset f^*F$ be the maximal destabilizing submodule. Because of its uniqueness, F'_Y is invariant under the action of G. By descent theory, there is a submodule $F' \subset F$ such that f^*F' is isomorphic to F'_Y in codimension 1, i.e. $F'_Y \cong f^*F'$ in $\mathrm{Coh}_{n,n-1}(Y)$. Thus F' destabilizes F.   □

This lemma can be extended to cover the case of polystable sheaves:

Lemma 3.2.3 — *Let F be an n-dimensional coherent sheaf on X. Then F is μ-polystable if and only if f^*F is μ-polystable.*

Proof. Again, we prove the 'if'-direction first. There is a natural trace map $tr : \mathcal{A} \to \mathcal{O}_X$. This map is obtained as the composition of the homomorphism $\mathcal{A} \to \mathcal{E}nd(\mathcal{A})$ given by the algebra structure of $\mathcal{A}$ and the trace map $\mathcal{E}nd(\mathcal{A}) \to \mathcal{O}_X$. The latter is first defined in the usual way over the maximal open set $U \subset X$ where $\mathcal{A}|_U$ is locally free: it extends uniquely over all of X, since X is normal. The homomorphism $\frac{1}{d}tr$ splits the inclusion morphism $i : \mathcal{O}_X \to \mathcal{A}$.

We may assume that F is torsion free. Suppose that F' is a non-trivial proper submodule with the property that the homomorphism $f^*F' \to f^*F$ splits. Such a splitting is given by an $\mathcal{O}_X$-homomorphism $\tilde{\sigma} : F \to \mathcal{A} \otimes F'$ which makes the diagram

$$
\begin{array}{ccccc}
F & \xrightarrow{\tilde{\sigma}} & \mathcal{A} & \otimes & F' \\
\uparrow & & & & \uparrow{\scriptstyle i \otimes 1_{F'}} \\
F' & \cong & \mathcal{O}_X & \otimes & F'
\end{array}
$$

commutative. Thus composing $\tilde{\sigma}$ with $\frac{1}{d}tr \otimes 1_{F'} : \mathcal{A} \otimes F' \to F'$ defines a splitting σ of the inclusion $F' \to F$. If $\tilde{\sigma}$ is defined outside a set of codimension 2, then the same is true for σ. Now if f^*F is μ-polystable, then F is μ-semistable by the previous lemma, and the arguments above show, that any destabilizing submodule in F is a direct summand (in codimension 1).

For the converse direction we may again assume that $f : Y \to X$ is a Galois covering because of the first part of the proof. Let F be μ-stable and let $E \subset f^*F$ be a destabilizing stable subsheaf. Then $E' := \sum_{g \in \mathrm{Gal}(Y/X)} g^*E \subset f^*F$ is a μ-polystable subsheaf which is invariant under the Galois action and is therefore the pull-back of a submodule $F' \subset F$. Since F is stable, we must have $F' = F$. Thus $E' \cong f^*F$. □

The next step is to relate semistability to ampleness. A vector bundle E on a projective k-scheme X is *ample* if the canonical line bundle $\mathcal{O}(1)$ on $\mathbb{P}(E)$ is ample. On curves a line bundle is ample if and only if its degree is positive. For arbitrary vector bundles of higher rank the degree condition is of course much too weak to imply any positivity properties. However, this is true if the vector bundle is semistable. Before we prove this result of Gieseker, recall some notions related to ampleness:

Let X be a projective k-scheme.

Definition 3.2.4 — *A Cartier divisor D on X is pseudo-ample, if for all integral closed subschemes $Y \subset X$ one has $Y.D^{\dim(Y)} \geq 0$. And D is called nef, if $Y.D \geq 0$ for all closed integral curves $Y \subset X$.*

This notion of pseudo-ampleness is justified in view of the following ampleness criterion:

Theorem 3.2.5 (Nakai Criterion) — *A Cartier divisor D on X is ample, if and only if for all integral closed subschemes $Y \subset X$ one has $Y.D^{\dim(Y)} > 0$.*

Proof. See [100]. □

The following theorem of Kleiman says that it is enough to test the nonnegativity of a divisor on curves in order to infer its pseudo-ampleness.

Theorem 3.2.6 — *A Cartier divisor D on X is pseudo-ample if and only if it is nef.* □

The analogous statement for ampleness is wrong. For counterexamples and a proof of the theorem see [100]. However, if D is contained in the interior of the cone dual to the cone generated by the integral curves, then D is indeed ample. This result due to Kleiman and references to the original papers can also be found in [100].

Theorem 3.2.7 — *Let X be a smooth projective curve over an algebraically closed field of characteristic zero and E a semistable vector bundle of rank r on X. Denote by $\pi :$ $\mathbb{P}(E) \to X$ the canonical projection and by $\mathcal{O}_\pi(1)$ the tautological line bundle on $\mathbb{P}(E)$.*
i) $\deg(E) \geq 0 \Leftrightarrow \mathcal{O}_\pi(1)$ *is pseudo-ample.*
ii) $\deg(E) > 0 \Leftrightarrow \mathcal{O}_\pi(1)$ *is ample.*

Proof. One direction is easy: assume that $\mathcal{O}_\pi(1)$ is pseudo-ample or ample. Then the self-intersection number $(\mathcal{O}_\pi(1))^r$ is ≥ 0 or > 0, respectively. But this number is the leading coefficient of the polynomial $\chi(\mathcal{O}_\pi(m))$. By the projection formula and the Riemann-Roch formula we get:

$$\chi(\mathcal{O}_\pi(m)) = \chi(\pi_*\mathcal{O}_\pi(m)) = \chi(S^m E) = \deg(S^m E) + \operatorname{rk}(S^m E)(1-g).$$

Now

$$\operatorname{rk}(S^m E) = \binom{m+r-1}{r-1} \text{ and } \det(S^m E) = \det(E)^{\otimes\binom{m+r-1}{r}}.$$

Thus the leading term is indeed $\deg(E)\frac{m^r}{r!}$. Now to the converse:

i) By Theorem 3.2.6, it suffices to show that $\mathcal{O}_\pi(1)$ is nef. Let $C' \subset \mathbb{P}(E)$ be any integral closed curve, $\nu : C \to C'$ its normalization and $f = \pi \circ \nu : C \to X$. If C' is mapped to a point by π then it is contained in a fibre. But the restriction of $\mathcal{O}_\pi(1)$ to any fibre is ample, hence $C'.\mathcal{O}_\pi(1) > 0$. Thus we may assume that $f : C \to X$ is a finite map of smooth curves. We have $C'.\mathcal{O}_\pi(1) = \deg(\nu^*(\mathcal{O}_\pi(1)))$ and a surjection $f^*E \to \nu^*\mathcal{O}_\pi(1)$. According to Lemma 3.2.2, f^*E is semistable. This implies $\deg(\nu^*\mathcal{O}_\pi(1)) \geq \deg(f^*E)/r = \deg(E) \cdot \deg(f)/r \geq 0$.

ii) Choose a finite morphism $f : Y \to X$ of smooth curves of degree $\deg(f) > r$, and let $P \in Y$ be a closed point. The vector bundle $E' = f^*E \otimes \mathcal{O}_Y(-P)$ still has positive degree:

$$\deg(E') = \deg(f) \cdot \deg(E) - \operatorname{rk}(E) > 0.$$

Moreover, E' is semistable by Lemma 3.2.2, so that by *i)* the line bundle $L' = \mathcal{O}_{\mathbb{P}(E')}(1)$ is pseudo-ample. Under the isomorphism

$$\mathbb{P}(E') \cong \mathbb{P}(f^*E) \cong Y \times_X \mathbb{P}(E)$$

L' can be identified with $L(-F)$, where $L = \mathcal{O}_{\mathbb{P}(f^*E)}(1)$ and F is the fibre over P. Now let V be any closed integral subscheme of $\mathbb{P}(f^*E)$ of dimension s. If V is contained in a fibre, then $V.L^s > 0$, since L is very ample on the fibres. If V is not contained in a fibre, it maps surjectively onto Y and has a proper intersection Z with F. Now

$$V.L^s = V.(L' + F)^s = V.(L')^s + s \cdot V.F.L^{s-1},$$

since $F.F = 0$ and $L|_F = L'|_F$. We know that L' is pseudo-ample. Therefore $V.(L')^s \geq 0$. Moreover, $L|_F$ is very ample on the fibre F. Therefore $V.F.L^{s-1} = Z.(L|_F)^{s-1} > 0$. Using the Nakai Criterion we conclude that L is ample. But L is the pull-back of $\mathcal{O}_\pi(1)$ under the finite map $\mathbb{P}(f^*E) \to \mathbb{P}(E)$. Therefore $\mathcal{O}_\pi(1)$ is ample, too. $\qquad\square$

We are now ready to prove the theorem on the μ-semistability of tensor products as stated in the previous section.

3.2.8 Proof of Theorem 3.1.4: — We may assume that $\mathcal{O}_X(1)$ is very ample. Let $F_1 \otimes F_2 \to Q$ be a torsion free destabilizing quotient, i.e. $\mathrm{rk}(Q) > 0$ and $\mu(F_1 \otimes F_2) = \mu(F_1) + \mu(F_2) > \mu(Q)$.

Step 1. Assume that $\mu(F_1) + \mu(F_2) - \mu(Q) > \deg(X).(\mathrm{rk}(F_1) + \mathrm{rk}(F_2) + 2)/2$. A general complete intersection of $\dim(X) - 1$ hyperplane sections is a smooth curve C, and the restrictions of the sheaves F_1, F_2, and Q to C are locally free. By the Corollary 3.1.6 to the Grauert-Mülich Theorem their Harder-Narasimhan factors satisfy $\mu(gr_j^{\mathrm{HN}}(F_i|_C)) \geq \mu(F_i) - \deg(X)(\mathrm{rk}(F_i) - 1)/2$. Define

$$n_i = \left\lceil \frac{\mu(F_i)}{\deg(X)} - \frac{\mathrm{rk}(F_i) - 1}{2} \right\rceil - 1.$$

Then

$$\mu\big(gr_j^{\mathrm{HN}} F_i(-n_i)|_C\big) \geq \mu(F_i) - \deg(X)(n_i + (\mathrm{rk}(F_i) - 1)/2) > 0.$$

Thus $gr_j^{\mathrm{HN}} F_i(-n_i)|_C$ is a semistable vector bundle of positive degree. By Theorem 3.2.7 it is ample. As Hartshorne shows [97], the tensor product of two ample vector bundles is again ample (in characteristic 0). Thus $gr_j^{\mathrm{HN}} F_1(-n_1) \otimes gr_k^{\mathrm{HN}} F_2(-n_2)$ is ample. Hence $(F_1 \otimes F_2)|_C(-n_1 - n_2)$ is an iterated extension of ample vector bundles and therefore itself ample, and finally, being a quotient of an ample vector bundle, $Q|_C(-n_1 - n_2)$ is ample as well. To get a contradiction it suffices to show that the slope of $Q(-n_1 - n_2)$ is negative. But

$$
\begin{aligned}
\mu(Q(-n_1 - n_2)) \;=\;& \mu(Q) - (n_1 + n_2)\deg(X) \\
<\;& \mu(F_1) + \mu(F_2) - \deg(X)(n_1 + n_2 + (\mathrm{rk}(F_1) + \mathrm{rk}(F_2) + 2)/2) \\
=\;& \mu(F_1) - \left(\frac{\mathrm{rk}(F_1) - 1}{2} + n_1 + 1\right)\deg(X) \\
& +\mu(F_2) - \left(\frac{\mathrm{rk}(F_2) - 1}{2} + n_2 + 1\right)\deg(X) \\
\leq\;& 0.
\end{aligned}
$$

Step 2. To reduce the general case to the situation of Step 1, we apply the following theorem which allows us to take 'roots' of line bundles.

Theorem 3.2.9 — *Let X be a projective normal variety over an algebraically closed field k of characteristic zero and let $\mathcal{O}_X(1)$ be a very ample invertible sheaf. For any positive integer d there exist a normal variety X' with a very ample invertible sheaf $\mathcal{O}_{X'}(1)$ and a finite morphism $f : X' \to X$ such that $f^*\mathcal{O}_X(1) \cong \mathcal{O}_{X'}(d)$. Moreover, if X is smooth, X' can be chosen to be smooth as well.*

Proof. See [27, 166, 260]. □

Using this theorem the proof proceeds as follows: choose a finite map $f : X' \to X$ as in the theorem with sufficiently large d. Observe that if slope and degree on X' are defined with respect to $\mathcal{O}_{X'}(1)$, then for any coherent sheaf F on X one has

$$\frac{\mu(f^*F)}{\deg(X')} = d \cdot \frac{\mu(F)}{\deg(X)}.$$

And according to Lemma 3.2.2, f^*F_1 and f^*F_2 are μ-semistable with respect to $\mathcal{O}_{X'}(1)$, and f^*Q destabilizes $f^*F_1 \otimes f^*F_2 = f^*(F_1 \otimes F_2)$. Choosing the degree d large enough it is easy to satisfy the condition

$$\frac{\mu(f^*F_1) + \mu(f^*F_2) - \mu(f^*Q)}{\deg(X')} = d \cdot \frac{\mu(F_1 \otimes F_2) - \mu(Q)}{\deg(X)} > \frac{\mathrm{rk}(F_1) + \mathrm{rk}(F_2) + 2}{2}$$

of Step 1. This finishes the proof. $\square$

Corollary 3.2.10 — *If F is a μ-semistable sheaf, then $\mathcal{E}nd(F)$, all exterior powers $\Lambda^\nu F$ and all symmetric powers $S^\nu F$ are again μ-semistable.*

Proof. $F^{\otimes \nu}$ is μ-semistable by Theorem 3.1.4. Since the characteristic of the base field is 0, $\Lambda^\nu F$ and $S^\nu F$ are direct summands of $F^{\otimes \nu}$ and therefore μ-semistable. Up to sheaves of codimension 2 one has $\mathcal{E}nd(F) \cong F^\vee \otimes F$, so again the assertion follows from the theorem. $\square$

Theorem 3.2.11 — *Let X be a smooth projective variety and $\mathcal{O}_X(1)$ an ample line bundle. The tensor product of any two μ-polystable locally free sheaves is again μ-polystable. In particular, the exterior and symmetric powers of a μ-polystable locally free sheaf are μ-polystable.*

Using transcendental methods, one can argue as follows: first reduce to the case $k = \mathbb{C}$. Then a complex algebraic vector bundle on X is polystable if and only if it admits a Hermite-Einstein metric with respect to the Kähler metric of X induced by $\mathcal{O}_X(1)$. (This deep theorem, known as the Kobayashi-Hitchin Correspondence, was proved in increasing generality by Narasimhan-Seshadri [202], Donaldson [44, 45] and Uhlenbeck-Yau [255]. For details see the books [127],[46] and [157].) Now, if E and F have a Hermite-Einstein metric, it is not difficult to see that the induced metric on the tensor product $E \otimes F$ is again Hermite-Einstein. The assertion of the theorem follows.

There are purely algebraic proofs of this fact, which however use different methods than the one developed here. Usually the paper [226] is cited, but the result is phrased in terms of quasistability of principal bundles. Another proof uses standard Tannakian arguments and was explained to us by Burt Totaro. Roughly, one forms the Tannakian category of semistable vector bundles lets say for simplicity of degree zero. The semisimple objects, i.e. direct sums of irreducible objects, of this category are precisely the polystable vector bundles. Then one uses that this Tannakian category is equivalent to the category of representations of a pro-algebraic group. Eventually one uses the fact that any representation of a reductive group is semisimple.

3.3 Boundedness II

The Grauert-Mülich Theorem allows one to give uniform bounds for the number of global sections of a μ-semistable sheaf in terms of its slope. This is made precise in a very elegant manner by the following theorem. Let $[x]_+ := \max\{0, x\}$ for any real number x.

Theorem 3.3.1 (Le Potier-Simpson) — *Let X be a projective scheme with a very ample line bundle $\mathcal{O}_X(1)$. For any purely d-dimensional coherent sheaf F of multiplicity $r(F)$ there is an F-regular sequence of hyperplane sections $H_1, \ldots, H_d$, such that*

$$\frac{h^0(X_\nu, F|_{X_\nu})}{r(F)} \leq \frac{1}{\nu!} \left[\hat{\mu}_{\max}(F) + r(F)^2 + \frac{1}{2}(r(F) + d) - 1 \right]_+^\nu,$$

for all $\nu = d, \ldots, 0$ and $X_\nu = H_1 \cap \ldots \cap H_{d-\nu}$.

We prove this theorem in several steps.

Lemma 3.3.2 — *Suppose that X is a normal projective variety of dimension d and that F is a torsion free sheaf of rank $\mathrm{rk}(F)$. Then for any F-regular sequence of hyperplane sections $H_1, \ldots, H_d$ and $X_\nu = H_1 \cap \ldots \cap H_{d-\nu}$ the following estimate holds for all $\nu = 1, \ldots, d$.*

$$\frac{h^0(X_\nu, F|_{X_\nu})}{\mathrm{rk}(F) \cdot \deg(X)} \leq \frac{1}{\nu!} \left[\frac{\mu_{\max}(F|_{X_1})}{\deg(X)} + \nu \right]_+^\nu.$$

Proof. Let $F_\nu = F|_{X_\nu}$ for brevity. The lemma is proved by induction on ν.

Let $\nu = 1$. Since $h^0(X_1, F_1) \leq \sum_i h^0(X_1, gr_i^{HN}(F_1))$ and since the right hand side of the estimate in the lemma is monotonously increasing with μ, we may assume without loss of generality that $\mu_{\max}(F_1) = \mu(F_1)$, i.e. that F_1 is μ-semistable. For $\ell \geq 0$ one gets estimates

$$h^0(X_1, F_1) \leq h^0(X_1, F_1(-\ell)) + \mathrm{rk}(F)\ell \deg(X). \tag{3.2}$$

But $h^0(X_1, F_1(-\ell)) = \hom(\mathcal{O}_{X_1}(\ell), F_1) = 0$ if $\ell > \mu(F_1)/\deg(X)$ by Proposition 1.2.7. Put $\ell := \lfloor \mu(F_1)/\deg(X) + 1 \rfloor$. Then (3.2) is the required bound in the case $\nu = 1$.

Suppose the inequality has been proved for $\nu - 1$, $\nu \geq 2$. From the standard exact sequences

$$0 \to F_\nu(-k-1) \to F_\nu(-k) \to F_{\nu-1}(-k) \to 0 \quad , k = 0, 1, 2 \ldots$$

one inductively derives estimates

$$h^0(X_\nu, F_\nu) \leq h^0(X_\nu, F_\nu(-\ell)) + \sum_{i=0}^{\ell-1} h^0(X_{\nu-1}, F_{\nu-1}(-i)) \leq \sum_{i=0}^{\infty} h^0(X_{\nu-1}, F_{\nu-1}(-i)).$$

Of course, the sum on the right hand side is in fact finite. Using the induction hypothesis and replacing the sum by an integral, we can write

$$\frac{h^0(X_\nu, F_\nu)}{\mathrm{rk}(F) \cdot \deg(X)} \leq \frac{1}{(\nu-1)!} \int_{-1}^{C} \left[\frac{\mu_{\max}(F_1)}{\deg(X)} + (\nu-1) - t \right]_+^{\nu-1} dt,$$

where C is the maximum of -1 and the smallest zero of the integrand. Evaluating the integral yields the bound of the lemma. $\square$

Corollary 3.3.3 — *Under the hypotheses of the lemma there is an F-regular sequence of hyperplane sections $H_1, \ldots, H_d$ such that*

$$\frac{h^0(X_\nu, F|_{X_\nu})}{\mathrm{rk}(F) \cdot \deg(X)} \leq \frac{1}{\nu!} \left[\frac{\mu_{\max}(F)}{\deg(X)} + \frac{\mathrm{rk}(F) - 1}{2} + \nu \right]_+^{\nu},$$

for all $\nu = 1, \ldots, d$.

Proof. Combine the lemma with Corollary 3.1.6. $\square$

Corollary 3.3.3 gives the assertion of the theorem in the case that F is torsion free on a normal projective variety. In order to reduce the general case to this situation we use the same trick that was already employed in the proof of Grothendieck's Lemma 1.7.9.

Proof of the theorem. Let $i : X \to \mathbb{P}^N$ be the closed embedding induced by the complete linear system of $\mathcal{O}_X(1)$. Let Z be the (d-dimensional) support of $\tilde{F} = i_* F$, and choose a linear subspace $L \subset \mathbb{P}^N$ of dimension $N - d - 1$ which does not intersect Z. Linear projection with center L then induces a finite map $\pi : Z \to Y \cong \mathbb{P}^d$ such that $\mathcal{O}_X(1)|_Z \cong \pi^* \mathcal{O}_Y(1)$. Since F is pure, $\pi_* F$ is also pure, i.e. torsion free. Moreover, $r(F) = \mathrm{rk}(\pi_* F)$ and $\hat{\mu}(F) = \hat{\mu}(\pi_* F) = \mu(\pi_* F) + \hat{\mu}(\mathcal{O}_Y) = \mu(\pi_* F) + (d+1)/2$. A $\pi_* F$-regular sequence of hyperplanes H_i' in Y induces an F-regular sequence of hyperplane sections H_i on X. If $Y_\nu = H_1' \cap \ldots \cap H_{d-\nu}'$, then $\pi_*(F|_{X_\nu}) = (\pi_* F)|_{Y_\nu}$ and hence $h^0(F|_{X_\nu}) = h^0(\pi_* F|_{Y_\nu})$. We need to relate $\hat{\mu}_{\max}(F)$ and $\mu_{\max}(\pi_* F)$. For that purpose consider the sheaf of algebras $\mathcal{A} := \pi_* \mathcal{O}_Z$.

Lemma 3.3.4 — $\mathcal{A}$ *is a torsion free sheaf with* $\mu_{\min}(\mathcal{A}) \geq -\mathrm{rk}(\mathcal{A}) \geq -\mathrm{rk}(\pi_*F)^2 = -r(F)^2.$

Proof. π_*F carries an $\mathcal{A}$-module structure. The corresponding algebra homomorphism $\mathcal{A} \to \mathcal{E}nd_{\mathcal{O}_Y}(\pi_*F)$ is injective, since by definition Z is the support of F. This implies that $\mathcal{A}$ is torsion free and has rank less or equal to $\mathrm{rk}(\pi_*F)^2 = r(F)^2$. By construction, Z is a closed subscheme of the geometric vector bundle

$$\mathbb{P}^N \setminus L \cong \operatorname{Spec} S^*W \longrightarrow Y,$$

where $W = \mathcal{O}_Y(-1)^{\oplus(N-d)}$. Hence, there is a surjection $\varphi : S^*W \to \mathcal{A}$. Consider the ascending filtration of $\mathcal{A}$ given by the submodules $F_p\mathcal{A} = \varphi(\mathcal{O} \oplus W \oplus \ldots \oplus S^pW)$. Since $\mathcal{A}$ is coherent, only finitely many factors $gr_p^F\mathcal{A}$ are nonzero. Moreover, since the multiplication map $W \otimes gr_p^F\mathcal{A} \to gr_{p+1}^F\mathcal{A}$ is surjective, it follows that, once $gr_p^F\mathcal{A}$ is torsion, the same is true for all $gr_{p+i}^F\mathcal{A}$, $i \geq 0$. In particular, if $gr_p^F(\mathcal{A})$ is not torsion then $p \leq \mathrm{rk}(\mathcal{A})$. In other words, the cokernel of $\varphi : \mathcal{O}_Y \oplus \ldots \oplus S^{\mathrm{rk}(\mathcal{A})}W \to \mathcal{A}$ is torsion. This implies that $\mu_{\min}(\mathcal{A}) \geq \mu_{\min}(S^{\mathrm{rk}(\mathcal{A})}W) = -\mathrm{rk}(\mathcal{A})$. $\qquad\square$

Lemma 3.3.5 — $\mu_{max}(\pi_*F) \leq \hat{\mu}_{\max}(F) + r(F)^2 - (d+1)/2.$

Proof. Let G be the maximal destabilizing submodule of π_*F, and let G' be the image of the multiplication map $\mathcal{A} \otimes G \to \mathcal{A} \otimes \pi_*F \to \pi_*F$, i.e. G' is the $\mathcal{A}$-submodule of π_*F generated by G. Then $G' \cong \pi_*G''$ for some $\mathcal{O}_X$-submodule $G'' \subset F$. It follows that

$$
\begin{aligned}
\hat{\mu}_{max}(F) &\geq \hat{\mu}(G'') = \hat{\mu}(G') = \mu(G') + \hat{\mu}(\mathcal{O}_Y) \\
&\geq \mu_{\min}(\mathcal{A} \otimes G) + \hat{\mu}(\mathcal{O}_Y) \\
&= \mu(G) + \mu_{\min}(\mathcal{A}) + \hat{\mu}(\mathcal{O}_Y) \qquad \text{because of 3.1.4} \\
&\geq \mu_{\max}(\pi_*F) - r(F)^2 + (d+1)/2,
\end{aligned}
$$

where for the last inequality we have used that $\mu(G) = \mu_{\max}(\pi_*F)$ by the choice of G, $\hat{\mu}(\mathcal{O}_Y) = (d+1)/2$, and $\mu_{\min}(\mathcal{A}) \geq -r(F)^2$ by the previous lemma. $\qquad\square$

As a consequence of Lemma 3.3.5 we have

$$\mu_{\max}(\pi_*F) + \nu + \frac{\mathrm{rk}(\pi_*F) - 1}{2} \leq \hat{\mu}_{\max}(F) + r(F)^2 + \frac{r(F)-1}{2} + \frac{d-1}{2}$$

for any $0 \leq \nu \leq d$. Applying Corollary 3.3.3 to π_*F and using this estimate we get the inequality of the theorem. $\qquad\square$

A slight modification of the proof of 3.3.2 gives the following proposition:

Proposition 3.3.6 — *Let X be a smooth projective surface and $\mathcal{O}_X(1)$ a globally generated ample line bundle. Let E and F be torsion free μ-semistable sheaves. Then*

$$\hom(E, F) \leq \frac{\mathrm{rk}(E)\mathrm{rk}(F)}{2\deg(X)} \left[\mu(F) - \mu(E) + \frac{\mathrm{rk}(E) + \mathrm{rk}(F) + 1}{2} \deg(X) \right]_+^2$$

To see this, apply Corollary 3.1.6 to both sheaves E and F, and use the same induction process as in Lemma 3.3.2. The bound of the proposition is slightly sharper than the one obtained by applying the theorem to $E^\vee \otimes F$, say in case E is locally free. □

As a major application of the Le Potier-Simpson Estimate we get the boundedness of the family of semistable sheaves:

Theorem 3.3.7 — *Let $f : X \to S$ be a projective morphism of schemes of finite type over k and let $\mathcal{O}_X(1)$ be an f-ample line bundle. Let P be a polynomial of degree d, and let μ_0 be a rational number. Then the family of purely d-dimensional sheaves on the fibres of f with Hilbert polynomial P and maximal slope $\hat{\mu}_{\max} \leq \mu_0$ is bounded. In particular, the family of semistable sheaves with Hilbert polynomial P is bounded.*

Proof. Covering S by finitely many open subschemes and replacing $\mathcal{O}_X(1)$ by an appropriate high tensor power, if necessary, we may assume that f factors through an embedding $X \to S \times \mathbb{P}^N$. Thus we may reduce to the case $S = \mathrm{Spec}(k)$, $X = \mathbb{P}^N$. According to Theorem 3.3.1 we can find for each purely d-dimensional coherent sheaf F a regular sequence of hyperplanes $H_1, \ldots, H_d$ such that $h^0(F|_{H_1 \cap \ldots \cap H_i}) \leq C$ for all $i = 0, \ldots, d$, where C is a constant depending only on the dimension and degree of X and the multiplicity and maximal slope of F. Since these are given or bounded by P and μ_0, respectively, the bound is uniform for the family in question. The assertion of the theorem follows from this and the Kleiman Criterion 1.7.8. □

For later reference we note the following variant of Theorem 3.3.1. Let X be a projective scheme with a very ample line bundle $\mathcal{O}_X(1)$. Let F_i, $i = 1 < \ldots < \ell$, be the factors of the Harder-Narasimhan filtration of a purely d-dimensional sheaf F, and let r_i and r denote the multiplicities of F_i and F. Then $h^0(F) \leq \sum_{i=1}^\ell h^0(F_i)$, and applying the Le Potier-Simpson Estimate to each factor individually and summing up, we get

$$\frac{h^0(F)}{r} = \sum_{i=1}^\ell \frac{r_i}{r} \cdot \frac{h^0(F_i)}{r_i} \leq \sum_{i=1}^\ell \frac{r_i}{r} \cdot \frac{1}{d!} \left[\hat{\mu}(F_i) + r_i^2 + \frac{1}{2}(r_i + d) - 1 \right]_+^d$$

Using $\hat{\mu}(F_i) \leq \hat{\mu}_{\max}(F)$ for $i = 1, \ldots, \ell - 1$, $\hat{\mu}(F_\ell) \leq \hat{\mu}(F)$ and $\hat{\mu}(F(m)) = \hat{\mu}(F) + m$, one finally gets:

Corollary 3.3.8 — *Let $C = r^2 + \frac{1}{2}(r + d)$. Then*

$$\frac{h^0(F(m))}{r} \leq \frac{r - 1}{r} \cdot \frac{1}{d!}\, [\hat{\mu}_{\max}(F) + C - 1 + m]_+^d + \frac{1}{r} \cdot \frac{1}{d!}\, [\hat{\mu}(F) + C - 1 + m]_+^d .$$

$\square$

3.4 The Bogomolov Inequality

Another application of Theorem 3.1.4 on the semistability of tensor products is the Bogomolov Inequality. This important result has manifold applications to the theory of vector bundles and to the geometry of surfaces. We begin with the definition of the discriminant of a sheaf.

Let F be a coherent sheaf on a smooth projective variety X with Chern classes c_i and rank r. The *discriminant* of F by definition is the characteristic class

$$\Delta(F) = 2rc_2 - (r - 1)c_1^2.$$

If X is a complex surface, we will denote the characteristic number obtained by evaluating $\Delta(F)$ on the fundamental class of X by the same symbol. (Warning: This definition of the discriminant differs from many other conventions in the literature, partly by the sign, partly by a multiple or a power of r, each of which has its own virtues.) Clearly, the discriminant of a line bundle vanishes. If F is locally free, then $\Delta(F^\vee) = \Delta(F)$. The Chern character of F is given by the series

$$ch(F) = r + c_1 + \frac{1}{2}(c_1^2 - 2c_2)\ldots .$$

Hence $ch(F)/r = 1 + y$ for some nilpotent element y. Following Drezet we write

$$\log ch(F) = \log r + \frac{c_1}{r} - \frac{\Delta(F)}{2r^2}\ldots .$$

The Chern character is a ring homomorphism from $K^0(X)$ to $H^*(X, \mathbb{Q})$, and the logarithm converts multiplicative relations into additive ones. From this it is clear that for locally free sheaves F' and F'' one has

$$\frac{\Delta(F' \otimes F'')}{r'^2 r''^2} = \frac{\Delta(F')}{r'^2} + \frac{\Delta(F'')}{r''^2}, \tag{3.3}$$

where $r' = \mathrm{rk}(F')$ and $r'' = \mathrm{rk}(F'')$. In particular, the discriminant of a coherent sheaf is invariant under twisting with a line bundle, and if F is locally free and n is a positive integer, then

$$\Delta(F^{\otimes n}) = nr^{2(n-1)}\Delta(F) \quad \text{and} \quad \Delta(\mathcal{E}nd(F)) = 2r^2\Delta(F), \tag{3.4}$$

The latter equation also implies the relation $\Delta(F) = c_2(\mathcal{E}nd(F))$.

Theorem 3.4.1 (Bogomolov Inequality) — *Let X be a smooth projective surface and H an ample divisor on X. If F is a μ-semistable torsion free sheaf, then*

$$\Delta(F) \geq 0.$$

Proof. Let r be the rank of F. The double dual $F^{\vee\vee}$ of F is still μ-semistable, and the discriminants of F and $F^{\vee\vee}$ are related by $\Delta(F) = \Delta(F^{\vee\vee}) + 2r\ell(F^{\vee\vee}/F) \geq \Delta(F^{\vee\vee})$. Hence replacing F by $F^{\vee\vee}$, if necessary, we may assume that F is locally free. Moreover, $\mathcal{E}nd(F)$ is also μ-semistable and $\Delta(\mathcal{E}nd(F)) = 2r^2\Delta(F)$, so that by replacing F by $\mathcal{E}nd(F)$ we may further reduce to the case that F has trivial determinant and is isomorphic to its dual $F^{\vee}$. Let k be a sufficiently large integer so that $k \cdot H^2 > H.K_X$ and that there is a smooth curve $C \in |kH|$. Recall that μ-semistable sheaves of negative slope have no global sections. The standard exact sequence $0 \to S^n F \otimes \mathcal{O}_X(-C) \to S^n F \to S^n F|_C \to 0$ and Serre duality lead to the estimates:

$$
\begin{aligned}
h^0(S^n F) &\leq h^0(S^n F(-C)) + h^0(S^n F|_C) = h^0(S^n F|_C) \\
h^2(S^n F) &= h^0((S^n F)^{\vee} \otimes K_X) = h^0(S^n F \otimes K_X) \\
&\leq h^0(S^n F \otimes K_X(-C)) + h^0(S^n F|_C \otimes K_X|_C) = h^0(S^n F|_C \otimes K_X|_C).
\end{aligned}
$$

Thus we can bound the Euler characteristic of $S^n F$ by

$$\chi(S^n F) \leq h^0(S^n F) + h^2(S^n F) \leq h^0(S^n F|_C) + h^0(S^n F|_C \otimes K_X|_C).$$

Now let $\pi : Y := \mathbb{P}(F) \to X$ be the projective bundle associated to F, $Y_C = Y \times_X C$, and consider the tautological line bundle $\mathcal{O}_\pi(1)$ on Y. Then for all $n \geq 0$ we have

$$\pi_* \mathcal{O}_\pi(n) = S^n F, \text{ and } R^i \pi_* \mathcal{O}_\pi(n) = 0, \text{ for all } i > 0.$$

In particular, $\chi(S^n F) = \chi(\mathcal{O}_\pi(n))$, and by the projection formula we get

$$h^0(C, S^n F|_C \otimes \mathcal{M}) = h^0(Y_C, \mathcal{O}_\pi(n)|_{Y_C} \otimes \pi^* \mathcal{M})$$

for any line bundle $\mathcal{M} \in \text{Pic}(C)$. Since $\dim(Y_C) = r$, there are constants $\gamma_{\mathcal{M}}$ such that

$$h^0(Y_C, \mathcal{O}_\pi(n)|_{Y_C} \otimes \pi^* \mathcal{M}) \leq \gamma_{\mathcal{M}} \cdot n^r \text{ for all } n > 0.$$

This shows that

$$\chi(S^n F) \leq (\gamma_{\mathcal{O}_C} + \gamma_{K_X|_C}) \cdot n^r \text{ for all } n > 0. \tag{3.5}$$

On the other hand, we can compute $\chi(S^n F)$ by the Hirzebruch-Riemann-Roch formula applied to the line bundle $\mathcal{O}_\pi(n)$:

$$\chi(S^n F) = \chi(\mathcal{O}_\pi(n)) = \int_Y \frac{(n\xi)^{r+1}}{(r+1)!} + \cdots, \tag{3.6}$$

where we have set $\xi = c_1(\mathcal{O}_\pi(1))$ and suppressed terms of lower order in n. The cohomology class ξ satisfies the relation $\xi^r - c_1(F) \cdot \xi^{r-1} + c_2(F) \cdot \xi^{r-2} = 0$. Since $c_1(F) = 0$, we get $\xi^{r+1} = -c_2(F) \cdot \xi^{r-1} = -\frac{\Delta(F)}{2r} \cdot \xi^{r-1}$. Finally, $\mathcal{O}_\pi(1)$ has degree 1 on the fibres of π, so that (3.6) yields:

$$\chi(S^n F) = \int_X -\frac{\Delta(F)}{2r} \cdot \frac{n^{r+1}}{(r+1)!} + \text{ terms of lower order in } n$$

If $\Delta(F)$ were negative, this would contradict (3.5). $\square$

Appendix to Chapter 3

3.A e-Stability and Some Estimates

Throughout this appendix let X be a smooth projective surface, K its canonical divisor, and $\mathcal{O}_X(H)$ a very ample line bundle. The following definition generalizes the notion of μ-stability.

Definition 3.A.1 — *Let e be a nonnegative real number. A coherent sheaf F of rank r is e-stable, if it is torsion free in codimension 1 and if*

$$\mu(F') < \mu(F) - \frac{e|H|}{r'}$$

for all subsheaves $F' \subset F$ of rank r', $0 < r' < r$.

The factor $|H| = (H.H)^{1/2}$ is thrown in to make the inequality invariant under rescaling $H \to \lambda H$. Obviously 0-stability is the same as μ-stability, and e-stability is stronger than e'-stability if $e > e'$. The same arguments as in the proof of Proposition 2.3.1 show that e-stability is an open property.

The following proposition due to O'Grady [209] is rather technical. It will be needed in Chapter 9 to give dimension bounds for the locus of μ-unstable sheaves in the moduli space of semistable sheaves. The main ingredients in the proof are the Hirzebruch-Riemann-Roch formula, the Le Potier-Simpson Estimate 3.3.1 for the number of global sections of μ-semistable sheaves and the Bogomolov Inequality 3.4.1.

Let F be a torsion free μ-semistable sheaf of rank r and slope μ which, however, is not e-stable for some $e \geq 0$. Let

$$0 = F_{(0)} \subset F_{(1)} \subset \ldots \subset F_{(n)} = F$$

be a filtration of F with factors $F_i = F_{(i)}/F_{(i-1)}$ of rank $r_i \geq 1$ and slope μ_i such that the following holds: all factors are torsion free and μ-semistable and satisfy the conditions

$$\mu - \mu_1 \leq \frac{e|H|}{r_1}, \quad \text{and } \mu_2 \geq \ldots \geq \mu_n,$$

i.e. $F_{(1)}$ is e-destabilizing, and $F_{(\bullet)}/F_{(1)}$ is a μ-Harder-Narasimhan filtration of $F/F_{(1)}$. For a filtered sheaf F we defined groups $\mathrm{Ext}^i_-(F, F)$ in the appendix to Chapter 2. We use ext^i_- for $\dim \mathrm{Ext}^i_-$ and $\chi(A, B)$ for the alternating sum of the dimensions $\mathrm{ext}^i(A, B)$ (cf. 6.1.1).

Proposition 3.A.2 — *There is a constant B depending on X, H and r such that the following holds: if F is a μ-semistable torsion free sheaf of rank r which is not e-stable, and* *if $F_{(\bullet)}$ is a filtration of F as above, then*

$$\mathrm{ext}^1_-(F,F) \leq \left(1 - \frac{1}{2r}\right)\Delta(F) + (3r-1)e^2 + \frac{r[K.H]_+}{2|H|}e + B.$$

Proof. First, the spectral sequence 2.A.4 for Ext_ and Serre duality allow us to write

$$
\begin{aligned}
\mathrm{ext}^1_-(F,F) &\leq \sum_{i \leq j} \mathrm{ext}^1(F_j, F_i) \\
&= \sum_{i \leq j} \left(\mathrm{ext}^0(F_j, F_i) + \mathrm{ext}^2(F_j, F_i) - \chi(F_j, F_i)\right) \\
&= \sum_{i \leq j} \left(\hom(F_j, F_i) + \hom(F_i, F_j \otimes K)\right) + \sum_{i > j} \chi(F_j, F_i) - \chi(F, F).
\end{aligned}
$$

By Le Potier-Simpson 3.3.6 we have

$$
\begin{aligned}
\hom(F_j, F_i) &\leq \frac{r_i r_j}{2H^2}\left(\mu_i - \mu_j + (r+1)H^2\right)^2 \\
\hom(F_i, F_j \otimes K) &\leq \frac{r_i r_j}{2H^2}\left(\mu_j - \mu_i + K.H + (r+1)H^2\right)^2,
\end{aligned}
$$

so that

$$
\begin{aligned}
\sum_{i \leq j} \hom(F_j, F_i) + \hom(F_i, F_j \otimes K) &\leq \sum_{i \leq j} \frac{r_i r_j}{H^2}(\mu_i - \mu_j)^2 \\
&\quad + K.H \sum_{i \leq j} \frac{r_i r_j}{H^2}(\mu_j - \mu_i) \\
&\quad + \frac{\left((r+1)H^2\right)^2 + \left((r+1)H^2 + K.H\right)^2}{2H^2} \sum_{i \leq j} r_i r_j.
\end{aligned}
$$

The Hirzebruch-Riemann-Roch formula yields

$$
\begin{aligned}
\chi(F,F) &= -\Delta + r^2 \chi(\mathcal{O}_X) \\
\chi(F_j, F_i) &= -\left(r_j \frac{\Delta_i}{2r_i} + r_i \frac{\Delta_j}{2r_j}\right) + r_i r_j \left(\frac{\xi_{ij}^2}{2} - \frac{\xi_{ij}K}{2} + \chi(\mathcal{O}_X)\right),
\end{aligned}
$$

where we have used the abbreviations

$$\Delta = \Delta(F), \ \Delta_i = \Delta(F_i) \qquad \text{and} \qquad \xi_{ij} = \frac{c_1(F_i)}{r_i} - \frac{c_1(F_j)}{r_j}.$$

Using the additivity of the Chern character and $2r \cdot ch_2 = r(c_1^2 - 2c_2) = c_1^2 - \Delta$, the following identities are easily verified:

$$\sum_i \frac{\Delta_i}{2r_i} - \frac{\Delta}{2r} = \sum_i \frac{c_1(F_i)^2}{2r_i} - \frac{c_1(F)^2}{2r},$$

and, clearly, $\xi_{ij}.H = \mu_i - \mu_j$. The Bogomolov Inequality implies $\Delta_i \geq 0$ for all i. Hence

$$
\Delta - \sum_{i>j}\left(r_j\frac{\Delta_i}{2r_i} + r_i\frac{\Delta_j}{2r_j}\right) = \Delta - \sum_i (r - r_i)\frac{\Delta_i}{2r_i} \leq \Delta - \sum_i \frac{\Delta_i}{2r_i}
$$

$$
= \left(1 - \frac{1}{2r}\right)\Delta - \sum_{i>j}\frac{r_ir_j}{2r}\xi_{ij}^2 .
$$

This shows:

$$
\sum_{i>j}\chi(F_j, F_i) - \chi(F, F) \leq \left(1 - \frac{1}{2r}\right)\Delta - \sum_{i\leq j} r_ir_j\chi(\mathcal{O}_X) + \sum_{i>j}\frac{r_ir_j}{2}\left(\frac{r-1}{r}\xi_{ij}^2 - K\xi_{ij}\right).
$$

The first term on the right hand side has already the required shape, the second one is clearly bounded by $r^2 \cdot [-\chi(\mathcal{O}_X)]_+$. For the third term we use quadratic completion and the Hodge Index Theorem, which says that $\xi^2 \leq (\xi.H)^2/H^2$ for any class ξ.

This leads to

$$
\sum_{i>j}\frac{r_ir_j}{2}\left(\frac{r-1}{r}\xi_{ij}^2 - K\xi_{ij}\right)
$$

$$
= \sum_{i>j}\frac{r_ir_j}{2}\frac{r-1}{r}\left(\xi_{ij} - \frac{rK}{2(r-1)}\right)^2 - \frac{rK^2}{8(r-1)}\sum_{i>j}r_ir_j
$$

$$
\leq \sum_{i>j}\frac{r_ir_j}{2H^2}\frac{r-1}{r}\left(\mu_i - \mu_j - \frac{rK.H}{2(r-1)}\right)^2 - \frac{rK^2}{8(r-1)}\sum_{i>j}r_ir_j
$$

$$
= \frac{r-1}{2H^2}\sum_{i>j}\frac{r_ir_j}{r}(\mu_i - \mu_j)^2 - \frac{rK.H}{2H^2}\sum_{i>j}\frac{r_ir_j}{r}(\mu_i - \mu_j)
$$

$$
+ \frac{r}{8(r-1)}\left(\frac{(K.H)^2}{H^2} - K^2\right)\sum_{i>j}r_ir_j .
$$

Note that the term in brackets in the last summand is nonnegative by the Hodge Index Theorem, so that the sum $\sum_{i>j}r_ir_j$ has to be bounded from above (its maximum value being $r(r-1)/2$).

Putting things together and using the abbreviation

$$
B(r, X, H) := \frac{r^2}{16}\left(\frac{(K.H)^2}{H^2} - K^2\right) + r^2[-\chi(\mathcal{O}_X)]_+
$$

$$
+ (r^2 - r + 1)\frac{\left((r+1)H^2\right)^2 + \left((r+1)H^2 + K.H\right)^2}{2H^2},
$$

we have proved so far that

$$
\text{ext}_-^1(F, F) \leq \left(1 - \frac{1}{2r}\right)\Delta + \frac{3r-1}{2H^2}a + \frac{rK.H}{2H^2}b + B(r, X, H),
$$

where a and b stand for $a = \sum_{i>j} \frac{r_i r_j}{r} (\mu_i - \mu_j)^2$ and $b = \sum_{i>j} \frac{r_i r_j}{r} (\mu_i - \mu_j)$, and we are left with the assertions $0 \leq a \leq 2e^2 H^2$ and $0 \leq b \leq e|H|$.

We have

$$0 \leq a = \sum_{i,j} \frac{r_i r_j}{2r} \left((\mu_i - \mu) - (\mu_j - \mu) \right)^2 = \sum_i r_i (\mu_i - \mu)^2,$$

since $\sum_i r_i (\mu_i - \mu) = 0$. Because of the μ-semistability of F and the assumptions on the μ_i, we have $\mu_1 - \mu \leq 0$, and $\mu_i - \mu \geq 0$ for all $i \geq 2$, and, fixing μ_1 for a moment, the problem is to maximize $\sum_{i \geq 2} r_i (\mu_i - \mu)^2$ subject to the conditions $\mu_i - \mu \geq 0$ and $\sum_{i \geq 2} r_i (\mu_i - \mu) = r_1 (\mu - \mu_1)$. A moment's thought yields:

$$a \leq r_1 (\mu - \mu_1)^2 + r_1^2 (\mu - \mu_1)^2 \leq 2e^2 H^2.$$

Now let $r_{(i)}$ and $\mu_{(i)}$ denote rank and slope of $F_{(i)}$, the i-th step in the filtration of F. Note that the following relations hold:

$$\mu_1 = \mu_{(1)} \leq \mu_{(2)} \leq \cdots \leq \mu_{(n)} = \mu \leq \mu_n \leq \cdots \leq \mu_2.$$

From this we get

$$b = \sum_i \frac{r_i}{r} (\mu_i - \mu_{(i-1)}) r_{(i-1)} \geq 0.$$

Moreover, $\mu_i - \mu_j$ is negative when $i > j \geq 2$. Hence

$$b \leq \sum_{i>1} \frac{r_1}{r} r_i (\mu_i - \mu_1) = \frac{r_1}{r} (r\mu - r_1 \mu_1) - \frac{r_1}{r} (r - r_1) \mu_1 = r_1 (\mu - \mu_1) \leq e|H|.$$

This finishes the proof of the proposition. □

Comments:

— In [20] Barth proves Theorem 3.0.1 for vector bundles of rank 2 and attributes it to Grauert and Mülich. This result was extended to vector bundles of arbitrary rank by Spindler [242]. As Schneider observed, Spindler's theorem together with results of Maruyama [162] implied the boundedness of the family of semistable vector bundles of fixed rank and Chern classes. For this result see also [54]. Shortly afterwards, Spindler's theorem was further extended to arbitrary projective manifolds by Maruyama [166] and Forster, Hirschowitz and Schneider [66]. The bound for $\delta\mu$ in terms of the minimal slope of the relative tangent bundle was given by Hirschowitz [102], based on Maruyama's results on tensor products of semistable sheaves.

— Lemma 3.2.2 and Theorem 3.2.7 are contained in [78]. In this paper Gieseker also gives an algebraic proof that symmetric powers of μ-semistable sheaves are again μ-semistable if the characteristic is zero. This had been proved in the curve case by Hartshorne [99] using the relation between stable bundles on a curve and representations of the fundamental groups established by Narasimhan and Seshadri [202]. The fact that tensor products of μ-semistable sheaves are again μ-semistable is due to Maruyama [166]. His proof uses Hartshorne's corresponding result on ampleness [97] and the techniques developed by Gieseker. More results on μ-stability in connection with unramified coverings can be found in Takemoto's article [245].

— The boundedness theorem for surfaces was proved by Takemoto [244] for sheaves of rank 2, and by Maruyama [160] and Gieseker [77] for semistable sheaves of arbitrary rank. Our approach via Theorem 3.3.1 follows the papers of Simpson [240] and Le Potier [145].

— Theorem 3.4.1 first appears in a special case in Reid's report [228]. A detailed account then was given by Bogomolov in [28]. In fact, he proves a stronger statement, that we will discuss in Section 7.3. In this stronger form the Bogomolov Inequality has interesting applications known as Reider's method. See Reider's paper [229] and the presentation in [139]. Gieseker gave a different proof of the Bogomolov Inequality in [78]. The proof given here is based on [180], where Miyaoka generalizes the Bogomolov Inequality to higher dimensions.

— The proof of the estimate in the appendix follows O'Grady [209]. Our coefficients differ slightly from his paper.

3.B Further comments (second edition)

I. Positive characteristic.

Boundedness of semistable sheaves in positive characteristic has been a long-standing open problem. It was completely solved by Langer in [356]. He shows that the family of purely d-dimensional sheaves on the geometric fibres of a projective morphism with fixed Hilbert polynomial and bounded maximal slope is bounded regardless of the characteristic of the base field. In the same paper, Langer also proves effective restriction theorems in the same setting. From the boundedness results he deduces the existence of moduli spaces of finite type in positive and mixed characteristic [356, 357]. These deductions require new arguments, too, as characteristic zero methods discussed in this book fail at various places. In particular, in [357], Langer gives a generalized Le Potier-Simpson type bound for the number of global sections of a torsion free sheaf.

One of the main problems in positive characteristic is that the Frobenius pull-back F^*E of a μ-semistable torsion free sheaf on a smooth projective variety X over an algebraically closed field k of $\mathrm{char}(k) = p > 0$ need not be μ-semistable again. The sheaf is called strongly μ-semistable if all Frobenius pull-backs $(F^n)^*E$ are again μ-semistable. One of the central results in [356] is the following theorem, which takes care:

For any torsion free sheaf E on X there exists a k_0 such that the Harder-Narasimhan factors of $(F^{k_0})^*E$ are strongly μ-semistable.

Langer also proves Theorem 3.1.4 in positive characteristic (see [356], Thm. 6.1). More precisely, he proves that the tensor product of two strongly μ-semistable sheaves is again strongly μ-semistable. He also remarks that the result had been observed earlier in [226]. Thm. 0.1 in [356] proves the Bogomolov Inequality for strictly μ-semistable sheaves (cf. Theorem 3.4.1). For comments on Langer's restriction theorem in positive characteristic see Appendix 7.A.

II. Algebraic holonomy and splitting of tensor products.

The results of the first two sections are nicely complemented by an article of Balaji and Kollár [272]. They introduce the *algebraic holonomy* of a μ-stable locally free sheaf E that governs the splitting type of all the tensor products of E and its dual. To be more precise, let E by a μ-stable locally free sheaf on a smooth projective variety over $\mathbb{C}$ with trivial determinant $\det(E) \cong \mathcal{O}_X$. Fix a closed point $x \in X$. Then the following result is shown:

There exists a reductive subgroup $H_x \subset \mathrm{SL}(E_x)$ such that the H_x-invariant subspaces of $E_x^{\otimes k} \otimes E_x^{\vee \otimes \ell}$ are in one-to-one correspondence with the direct summands $F \subset E^{\otimes k} \otimes E^{\vee \otimes \ell}$.

The correspondence is given by mapping a direct summand F to its fibre in x. The subgroup H_x is obtained by considering stable restrictions to curves. By the theorem of

Narasimhan-Seshadri, which was only alluded to in the text, a stable bundle with trivial determinant on a curve C is associated to a unitary representation of $\pi_1(C)$. The subgroup H_x contains by construction the images of all representations $\pi_1(C) \to \mathrm{SL}(E_x)$ obtained from all curves C passing through $x \in X$ with $E|_C$ stable.

This approach was pursued in [273] where the authors construct, for a smooth projective variety over an algebraically closed field of characteristic zero, a neutral Tannakian category containing all μ-stable locally free sheaves, which is very much in the spirit of Nori's fundamental group scheme.

In the main body of the text we have alluded to the Kobayashi-Hitchin correspondence whose difficult part shows the existence of a Hermite-Einstein metric on any μ-stable locally free sheaf on a smooth projective variety over $\mathbb{C}$.

As it turns out, also sheaves that are (Gieseker) stable in the sense of Definition 1.2.4 have distinguished metrical properties. In [440] it was shown that stability of a locally free sheaf E is equivalent to the existence of a *balanced* embedding into a Grassmannian, where the embedding is given by the universality of the Grassmannian and the surjective evaluation map $H^0(X, E(k)) \otimes \mathcal{O} \twoheadrightarrow E(k)$ for $k \gg 0$. This then was used in [441] to write down a weak Hermite-Einstein equation for stable locally free sheaves.

4

Moduli Spaces

The goal of this chapter is to give a geometric construction for moduli spaces of semi-stable sheaves, the central object of study in these notes, and some of the properties that follow from the construction. As the chapter has grown a bit out of size, here is a short introduction:

Intuitively, a moduli space of semistable sheaves is a scheme whose points are in some 'natural bijection' to equivalence classes of semistable sheaves on some fixed polarized projective scheme (X, H). The phrase 'natural bijection' can be given a rigorous meaning in terms of corepresentable functors. The correct notion of 'equivalence' turns out to be S-equivalence. This is done in Section 4.1.

The moduli space can be constructed as a quotient of a certain Quot-scheme by a natural group action: instead of sheaves F one first considers pairs consisting of a sheaf F *and* a basis for the vector space $H^0(X, F(m))$ for some fixed large integer m. If m is large enough, such a basis defines a surjective homomorphism $\mathcal{H} := \mathcal{O}_X(-m)^{h^0(F(m))} \to F$ and hence a point in the Quot-scheme $\mathrm{Quot}(\mathcal{H}, P(F))$. An arbitrary point $[\rho : \mathcal{H} \to F]$ in this Quot-scheme is of this particular form if and only if F is semistable and ρ induces an isomorphism $k^{P(F,m)} \to H^0(F(m))$. The subset $R \subset \mathrm{Quot}(\mathcal{H}, P(F))$ of all points satisfying both conditions is open. The passage from R to the moduli space M consists in dividing out the ambiguity in the choice of the basis of $H^0(F(m))$. We collect the necessary terminology and results from Geometric Invariant Theory in Section 4.2. The construction itself is carried out in Section 4.3 following a method due to Simpson. In fact, the proofs of the more technical theorems are confined to a separate section.

The infinitesimal structure of the moduli space is described in Section 4.5. It also contains upper and lower bounds for the dimension of the moduli space. Once the existence of the moduli space is established, the question arises as to what can be said about universal families of semistable sheaves parametrized by the moduli space. Section 4.6 gives partial answers to this problem.

This chapter has three appendices. In the first we sketch an alternative and historically earlier construction of the moduli space due to Gieseker and Maruyama, which has the virtue of showing that a certain line bundle on the moduli space is ample relative to the Picard scheme of X. The second contains a short report about 'decorated sheaves', and in the third we state some results about the dependence of the moduli space on the polarization of the base scheme.

4.1 The Moduli Functor

Let $(X, \mathcal{O}_X(1))$ be a polarized projective scheme over an algebraically closed field k. For a fixed polynomial $P \in \mathbb{Q}[z]$ define a functor

$$\mathcal{M}' : (Sch/k)^{\circ} \to (Sets)$$

as follows. If $S \in \mathrm{Ob}(Sch/k)$, let $\mathcal{M}'(S)$ be the set of isomorphism classes of S-flat families of semistable sheaves on X with Hilbert polynomial P. And if $f : S' \to S$ is a morphism in (Sch/k), let $\mathcal{M}'(f)$ be the map obtained by pulling-back sheaves via $f_X = f \times \mathrm{id}_X$:

$$\mathcal{M}'(f) : \mathcal{M}'(S) \to \mathcal{M}'(S'), \quad [F] \to [f_X^* F].$$

If $F \in \mathcal{M}'(S)$ is an S-flat family of semistable sheaves, and if L is an arbitrary line bundle on S, then $F \otimes p^* L$ is also an S-flat family, and the fibres F_s and $(F \otimes p^* L)_s = F_s \otimes_{k(s)} L(s)$ are isomorphic for each point $s \in S$. It is therefore reasonable to consider the quotient functor $\mathcal{M} = \mathcal{M}' / \sim$, where $\sim$ is the equivalence relation:

$$F \sim F' \text{ for } F, F' \in \mathcal{M}'(S) \text{ if and only if } F \cong F' \otimes p^* L \text{ for some } L \in \mathrm{Pic}(S).$$

If we take families of geometrically stable sheaves only, we get open subfunctors $(\mathcal{M}')^s \subset \mathcal{M}'$ and $\mathcal{M}^s \subset \mathcal{M}$. In 2.2.1 we explained the notion of a scheme corepresenting a functor.

Definition 4.1.1 — *A scheme M is called a moduli space of semistable sheaves if it corepresents the functor $\mathcal{M}$.*

Recall that this characterizes M up to unique isomorphism. We will write $M_{\mathcal{O}_X(1)}(P)$ and $\mathcal{M}_{\mathcal{O}_X(1)}(P)$ instead of M and $\mathcal{M}$, if the dependence on the polarization and the Hilbert polynomial is to be emphasized.

If A is a local k-algebra of finite type, then any invertible sheaf on A is trivial. Hence the map $\mathcal{M}'(\mathrm{Spec}(A)) \to \mathcal{M}(\mathrm{Spec}(A))$ is a bijection. This implies that any scheme corepresenting $\mathcal{M}$ would also corepresent $\mathcal{M}'$ and conversely. We will see that there always is a projective moduli space for $\mathcal{M}$. In general, however, there is no hope that $\mathcal{M}$ can be represented.

Lemma 4.1.2 — *Suppose* M *corepresents* $\mathcal{M}$. *Then S-equivalent sheaves correspond to identical closed points in* M. *In particular, if there is a properly semistable sheaf* F, *(i.e. semistable but not stable), then* $\mathcal{M}$ *cannot be represented.*

Proof. Let $0 \to F' \to F \to F'' \to 0$ be a short exact sequence of semistable sheaves with the same reduced Hilbert polynomial. Then it is easy to construct a flat family $\mathcal{F}$ of semistable sheaves parametrized by the affine line $\mathbb{A}^1$, such that (cf. 2.1.12)

$$\mathcal{F}_0 \cong F' \oplus F'' \text{ and } \mathcal{F}_t \cong F \text{ for all } t \neq 0.$$

Either take $\mathcal{F}$ to be the tautological extension which is parametrized by the affine line in $\text{Ext}^1(F'', F')$ through the point given by the extension above, or, what amounts to the same, let $\mathcal{F}$ be the kernel of the surjection

$$q^* F \longrightarrow i_* F'',$$

where $q : \mathbb{A}^1 \times X \to X$ is the projection and $i : X \cong \{0\} \times X \to \mathbb{A}^1 \times X$ is the inclusion. Since over the punctured line $\mathbb{A}^1 \setminus \{0\}$ the modified family $\mathcal{F}$ and the constant family $\mathcal{O}_{\mathbb{A}^1} \times_k F$ are isomorphic, the morphism $\mathbb{A}^1 \longrightarrow M$ induced by $\mathcal{F}$ must be constant on $\mathbb{A}^1 \setminus \{0\}$, hence everywhere. This means that F and $F' \oplus F''$, or more generally all sheaves which are S-equivalent to F, correspond to the same closed point in M. Hence M does not represent $\mathcal{M}$. □

Such phenomena cannot occur for the subfunctor $\mathcal{M}^s$ of stable families. The question whether $\mathcal{M}^s$ is representable will be considered in Section 4.6.

4.2 Group Actions

In this section we briefly recall the notions of an algebraic group and a group action, various notions of quotients for group actions and linearizations of sheaves. We then list without proof results from Geometric Invariant Theory, which will be needed in the construction of moduli spaces. For text books on Geometric Invariant Theory we refer to the books of Mumford et al. [195], Newstead [203] and, in particular, Kraft et al. [131].

Let k be an algebraically closed field of characteristic zero.

Group Actions and Linearizations

An *algebraic group* over k is a k-scheme G of finite type together with morphisms

$$\mu : G \times G \to G, \quad \varepsilon : \text{Spec}(k) \to G \quad \text{and} \quad \iota : G \to G$$

defining the group multiplication, the unit element and taking the inverse, and satisfying the usual axioms for groups. This is equivalent to saying that the functor $\underline{G} : (Sch/k) \to$

(*Sets*) factors through the category of (abstract) groups. Since the characteristic of the base field is assumed to be zero, any such group is smooth by a theorem of Cartier. An algebraic group is affine if and only if it is isomorphic to a closed subgroup of some $\mathrm{GL}(N)$.

A (right) *action* of an algebraic group G on a k-scheme X is a morphism $\sigma : X \times G \to X$ which satisfies the usual associativity rules. Again this is equivalent to saying that for each k-scheme T there is a an action of the group $\underline{G}(T)$ on the set $\underline{X}(T)$ and that this action is functorial in T. A morphism $\varphi : X \to Y$ of k-schemes with G-actions σ_X and σ_Y, respectively, is *G-equivariant*, if $\sigma_Y \circ (\varphi \times \mathrm{id}_G) = \varphi \circ \sigma_X$. In the special case that G acts trivially on Y, i.e. if $\sigma_Y : Y \times G \to Y$ is the projection onto the second factor, an equivariant morphism $f : X \to Y$ is called *invariant*.

Let $\sigma : X \times G \to X$ be a group action as above, and let $x \in X$ be a closed point. Then the *orbit* of x is the image of the composite $\sigma_x : G \cong \{x\} \times G \subset X \times G \xrightarrow{\sigma} X$. It is a locally closed smooth subscheme of X, since G acts transitively on its closed points. The fibre $\sigma_x^{-1}(x) =: G_x \subset G$ is a subgroup of G and is called the *isotropy subgroup* or the *stabilizer* of x in G. If V is a G-representation space, let V^G denote the linear subspace of invariant elements.

Definition 4.2.1 — *Let $\sigma : X \times G \to X$ be a group action. A categorical quotient for σ is a k-scheme Y that corepresents the functor*

$$\underline{X/G} : (Sch/k)^o \to (Sets), \quad T \mapsto \underline{X}(T)/\underline{G}(T).$$

If Y universally corepresents $\underline{X/G}$, it is said to be a universal categorical quotient.

Suppose that Y corepresents the functor $\underline{X/G}$. The image of $[\mathrm{id}_X] \in \underline{X/G}(X)$ in $\underline{Y}(X)$ corresponds to a morphism $\pi : X \to Y$. This morphism has the following universal property: π is invariant, and if $\pi' : X \to Y'$ is any other G-invariant morphism of k-schemes then there is a unique morphism $f : Y \to Y'$ such that $\pi' = f \circ \pi$. Indeed, it is straightforward to check that this characterizes Y as a categorical quotient.

Even if a categorical quotient exists, it can be far from being an 'orbit space': let the multiplicative group $\mathbb{G}_m \cong \mathrm{Spec}(k[T, T^{-1}])$ act on $\mathbb{A}^n$ by homotheties. Then the projection $\mathbb{A}^n \to \mathrm{Spec}(k)$ is a categorical quotient. However, clearly, it is not an orbit space. We will need notions which are closer to the intuitive idea of a quotient:

Definition 4.2.2 — *Let G an affine algebraic group over k acting on a k-scheme X. A morphism $\varphi : X \to Y$ is a good quotient, if*
- *φ is affine and invariant.*
- *φ is surjective, and $U \subset Y$ is open if and only if $\varphi^{-1}(U) \subset X$ is open.*
- *The natural homomorphism $\mathcal{O}_Y \to (\varphi_* \mathcal{O}_X)^G$ is an isomorphism.*

• *If W is an invariant closed subset of X, then $\varphi(W)$ is a closed subset of Y. If W_1 and W_2 are disjoint invariant closed subsets of X, then $\varphi(W_1) \cap \varphi(W_2) = \emptyset$.*

The morphism φ is said to be a geometric quotient if the geometric fibres of φ are the orbits of geometric points of X. Finally, φ is a universal good (geometric) quotient if $Y' \times_Y X \to Y'$ is a good (geometric) quotient for any morphism $Y' \to Y$ of k-schemes.

Any (universal) good quotient is in particular a (universal) categorical quotient. If $\varphi : X \to Y$ is a good quotient and if X is irreducible, reduced, integral, or normal, then the same holds for Y. We will denote a good quotient of X, if it exists, by $X /\!\!/ G$.

Let G be an algebraic group and let $\pi : X \to Y$ be an invariant morphism of G-schemes. π is said to be a *principal G-bundle*, if there exists a surjective étale morphism $Y' \to Y$ and a G-equivariant isomorphism $Y' \times G \to Y' \times_Y X$, i.e. X is locally (in the étale toplogy) isomorphic as a G-scheme to the product $Y \times G$. Principal bundles are universal geometric quotients. Conversely, if $\pi : X \to Y$ is a flat geometric quotient and if the morphism $(\sigma, p_1) : X \times G \to X \times_Y X$ is an isomorphism, then π is a principal G-bundle.

Let $\pi : X \to Y$ be a principal G-bundle, and let Z be a k-scheme of finite type with a G-action. Then there is a geometric quotient for the diagonal action of G on $X \times Z$. It is a bundle (in the étale topology) over Y with typical fibre Z, and is denoted by $X \times^G Z$.

Example 4.2.3 — Let Y be a k-scheme of finite type, let F be a locally free $\mathcal{O}_Y$-module of rank r and let $\mathbb{H}om(\mathcal{O}_Y^r, F) := \operatorname{Spec} S^*(\mathcal{H}om(\mathcal{O}_Y^r, F)^\vee) \longrightarrow Y$ be the geometric vector bundle that parametrizes homomorphisms from $\mathcal{O}_Y^r$ to F. Let $X := \mathbb{I}som(\mathcal{O}_Y^r, F) \subset \mathbb{H}om(\mathcal{O}_Y^r, F)$ be the open subscheme corresponding to isomorphisms, and let $\pi : X \to Y$ be the natural projection. X is called the *frame bundle* associated to F. The group $\mathrm{GL}(r)$ acts naturally on X by composition: if $y \in \underline{Y}(k)$, $g \in \underline{\mathrm{GL}(r)}(k)$, and if $f : k(y)^r \to F(y)$ is an isomorphism, then $\sigma(f, g) := f \circ g$. Then $\pi : X \to Y$ is a principal $\mathrm{GL}(r)$-bundle, which is locally trivial even in the Zariski topology. (In fact, as Serre shows in [234], any principal $\mathrm{GL}(r)$-bundle is locally trivial in the Zariski topology.) Similarly, we can construct a principal $\mathrm{PGL}(r)$-bundle by taking the image X' of X in $\operatorname{Proj}(S^*(\mathcal{H}om(\mathcal{O}_Y^r, F)^\vee))$. We will refer to X' as the *projective frame bundle* associated to F. □

Example 4.2.4 — Let G be an algebraic group and $H \subset G$ a closed subgroup. Then there is a geometric quotient $\pi : G \to H \backslash G$ for the natural (left) action of H on G, which in fact is a (left) principal H-bundle. □

The following gives the precise definition for a group action on a sheaf that is compatible with a given group action on the supporting scheme.

Definition 4.2.5 — *Let X a k-scheme of finite type, G an algebraic k-group and σ : $X \times G \to X$ a group action. A G-linearization of a quasi-coherent $\mathcal{O}_X$-sheaf F is an isomorphism of $\mathcal{O}_{X \times G}$-sheaves $\Phi : \sigma^* F \to p_1^* F$, where $p_1 : X \times G \to X$ is the projection, such that the following cocycle condition is satisfied:*

$$(\mathrm{id}_X \times \mu)^* \Phi = p_{12}^* \Phi \circ (\sigma \times \mathrm{id}_G)^* \Phi,$$

where $p_{12} : X \times G \times G \to X \times G$ is the projection onto the first two factors.

Intuitively this means the following: if g and x are k-rational points in G and X, respectively, and if we write xg for $\sigma(x, g)$, then Φ provides an isomorphism of fibres of F

$$\Phi_{x,g} : F(xg) \to F(x).$$

And the cocycle condition translates into

$$\Phi_{x,g} \circ \Phi_{xg,h} = \Phi_{x,gh} : F(xgh) \to F(x).$$

Note that a given $\mathcal{O}_X$-sheaf might be endowed with different G-linearizations. A homomorphism $\phi : F \to F'$ of G-linearized quasi-coherent $\mathcal{O}_X$-sheaves is a homomorphism of $\mathcal{O}_X$-sheaves which commutes with the G-linearizations Φ and Φ' of F and F', respectively, in the sense that $\Phi' \circ \sigma^* \phi = p_1^* \phi \circ \Phi$. Kernels, images, cokernels of homomorphisms of G-linearized sheaves as well as tensor products, exterior or symmetric powers of G-linearized sheaves inherit G-linearizations in a natural way. Similarly, if $f : X \to Y$ is an equivariant morphism of G-schemes, then the pull-back $f^* F$ and the derived direct images $R^i f_* F'$, $i \geq 0$ of any G-linearized sheaves F and F' on Y and X, respectively, inherit natural linearizations. In the case of the derived direct image functor this follows from the fact that a group action $\sigma : X \times G \to X$ is flat and that taking direct images commutes with flat base change. In particular, the space of global sections of a linearized sheaf on a projective scheme naturally has the structure of a G-representation.

A G-linearization on a sheaf induces an 'ordinary' action on all schemes which are functorially constructed from the sheaf: Let X be a k-scheme with an action by an algebraic group G and let $\mathcal{A}$ be a quasi-coherent sheaf of commutative $\mathcal{O}_X$-algebras with a G-linearization Φ that respects the $\mathcal{O}_X$-algebra structure. Let $\pi : A := \mathrm{Spec}\,(\mathcal{A}) \to X$ be the associated X-scheme. Then Φ induces a morphism

$$\sigma_A : A \times_k G = A \times_{X,p_1} (X \times G) \to A \times_{X,\sigma} (X \times G) \to A$$

such that the diagram

$$
\begin{array}{ccc}
A \times G & \xrightarrow{\sigma_A} & A \\
\downarrow & & \downarrow \\
X \times G & \xrightarrow{\sigma} & X
\end{array}
$$

commutes. The cocycle condition for Φ implies that σ_A is group action of G on A, and the commutativity of the diagram says that $\pi : A \to X$ is equivariant. Similarly, if $\mathcal{A}$ is a G-linearized $\mathbb{Z}$-graded algebra, then $\mathrm{Proj}(\mathcal{A})$ inherits a natural G-action that makes the projection $\mathrm{Proj}(\mathcal{A}) \to X$ equivariant. Typically, $\mathcal{A}$ will be the symmetric algebra S^*F of a linearized coherent sheaf F.

Apply this to the following special situation: suppose that X is a projective scheme with a G-action and that L is a G-linearized very ample line bundle. Then G acts naturally on the vector space $H^0(X, L)$, the natural homomorphism $H^0(X, L) \otimes_k \mathcal{O}_X \to L$ is equivariant and induces a G-equivariant embedding $X \cong \mathbb{P}(L) \to \mathbb{P}(H^0(X, L))$. Thus the G-linearization of L *linearizes* the action on X in the sense that this action is induced by the projective embedding given by L and a *linear* representation on $H^0(X, L)$.

Example 4.2.6 — Let Y be a k-scheme of finite type, F a locally free $\mathcal{O}_Y$-module of rank r, and let $\pi : X \to Y$ be the associated frame bundle (cf. 4.2.3). It follows from the definition of X that there is an isomorphism $\varphi : \mathcal{O}_X^r \to \pi^*F$, the *universal trivialization* of F. If we give F the trivial linearization and $\mathcal{O}_X^r$ the linearization which is induced by the standard representation of $\mathrm{GL}(r)$ on k^r, then φ is equivariant. Similarly, there is a $\mathrm{GL}(r)$-equivariant isomorphism $\tilde{\varphi} : \mathcal{O}_{\tilde{X}}^r \to \tilde{\pi}^*F \otimes \mathcal{O}_{\tilde{X}}(1)$ of sheaves on the projective frame bundle $\tilde{\pi} : \tilde{X} \to Y$ associated to F.

Geometric Invariant Theory (GIT)

In general, good quotients for group actions do not exist. The situation improves if we restrict to a particular class of groups, which fortunately contains those groups we are most interested in.

Definition 4.2.7 — *An algebraic group G is called reductive, if its unipotent radical, i.e. its maximal connected unipotent subgroup, is trivial.*

For the purposes of these notes it suffices to notice that all tori $\mathbb{G}_m^N$ and the groups $\mathrm{GL}(n)$, $\mathrm{SL}(n)$, $\mathrm{PGL}(n)$ are reductive.

The main reason for considering reductive groups is the following theorem:

Theorem 4.2.8 — *Let G be a reductive group acting on an affine k-scheme X of finite type. Let $A(X)$ be the affine coordinate ring of X and let $Y = \mathrm{Spec}(A(X)^G)$. Then $A(X)^G$ is finitely generated over k, so that Y is of finite type over k, and the natural map $\pi : X \to Y$ is a universal good quotient for the action of G.*

Proof. See Thm. 1.1 in [195] or Thm. 3.4 and Thm. 3.5 in [203] $\qquad\square$

Assume that X is a projective scheme with an action of a reductive group G and that L is a G-linearized ample line bundle on X. Let $R = \bigoplus_{n \geq 0} H^0(X, L^{\otimes n})$ be the associated homogeneous coordinate ring. Then R^G is a finitely generated $\mathbb{Z}$-graded k-algebra as well. Let $Y = \mathrm{Proj}(R^G)$. The inclusion $R^G \subset R$ induces a rational map $X \to Y$ which is defined on the complement of the closed subset $V(R_+^G \cdot R) \subset \mathrm{Proj}(R) = X$, i.e. on all points x for which there is an integer n and a G-invariant section $s \in H^0(X, L^{\otimes n})$ with $s(x) \neq 0$. This property is turned into a definition:

Definition 4.2.9 — *A point $x \in X$ is semistable with respect to a G-linearized ample line bundle L if there is an integer n and an invariant global section $s \in H^0(X, L^{\otimes n})$ with $s(x) \neq 0$. The point x is stable if in addition the stabilizer G_x is finite and the G-orbit of x is closed in the open set of all semistable points in X.*

A point is called *properly semistable* if it is semistable but not stable. The sets $X^s(L)$ and $X^{ss}(L)$ of stable and semistable points, respectively, are open G-invariant subsets of X, but possibly empty.

Theorem 4.2.10 — *Let G be a reductive group acting on a projective scheme X with a G-linearized ample line bundle L. Then there is a projective scheme Y and a morphism $\pi : X^{ss}(L) \to Y$ such that π is a universal good quotient for the G-action. Moreover, there is an open subset $Y^s \subset Y$ such that $X^s(L) = \pi^{-1}(Y^s)$ and such that $\pi : X^s(L) \to Y^s$ is a universal geometric quotient. Finally, there is a positive integer m and a very ample line bundle M on Y such that $L^{\otimes m}|_{X^{ss}(L)} \cong \pi^{-1}(M)$.*

Proof. Indeed, $Y = \mathrm{Proj}\left(\bigoplus_{n \geq 0} H^0(X, L^{\otimes n})\right)^G$. For details see Thm. 1.10 and the remarks following 1.11 in [195] or Thm. 3.21 in [203]. $\square$

Suppose we are in the set-up of the theorem. The problem arises how to decide whether a given point x is semistable or stable. A powerful method is provided by the Hilbert-Mumford criterion. Let $\lambda : \mathbb{G}_m \to G$ be a non-trivial one-parameter subgroup of G. Then the action of G on X induces an action of $\mathbb{G}_m$ on X. Since X is projective, the orbit map $\mathbb{G}_m \to X, t \mapsto \sigma(x, \lambda(t))$ extends in a unique way to a morphism $f : \mathbb{A}^1 \to X$ such that the diagram

$$
\begin{array}{ccc}
\mathbb{G}_m & \xrightarrow{\lambda} & G, \qquad g \\
\downarrow & & \downarrow \qquad \downarrow \\
\mathbb{A}^1 & \xrightarrow{f} & X, \quad \sigma(x, g)
\end{array}
$$

commutes, where $\mathbb{G}_m = \mathbb{A}^1 \setminus \{0\} \to \mathbb{A}^1$ is the inclusion. We write symbolically

$$\lim_{t \to 0} \sigma(x, \lambda(t)) := f(0).$$

Now $f(0)$ is a fixed point of the action of $\mathbb{G}_m$ on X via λ. In particular, $\mathbb{G}_m$ acts on the fibre of $L(f(0))$ with a certain weight r, i.e. if Φ is the linearization of L, then $\Phi(f(0), \lambda(t)) = t^r \cdot \mathrm{id}_{L(f(0))}$. Define the number $\mu^L(x, \lambda) := -r$.

Theorem 4.2.11 (Hilbert-Mumford Criterion) — *A point $x \in X$ is semistable if and only if for all non-trivial one-parameter subgroups $\lambda : \mathbb{G}_m \to G$, one has*

$$\mu^L(x, \lambda) \geq 0.$$

And x is stable if and only if strict inequality holds for all non-trivial λ.

Proof. See Thm. 2.1 in [195] or Thm. 4.9 in [203]. □

Once a good quotient is constructed, one wants to know about its local structure.

Theorem 4.2.12 (Luna's Etale Slice Theorem) — *Let G be a reductive group acting on a k-scheme X of finite type, and let $\pi : X \to X/\!/G$ be a good quotient. Let $x \in X$ be a point with a closed G-orbit and therefore reductive stabilizer G_x. Then there is a G_x-invariant locally closed subscheme $S \subset X$ through x such that the multiplication $S \times G \to X$ induces a G-equivariant étale morphism $\psi : S \times^{G_x} G \to X$. Moreover, ψ induces an étale morphism $S/\!/G_x \to X/\!/G$, and the diagram*

$$
\begin{array}{ccc}
S \times^{G_x} G & \to & X \\
\downarrow & & \downarrow \\
S/\!/G_x & \to & X/\!/G
\end{array}
$$

is cartesian. Moreover, if X is normal or smooth, then S can be taken to be normal or smooth as well.

Proof. See the Appendix to Chapter 1 in [195] or [131]. □

Corollary 4.2.13 — *If the stabilizer of x is trivial, then $\pi : X \to Y$ is a principal G-bundle in a neighbourhood of $\pi(x)$.* □

Some Descent Results

Let G be a reductive algebraic group over a field k that acts on a k-scheme of finite type. Assume that there is a good quotient $\pi : X \to Y$. Let F be a G-linearized coherent sheaf on X. We say that F descends to Y, if there is a coherent sheaf E on Y such that there is an isomorphism $F \cong \pi^* E$ of G-linearized sheaves.

Theorem 4.2.14 — *Let $\pi : X \to Y$ be a principal G-bundle, and let F be a G-linearized coherent sheaf. Then F descends. Moreover, an equivariant homomorphism $F \to F'$ between G-linearized coherent sheaves F and F' descends to a homomorphism on Y.*

Proof. If π is a principal bundle then there is an isomorphism $X \times G \to X \times_Y X$. Under this isomorphism the G-linearization of F induces an isomorphism $p_1^* F \cong p_2^* F$, where $p_1, p_2 : X \times_Y X \to X$ are the two projections. Moreover, the cocyle condition for the linearization translates precisely into the cocycle condition for usual descent theory for faithfully flat quasi-compact morphisms (cf. Thm. 2.23 in [178]).

For the second assertion write $F = \pi^* F_0$ and $F' = \pi^* F_0'$. Taking the direct image of a homomorphism $\varphi : \pi^* F_0 \to \pi^* F_0'$ and applying the projection formula yields a homomorphism $F_0 \otimes \pi_* \mathcal{O}_X \to F_0' \otimes \pi_* \mathcal{O}_X$. If φ was equivariant, then this homomorphism is equivariant with respect to the natural G-action on $\pi_* \mathcal{O}_X$ and hence compatible with taking invariants. Since $(\pi_* \mathcal{O}_X)^G = \mathcal{O}_Y$, this yields a homomorphism $F_0 \to F_0'$. It is easy to see that its pull-back to X gives back φ. $\qquad\square$

In other words, the theorem asserts an equivalence between the category of coherent sheaves on Y and the category of G-linearized coherent sheaves with equivariant homomorphisms on X. In general, we only have the following

Theorem 4.2.15 — *Let $\pi : X \to Y$ be a good quotient, and let F be a G-linearized locally free sheaf on X. A necessary and sufficient condition for F to descend is that for any point $x \in X$ in a closed G-orbit the stabilizer G_x of x acts trivially on the fibre $F(x)$.*

Proof. See 'The Picard Group of a G-Variety' in [131]. $\qquad\square$

Let $\mathrm{Pic}^G(X)$ denote the group of all isomorphism classes of G-linearized line bundles on X, the group structure being given by the tensor product of two line bundles.

Theorem 4.2.16 — *Let $\pi : X \to X /\!\!/ G$ be a good quotient. Then the natural homomorphism $\pi^* : \mathrm{Pic}(X /\!\!/ G) \to \mathrm{Pic}^G(X)$ is injective.*

Proof. Loc. cit. $\qquad\square$

4.3 The Construction — Results

Let X be a connected projective scheme over an algebraically closed field k of characteristic zero and let $\mathcal{O}_X(1)$ be an ample line bundle on X. If we fix a polynomial $P \in \mathbb{Q}[z]$, then according to Theorem 3.3.7 the family of semistable sheaves on X with Hilbert polynomial equal to P is bounded. In particular, there is an integer m such that any such sheaf

F is m-regular. Hence, $F(m)$ is globally generated and $h^0(F(m)) = P(m)$. Thus if we let $V := k^{\oplus P(m)}$ and $\mathcal{H} := V \otimes_k \mathcal{O}_X(-m)$, then there is a surjection

$$\rho : \mathcal{H} \longrightarrow F$$

obtained by composing the canonical evaluation map $H^0(F(m)) \otimes \mathcal{O}_X(-m) \to F$ with an isomorphism $V \to H^0(F(m))$. This defines a closed point

$$[\rho : \mathcal{H} \to F] \in \mathrm{Quot}(\mathcal{H}, P).$$

In fact, this point is contained in the open subset $R \subset \mathrm{Quot}(\mathcal{H}, P)$ of all those quotients $[\mathcal{H} \to E]$, where E is semistable and the induced map

$$V \to H^0(\mathcal{H}(m)) \to H^0(E(m))$$

is an isomorphism. (Note that the first map is bijective if X is reduced.) The first condition is open according to 2.3.1 and the second because of the semicontinuity theorem for cohomology. Moreover, let $R^s \subset R$ denote the open subscheme of those points which parametrize geometrically stable sheaves F.

Thus R parametrizes all semistable sheaves with Hilbert polynomial P but with an ambiguity arising from the choice of a basis of the vector space $H^0(F(m))$. The group $\mathrm{GL}(V)$ acts on $\mathrm{Quot}(\mathcal{H}, P)$ from the right by composition:

$$[\rho] \cdot g := [\rho \circ g]$$

for any two S-valued points ρ and g in $\mathrm{Quot}(\mathcal{H}, P)$ and $\mathrm{GL}(V)$, respectively. Clearly, R is invariant under this action, and isomorphism classes of semistable sheaves are given by the set $R(k)/\mathrm{GL}(V)(k)$. Let $\mathcal{M}' = \mathcal{M}'(P)$ be the functor defined in Section 4.1. The next lemma relates the moduli problem with the problem of finding a quotient for the group action.

Lemma 4.3.1 — *If $R \to M$ is a categorical quotient for the $\mathrm{GL}(V)$-action then M corepresents the functor $\mathcal{M}'$. Conversely, if M corepresents $\mathcal{M}'$ then the morphism $R \to M$, induced by the universal quotient module on $R \times X$, is a categorical quotient. Similarly, $R^s \to M^s$ is a categorical quotient if and only if M^s corepresents $\mathcal{M}^s$.*

Proof. Suppose that S is a Noetherian k-scheme and $\mathcal{F}$ a flat family of m-regular $\mathcal{O}_X$-sheaves with Hilbert polynomial P which is parametrized by S. Then $V_\mathcal{F} := p_*(\mathcal{F} \otimes q^*\mathcal{O}_X(m))$ is a locally free $\mathcal{O}_S$-sheaf of rank $P(m)$, and there is a canonical surjection

$$\varphi_\mathcal{F} : p^*V_\mathcal{F} \otimes q^*\mathcal{O}_X(-m) \longrightarrow \mathcal{F}.$$

Let $R(\mathcal{F}) := \mathbb{I}\mathrm{som}(V \otimes \mathcal{O}_S, V_{\mathcal{F}})$ be the frame bundle associated to $V_{\mathcal{F}}$ (cf. 4.2.3) with the natural projection $\pi : R(\mathcal{F}) \to S$. Composing $\varphi_{\mathcal{F}}$ with the universal trivialization of $V_{\mathcal{F}}$ on $R(\mathcal{F})$ we obtain a canonically defined quotient

$$\tilde{q}_{\mathcal{F}} : \mathcal{O}_{R(\mathcal{F})} \otimes_k \mathcal{H} \longrightarrow \pi_X^* \mathcal{F}$$

on $R(\mathcal{F}) \times X$. This quotient $\tilde{q}_{\mathcal{F}}$ gives rise to a classifying morphism

$$\widetilde{\Phi}_{\mathcal{F}} : R(\mathcal{F}) \to \mathrm{Quot}(\mathcal{H}, P).$$

As discussed earlier, the group $\mathrm{GL}(V)$ acts on $R(\mathcal{F})$ from the right by composition, so that $\pi : R(\mathcal{F}) \to S$ becomes a principal $\mathrm{GL}(V)$-bundle. The morphism $\widetilde{\Phi}_{\mathcal{F}}$ is clearly equivariant. It follows directly from the construction that $\widetilde{\Phi}_{\mathcal{F}}^{-1}(R) = \pi^{-1}(S^{ss})$, where $S^{ss} = \{s \in S | \mathcal{F}_s \text{ is semistable}\}$. In particular, if S parametrizes semistable sheaves only, then $\widetilde{\Phi}_{\mathcal{F}}(R(\mathcal{F})) \subset R$. In this case, the morphism $\widetilde{\Phi}_{\mathcal{F}} : R(\mathcal{F}) \to R$ induces a transformation of functors $\underline{R(\mathcal{F})/\mathrm{GL}(V)} \to \underline{R/\mathrm{GL}(V)}$ and, since $R(\mathcal{F}) \to S$ is a principal bundle and therefore a categorical quotient as well, defines an element in $\underline{R/\mathrm{GL}(V)}(S)$. In this way, we have constructed a transformation $\mathcal{M}' \to \underline{R/\mathrm{GL}(V)}$. The universal family on R yields an inverse transformation. Hence, indeed it amounts to the same to corepresent $\mathcal{M}'$ and to corepresent $\underline{R/\mathrm{GL}(V)}$. $\square$

The construction used in the proof is functorial in the following sense: if $f : S' \to S$ is a morphism of finite type of Noetherian schemes and if we set $\mathcal{F}' = f_X^* \mathcal{F}$ then there is a canonical $\mathrm{GL}(V)$-equivariant morphism $\tilde{f} : R(\mathcal{F}') \to R(\mathcal{F})$ commuting with f and the projections to S' and S, respectively, such that $\widetilde{\Phi}_{\mathcal{F}'} = \widetilde{\Phi}_{\mathcal{F}} \circ \tilde{f}$. As a consequence, if S and $\mathcal{F}$ carry in addition compatible G-actions for some algebraic group G, then $R(\mathcal{F})$ inherits a natural G-structure commuting with the action of $\mathrm{GL}(V)$ such that π is equivariant and $\widetilde{\Phi}_{\mathcal{F}}$ is invariant.

Lemma 4.3.2 — *Let $[\rho : \mathcal{H} \to F] \in \mathrm{Quot}(\mathcal{H}, P)$ be a closed point such that $F(m)$ is globally generated and such that the composition of $H^0(\rho(m)) : H^0(\mathcal{H}(m)) \to H^0(F(m))$ with $V \to H^0(\mathcal{H}(m))$ is an isomorphism $V \xrightarrow{\sim} H^0(F(m))$. Then there is a natural injective homomorphism $\mathrm{Aut}(F) \to \mathrm{GL}(V)$ whose image is precisely the stabilizer subgroup $\mathrm{GL}(V)_{[\rho]}$ of the point $[\rho]$.*

Proof. Consider the map $\mathrm{Aut}(F) \to \mathrm{GL}(V)$ defined by

$$\varphi \mapsto H^0(\rho(m))^{-1} \circ H^0(\varphi(m)) \circ H^0(\rho(m)).$$

Since $F(m)$ is globally generated, this map is injective. By the definition of equivalence for two surjective homomorphisms representing the same quotient, an element $g \in \mathrm{GL}(V)$ is in the stabilizer $\mathrm{GL}(V)_{[\rho]}$ if and only if there is an automorphism φ of F such that $\rho \circ g = \varphi \circ \rho$. $\square$

The lemma implies that the center $Z \subset \mathrm{GL}(V)$ is contained in the stabilizer of any point in $\mathrm{Quot}(\mathcal{H}, P)$. Instead of the action of $\mathrm{GL}(V)$ we will therefore consider the actions of $\mathrm{PGL}(V)$ and $\mathrm{SL}(V)$. There is no difference in the action of these groups on $\mathrm{Quot}(\mathcal{H}, P)$, since the natural map $\mathrm{SL}(V) \to \mathrm{PGL}(V)$ is a finite surjective homomorphism. But it is a little easier to find linearized line bundles for the action of $\mathrm{SL}(V)$ than for the action of $\mathrm{PGL}(V)$, though not much: if L carries a $\mathrm{PGL}(V)$-linearization, then *a fortiori* it is also $\mathrm{SL}(V)$-linearized. Conversely, if L carries an $\mathrm{SL}(V)$-linearization, then all that could prevent it from being $\mathrm{PGL}(V)$-linearized is the action of the group of units $Z \cap \mathrm{SL}(V)$ of order $\dim(V)$. But this action becomes trivial if we pass to the tensor power $L^{\otimes \dim(V)}$.

The next step, before we can apply the methods of Geometric Invariant Theory as described in the previous section, is to find a linearized ample line bundle on R:

Let $\tilde\rho : q^*\mathcal{H} \to \widetilde{F}$ be the universal quotient on $\mathrm{Quot}(\mathcal{H}, P) \times X$, and let $\tau : V \otimes \mathcal{O}_{\mathrm{GL}(V)} \to V \otimes \mathcal{O}_{\mathrm{GL}(V)}$ be the 'universal automorphism' of V parametrized by $\mathrm{GL}(V)$. Let p_1 and p_2 denote the projection from $\mathrm{Quot}(\mathcal{H}, P) \times \mathrm{GL}(V)$ to the first and the second factor, respectively. The composition $q^*\mathcal{H} \xrightarrow{p_2^*\tau} q^*\mathcal{H} \xrightarrow{p_{1,X}^*\tilde\rho} p_{1,X}^*\widetilde{F}$ is a family of quotients parametrized by $\mathrm{Quot}(\mathcal{H}, P) \times \mathrm{GL}(V)$, whose classifying morphism

$$\sigma : \mathrm{Quot}(\mathcal{H}, P) \times \mathrm{GL}(V) \longrightarrow \mathrm{Quot}(\mathcal{H}, P),$$

is, of course, just the $\mathrm{GL}(V)$-action on $\mathrm{Quot}(\mathcal{H}, P)$, which we defined earlier in terms of point functors. By the definition of the classifying morphism, the epimorphisms $\sigma_X^*\tilde\rho$ and $p_{1,X}^*\tilde\rho \circ p_2^*\tau$ yield equivalent quotients. This means that there is an isomorphism $\Lambda : \sigma_X^*\widetilde{F} \to p_{1,X}^*\widetilde{F}$ such that the diagram

$$
\begin{array}{ccc}
q^*\mathcal{H} & \xrightarrow{\;p_{1,X}^*\tilde\rho\;} & p_{1,X}^*\widetilde{F} \\[2pt]
{\scriptstyle p_2^*\tau}\big\uparrow & & \big\uparrow{\scriptstyle \Lambda} \\[2pt]
q^*\mathcal{H} & \xrightarrow{\;\sigma_X^*\tilde\rho\;} & \sigma_X^*\widetilde{F}
\end{array}
$$

commutes. It is not difficult to check that Λ satisfies the cocyle condition 4.2.5. Thus Λ is a natural $\mathrm{GL}(V)$-linearization for the universal quotient sheaf $\widetilde{F}$. We saw in Chapter 2, (cf. Proposition 2.2.5), that the line bundle

$$L_\ell := \det(p_*(\widetilde{F} \otimes q^*\mathcal{O}_X(\ell)))$$

on $\mathrm{Quot}(\mathcal{H}, P)$ is very ample if ℓ is sufficiently large. Since the definition of L_ℓ commutes with base change (if ℓ is sufficiently large), Λ induces a natural $\mathrm{GL}(V)$-linearization on L_ℓ.

Thus we can speak of semistable and stable points in the closure $\overline{R}$ of R in $\mathrm{Quot}(\mathcal{H}, P)$ with respect to L_ℓ and the $\mathrm{SL}(V)$-action (!). Remember that the definition of the whole set-up depended on the integer m.

Theorem 4.3.3 — *Suppose that m, and for fixed m also ℓ, are sufficiently large integers. Then $R = \overline{R}^{ss}(L_\ell)$ and $R^s = \overline{R}^s(L_\ell)$. Moreover, the closures of the orbits of two points $[\rho_i : \mathcal{H} \to F_i]$, $i = 1, 2$, in R intersect if and only if $gr^{JH}(F_1) \cong gr^{JH}(F_2)$. The orbit of a point $[\rho : \mathcal{H} \to F]$ is closed in R if and only if F is polystable.*

The proof of this theorem will take up Section 4.4. Together with Lemma 4.3.1 and Theorem 4.2.10 it yields:

Theorem 4.3.4 — *There is a projective scheme $M_{\mathcal{O}_X(1)}(P)$ that universally corepresents the functor $\mathcal{M}_{\mathcal{O}_X(1)}(P)$. Closed points in $M_{\mathcal{O}_X(1)}(P)$ are in bijection with S-equivalence classes of semistable sheaves with Hilbert polynomial P. Moreover, there is an open subset $M^s_{\mathcal{O}_X(1)}(P)$ that universally corepresents the functor $\mathcal{M}^s_{\mathcal{O}_X(1)}(P)$.* □

More precisely, Theorem 4.3.3 tells us that $\pi : R \to M := M_{\mathcal{O}_X(1)}(P)$ is a good quotient, and that $\pi : R^s \to M^s := M^s_{\mathcal{O}_X(1)}(P)$ is a geometric quotient, since the orbits of stable sheaves are closed. According to Lemma 4.3.2 the stabilizer in $\mathrm{PGL}(V)$ of a closed point in R^s is trivial. Thus:

Corollary 4.3.5 — *The morphism $\pi : R^s \to M^s$ is a principal $\mathrm{PGL}(V)$-bundle.*

Proof. This follows from Theorem 4.3.3 and Luna's Etale Slice Theorem 4.2.12. □

Example 4.3.6 — Let X be a projective scheme over k, and let $S^n(X)$ be its n-th symmetric product, i.e. the quotient of the product $X \times \ldots \times X$ of n copies of X by the permutation action of the symmetric group S_n. And let M_n denote the moduli space of zero-dimensional coherent sheaves of length n on X. It is easy to see that any zero-dimensional sheaf F of length n is semistable. Moreover, if $n_x = \mathrm{length}(F_x)$ for each $x \in X$, then F is S-equivalent to the direct sum $\bigoplus_{x \in X} k(x)^{\oplus n_x}$ of skyscraper sheaves. This shows that the following morphism $f : S^n(X) \to M_n$ is bijective. Consider the structure sheaf $\mathcal{O}_\Delta$ of the diagonal $\Delta \subset X \times X$ as a family of sheaves of length one on X parametrized by X. Thus on $(X \times \ldots \times X) \times X$ we can form the family $\bigoplus_{i=1}^n p_i^* \mathcal{O}_\Delta$, where $p_i : (X \times \ldots \times X) \times X \to X \times X$ is the projection onto the product of the i-th and the last factor. This family induces a morphism $\tilde{f} : X \times \ldots \times X \to M_n$ which is obviously S_n-invariant and therefore descends to a morphism $f : S^n(X) \to M_n$. In fact, f is an isomorphism. In order to see this, we shall construct an inverse morphism $g : M_n \to S^n(X)$. In general, there is no universal family on M_n which we could use. Instead, we construct a natural transformation $g : \mathcal{M}_n \to \underline{S^n(X)}$ for the moduli functor corepresented by M_n. Let F be a flat family of zero-dimensional sheaves of length n on X parametrized by a scheme S. Let $s \in S$ be a closed point representing a sheaf F_s on X. Since the support of F_s is finite, and since X is projective, there is an open affine subset $U = \mathrm{Spec}(B) \subset X$ containing the support of F_s. Then there is an open affine neighbourhood $V = \mathrm{Spec}(A) \subset S$ of s such that $\mathrm{Supp}(F_t) \subset U$ for all $t \in V$. Moreover, making V smaller if necessary, we may assume that $H := p_* F|_V$ is free of rank n. Choose a basis of sections. Then the $\mathcal{O}_U$-module structure of F_V is determined by a k-algebra homomorphism $\rho : B \to E$, where $E = \mathrm{End}_A(H)$ is isomorphic to the ring of $n \times n$-matrices with values in A.

Recall the notion of the *linear determinant*: there is a natural equivariant identification $\Phi : E^{\otimes n} \cong \mathrm{End}_A(H^{\otimes n})$ with respect to the actions of the symmetric group S_n on $H^{\otimes n}$ and $E^{\otimes n}$. Hence $(E^{\otimes n})^{S_n} \subset E^{\otimes n}$ is the subalgebra of those endomorphisms of $H^{\otimes n}$ which commute with the action of S_n. In particular, $(E^{\otimes n})^{S_n}$ commutes with the anti-symmetrization operator

$$a : H^{\otimes n} \to H^{\otimes n}, h_1 \otimes \ldots \otimes h_n \mapsto \sum_{\pi \in S_n} \mathrm{sgn}(\pi) h_{\pi(1)} \otimes \ldots \otimes h_{\pi(n)}$$

and therefore acts naturally on the image of a, which is $\Lambda^n H$ and, hence, free of rank 1. This gives a ring homomorphism $ld : (E^{\otimes n})^{S_n} \to A$. An equivalent description is the following: let $b : E \otimes \ldots \otimes E \to A$ be the polar form of the determinant. Then b restricted to symmetric tensors is formally divisible by $n!$, and $ld = b/n!$.

Using the linear determinant we can finish our argument: let $g(F) : V \to S^n U \subset S^n(X)$ be the morphism induced by the ring homomorphism

$$(B^{\otimes n})^{S_n} \xrightarrow{\rho^n} (E^{\otimes n})^{S_n} \xrightarrow{ld} A.$$

Check that the morphisms thus obtained for an open cover of S glue to give a morphism $S \to S^n(X)$, that this construction is functorial, and that the natural transformation g constructed in this way provides an inverse of f.

Consider now the Hilbert scheme $\mathrm{Hilb}^n(X)$ of zero-dimensional subschemes of X of length n. The structure sheaf $\mathcal{O}_Z$ of the universal subscheme $Z \subset \mathrm{Hilb}^n(X) \times X$ induces a morphism $\mathrm{Hilb}^n(X) \to M_n$. Using the above identification, we obtain the Hilbert-to-Chow morphism $\mathrm{Hilb}^n(X) \to S^n(X)$, which associates to any cycle in X its support counted with the correct multiplicity.

Assume now that X is a smooth projective surface, and let $M_X(1, \mathcal{O}_X, n)$ denote the moduli space of rank one sheaves with trivial determinant and second Chern number n. Then there is a canonical isomorphism $\mathrm{Hilb}^n(X) \cong M_X(1, \mathcal{O}_X, n)$ obtained by sending a subscheme $Z \subset X$ to its ideal sheaf $\mathcal{I}_Z$. In this context the morphism $M_X(1, \mathcal{O}_X, n) \to S^n(X)$ appears as a particular case of the 'Gieseker-to-Donaldson' morphism which we will discuss later (cf. 8.2.8 and 8.2.17). $\qquad\square$

Occasionally, one also needs to consider relative moduli spaces, i.e. moduli spaces of semistable sheaves on the fibres of a projective morphism $X \to S$. It is easy to generalize the previous construction to this case.

Theorem 4.3.7 — *Let $f : X \to S$ be a projective morphism of k-schemes of finite type with geometrically connected fibres, and let $\mathcal{O}_X(1)$ be a line bundle on X very ample relative to S. Then for a given polynomial P there is a projective morphism $M_{X/S}(P) \to S$ which universally corepresents the functor*

$$\mathcal{M}_{X/S} : (Sch/S)^o \to (Sets),$$

which by definition associates to an S-scheme T of finite type the set of isomorphism classes of T-flat families of semistable sheaves on the fibres of the morphism $X_T := T \times_S X \to T$ with Hilbert polynomial P. In particular, for any closed point $s \in S$ one has $M_{X/S}(P)_s \cong M_{X_s}(P)$. Moreover, there is an open subscheme $M^s_{X/S}(P) \subset M_{X/S}(P)$ that universally corepresents the subfunctor $\mathcal{M}^s_{X/S} \subset \mathcal{M}_{X/S}$ of families of geometrically stable sheaves.

Proof. Because of the assertion that $M_{X/S}$ universally corepresents $\mathcal{M}_{X/S}$, the statement of the theorem is local in S. We may therefore assume that S is quasi-projective. The family of semistable sheaves on the fibres of f with given Hilbert polynomial is bounded and hence m-regular for some integer m. As in the absolute case, let $\mathcal{H} := \mathcal{O}_X(-m)^{P(m)}$ and let $R \subset \mathrm{Quot}_{X/S}(\mathcal{H}, P)$ denote the open subset of all points $[\rho : \mathcal{H}_s \to F]$ where F is a semistable sheaf on X_s, $s \in S$, and ρ induces an isomorphism $H^0(X_s, \mathcal{H}_s(m)) \to H^0(X_s, F(m))$. If $\mathcal{O}_R \otimes_{\mathcal{O}_S} \mathcal{H} \to \widetilde{F}$ denotes the universal quotient family, $L_\ell := \det(p_*(\widetilde{F} \otimes q^* \mathcal{O}_X(\ell)))$ is well-defined and very ample relative to S for sufficiently large ℓ. For any such ℓ there is a very ample line bundle B_ℓ on S such that $L_\ell \otimes g^* B_\ell$ is very ample on R (where $g : R \to S$ is the structure morphism). Then the following statements about a closed point $[\rho : \mathcal{H}_s \to F]$ in the fibre R_s over $s \in S$ are equivalent:

1. $[\rho]$ is a (semi)stable point in $\overline{R}$ with respect to the linearization of $L_\ell \otimes q^* B_\ell$.
2. $[\rho]$ is a (semi)stable point in $\overline{R}_s$ with respect to the linearization of L_ℓ.

This follows either directly from the definition of semistable points (4.2.9), or can be deduced by means of the Hilbert-Mumford Criterion 4.2.11. The theorem then is a consequence of this easy fact, Theorem 4.3.3 and the fact that $M_{X/S}(P) := R/\!/\mathrm{SL}(P(m))$ is a *universal* good quotient (Theorem 4.2.10). We omit the details.   □

4.4 The Construction — Proofs

The proof of Theorem 4.3.3 has two parts: in order to determine whether a given point $[\rho : V \otimes \mathcal{O}_X(-m) \to F]$ in $\overline{R}$ is semistable or stable by means of the Hilbert-Mumford

Criterion we must compute the weight of a certain action of $\mathbb{G}_m$. In this way we shall obtain a condition for the semistability of ρ (in the sense of Geometric Invariant Theory) in terms of numbers of global sections of subsheaves F' of F, which then must be related to the semistability of F. We begin with the second problem and prove a theorem due to Le Potier that makes this relation precise.

Theorem 4.4.1 — *Let p be a polynomial of degree d, and let r be a positive integer. Then for all sufficiently large integers m the following properties are equivalent for a purely d-dimensional sheaf F of multiplicity r and reduced Hilbert polynomial p.*

(1) *F is (semi)stable*

(2) *$r \cdot p(m) \leq h^0(F(m))$, and $h^0(F'(m)) (\leq) r' \cdot p(m)$ for all subsheaves $F' \subset F$ of multiplicity r', $0 < r' < r$.*

(3) *$r'' \cdot p(m) (\leq) h^0(F''(m))$ for all quotient sheaves $F \to F''$ of multiplicity r'', $r > r'' > 0$.*

Moreover, for sufficiently large m, equality holds in (2) and (3) if and only if F' or F'', respectively, are destabilizing.

Proof. (1) $\Rightarrow$ (2): The family of semistable sheaves with Hilbert polynomial equal to $r \cdot p$ is bounded by 3.3.7. Therefore, if m is sufficiently large, any such sheaf F is m-regular, and $r \cdot p(m) = h^0(F(m))$. Let $F' \subset F$ be a subsheaf of multiplicity r', $0 < r' < r$. In order to show (2) we may assume that F' is saturated in F. We distinguish two cases:

A. $\hat{\mu}(F') < \hat{\mu}(F) - r \cdot C$

B. $\hat{\mu}(F') \geq \hat{\mu}(F) - r \cdot C$,

where $C := r^2 + (r + d)/2$ is the constant that appears in Corollary 3.3.8. The family of (saturated!) subsheaves F' of type B is bounded according to Grothendieck's Lemma 1.7.9. Thus for large m, any such sheaf F' is m-regular, implying $h^0(F'(m)) = P(F', m)$, and, moreover, since the set of Hilbert polynomials $\{P(F')\}$ is finite, we can assume that

$$P(F', m) (\leq) r' \cdot p(m) \quad \Leftrightarrow \quad P(F') (\leq) r' \cdot p.$$

For subsheaves of type A we use estimate 3.3.8 to bound the number of global sections directly. Note that $\hat{\mu}_{\max}(F') \leq \hat{\mu}(F)$ by the semistability of F, and $\hat{\mu}(F') < \hat{\mu}(F) - r \cdot C$, since F' is of type A. Thus

$$
\begin{aligned}
\frac{h^0(F'(m))}{r'} &\leq \frac{r' - 1}{r'} \cdot \frac{1}{d!} \left[\hat{\mu}_{\max}(F') + C - 1 + m \right]_+^d + \frac{1}{r'} \cdot \frac{1}{d!} \left[\hat{\mu}(F') + C - 1 + m \right]_+^d \\
&\leq \frac{r - 1}{r} \cdot \frac{1}{d!} \left[\hat{\mu}(F) + C - 1 + m \right]_+^d + \frac{1}{r} \cdot \frac{1}{d!} \left[\hat{\mu}(F) - (r - 1) \cdot C - 1 + m \right]_+^d
\end{aligned}
$$

Hence for large m we get

$$\frac{h^0(F'(m))}{r'} \leq \frac{m^d}{d!} + \frac{m^{d-1}}{(d-1)!} \cdot (\hat{\mu}(F) - 1) + \dots, \tag{4.1}$$

where ... stands for monomials in m of degree smaller than $d - 1$ with coefficients that depend only on r, d, C and $\hat{\mu}(F)$, but not on F'. Since $p(m) = \frac{m^d}{d!} + \frac{m^{d-1}}{(d-1)!} \cdot \hat{\mu}(F) + \ldots$, the right hand side of (4.1) is strictly smaller than $p(m)$ for sufficiently large m.

(2) $\Rightarrow$ (3): Let F' be the kernel of a surjection $F \to F''$ and let r' and r'' be the multiplicities of F' and F'', respectively. Then (2) implies:

$$h^0(F''(m)) \geq h^0(F(m)) - h^0(F'(m)) \ (\geq) \ p(m) \cdot r - p(m) \cdot r' = p(m) \cdot r''.$$

(3) $\Rightarrow$ (1): Apply (3) to the minimal destabilizing quotient sheaf F'' of F. Then, by Corollary 3.3.8

$$p(m) \ (\leq) \ \frac{h^0(F''(m))}{r''} \leq \frac{1}{d!} \ [\hat{\mu}(F'') + C - 1 + m]_+^d \ .$$

This shows that $\hat{\mu}_{\min}(F) = \hat{\mu}(F'')$ is bounded from below and consequently $\hat{\mu}_{\max}(F)$ is bounded uniformly from above. Hence by 3.3.7 the family of sheaves F satisfying (3) is bounded. Now let F'' be any purely d-dimensional quotient of F. Then either $\hat{\mu}(F) < \hat{\mu}(F'')$ and F'' is far from destabilizing F, or indeed, $\hat{\mu}(F) \geq \hat{\mu}(F'')$. But according to Grothendieck's Lemma 1.7.9, the family of such quotients F'' is bounded. As before, this implies that for large m one has $h^0(F''(m)) = P(F'', m)$ and

$$P(F'', m) \ (\geq) \ r'' \cdot p(m) \quad \Leftrightarrow \quad P(F'') \ (\geq) \ r'' \cdot p.$$

Hence indeed, (3) $\Rightarrow$ (1) $\square$

This theorem works for pure sheaves only. The following proposition allows us to make the passage to a more general class of sheaves:

Proposition 4.4.2 — *If F is a coherent module of dimension d which can be deformed to a pure sheaf, then there exists a pure sheaf E with $P(E) = P(F)$ and a homomorphism $\varphi : F \to E$ with $\ker(\varphi) = T_{d-1}(F)$.*

Proof. If F itself is pure there is nothing to show. Hence, assume that $T_{d-1}(F)$ is non-trivial. The condition on F means that there is a smooth connected curve C and a C-flat family $\mathcal{F}$ of d-dimensional sheaves on X such that $\mathcal{F}_0 \cong F$ for some closed point $0 \in C$ and such that $\mathcal{F}_s$ is pure for all $s \in C \setminus \{0\}$. (Note that this implies that $\mathcal{F}$ is pure of dimension $d + 1$: any torsion subsheaf supported on a fibre would contradict flatness, and any other torsion subsheaf could be detected in the restriction of $\mathcal{F}$ to the fibre over a point in $C \setminus \{0\}$).

Let t be a uniformizing parameter in the local ring $\mathcal{O}_{C,0}$. Consider the action of t on the cokernel N of the natural homomorphism $\mathcal{F} \to \mathcal{F}^{DD}$ from $\mathcal{F}$ to its reflexive hull (cf. 1.1.9). Since $\mathcal{F}$ is pure, this homomorphism is injective (cf. Proposition 1.1.10). (Note that in order to apply the results of Section 1.1 we tacitly embed X into a smooth ambient variety X' and consider the reflexive hull of $\mathcal{F}$ on $C \times X'$. By the local description of the reflexive hull it is clear that $\mathcal{F}^{DD}$ is again a sheaf on X.)

The kernels N_n of the multiplication maps $t^n : N \to N$ form an increasing sequence of submodules and hence stabilize. Let N' be the union of all N_n. Then t is injective on N/N', which is equivalent to saying that N/N' is C-flat. Let $\mathcal{E}$ be the kernel of $\mathcal{F}^{DD} \to N/N'$. Thus we get the following commutative diagram with exact columns and rows:

$$
\begin{array}{ccccccccc}
& & & & N/N' & = & N/N' & & \\
& & & & \uparrow & & \uparrow & & \\
0 & \to & \mathcal{F} & \to & \mathcal{F}^{DD} & \to & N & \to & 0 \\
& & \| & & \uparrow & & \uparrow & & \\
0 & \to & \mathcal{F} & \to & \mathcal{E} & \to & N' & \to & 0.
\end{array}
$$

$\mathcal{F}^{DD}$ is reflexive and therefore pure of dimension $d+1$, and the same holds for $\mathcal{E}$. In particular, both sheaves as well as N/N' are C-flat. Restricting the middle column to the special fibre $\{0\} \times X$ we get an exact sequence $0 \to \mathcal{E}_0 \to (\mathcal{F}^{DD})_0 \to (N/N')_0 \to 0$. By Corollary 1.1.14 the sheaf $(\mathcal{F}^{DD})_0$ is pure, and being a subsheaf of a pure sheaf, $E := \mathcal{E}_0$ is pure as well. Since N' has support in $\{0\} \times X$, $\mathcal{F}$ and $\mathcal{E}$ are isomorphic over $C \setminus \{0\}$, and since both are C-flat, they have the same Hilbert polynomial: $P(F) = P(E)$. Note that $\dim(N) \leq \dim(\mathcal{F}^{DD}) - 2 = d - 1$ (cf. 1.1.8). This implies that $\varphi : F \to E$ has at most $(d-1)$-dimensional cokernel and kernel. In particular, $\ker(\varphi)$ is precisely the torsion submodule of F. $\qquad\square$

After these preparations we can concentrate on the geometric invariant theoretic part of the proof. Let $[\rho : V \otimes \mathcal{O}_X(-m) \to F]$ be a closed point in $\overline{R}$. In order to apply the Hilbert-Mumford Criterion we need to determine the limit point $\lim_{t \to 0}[\rho] \cdot \lambda(t)$ for the action of any one-parameter subgroup $\lambda : \mathbb{G}_m \to SL(V)$ on $[\rho]$. Now λ is completely determined by the decomposition $V = \bigoplus_{n \in \mathbb{Z}} V_n$ of V into weight spaces V_n, $n \in \mathbb{Z}$, of weight n, i.e. $v \cdot \lambda(t) = t^n \cdot v$ for all $v \in V_n$. Of course, $V_n = 0$ for almost all n. Define ascending filtrations of V and F by

$$
V_{\leq n} = \bigoplus_{\nu \leq n} V_\nu \quad \text{and} \quad F_{\leq n} = \rho(V_{\leq n} \otimes \mathcal{O}_X(-m)).
$$

Then ρ induces surjections $\rho_n : V_n \otimes \mathcal{O}_X(-m) \to F_n := F_{\leq n}/F_{\leq n-1}$. Summing up over all weights we get a closed point

$$
\left[\bar{\rho} := \oplus_n \rho_n : V \otimes \mathcal{O}_X(-m) \longrightarrow \overline{F} := \bigoplus_n F_n \right]
$$

in $\mathrm{Quot}(\mathcal{H}, P)$.

Lemma 4.4.3 — $[\bar{\rho}] = \lim_{t \to 0} [\rho] \cdot \lambda(t)$.

Proof. We will explicitly construct a quotient $\theta : V \otimes \mathcal{O}_X(-m) \otimes k[T] \to \mathcal{F}$ parametrized by $\mathbb{A}^1 = \mathrm{Spec}(k[T])$ such that $[\theta_0] = [\bar{\rho}]$ and $[\theta_\alpha] = [\rho] \cdot \lambda(\alpha)$ for all $\alpha \neq 0$. The assertion follows from this. Let

$$\mathcal{F} := \bigoplus_n F_{\leq n} \otimes T^n \subset F \otimes_k k[T, T^{-1}].$$

Only finitely many summands with negative exponent n are nonzero, so that $\mathcal{F}$ can be considered as a coherent sheaf on $\mathbb{A}^1 \times X$. Indeed, let N be a positive integer such that $V_n = 0$ and $F_n = 0$ for all $n \leq -N$. Then $\mathcal{F} \subset F \otimes_k T^{-N} k[T]$. Similarly, define a module

$$\mathcal{V} := \bigoplus_n V_{\leq n} \otimes \mathcal{O}_X(-m) \otimes T^n \subset V \otimes_k \mathcal{O}_X(-m) \otimes_k T^{-N} k[T].$$

Clearly, ρ induces a surjection $\rho' : \mathcal{V} \to \mathcal{F}$ of $\mathbb{A}^1$-flat coherent sheaves on $\mathbb{A}^1 \times X$. Finally, define an isomorphism (!) $\gamma : V \otimes_k k[T] \to \bigoplus_n V_{\leq n} \otimes T^n$ by $\gamma|_{V_\nu} = T^\nu \cdot \mathrm{id}_{V_\nu}$ for all ν; and let θ be the surjection that makes the following diagram commutative:

$$
\begin{array}{ccccc}
\bigoplus_n F_{\leq n} \otimes T^n & = & \mathcal{F} & \longrightarrow & F \otimes T^{-N} k[T] \\
\uparrow{\scriptstyle\theta} & & \uparrow{\scriptstyle\rho'} & & \uparrow{\scriptstyle\rho \otimes 1} \\
V \otimes \mathcal{O}_X(-m) \otimes k[T] & \xrightarrow{\;\gamma\;} & \mathcal{V} & \longrightarrow & V \otimes \mathcal{O}_X(-m) \otimes T^{-N} k[T]
\end{array}
$$

First, restrict to the special fibre $\{0\} \times X$: it is easy to see that $\theta_0 = \oplus_n \rho_n$; for we have $\mathcal{F}_0 = \mathcal{F}/T \cdot \mathcal{F} = \bigoplus_n F_n$ etc. Restricting to the open complement $\mathbb{A}^1 \setminus \{0\}$ corresponds to inverting the variable T: all horizontal arrows in the diagram above become isomorphisms. Thus we get:

$$
\begin{array}{ccc}
\mathcal{F} \otimes_{k[T]} k[T, T^{-1}] & \xrightarrow{\;\cong\;} & F \otimes_k k[T, T^{-1}] \\
\theta \uparrow & & \uparrow{\scriptstyle\rho \otimes 1} \\
V \otimes_k \mathcal{O}_X(-m) \otimes_k k[T, T^{-1}] & \xrightarrow{\;\gamma\;} & V \otimes_k \mathcal{O}_X(-m) \otimes_k k[T, T^{-1}]
\end{array}
$$

Note that γ describes precisely the action of λ! Hence θ has the required properties, and we are done. $\square$

Lemma 4.4.4 — *The weight of the action of $\mathbb{G}_m$ via λ on the fibre of L_ℓ at the point $[\bar{\rho}]$ is given by*

$$\sum_{n \in \mathbb{Z}} n \cdot P(F_n, \ell).$$

Proof. $\bar{F} = \bigoplus F_n$ decomposes into a direct sum of subsheaves on which $\mathbb{G}_m$ acts via a character of weight n. Hence for each integer n the group $\mathbb{G}_m$ acts with weight n on the complex which defines the cohomology groups $H^i(F_n(\ell))$, $i \geq 0$, (cf. Section 2.1). This complex has (virtual) total dimension $P(F_n, \ell)$, so that $\mathbb{G}_m$ acts on its determinant with weight $n \cdot P(F_n, \ell)$). Since $L_\ell([\bar{\rho}]) = \bigotimes_n \det(H^*(F_n(\ell)))$, the weight of the action on $L_\ell([\bar{\rho}])$ is indeed $\sum_n n \cdot P(F_n, \ell)$. $\qquad \square$

We can rewrite this weight in the following form: use the fact that $\sum_n n \cdot \dim(V_n) = 0$ since the determinant of λ is 1, and that in both sums only finitely many summands are nonzero:

$$\sum_{n \in \mathbb{Z}} n \cdot P(F_n, \ell) = \frac{1}{\dim(V)} \sum_{n \in \mathbb{Z}} n \cdot (\dim(V)P(F_n, \ell) - \dim(V_n)P(F, \ell))$$

$$= -\frac{1}{\dim(V)} \sum_{n \in \mathbb{Z}} (\dim(V)P(F_{\leq n}, \ell) - \dim(V_{\leq n})P(F, \ell))$$

Lemma 4.4.5 — *A closed point $[\rho : \mathcal{H} \to F] \in \overline{R}$ is (semi)stable if and only if for all non-trivial proper linear subspaces $V' \subset V$ and the induced subsheaf $F' := \rho(V' \otimes \mathcal{O}_X(-m)) \subset F$, the following inequality holds:*

$$\dim(V) \cdot P(F', \ell) \, (\geq) \, \dim(V') \cdot P(F, \ell). \tag{4.2}$$

Proof. Define a function θ on the set of subspaces of V by

$$\theta(V') := \dim(V) \cdot P(F', \ell) - \dim(V') \cdot P(F, \ell).$$

Then, with the notations of Lemma 4.4.4, we have

$$\mu(x, \lambda) = -\sum_{n \in \mathbb{Z}} n \cdot P(F_n, \ell) = \frac{1}{\dim(V)} \sum_{n \in \mathbb{Z}} \theta(V_{\leq n}).$$

Hence, according to the Hilbert-Mumford Criterion 4.2.11, a point $[\rho]$ is (semi)stable, if for any non-trivial weight decomposition $V = \bigoplus V_n$, the condition $\sum_n \theta(V_{\leq n}) \, (\geq) \, 0$ is satisfied. Hence if $\theta(V') \, (\geq) \, 0$ for any non-trivial proper subspace $V' \subset V$, then $[\rho]$ is (semi)stable. Conversely, if $V' \subset V$ is a subspace with $\theta(V') \, (<) \, 0$ and $V'' \subset V$ is any complement of V', define a weight decomposition of V by

$$V_{-\dim(V'')} = V', \quad V_{\dim(V')} = V'', \quad \text{and } V_n = 0 \text{ else.}$$

Then $\sum_n \theta(V_{\leq n}) = \dim(V) \cdot \theta(V') \, (<) \, 0$. This proves the converse. $\qquad \square$

Lemma 4.4.6 — *If ℓ is sufficiently large, a closed point $[\rho : \mathcal{H} \to F] \in \overline{R}$ is (semi)stable if and only if for all coherent subsheaves $F' \subset F$ and $V' = V \cap H^0(F'(m))$ the following inequality holds:*

$$\dim(V) \cdot P(F') \, (\geq) \, \dim(V') \cdot P(F). \tag{4.3}$$

Here and in the following we use the more suggestive notation $V \cap H^0(F'(m))$ instead of $H^0(\rho(m))^{-1}(H^0(F'(m)))$.

Proof. If $V' \subset V$ runs through the linear subsets of V then the family of subsheaves $F' \subset F$ generated by V' is bounded. Hence, the set of polynomials $\{P(F')\}$ is finite, and if ℓ is large, the conditions (4.2) and (4.3) are equivalent (with F' still denoting the subsheaf generated by V'). Moreover, if F' is generated by V', then $V' \subset V \cap H^0(F'(m))$, and conversely, if F' is an arbitrary subsheaf of F and $V' = V \cap H^0(F'(m))$, then the subsheaf of F generated by V' is contained in F'. This shows that the condition of Lemma 4.4.6 is equivalent to the condition of Lemma 4.4.5. $\square$

Corollary 4.4.7 — *For $[\rho]$ to be semistable, a necessary condition is that the induced homomorphism $V \to H^0(F(m))$ is injective and that for any submodule $F' \subset F$ of dimension $\leq d - 1$ one has $H^0(X, F'(m)) \cap V = 0$.*

Proof. Indeed, if this were false, let $V' \subset V$ be a non-trivial linear subspace such that the subsheaf $F'(m) \subset F(m)$ generated by V' is trivial or torsion. Then $P(F')$ has degree less than d and we get a contradiction to (4.3). $\square$

Note that the arguments below actually show $H^0(X, F'(m)) = 0$ for any submodule $F' \subset F$ of dimension $\leq d - 1$.

4.4.8 Proof of Theorem 4.3.3 — Let m be large enough in the sense of Theorem 4.4.1 and such that any semistable sheaf with multiplicity $\rho \leq r$ and Hilbert polynomial $\rho \cdot p$ is m-regular. Moreover, let ℓ be large enough in the sense of Lemma 4.4.6.

First assume that $[\rho : \mathcal{H} \to F]$ is a closed point in R. By definition of R, the map $V \to H^0(F(m))$ is an isomorphism. Let $F' \subset F$ be a subsheaf of multiplicity $0 < r' < r$ and let $V' = V \cap H^0(F'(m))$. According to Theorem 4.4.1 one has either

- $p(F') = p(F)$, i.e. $P(F') \cdot r = P(F) \cdot r'$, or
- $h^0(F'(m)) < r' \cdot p(m)$.

In the first case F' is m-regular, and we get $\dim(V') = h^0(F'(m)) = r' \cdot p(m)$ and therefore

$$\dim(V') \cdot P(F) = (r'p(m)) \cdot (rp) = (rp(m)) \cdot (r'p) = \dim(V) \cdot P(F').$$

In the second case

$$\dim(V) \cdot r' = r \cdot r' \cdot p(m) > h^0(F'(m)) \cdot r = \dim(V') \cdot r.$$

These are the leading coefficients of the two polynomials appearing in (4.3), so that indeed $\dim(V) \cdot P(F') > \dim(V') \cdot P(F)$ and hence Criterion (4.3) is satisfied. This proves: $[\rho] \in R^s \Rightarrow [\rho] \in \overline{R}^s$ and $[\rho] \in R \setminus R^s \Rightarrow [\rho] \in \overline{R}^{ss} \setminus \overline{R}^s$.

Conversely, suppose that $[\rho : V \otimes \mathcal{O}_X(-m) \to F] \in \overline{R}$ is semistable in the GIT sense. Because of the first part of the proof it suffices to show that $[\rho] \in R$. By Lemma 4.4.6 we have an inequality

$$\dim(V) \cdot P(F') \geq \dim(V') \cdot P(F)$$

for any $F' \subset F$ and $V' = V \cap H^0(F'(m))$. Passing to the leading coefficients of the polynomials we get

$$p(m) \cdot r \cdot r' = \dim(V) \cdot r' \geq \dim(V') \cdot r. \tag{4.4}$$

As $[\rho]$ is in the closure of R by assumption, the sheaf F can be deformed into a semistable sheaf, hence a fortiori into a pure sheaf. Thus we can apply Proposition 4.4.2 and conclude that there exists a generically injective homomorphism $\varphi : F \to E$ to a pure sheaf E with $P(E) = P(F)$ and whose kernel is the torsion of F. According to Corollary 4.4.7 the composite map $V \to H^0(F(m)) \to H^0(E(m))$ is injective, since any element in the kernel would give a section of $T_{d-1}(F)(m)$. Let E'' be any quotient module of E of multiplicity r'', $r > r'' > 0$. Let F' be the kernel of the composite map $F \to E \to E''$ and $V' = V \cap H^0(F'(m))$. Using inequality (4.4) we get:

$$
\begin{aligned}
h^0(E''(m)) &\geq h^0(F(m)) - h^0(F'(m)) \\
&\geq \dim(V) - \dim(V') \\
&\geq r \cdot p(m) - r' \cdot p(m) = r'' \cdot p(m).
\end{aligned}
$$

Thus E is semistable by Theorem 4.4.1. In particular, $h^0(E(m)) = \dim(V)$. Since $V \to H^0(E(m))$ is injective, it is in fact an isomorphism, and V generates E. But the map $V \otimes \mathcal{O}_X(-m) \to E$ factors through F, forcing the homomorphism $\varphi : F \to E$ to be surjective. Since E and F have the same Hilbert polynomial, φ must be an isomorphism. Hence, F is semistable and $V \to H^0(F(m))$ is bijective. This means that $[\rho]$ is a point in R^{ss}.

Remark 4.4.9 — The last paragraph is the only place where have used the fact that the given semistable point $[\rho]$ lies in $\overline{R}$ rather than just in $\text{Quot}(\mathcal{H}, P)$. Sometimes this restriction is not necessary: Suppose that X is a smooth curve. Then any torsion submodule of F is zero-dimensional and can therefore be detected by its global sections. Hence Corollary 4.4.7 implies that F is torsion free if $[\rho : \mathcal{H} \to F]$ is semistable. $\qquad\square$

We are almost finished with the proof of 4.3.3. What is left to prove is the identification of closed orbits. Observe first that we can read the proof of Lemma 4.4.3 backwards: let $[\rho : \mathcal{H} \to F]$ be a point in R and $JH_\bullet F$ a Jordan-Hölder filtration of F. Let $V_{\leq n} = H^0(JH_n F(m)) \cap V$ for all n, and choose linear subspaces $V_n \subset V_{\leq n}$ which split the filtration. Summing up the induced surjections $V_n \otimes \mathcal{O}_X(-m) \to gr_n^{JH} F$ one gets a point $[\bar{\rho} : \mathcal{H} \to gr^{JH}(F)]$, *and* a one-parameter subgroup λ such that $\lim_{t \to 0}[\rho] \cdot \lambda(t) = [\bar{\rho}]$. Thus, loosely speaking, any semistable sheaf contains its associated polystable sheaf in the closure of its orbit. Now π is a good quotient and separates closed invariant subschemes. It therefore suffices to show that the orbit of a point $[\rho : \mathcal{H} \to F]$ is closed in R if F is polystable. Suppose $[\rho' : \mathcal{H} \to F'] \in R$ is in the closure of the orbit of $[\rho]$. It suffices to show that in this case $F' \cong F$. The assumption implies that there is a smooth curve C parametrizing a flat family $\mathcal{E}$ of sheaves on X such that $\mathcal{E}_0 \cong F'$ for some closed point $0 \in C$ and $\mathcal{E}_{C \setminus \{0\}} \cong \mathcal{O}_{C \setminus \{0\}} \otimes F$. Let $F = \bigoplus_i F_i^{n_i}$ be the (unique) decomposition of F into isotypical components. Formally, we can think of F_i as running through a complete set of representatives of isomorphism classes of stable sheaves with reduced Hilbert polynomial p, where the n_i are given by $\hom(F_i, F)$. Since the family $\mathcal{E}$ is flat, the function

$$C \longrightarrow \mathbb{N}_0, \quad t \mapsto \hom(F_i, \mathcal{E}_t)$$

is semicontinuous for each i and equals n_i for all $t \neq 0$. Thus $n_i' = \hom(F_i, F') \geq n_i$. The image of the homomorphism $\psi_i : F_i \otimes_k \operatorname{Hom}(F_i, F') \to F'$ is polystable with all summands isomorphic to F_i. Moreover, ψ_i must be injective. Finally, the sum $\sum_i F_i^{n_i'} \subset F'$ must be direct. This is possible only if $n_i' = n_i$ and $F' \cong \bigoplus_i F_i^{n_i} \cong F$. We are done.$\square$

4.5 Local Properties and Dimension Estimates

In this section we want to derive some easy bounds for the dimension of the moduli spaces of stable sheaves on a projective scheme X. This is done by showing that at stable points the moduli space pro-represents the local deformation functor. From this we get a smoothness criterion and dimension bounds by applying Mori's result 2.A.11. If the scheme X is a smooth variety these results can be refined by exploiting the determinant map from the moduli space to the Picard scheme of X.

Theorem 4.5.1 — *Let F be a stable sheaf on X represented by a point $[F] \in M$. Then the completion of the local ring $\mathcal{O}_{M,[F]}$ pro-represents the deformation functor $\mathcal{D}_F$. (cf. 2.A.5).*

Proof. Clearly, there is a natural map of functors $\mathcal{D}_F \to \hat{\underline{\mathcal{O}}}_{M,[F]}$ by the openness of stability 2.3.1 and the universal property of M. To get an inverse consider the geometric quotient $\pi : R^s \to M^s$ constructed in Section 4.3. Let $[q : \mathcal{H} \to F] \in R^s$ be a point in the fibre over $[F]$. By Luna's Etale Slice Theorem there is a subscheme $S \subset R^s$ through the closed point $[q]$ such that the projection $S \to M$ is étale near $[q]$. Then $\hat{\underline{\mathcal{O}}}_{S,[q]} \cong \hat{\underline{\mathcal{O}}}_{M,[F]}$ as functors on $(Artin/k)$, and the universal family on $R^s \times X$, restricted to $S \times X$, induces a map $\hat{\mathcal{O}}_{S,[q]} \to \mathcal{D}_F$ which yields the required inverse. $\qquad \square$

As a consequence of this theorem and Proposition 2.A.11 we get

Corollary 4.5.2 — *Let F be a stable point. Then the Zariski tangent space of M at $[F]$ is canonically given by $T_{[F]}M \cong \mathrm{Ext}^1(F, F)$. If $\mathrm{Ext}^2(F, F) = 0$, then M is smooth at $[F]$. In general, there are bounds*

$$\mathrm{ext}^1(F, F) \geq \dim_{[F]} M \geq \mathrm{ext}^1(F, F) - \mathrm{ext}^2(F, F).$$

$\qquad \square$

If X is smooth, these estimates can be improved. Recall that to any flat family of sheaves on X parametrized by a scheme S we can associate the family of determinant line bundles which in turn induces a morphism $S \to \mathrm{Pic}(X)$. By the universal property of the moduli space we also obtain a morphism

$$\det : M \to \mathrm{Pic}(X),$$

which coincides with the morphism induced by a universal family on $M \times X$ in case such a family exists. Similarly, if F is a stable sheaf, there is a natural map of functors $\mathcal{D}_F \to \mathcal{D}_{\det(F)}$ from deformations of F to deformations of its determinant. We want to relate the obstruction spaces for these functors and their tangent spaces.

If E is a locally free sheaf, then the trace map $tr : \mathcal{E}nd(E) \to \mathcal{O}_X$ induces maps $tr : \mathrm{Ext}^i(E, E) \cong H^i(\mathcal{E}nd(E)) \to H^i(\mathcal{O}_X)$. We shall see later (cf. Section 10.1) how to construct natural maps $tr : \mathrm{Ext}^i(F, F) \to H^i(\mathcal{O}_X)$ for sheaves F which are not necessarily locally free. These homomorphisms are surjective if the rank of F is nonzero as an element of the base field k. Let $\mathrm{Ext}^i(F, F)_0$ denote the kernel of tr^i, and let $\mathrm{ext}^i(F, F)_0$ be its dimension.

Theorem 4.5.3 — *Let F be a stable sheaf. The tangent map of $\det : M \to \mathrm{Pic}(X)$ at $[F]$ is given by*

$$tr : T_{[F]}M \cong \mathrm{Ext}^1(F, F) \to H^1(\mathcal{O}_X) \cong T_{[\det(F)]}\mathrm{Pic}(X).$$

Moreover, if $\sigma : A' \to A$ is an extension in $(Artin/k)$ with $\mathfrak{m}_{A'} \cdot \ker(\sigma) = 0$, and if $F_A \in \mathcal{D}_F(A)$, then the homomorphism

$$tr : \mathrm{Ext}^2(F, F) \to H^2(\mathcal{O}_X) \cong \mathrm{Ext}^2(\det(F), \det(F))$$

maps the obstruction $\mathfrak{o}(F_A, \sigma)$ to extend F_A to A' onto the obstruction $\mathfrak{o}(\det(F_A), \sigma)$ to extend the determinant.

Proof. The proof of this theorem requires a description of the deformation obstruction which differs from the one we gave, and a cocycle computation. We refer to Artamkin's paper [5] and in particular to Friedman's book [69].   □

Now $\text{Pic}(X)$ naturally has the structure of an algebraic group scheme: the multiplication being given by tensorizing two line bundles. A theorem of Cartier asserts that (in characteristic zero) such a group scheme must be smooth (cf. II.6, no 1, 1.1, in [43]). In particular, all obstructions for extending the determinant of a sheaf F vanish.

Theorem 4.5.4 — *Let X be a smooth projective variety and let F be a stable $\mathcal{O}_X$-module of rank $r > 0$ and determinant bundle $\mathcal{Q}$. Let $M(\mathcal{Q})$ be the fibre of the morphism* $\det :$ $M \to \text{Pic}(X)$ *over the point* $[\mathcal{Q}]$. *Then* $T_{[F]}M(\mathcal{Q}) \cong \text{Ext}^1(F, F)_0$. *If* $\text{Ext}^2(F, F)_0 = 0$, *then M and $M(\mathcal{Q})$ are smooth at $[F]$. Moreover,*

$$\text{ext}^1(F, F)_0 \geq \dim_{[F]} M(\mathcal{Q}) \geq \text{ext}^1(F, F)_0 - \text{ext}^2(F, F)_0.$$

Proof. Tensorizing a sheaf E or rank r by a line bundle B twists the determinant bundle $\det(E)$ by B^r. Moreover, if B is numerically trivial, $E \otimes B$ is semistable or stable if and only if E is semistable or stable, respectively. It follows from this that $\det : M \to \text{Pic}(X)$ is surjective in a neighbourhood of $[\mathcal{Q}]$ and is, in fact, a fibre bundle with fibre $M(\mathcal{Q})$ in an étale neighbourhood of $[\mathcal{Q}]$. Then 4.5.3 implies that the tangent space of $M(\mathcal{Q})$ at $[F]$ is the kernel of the trace homomorphism $tr : \text{Ext}^1(F, F) \to H^1(\mathcal{O}_X)$, and moreover, that F has an obstruction theory with values in $\text{Ext}^2(F, F)_0$. Thus the vanishing of $\text{Ext}^2(F, F)_0$ implies smoothness for M and hence for $M(\mathcal{Q})$. Finally, the estimates stated in the theorem follow as above from Proposition 2.A.11.   □

Corollary 4.5.5 — *Let C be a smooth projective curve of genus $g \geq 2$. Then the moduli space of stable $\mathcal{O}_C$-sheaves of rank r with fixed determinant bundle is smooth of dimension* $(r^2 - 1)(g - 1)$.

Proof. As C is one-dimensional, $\text{ext}^2(F, F)_0 = 0$ for any coherent sheaf F. Thus the moduli space is smooth according to the theorem, and, using the Riemann-Roch formula, its dimension is given by $\text{ext}^1(F, F)_0 = -\chi(F, F) + \chi(\mathcal{O}_C) = (r^2 - 1)(g - 1)$.   □

As a matter of fact, for a smooth projective curve the moduli space of stable sheaves is irreducible and dense in the moduli space of semistable sheaves [236].

If X is a smooth surface we can make the dimension bound more explicit: Note that for a stable sheaf F

$$\text{ext}^1(F, F)_0 - \text{ext}^2(F, F)_0 = \chi(\mathcal{O}_X) - \sum_{i=0}^{2}(-1)^i\text{ext}^i(F, F),$$

which by the Hirzebruch-Riemann-Roch formula is equal to

$$\Delta(F) - (r^2 - 1) \cdot \chi(\mathcal{O}_X).$$

(Recall that $\Delta(F) = 2rc_2(F) - (r-1)c_1(F)^2$.)

Definition 4.5.6 — *The number*

$$\exp \dim(M(\mathcal{Q})) := \Delta(F) - (r^2 - 1)\chi(\mathcal{O}_X)$$

is called the expected dimension of $M(\mathcal{Q})$.

Lemma 4.5.7 — *Let X be a smooth polarized projective surface and let r be a positive integer. There is a constant β_∞ depending only on X and r such that for any semistable sheaf F of rank $r > 0$ on X one has*

$$\mathrm{ext}^2(F, F)_0 \le \beta_\infty.$$

Proof. By Serre Duality, $\mathrm{ext}^2(F, F)_0 = \hom(F, F \otimes \omega_X) - h^0(\omega_X)$. Applying Proposition 3.3.6 we get $\hom(F, F \otimes \omega_X) \le \frac{r^2}{2 \deg(X)} \left[\mu(\omega_X) + (r + \frac{1}{2}) \cdot \deg(X) \right]_+^2$. Obviously, the right hand side depends only on X and r. $\square$

Thus we can state

Theorem 4.5.8 — *Let X be a smooth polarized projective surface and let F be a stable sheaf of rank $r > 0$ and determinant $\mathcal{Q}$. Then*

$$\exp \dim(M(\mathcal{Q})) \le \dim_{[F]} M(\mathcal{Q}) \le \exp \dim(M(\mathcal{Q})) + \beta_\infty.$$

If $\exp \dim(M(\mathcal{Q})) = \dim_{[F]} M(\mathcal{Q})$ then $M(\mathcal{Q})$ is a local complete intersection at $[F]$ (cf. 2.A.12). $\square$

What can be said about points in $M \setminus M^s$? In this case the picture is blurred because of the existence of non-scalar automorphisms. At this stage we only prove a lower bound for the dimension of R, that will be needed later in Chapter 9. The setting is that of Section 4.3 for the case of a smooth projective surface: m is a sufficiently large integer, V a vector space of dimension $P(m)$ and $\mathcal{H} := V \otimes_k \mathcal{O}_X(-m)$. Let $R \subset \mathrm{Quot}(\mathcal{H}, P)$ be the open subscheme of those quotients $[\rho : \mathcal{H} \to F]$ where F is semistable and $V \to H^0(F(m))$ is an isomorphism. Let $[\rho : \mathcal{H} \to F] \in R$ be fixed. As F is m-regular, it follows from the properties of $[\rho]$, that $\mathrm{End}(\mathcal{H}) \cong \mathrm{Hom}(\mathcal{H}, F)$ and $\mathrm{Ext}^i(\mathcal{H}, F) = 0$ for all $i > 0$. Let K be the kernel of ρ. Then there is an exact sequence

$$0 \longrightarrow \mathrm{End}(F) \longrightarrow \mathrm{Hom}(\mathcal{H}, F) \overset{T}{\longrightarrow} \mathrm{Hom}(K, F) \longrightarrow \mathrm{Ext}^1(F, F) \longrightarrow 0$$

and isomorphisms $\text{Ext}^i(K, F) \cong \text{Ext}^{i+1}(F, F)$ for $i > 0$. Recall that the boundary map $\text{Ext}^1(K, F) \to \text{Ext}^2(F, F)$ maps the obstruction to extend $[\rho]$ onto the obstruction to extend $[F]$ (cf. 2.A.8), and that the latter is contained in the subspace $\text{Ext}^2(F, F)_0$. This leads to the dimension bound

$$
\begin{aligned}
\dim_{[\rho]} R \;\; &\geq \;\; \hom(K, F) - \text{ext}^2(F, F)_0 \\
&= \;\; \hom(\mathcal{H}, F) + \text{ext}^1(F, F) - \text{ext}^0(F, F) - \text{ext}^2(F, F)_0 \\
&= \;\; \text{end}(\mathcal{H}) - 1 + h^1(\mathcal{O}_X) + \exp \dim(M(\mathcal{Q})),
\end{aligned}
$$

where $\mathcal{Q} = \det(F)$ as before. Consider the map $\det : R \to \text{Pic}(X)$ induced by the universal quotient on $R \times X$. This map is surjective onto a neighbourhood of $[\mathcal{Q}]$, and since $\dim(\text{Pic}(X)) = h^1(\mathcal{O}_X)$, we finally get the following dimension bound for the fibre $R(\mathcal{Q}) = \det^{-1}([\mathcal{Q}])$:

Proposition 4.5.9 — $\dim_{[\rho]} R(\mathcal{Q}) \geq \exp \dim(M(\mathcal{Q})) + \text{end}(\mathcal{H}) - 1.$ $\qquad\square$

Example 4.5.10 — Let X be a smooth projective surface, and consider the Hilbert scheme $\text{Hilb}^\ell(X) = \text{Quot}(\mathcal{O}_X, \ell)$ of zero-dimensional subschemes in X of length $\ell \geq 0$. It is easy to see that $\text{Hilb}^1(X) = X$ and that $\text{Hilb}^2(X)$ is the quotient of the blow-up of $X \times X$ along the diagonal by the action of $\mathbb{Z}/2$ that flips the two components. In fact, $\text{Hilb}^\ell(X)$ is a smooth projective variety of dimension 2ℓ for all $\ell \geq 1$. We give two arguments: first, let $\mathcal{I}$ denote the ideal sheaf of the universal family in $\text{Hilb}^\ell(X) \times X$. Let $(Z, x) \in \text{Hilb}^\ell(X) \times X$ be an arbitrary point. For any surjection $\lambda : \mathcal{I}_Z(x) \to k(x)$ we can consider the kernel $\mathcal{I}'$ of $\mathcal{I}_Z \to \mathcal{I}_Z(x) \to k(x)$ and the associated point $Z' \in \text{Hilb}^{\ell+1}(X)$. This construction yields a surjective morphism $\mathbb{P}(\mathcal{I}) \to \text{Hilb}^{\ell+1}(X)$. Note that the fibres of $\mathbb{P}(\mathcal{I}) \to \text{Hilb}^\ell(X) \times X$ are projective spaces and hence connected. In particular, by induction we see that $\text{Hilb}^\ell(X)$ is connected.

Now $\text{Hilb}^\ell(X)$ always contains the following 2ℓ-dimensional smooth variety as an open subset: let U be the quotient of the open subset $\{(x_1, \dots, x_\ell) | x_i \neq x_j \forall i \neq j\}$ in X^ℓ by the permutation action of the symmetric group. Thus, if we can show that the dimension of the Zariski tangent space at every point in $\text{Hilb}^\ell(X)$ is 2ℓ we are done: for then the closure of U is smooth and cannot meet any other irreducible component of $\text{Hilb}^\ell(X)$, hence is all of $\text{Hilb}^\ell(X)$ as the latter is connected. Let $Z \subset X$ be a closed point in $\text{Hilb}^\ell(X)$. Recall that $T_Z \text{Hilb}^\ell(X) \cong \text{Hom}(\mathcal{I}_Z, \mathcal{O}_Z)$. Moreover,

$$
\begin{aligned}
\hom(\mathcal{I}_Z, \mathcal{O}_Z) &= \text{ext}^1(\mathcal{O}_Z, \mathcal{O}_Z) + \hom(\mathcal{O}_X, \mathcal{O}_Z) - \hom(\mathcal{O}_Z, \mathcal{O}_Z) \\
&= \hom(\mathcal{O}_X, \mathcal{O}_Z) - \chi(\mathcal{O}_Z, \mathcal{O}_Z) + \text{ext}^2(\mathcal{O}_Z, \mathcal{O}_Z) \\
&= \hom(\mathcal{O}_X, \mathcal{O}_Z) + \hom(\mathcal{O}_Z, \mathcal{O}_Z) \\
&= 2 \cdot \text{length}(\mathcal{O}_Z) = 2\ell.
\end{aligned}
$$

Using Theorem 4.5.4 we can give a shorter proof: observe that we can identify $\text{Hilb}^\ell(X) \cong M(1, \mathcal{O}_X, \ell)$ by sending a subscheme $Z \subset X$ to its ideal sheaf $\mathcal{I}_Z$. In order to conclude smoothness it suffices to check that $\text{Ext}^2(\mathcal{I}_Z, \mathcal{I}_Z)_0 = 0$. But

$$
\text{ext}^2(\mathcal{I}_Z, \mathcal{I}_Z)_0 = \hom(\mathcal{I}_Z, \mathcal{I}_Z \otimes K_X)_0 = \hom(\mathcal{O}_X, K_X)_0 = 0.
$$

See also 6.A.1 for a generalization of this example. □

4.6 Universal Families

We now turn to the question under which hypotheses the functor $\mathcal{M}^s$ is represented by the moduli space M^s. If this is the case M^s is sometimes called a *fine* moduli space.

Let X be a polarized projective scheme. Recall our convention that whenever we speak about a family of sheaves on X parametrized by a scheme S, p and q denote the projections $S \times X \to S$ and $S \times X \to X$, respectively.

Definition 4.6.1 — *A flat family $\mathcal{E}$ of stable sheaves on X parametrized by M^s is called universal, if the following holds: if F is an S-flat family of stable sheaves on X with Hilbert polynomial P and if $\Phi_F : S \to M^s$ is the induced morphism, then there is a line bundle L on S such that $F \otimes p^* L \cong \Phi_F^* \mathcal{E}$. An M^s-flat family $\mathcal{E}$ is called quasi-universal, if there is a locally free $\mathcal{O}_S$-module W such that $F \otimes p^* W \cong \Phi_F^* \mathcal{E}$.*

Clearly, M^s represents the moduli functor $\mathcal{M}^s$ if and only if a universal family exists. Though this will in general not be the case, quasi-universal families always exist. Recall that the centre Z of $\text{GL}(V)$ acts trivially on R. Therefore the fibre over any point $[\rho] \in R$ or $([\rho], x) \in R \times X$ of any $\text{GL}(V)$-linearized sheaf on R or $R \times X$, respectively, such as the universal quotient $\widetilde{F}$ on $R \times X$, has the structure of a Z-representation and decomposes into weight spaces. We say that a sheaf or a particular fibre of a sheaf has Z-weight ν, if $t \in Z \cong \mathbb{G}_m$ acts via multiplication by t^ν.

Proposition 4.6.2 — *There exist $\text{GL}(V)$-linearized vector bundles on R^s with Z-weight 1. If A is any such vector bundle then $\mathcal{H}om(p^* A, \widetilde{F})$ descends to a quasi-universal family $\mathcal{E}$, and any quasi-universal family arises in this way. If A is a line bundle then $\mathcal{E}$ is universal.*

In the proof of the proposition we will need the following observation, which is the relative version of 1.2.8.

Lemma 4.6.3 — *Let F be a flat family of stable sheaves on a projective scheme X, parametrized by a scheme S. Then the natural homomorphism $\mathcal{O}_S \to p_*\mathcal{E}nd(F)$ is an isomorphism.*

Proof. The homomorphism $\mathcal{O}_S \to p_*\mathcal{E}nd(F)$ is given by scalar multiplication of $\mathcal{O}_S$ on F. The assumption that F is S-flat implies that the homomorphism is injective. Now, for each k-rational point $s \in S$ the fibre F_s is stable and therefore simple, i.e. $\mathrm{End}(F_s) \cong k(s)$, so that the composite homomorphism $k(s) \to p_*\mathcal{E}nd(F)(s) \to \mathrm{End}(F_s)$ is surjective. Hence $p_*\mathcal{E}nd(F)(s) \to \mathrm{End}(F_s)$ is surjective and therefore even isomorphic by the semicontinuity theorems for the functors $\mathrm{Ext}_p^\bullet$. This means that $\mathcal{O}_S \to p_*\mathcal{E}nd(F)$ is surjective as well. $\square$

Proof of the proposition. If n is sufficiently large then $A_n = p_*(\widetilde{F} \otimes q^*\mathcal{O}_X(n))$ is a locally free sheaf on R^s of rank $P(n)$ and carries a natural $\mathrm{GL}(V)$-linearization of Z-weight 1. Let A be any $\mathrm{GL}(V)$-linearized vector bundle on R^s with Z-weight 1. Then Z acts trivially on the bundle $\mathcal{H}om(p^*A, \widetilde{F})$, which therefore carries a $\mathrm{PGL}(V)$-linearization and descends to a family $\mathcal{E}$ on $M^s \times X$ by 4.2.14. We claim that $\mathcal{E}$ is quasi-universal. Suppose that F is a family of stable sheaves on X parametrized by a scheme S with Hilbert polynomial P. Let $R(F)$ be the associated frame bundle and consider the commutative diagram

$$
\begin{array}{ccc}
R(F) & \xrightarrow{\tilde{\Phi}_F} & R^s \\
\pi \downarrow & & \downarrow \pi \\
S & \xrightarrow{\Phi_F} & M^s.
\end{array}
$$

Then $\pi_X^* F \cong \tilde{\Phi}_{F,X}^* \widetilde{F}$ and we thus have equivariant isomorphisms

$$
\begin{aligned}
\pi_X^* \Phi_{F,X}^* \mathcal{E} & \cong \tilde{\Phi}_{F,X}^* \pi_X^* \mathcal{E} \cong \tilde{\Phi}_{F,X}^* \mathcal{H}om(p^*A, \widetilde{F}) \\
& \cong \mathcal{H}om(\tilde{\Phi}_{F,X}^* p^*A, \tilde{\Phi}_{F,X}^* \widetilde{F}) \\
& \cong \mathcal{H}om(p^* \tilde{\Phi}_F^* A, \pi_X^* F).
\end{aligned}
$$

Now $\tilde{\Phi}_F^* A$ is linearized in a natural way and $\pi : R(F) \to S$ is a $\mathrm{GL}(V)$-principal bundle so that $\tilde{\Phi}_F^* A \cong \pi^* B$ for some vector bundle B on S. It follows that

$$
\pi_X^* \Phi_{F,X}^* \mathcal{E} \cong \mathcal{H}om(\pi_X^* p^* B, \pi_X^* F) \cong \pi_X^* \mathcal{H}om(p^* B, F),
$$

which is equivariant and therefore (cf. 4.2.14) descends to an isomorphism

$$
\Phi_{F,X}^* \mathcal{E} \cong \mathcal{H}om(p^* B, F) \cong p^* B^\vee \otimes F.
$$

Thus $\mathcal{E}$ is indeed a quasi-universal family. Conversely, let $\mathcal{E}$ be a quasi-universal family on $M^s \times X$. Then applying the universal property of $\mathcal{E}$ to the family $\widetilde{F}$ on $R^s \times X$ we find a vector bundle A on R^s such that $\pi_X^* \mathcal{E} \cong \mathcal{H}om(p^* A, \widetilde{F})$. Then $p_*(\mathcal{H}om(\pi_X^* \mathcal{E}, \widetilde{F})) = p_* \mathcal{H}om(p^* A^{\vee} \otimes \widetilde{F}, \widetilde{F}) = A \otimes_{\mathcal{O}_{R^s}} p_* \mathcal{E}nd(\widetilde{F}) \cong A$ by the previous lemma. This description shows that A carries a $GL(V)$-linearization of Z-weight 1 which is compatible with the isomorphism $\pi_X^* \mathcal{E} \cong \mathcal{H}om(p^* A, \widetilde{F})$. It is clear that $\mathcal{E}$ is universal if and only if A is a line bundle. $\qquad\square$

Exercise 4.6.4 — Show by the same method: if $\mathcal{E}'$ and $\mathcal{E}''$ are two quasi-universal families, then there are locally free sheaves W' and W'' on M^s such that $\mathcal{E}' \otimes p^* W'' \cong \mathcal{E}'' \otimes p^* W'$.

For the remaining part of this section let X be a smooth projective variety. Let c be a fixed class in $K_{\mathrm{num}}(X)$, let P be the associated Hilbert polynomial, and let $M(c)^s \subset M^s$ and $R(c)^s \subset R^s$ be the open and closed parts that parametrize stable sheaves of numerical class c (see also Section 8.1).

Suppose B is a locally free sheaf on X. Then the line bundle

$$\lambda(B) := \det p_!(\widetilde{F} \otimes q^* B) \in \mathrm{Pic}(R(c))$$

as defined in Section 2.1 carries a natural linearization of weight $\chi(c \otimes B)$. If B is not locally free, we can still choose a finite locally free resolution $B_{\bullet} \to B$ and define $\lambda(B) := \bigotimes_i \lambda(B_i)^{(-1)^i}$. Then $\lambda(B)$ has weight $\sum_i (-1)^i \chi(c \otimes B_i) =: \chi(c \cdot B)$.

Theorem 4.6.5 — *If the greatest common divisor of all numbers $\chi(c \cdot B)$, where B runs through some family of coherent sheaves on X, equals 1, then there is a universal family on $M(c)^s \times X$.*

Proof. Suppose there are sheaves $B_1, \ldots, B_{\ell}$ and integers $w_1, \ldots, w_{\ell}$ such that $1 = \sum_i w_i \chi(c \cdot B_i)$, then $A := \bigotimes_i \lambda(B_i)^{w_i}$ is a line bundle of Z-weight 1. Hence the theorem follows from the proposition. $\qquad\square$

Recall that the Hilbert polynomial P can be written in the form

$$P(n) := \sum_{i=0}^{d} a_i \binom{n+i-1}{i}$$

with integral coefficients $a_0, \ldots, a_d$, where $d = \dim(F)$.

Corollary 4.6.6 — *If* $\mathrm{g.c.d.}(a_0, \ldots, a_d) = 1$ *then there is a universal family on $M^s \times X$.*

Proof. Apply the previous theorem to the sheaves $\mathcal{O}_X(0), \ldots, \mathcal{O}_X(d)$. It suffices to check that the g.c.d.$(a_0, \ldots, a_d)$ = g.c.d.$(P(0), \ldots, P(d))$. But this follows from the observation that the matrix

$$\left(\binom{\nu + i - 1}{i} \right)_{i, \nu = 0, \ldots, d}$$

is invertible over the integers. $\qquad\square$

Corollary 4.6.7 — *Let X be a smooth surface. Let r, c_1, c_2 be the rank and the Chern classes corresponding to c. If g.c.d.$(r, c_1.H, \frac{1}{2}c_1.(c_1 - K_X) - c_2) = 1$, then there is a universal family on $M(c)^s \times X$.*

Proof. Apply the previous theorem to the sheaves $\mathcal{O}_X$, $\mathcal{O}_X(1)$, and the structure sheaf $\mathcal{O}_P$ of a point $P \in X$. The assertion then follows by expressing $P(0)$ and $P(1)$ in terms of Chern classes and using that $\chi(c \otimes \mathcal{O}_P) = r$. $\qquad\square$

Remark 4.6.8 — The condition of Corollary 4.6.7 is also sufficient to ensure that there are no properly semistable sheaves, in other words that $M(c)^s = M(c)$. Namely, suppose that F is a semistable sheaf of class c admitting a destabilizing subsheaf F' of rank $r' < r$ and Chern classes c_1' and c_2'. Then we have the relations:

$$r \cdot (c_1'.H) = r' \cdot (c_1.H)$$

and

$$r \cdot (c_1'(c_1' - K_X) - 2c_2')/2 = r' \cdot (c_1(c_1 - K_X) - 2c_2)/2.$$

If α, β and γ are integers with $\alpha \cdot r + \beta \cdot (c_1.H) + \gamma \cdot (c_1(c_1 - K_X) - 2c_2)/2 = 1$, then $r \cdot (\alpha \cdot r' + \beta \cdot (c_1'.H) + \gamma \cdot (c_1'(c_1' - K_X) - 2c_2')/2) = r'$, obviously a contradiction.

Appendix to Chapter 4

4.A Gieseker's Construction

The first construction of a moduli space for semistable torsion free sheaves on a smooth projective surface was given by Gieseker [77]. We briefly sketch his approach, at least where it differs from the construction discussed before, for the single reason that it gives a bit more: namely, the ampleness of a certain line bundle on the moduli space relative to the Picard variety of X.

Let $(X, \mathcal{O}_X(1))$ be a polarized smooth projective variety over an algebraically closed field of characteristic zero. Let P be a polynomial of degree equal to the dimension of X (i.e. we consider torsion free sheaves only) and let r be the rank determined by P. Recall the notations of Section 4.3: m is a sufficiently large integer, V a vector space of dimension $P(m)$, and $\mathcal{H} := V \otimes \mathcal{O}_X(-m)$. Let $R \subset \mathrm{Quot}(\mathcal{H}, P)$ be the subscheme of all quotients $[\rho : \mathcal{H} \to F]$ such that F is semistable torsion free and $V \to H^0(F(m))$ is an isomorphism. Let $\overline{R}$ be the closure of R in $\mathrm{Quot}(\mathcal{H}, P)$. The universal quotient $\tilde{\rho} : \mathcal{H} \otimes \mathcal{O}_{\overline{R}} \to \widetilde{F}$ induces an invariant morphism

$$\det : \overline{R} \to \mathrm{Pic}(X)$$

such that $\det(\widetilde{F}) = \det_X^*(\mathcal{P}) \otimes p^*\mathcal{A}$, where $\mathcal{P}$ denotes the Poincaré line bundle on $\mathrm{Pic}(X) \times X$ and $\mathcal{A}$ is some line bundle on $\overline{R}$. We may assume that m was chosen large enough so that any line bundle represented by a point in the image $\det(\overline{R}) \subset \mathrm{Pic}(X)$ is m-regular. From $\tilde{\rho} : \mathcal{H} \otimes \mathcal{O}_{\overline{R}} \to \widetilde{F}$ we get homomorphisms $\Lambda^r V \otimes \mathcal{O}_{\overline{R} \times X} \longrightarrow \det(\widetilde{F} \otimes q^* \mathcal{O}_X(m))$ and

$$\Lambda^r V \otimes \mathcal{O}_{\overline{R}} \longrightarrow p_* \det(\widetilde{F}(m)) = \det^* p_*(\mathcal{P}(rm)) \otimes \mathcal{A}$$

which is adjoint to

$$\tilde{\zeta} : \det^*(\mathcal{H}om(\Lambda^r V, p_* \mathcal{P}(rm))^\vee) \longrightarrow \mathcal{A}.$$

Note that $\tilde{\zeta}$ is everywhere surjective and therefore defines a morphism

$$\zeta : \overline{R} \longrightarrow Z := \mathbb{P}(\mathcal{H}om(\Lambda^r V, p_* \mathcal{P}(rm))^\vee)$$

of schemes over $\mathrm{Pic}(X)$, such that $\zeta^* \mathcal{O}_Z(1) \cong \mathcal{A}$. Moreover, ζ is clearly equivariant for the obvious action of $\mathrm{SL}(V)$ on Z. Observe, that if F is torsion free then $\rho : V \otimes \mathcal{O}_X(-m) \to F$ is, as a quotient, completely determined by the homomorphism $\Lambda^r V \to H^0(\det(F(m)))$. This means that $\zeta|_R$ is injective, hence $\mathcal{A}|_R$ is ample relative to $\mathrm{Pic}(X)$.

Theorem 4.A.1 — ζ *maps R to the subscheme Z^{ss} of semistable points in Z with respect to the $\mathrm{SL}(V)$-action and the linearization of $\mathcal{O}_Z(1)$, and $R^s = \zeta^{-1}(Z^s)$. Moreover, as*

$\zeta : R \rightarrow Z^{ss}$ *is finite, good quotients of R and R^s exist, and some tensor power of $\mathcal{A}$ descends to a line bundle on the moduli space M which is very ample relative to* $\mathrm{Pic}(X)$.

We sketch a proof of the first statements. The last assertion then follows from these and general principles of GIT quotients.

Let $[\rho : \mathcal{H} \rightarrow F]$ be a point in R and let $\lambda : \mathbb{G}_m \rightarrow \mathrm{SL}(V)$ be a one-parameter group given by a weight decomposition $V = \oplus_n V_n$. As in the proof of 4.3.3, let $V_{\leq n} = \oplus_{\nu \leq n} V_\nu$ be the induced filtration on V, but define $F_{\leq n}$ as the saturation of $\rho(V_{\leq n} \otimes \mathcal{O}_X(-m))$ in F, and let r_n be the rank of $F_n = F_{\leq n}/F_{\leq n-1}$. Then $\det(F) \cong \otimes_n \det(F_n)$ and

$$\lim_{t \to 0} \tilde{\zeta}([\rho]\lambda(t)) = \left[\Lambda^r V \rightarrow \bigotimes_n \Lambda^{r_n} V_n \rightarrow H^0 \left(\bigotimes_n \det(F_n(m)) \right) \right].$$

Hence the weight of the action at the limit point is, up to some constant, given by

$$\dim(V) \cdot \sum_n n r_n = \sum_n n \cdot (r_n \cdot \dim(V) - r \cdot \dim(V_n)) = - \sum_n (r_{\leq n} \cdot \dim(V) - r \cdot \dim(V_{\leq n})).$$

The same reduction as in the proof of 4.3.3 shows that $[\rho]$ is (semi)stable, if and only if the following holds: If V' is any non-trivial proper subspace of V and if r' is the rank of the subsheaf in F generated by V', then

$$\dim(V') \cdot r \; (\leq) \; \dim(V) \cdot r'.$$

At this point we can re-enter the first half of the proof of 4.3.3 and conclude literally in the same way. $\qquad\square$

4.B Decorated Sheaves

So far we have encountered two different types of moduli spaces: the Grothendieck Quot-scheme and the moduli space of semistable sheaves. The Grothendieck Quot-scheme parametrizes all quotients of a sheaf, i.e. sheaves together with a surjection from a fixed one. In this spirit, one could, more generally, consider sheaves endowed with an additional structure such as a homomorphism to or from a fixed sheaf, a filtration or simply a global section. For many types of such 'decorated' sheaves one can set up a natural stability condition and then formulate the appropriate moduli problem. (Recall, there is no stability condition quotients parametrized by the Quot-scheme have to satisfy.) We are going to describe a moduli problem that is general enough to comprise various interesting examples. To a large extent the theory is modelled on things we have been explaining in the last sections. In particular, the boundedness and the actual construction of the moduli space, though involving some extra technical difficulties, are dealt with quite similarly. However, two things in the theory of decorated sheaves are different. First, the stability condition is usually slightly more complicated and depends on extra parameters, which can be varied. Second, by adding the additional structure we make the automorphism group of the objects in question smaller. This can be used to construct fine moduli spaces in many instances.

Let X be a smooth projective variety over an algebraically closed field k of characteristic zero. Fix an ample invertible sheaf $\mathcal{O}_X(1)$ and a non-trivial coherent sheaf E. Furthermore, let $\delta \in \mathbb{Q}[t]$ be a positive polynomial. A *framed module* is a pair (F, α) consisting of a coherent sheaf F and a homomorphism $\alpha : F \to E$. Its Hilbert polynomial is by definition $P(F, \alpha) := P(F) - \varepsilon(\alpha) \cdot \delta$, where $\varepsilon(\alpha) = 1$ if $\alpha \neq 0$ and $\varepsilon(\alpha) = 0$ otherwise. For simplicity we give the stability condition only for framed modules of dimension $\dim(X)$:

Definition 4.B.1 — *A framed module (F, α) of rank r is (semi)stable if for all framed submodules $(F', \alpha') \subset (F, \alpha)$, i.e. $F' \subset F$ and $\alpha' = \alpha|_{F'}$, one has $r \cdot P(F', \alpha')(\leq) \mathrm{rk}(F') \cdot P(F, \alpha)$.*

Remark 4.B.2 — *i)* If $\alpha = 0$, then this stability condition coincides with the stability condition for sheaves. If $\alpha \neq 0$, then the stability condition splits into the following two conditions: for subsheaves $F' \subset \ker(\alpha)$ one requires $rP(F') (\leq) \mathrm{rk}(F')P(F') - \mathrm{rk}(F')\delta$ and for arbitrary $F' \subset F$ only the weaker inequality $rP(F') (\leq) \mathrm{rk}(F')P(F') + (r - \mathrm{rk}(F'))\delta$.

ii) If (F, α) is a semistable framed module then α embeds the torsion of F into E.

iii) If (F, α) is semistable and $\alpha \neq 0$, then $\deg(\delta) \leq \dim(X)$. Moreover, if $\deg(\delta) = \dim(X)$ and (F, α) is semistable then α is injective. Thus, for $\deg(\delta) = \dim(X)$ all framed modules are just subsheaves of E. Since this case is covered by the Grothendieck Quot-scheme, we will henceforth assume $\deg(\delta) < \dim(X)$.

iv) By definition the stability of a framed module (F, α) with $\alpha \neq 0$ depends on the polynomial δ, but for 'generic' δ the stability condition is invariant under small changes of δ. Only when δ crosses certain critical values the stability condition actually changes. Moreover, for generic δ semistability and stability coincide.

v) For δ small and generic, e.g. δ is a positive constant close to zero, the underlying sheaf F of a semistable framed module (F, α) is semistable. Conversely, if F is a stable sheaf and $\alpha : F \to E$ is non-trivial, then (F, α) is stable with respect to small δ.

Example 4.B.3 — Let E be a sheaf supported on a divisor $D \subset X$. Then a sheaf F on X together with an isomorphism $F|_D \cong E$ (a 'framing') gives rise to a framed module (F, α) in our sense with $\alpha : F \to F|_D \cong E$. Here, E is considered as a sheaf on X with support on D. In the case of a curve X and a point $D = \{x\}$ these objects are also called bundles with a level structure. Next, let E be the trivial invertible sheaf $\mathcal{O}_X$. In this case, the underlying sheaf of a semistable framed module must be torsion free (this is true whenever E is torsion free). Thus, on a curve X semistable framed modules $(F, \alpha : F \to \mathcal{O}_X)$ are locally free and, therefore, there is no harm in dualizing, i.e. instead of considering (F, α) we could consider $(F^\vee, \varphi := \alpha^\vee \in H^0(F^\vee))$. This gives an equivalence between semistable framed modules and semistable pairs, i.e. bundles with a global section. This correspondence holds also true in higher dimensions if we restrict to locally free sheaves. There are of course interesting types of decorations that are not covered by framed modules. Most important, parabolic sheaves and Higgs sheaves.

Let us now introduce the corresponding moduli functor. As before, we fix a positive polynomial δ of degree less than $\dim(X)$, a coherent sheaf E, an ample invertible sheaf $\mathcal{O}_X(1)$ and a polynomial $P \in \mathbb{Q}[z]$ of degree $\dim(X)$. Then the moduli functor

$$\mathcal{M} : (Sch/k)^\circ \to (Sets)$$

maps $S \in \mathrm{Ob}(Sch/k)$ to the set of isomorphism classes of S-flat families $(F, \alpha : F \to E_S)$ of semistable framed modules with $P(F_t) = P$ and $P(F_t, \alpha_t) = P - \delta$ for any closed point $t \in S$. By $\mathcal{M}^s$ we denote the open subfunctor of geometrically stable framed modules.

Theorem 4.B.4 — *There exists a projective scheme $M_{\mathcal{O}_X(1)}(P, E, \delta)$ that universally corepresents the functor $\mathcal{M}$. Moreover, there is an open subscheme M^s of $M_{\mathcal{O}_X(1)}(P, E, \delta)$ that universally represents $\mathcal{M}^s$.* $\square$

Analogously to the case of semistable sheaves, the closed points of $M_{\mathcal{O}_X(1)}(P, E, \delta)$ parametrize S-equivalence classes of framed modules. We leave it as an exercise to find the right definition of S-equivalence in this context. Note that the second statement is stronger than the corresponding one in Theorem 4.3.4. It says that on the moduli space of stable framed modules (F, α) with $\alpha \neq 0$ there exists a universal family. This will be essentially used in the proof of the following proposition.

Proposition 4.B.5 — *Let $M = M_{\mathcal{O}_X(1)}(P)$ be the moduli space of semistable sheaves with Hilbert polynomial P and let M^s be the open subscheme of stable sheaves. Then there exists a projective scheme $\tilde{M}$, a morphism $\psi : \tilde{M} \to M$ and an $\tilde{M}$-flat family $\mathcal{E}$ such that:*
i) ψ is birational over M^s,
ii) on the open set over M^s where ψ is an isomorphism the family $\mathcal{E}$ is quasi-universal,
iii) if F is the sheaf corresponding to $\psi(t)$ for a closed point $t \in \tilde{M}$, then $\mathcal{E}_t$ is S-equivalent to $F^{\oplus b}$, where $b = \mathrm{rk}(\mathcal{E})/\mathrm{rk}(F)$.

Of course, if $M = M^s$ this is just the existence of a quasi-universal family (cf. 4.6.2). In general, a quasi-universal family cannot be extended to a family on the projective scheme M. The projective variety $\tilde{M}$ together with $\mathcal{E}$ is a replacement for this. As it turns out, for many purposes this is enough. Note, if M^s is reduced, by desingularizing $\tilde{M}$ and pulling-back $\mathcal{E}$ one can assume that $\tilde{M}$ is in fact smooth.

Proof. Let $\mathcal{E}$ be a quasi-universal family on $M^s \times X$ with Hilbert polynomial $b \cdot P$. (For the existence of $\mathcal{E}$ see Proposition 4.6.2.) Let $\mathcal{M}(P)$ denote the moduli functor of semistable sheaves with Hilbert polynomial P. We define a natural functor transformation $\mathcal{M}(P) \to \mathcal{M}(bP)$ by $[F] \mapsto [F^{\oplus b}]$. If $b > 1$, the image is contained in $\mathcal{M}(bP) \setminus \mathcal{M}^s(bP)$. The induced morphism $M = M(P) \to M(bP)$ is a closed immersion (see Lemma 4.B.6 below). Next, consider the moduli space $M(bP, \mathcal{O}(n), \delta)$ of framed modules $(F, F \to \mathcal{O}(n))$. For generic δ there exists a universal framed module $(\mathcal{F}, \mathcal{F} \to q^*\mathcal{O}(n))$ on the product $M(bP, \mathcal{O}(n), \delta) \times X$ and for small generic δ the map $[(F, F \to \mathcal{O}(n))] \mapsto [F]$ defines a morphism $M(bP, \mathcal{O}(n), \delta) \to M(bP)$. Let $N := M \times_{M(bP)} M(bP, \mathcal{O}(n), \delta)$. It suffices to construct a section of $N \to M$ over a dense open subset of M^s. Indeed, the closure $\tilde{M}$ of this section in N together with the pull-back of $\mathcal{F}$ under $\tilde{M} \subset N \to M(bP, \mathcal{O}(n), \delta)$ satisfies *i)*, *ii)*, and *iii)*. The construction of the section over a dense open subset of M^s goes as follows. For $n \gg 0$ and any sheaf $[F] \in M^s$ there exists a non-trivial homomorphism $F^{\oplus b} \to \mathcal{O}(n)$. Moreover, the generic homomorphism gives rise to a stable framed module, i.e. a point in $M(bP, \mathcal{O}(n), \delta)$ and hence in N. The sheaf $p_* \mathcal{H}om(\mathcal{E}, q^*\mathcal{O}(n))$ is free over a dense open subscheme $U \subset M^s$. A generic non-vanishing section of this free sheaf induces a section of $N \to M$ over U (we might have to shrink U slightly in order to make all framed modules semistable). $\qquad\square$

Lemma 4.B.6 — *The canonical morphism $j_b : M(P) \to M(bP)$ is a closed immersion.*

Proof. As points in $M(P)$ are in bijection with polystable sheaves $F = \oplus_i F_i \otimes W_i$, where F_i are pairwise non-isomorphic stable sheaves and $W_i = \mathrm{Hom}(F_i, F)$, and since the morphism j_b is given by $F \mapsto \oplus_i F_i \otimes (W_i \otimes k^b)$, it is clearly injective.

Let m be a sufficiently large integer, and let $R \subset \mathrm{Quot}(\mathcal{O}(-m)^{P(m)}, P)$ and $R' \subset \mathrm{Quot}(\mathcal{O}(-m)^{bP(m)}, bP)$ be the open subsets as defined in Section 4.3. Then j_b is covered by a natural morphism $\hat{\jmath} : R \to R'$, $[\rho] \mapsto [\rho^{\oplus b}]$. Let F be a polystable sheaf as above and $[\rho : \mathcal{O}(-m)^{P(m)} \to F]$ a point in the fibre of $\pi : R \to M(P)$ over $[F]$. The stabilizer subgroups of $\mathrm{GL}(P(m))$ and $\mathrm{GL}(bP(m))$ at the points $[\rho]$ and $\hat{\jmath}[\rho]$ are given by

$$G = \prod_i \mathrm{GL}(W_i) \quad \text{and} \quad G' = \prod_i \mathrm{GL}(W_i \otimes k^b),$$

respectively. The normal directions to the orbits of $[\rho]$ in R and $[\rho^{\oplus b}]$ in R' at these points are

$$E = \bigoplus_{i,j} \mathrm{Ext}^1(F_i, F_j) \otimes \mathrm{Hom}(W_i, W_j)$$

and

$$E' = \bigoplus_{i,j} \mathrm{Ext}^1(F_i, F_j) \otimes (\mathrm{Hom}(W_i, W_j) \otimes \mathrm{End}(k^{\oplus b})),$$

on which G and G' act by conjugation. By Luna's Etale Slice Theorem, an étale neighbourhood of $[F]$ in $M(P)$ embeds into $E /\!\!/ G$. Therefore it suffices to show that $E /\!\!/ G$ embeds into $E' /\!\!/ G'$, or equivalently, that the diagonal embedding $\delta = \mathrm{id}_E \otimes 1 : E \to E'$ induces a surjective homomorphism $\mathcal{O}_{E'}^{G'} \to \mathcal{O}_E^G$. In fact, one can check that the partial trace map $E' = E \otimes \mathrm{End}(k^{\oplus b}) \to E$ induces a splitting. $\square$

The proposition above is one application of moduli spaces of framed modules. They also provide a framework for the comparison of different moduli spaces, e.g. the moduli space of rank two sheaves on a surface and the Hilbert scheme. For simplicity we have avoided the extensive use of framed modules in these notes, but some of the results in Chapters 5, 6, and 11 could be conveniently and sometimes more conceptually formulated in this language.

4.C Change of Polarization

The definition of semistability depends on the choice of a polarization. The changes of the moduli space that occur when the polarization varies have been studied by several people in greater detail. We only touch upon this problem and formulate some general results that will be needed later on.

Let X be a smooth projective surface over an algebraically closed field of characteristic 0. Let $\equiv$ denote numerical equivalence on $\mathrm{Pic}(X)$, and let $\mathrm{Num}(X) = \mathrm{Pic}(X)/\equiv$. This is a free $\mathbb{Z}$-module equipped with an intersection pairing

$$\mathrm{Num}(X) \times \mathrm{Num}(X) \longrightarrow \mathbb{Z}.$$

The Hodge Index Theorem says that, over $\mathbb{R}$, the positive definite part is 1-dimensional. In other words, $\mathrm{Num}_\mathbb{R}$ carries the Minkowski metric. For any class $u \in \mathrm{Num}_\mathbb{R}$ let $|u| = |u^2|^{1/2}$. This is not a norm! Recall that the positive cone is defined as

$$K^+ := \{x \in \mathrm{Num}_\mathbb{R}(X) | x^2 > 0 \text{ and } x.H > 0 \text{ for some ample divisor } H\}.$$

It contains as an open subcone the cone A spanned by ample divisors. A polarization of X is a ray $\mathbb{R}_{>0}.H$, where $H \in A$. Let $\mathcal{H}$ denote the set of rays in K_+. This set can be identified with the hyperbolic manifold $\{H \in K_+ | |H| = 1\}$. The hyperbolic metric β is defined as follows: for points $[H], [H'] \in \mathcal{H}$ let

$$\beta([H], [H']) = \text{arcosh}\left(\frac{H.H'}{|H|.|H'|}\right).$$

Recall that arcosh is the inverse function of the hyperbolic cosine.

Definition 4.C.1 — *Let $r \geq 2$ and $\Delta > 0$ be integers. A class $\xi \in \text{Num}(X)$ is of type (r, Δ) if $-\frac{r^2}{4}\Delta \leq \xi^2 < 0$. The wall defined by ξ is the real 1-codimensional submanifold*

$$W_\xi = \{[H] \in \mathcal{H}|\xi.H = 0\} \subset \mathcal{H}.$$

Lemma 4.C.2 — *Fix r and Δ as in the definition above. Then the set of walls of type (r, Δ) is locally finite in $\mathcal{H}$.*

Proof. The lemma states that every point $[H]$ in $\mathcal{H}$ has an open neighbourhood intersecting only finitely many walls of type (r, Δ). Let $H \in [H]$ be the class of length 1. Then $\text{Num}_{\mathbb{R}} = \mathbb{R}.H \oplus H^{\perp}$, and any class u decomposes as $u = a.H + u_0$ with $a \in \mathbb{R}$ and $u_0.H = 0$. Define a norm on $\text{Num}_{\mathbb{R}}$ by $\|u\| = (a^2 + |u_0|^2)^{1/2}$. Let $\xi = b.H + \xi_0$ be a class of type (r, Δ) and let β_0 be a positive number. $B([H], \beta_0)$ is the open ball in $\mathcal{H}$ with centre $[H]$ and radius β_0. Suppose that $[H'] \in W_\xi \cap B([H], \beta_0)$. Write $H' = H + H'_0$ with $H'_0.H = 0$. Let $\beta' = \beta([H], [H']) < \beta_0$. Check that $|H'_0| = \tanh(\beta')$. Then $0 = H'.\xi = b + \xi_0.H'_0$ and $b^2 = |\xi_0.H'_0|^2 \leq |\xi_0|^2.|H'_0|^2 = \tanh^2(\beta')|\xi_0|^2$. Moreover, $\frac{r^2}{4}\Delta \geq |\xi|^2 = |\xi_0^2|-b^2 \geq (1-\tanh^2(\beta'))|\xi_0^2|$ and $\|\xi\|^2 = |\xi_0|^2+b^2 \leq (1+\tanh^2(\beta'))|\xi_0^2|$. Hence

$$\|\xi\|^2 \leq \frac{1 + \tanh^2(\beta')}{1 - \tanh^2(\beta')} \cdot \frac{r^2}{4}\Delta \leq \cosh(2\beta_0)\frac{r^2}{4}\Delta.$$

Thus ξ is contained in a bounded, discrete and therefore finite set. This proves that the set $\{\xi|W_\xi \cap B([H], \beta_0) \neq \emptyset\}$ is finite. $\square$

Theorem 4.C.3 — *Let H be an ample divisor, F a μ_H-semistable coherent sheaf of rank r and discriminant Δ, and let $F' \subset F$ be a subsheaf of rank r', $0 < r' < r$, with $\mu_H(F') = \mu_H(F)$. Then $\xi := r.c_1(F') - r'.c_1(F)$ satisfies:*

$$\xi.H = 0 \quad and \quad -\frac{r^2}{4}\Delta \leq \xi^2 \leq 0,$$

and $\xi^2 = 0$ if and only if $\xi = 0$.

In particular, if $c_1 \in \text{Num}(X)$ is indivisible, and if H is not on a wall of type (r, Δ), then a torsion free sheaf of rank r, first Chern class c_1 and discriminant Δ is μ_H-semistable if and only if it is μ_H-stable.

Proof. We may assume that F' is saturated. Then $F'' = F/F'$ is torsion free and μ_H-semistable of rank $r'' = r - r'$. Since $H.\xi = 0$ it follows from the Hodge Index Theorem that $\xi^2 \leq 0$ with equality if and only if $\xi = 0$. Moreover, the following identity holds:

$$\Delta - \frac{r}{r'}\Delta(F') - \frac{r}{r''}\Delta(F'') = -\frac{\xi^2}{r'r''}.$$

By the Bogomolov Inequality (3.4.1) one has $\Delta(F'), \Delta(F'') \geq 0$ and therefore

$$-\xi^2 \leq r'r''\Delta \leq \frac{r^2}{4}\Delta.$$

If c_1 is not divisible, then $\xi \neq 0$, hence $\xi^2 < 0$. Thus, if a subsheaf F' as above exists, then H lies on a wall of type (r, Δ). $\square$

Remark 4.C.4 — The assumption of the theorem that H be ample is too strong: if $H \in K^+$, it makes still sense to speak of μ_H-(semi)stable sheaves in the sense that $(rc_1(F') - r'c_1(F)).H (\leq) 0$ for all saturated subsheaves $F' \subset F$ of rank r', $0 < r' < r$. The proof of the theorem goes through except for the following point: in order to conclude that $\Delta(F') \geq 0$ and $\Delta(F'') \geq 0$ we need the Bogomolov Inequality in a stronger form (7.3.3) than proved so far (3.4.1). Chapter 7 is independent of this appendix. In the following we will therefore make use of the theorem in this form, since in the applications we have in mind H will be the canonical divisor K of a smooth minimal surface of general type, which is big and nef but in general not ample.

It is clear from the proof that the conditions on classes ξ whose walls could possibly affect the stability notion are more restrictive than just being of type (r, Δ) as defined in 4.C.1. Since we have no need for a more detailed analysis here, we leave the definition as it stands.

Recall that $\overline{A}$ is the closure of the ample cone. If H and H' are elements in Num, we write

$$[H, H'] := \{tH + (1 - t)H' \mid t \in [0, 1]\}.$$

Lemma 4.C.5 — *Let H be an ample divisor and $H' \in \overline{A} \cap K^+$. Let F be a torsion free sheaf which is μ_H-stable but not $\mu_{H'}$-stable. Then there is a divisor $H_0 \in [H, H']$ and a subsheaf $F_0 \subset F$ such that $\mu_{H'}(F_0) \geq \mu_{H'}(F)$, and F and F_0 are μ_{H_0}-semistable of the same slope.*

Proof. If F is $\mu_{H'}$-semistable we can choose $H_0 = H'$ and there is nothing to prove. Hence we may assume that F is not even $\mu_{H'}$-semistable. Then there exists a saturated subsheaf $F_0 \subset F$ with $\mu_{H'}(F_0) > \mu_{H'}(F)$. If F' is any saturated subsheaf with this property, let

$$t(F') := \frac{\mu_H(F) - \mu_H(F')}{\mu_{H'}(F') - \mu_{H'}(F)},$$

so that $\mu_{H+t(F')H'}(F') = \mu_{H+t(F')H'}(F)$. Note that $H_0 := H + t(F_0)H'$ is ample. If $t(F') < t(F_0)$, then $\mu_{H_0}(F') > \mu_{H_0}(F)$. By Grothendieck's Lemma 1.7.9, the family of saturated subsheaves F' with this property is bounded. This implies that there are only finitely many numbers $t(F')$ which are smaller than $t(F_0)$. In fact, we may assume that F_0 was chosen in such a way that $t(F_0)$ is minimal. Then F_0 and H_0 have the properties stated in the lemma. □

For the definition of e-stability see 3.A.1.

Proposition 4.C.6 — *Let H be an ample divisor, $H' \in \overline{A} \cap K^+$. Let $r \geq 2$ and $\Delta \geq 0$ be integers and put $e := \sqrt{\Delta/4} \sinh \beta([H], [H'])$. Suppose that F is a coherent sheaf of rank r and discriminant Δ. If F is e-stable with respect to H then F is $\mu_{H'}$-stable.*

Proof. Suppose that F is μ_H-stable but not $\mu_{H'}$-stable. By the previous lemma there exists a divisor $H_0 \in [H, H']$ and a subsheaf F_0 such that $\xi.H_0 = 0$ for $\xi := r.c_1(F_0) - \mathrm{rk}(F_0)c_1(F)$. Let $\beta_0 = \beta([H], [H_0])$. Note that $\beta_0 \leq \beta([H], [H'])$. Write $\xi = a.H + \tilde{\xi}$ and $H_0 = b.H + \tilde{H}_0$ with $a, b \in \mathbb{R}$ and $\tilde{\xi}, \tilde{H}_0 \perp H$. Then $\tanh \beta_0 = \frac{|\tilde{H}_0|}{b|H|}$. Moreover, $0 = \xi.H_0 = ab|H|^2 + \tilde{\xi}.\tilde{H}_0$ and therefore

$$|\tilde{\xi}| \geq \tilde{\xi}.\tilde{H}_0/|\tilde{H}_0| = -a|H|/\tanh(\beta_0).$$

Furthermore, by 4.C.3 and 4.C.4 the inequality

$$r^2 \Delta/4 \geq |\xi|^2 = -a^2|H|^2 + |\tilde{\xi}|^2 \geq a^2|H|^2/\sinh^2(\beta_0)$$

holds, implying that $(-a)|H| \leq r \cdot \sinh(\beta_0)\sqrt{\Delta}/2$. Finally we get:

$$\mu_H(F) - \mu_H(F_0) = -\frac{\xi.H}{r \cdot \mathrm{rk}(F_0)} = \frac{a.|H|^2}{r \cdot \mathrm{rk}(F_0)} \geq -\frac{\sqrt{\Delta}\sinh(\beta_0)}{2 \cdot \mathrm{rk}(F_0)}|H| \geq -\frac{|H|}{\mathrm{rk}(F_0)}e.$$

This means that F is not e-stable, contradicting the assumption of the proposition. □

Theorem 4.C.7 — *Let H and H' be ample divisors. If $\Delta \gg 0$, the moduli spaces*

$$M_H(r, c_1, \Delta) \sim M_{H'}(r, c_1, \Delta)$$

are birational.

Proof. We may assume that H and H' are very ample. Recall that we have an estimate for the e-unstable locus of M_H:

$$\dim M_H(e) \le \left(1 - \frac{1}{2r}\right) \Delta + (3r - 1)e^2 + \frac{r(K_X.H)_+}{2|H|}e + B(X, H).$$

Inserting $e = \sqrt{\Delta} \sinh(\beta_0)/2$ for some positive number β_0, the coefficient of Δ on the right hand side is $(1 - \frac{1}{2r} + \frac{1}{4}\sinh^2 \beta_0)$, and this coefficient is strictly smaller than 1 if $\sinh^2 \beta_0 < \frac{2}{r}$. Fix $\beta_0 = \text{arsinh}\frac{1}{r}$. Subdivide the line in $\mathcal{H}$ connecting $[H]$ and $[H']$ into finitely many sections such that the division points have mutual distances $< \beta_0$ and have very ample integral representatives $H = H_1, H_2, \dots, H_N = H'$. Now choose Δ large enough such that $M_{H_i}(r, c_1, \Delta)$ is a normal scheme of expected dimension (cf. Theorem 9.3.3) and such that $\dim M_{H_i}(e) < \dim M_{H_i}$ for each $i = 1, \dots, N$. This is possible since by our choice of β_0 the dimension of M_{H_i} grows faster than the dimension of $M_{H_i}(e)$ considered as functions of Δ. By the proposition only the e-unstable sheaves in M_{H_i} can be unstable with respect to H_{i-1} or H_{i+1}. Therefore the dimension estimate just derived shows $M_{H_1} \sim \dots \sim M_{H_N}$.   □

Comments:
— The notion of S-equivalence is due to C. S. Seshadri [235]. He constructs a projective moduli space for *semistable* vector bundles on a smooth curve, which compactifies the moduli space of stable bundles constructed by Mumford [191]. There exists an intensive literature on moduli of vector bundles on curves. We refer to Seshadri's book [236] and the references given there. In the curve case, G. Faltings [61] gave a construction without using GIT, see also the expository paper by Seshadri [237].

— The main reference for Geometric Invariant Theory is Mumford's book [195]. The lecture notes of Newstead [203] explain the material on a more elementary level. We also recommend the seminar notes of Kraft, Slodowy and Springer [131]. Theorem 4.2.15 is due to G. Kempf. For a proof see Thm. 2.3. in [52] or Prop. 4.2 in part 4 of [131].

— General constructions of moduli spaces of sheaves on higher dimensional varieties have been given by D. Gieseker [77] for surfaces and M. Maruyama [162, 163]. This approach has been sketched in Appendix 4.A. Our presentation in 4.3 and 4.4 follows the method of C. Simpson [240] and the J. Le Potier's exposé [145]. Theorem 4.4.1 and the proof of 4.4.2 are taken from Le Potier's exposé [145]. The observation of the dichotomy of the cases A and B in the proof of 4.4.1 as well as the statement of 4.4.2 are due to Simpson. This is one of the main technical improvements of his approach. Observe how well suited the Le Potier-Simpson estimate is to mediate between the Euler characteristic and the number of global sections of a sheaf. Proposition 4.4.2 is Lemma 1.17 in [240], where it is proved in a slightly different way. In a sense, this theorem is responsible for the projectivity of the moduli space of semistable sheaves. Thus in Gieseker's construction its rôle is played by Lemmas 4.2 and 4.5 in [77]. In a certain sense, the properness of the moduli space had been proved by Langton [135] before the moduli space itself was constructed. See Appendix 2.B. In the proof of 4.4.5 we used a pleasant technical device we learned from A. King [123].

— The smoothness of Hilbert schemes of points on surfaces (Example 4.5.10) is due to Fogarty [65]. His argument for smoothness is the first one given in the example, whereas his proof of the connectivity is quite different and very interesting. He shows that the *punctual* Hilbert scheme, i.e. the closed subscheme in $\text{Hilb}^\ell(X)$ of those cycles which are supported in a single, fixed point in X can be considered as the set of fixed points for the action of a unipotent algebraic group on a Grassmann variety and therefore must be connected. The existence of natural morphisms $\text{Hilb}^n(X) \to M_n \to S^n(X)$ as discussed in Example 4.3.6 is asserted by Grothendieck [93] though without proof and using a different terminology. Our presentation of the linear determinant follows Iversen [117].

— Deformations of coherent sheaves are discussed in the papers of Mukai [187], Elencwajg and Forster [55] and Artamkin [5, 7]. See also the book of Friedman [69]. Theorem 4.5.1 was proved by Wehler in [261].

— The existence of universal families was already discussed by Maruyama [163]. The notion of quasi-universal families is due to Mukai [188].

— Theorem 4.B.4 can be found in [115]. For other constructions of similar moduli spaces see [147], [236], [246], [156]. The probably most spectacular application of stable pairs is Thaddeus' proof of the Verlinde formula for rank two bundles [246]. Recently, moduli spaces of stable pairs on surfaces have found applications in non-abelian Seiberg-Witten theory.

— The changes that moduli spaces undergo when the ample divisor H on X crosses a wall have been studied by several authors, often with respect to their relation to gauge theory and the computation of Donaldson polynomials. We refer to the papers of Qin, Göttsche [84], Ellingsrud and Göttsche [57], Friedman and Qin [72], Matsuki and Wentworth [171].

— We also wish to draw the reader's attention to the papers of Altman, Kleiman [3] and Kosarew, Okonek [130]. In these papers, moduli spaces of simple coherent sheaves are considered. In [3] the moduli space of simple coherent sheaves on a projective variety is shown to be an algebraic space in the sense of Artin. In general, however, it is neither of finite type nor separated. The phenomenon of non-separated points in the moduli space was investigated in [204] and [130].

4.D Further comments (second edition)

I. Moduli space constructions via quiver representations.

In the very inspiring article [267] Álvarez-Cónsul and King have developed a new approach to the construction of moduli spaces of sheaves. Recall once more the three steps in the construction of the moduli space of semistable sheaves with Hilbert polynomial P as explained in this chapter:

i) Twist all semistable sheaves F in a uniform manner by $\mathcal{O}_X(n)$, $n \gg 0$, such that they can all be written as quotients $V \otimes \mathcal{O}_X(-n) \twoheadrightarrow F$ with V a vector space of dimension $P(n)$. This then defines a point in the Quot-scheme of quotients of $V \otimes \mathcal{O}_X(-n)$.

ii) Apply a second high twist and send a quotient $V \otimes \mathcal{O}_X(-n) \twoheadrightarrow F$ to the quotient of vector spaces $V \otimes H^0(\mathcal{O}_X(m-n)) \twoheadrightarrow H^0(F(m))$. This corresponds to embedding the Quot-scheme into a certain Grassmannian.

iii) Divide out by $\mathrm{GL}(V)$ to make the construction independent of the choice $V \cong H^0(X, F(n))$.

The main idea in [267] is now to combine these steps and embed the moduli functor of semistable sheaves directly into the moduli functor of representations of a certain Kronecker quiver. To make this a little more precise we fix the following notations. Let $m \gg n \gg 0$ as before, so in particular one relies on the boundedness results proven in earlier chapters. Let $H := H^0(\mathcal{O}_X(m-n))$ and let A be the finite-dimensional algebra that can be best written as

$$A = \begin{pmatrix} k & H \\ 0 & k \end{pmatrix}.$$

Note that A-modules correspond to representations of the quiver with two vertices P_1, P_2 and $\dim(H)$ arrows $P_1 \to P_2$. Then in [267] a sheaf F is sent to the A-module given by $H^0(F(n)) \otimes H \to H^0(F(m))$. In more functorial terms, consider $T = \mathcal{O}_X(-m) \oplus \mathcal{O}_X(-n)$ and define

$$\Phi = \mathrm{Hom}(T, \ ) : \mathrm{Coh}(X) \to \mathrm{mod}{-}A$$

from the category of coherent sheaves into the category of finite-dimensional right A-modules.

One of the key facts proved in [267] is then:

Suppose $\mathcal{O}_X(m-n)$ is regular. Then the functor Φ induces a fully faithful embedding

$$\mathcal{M}_X^{\mathrm{reg}} \hookrightarrow \mathcal{M}_A$$

of the moduli functor of n-regular sheaves into the functor of A-modules. Moreover, the image is locally closed.

The latter fact is used to prove that $\mathcal{M}_X^{\mathrm{reg}}$ is locally a quotient functor given by a locally closed subset of a finite-dimensional vector space by a reductive group.

Also for A-modules one can introduce the notion of (semi)stability and the corresponding moduli spaces had been constructed earlier in broad generality by King in [123]. Coarse
moduli spaces for $\mathcal{M}_A^s \subset \mathcal{M}_A^{ss}$ exist as quasi-projective respectively projective schemes.
Closed points in the latter parametrize S-equivalence classes of A-modules. The two con-
cepts of stability coincide due to the following result [267]:

*The sheaf F is semistable if and only if F is pure, n-regular and $\Phi(F)$ is a semistable
A-module.*

Moreover, the A-modules arising in the image of Φ can be described by properties of
the adjoint functor. Combining these results with Langton's criterion Álvarez-Cónsul and
King eventually prove:

*The embedding $\mathcal{M}_X^{\mathrm{reg}} \hookrightarrow M_A$ induces locally closed embeddings $\mathcal{M}_X^s \hookrightarrow \mathcal{M}_A^s$ and
$\mathcal{M}_X^{ss} \hookrightarrow \mathcal{M}_A^{ss}$. Moreover, the corresponding locally closed subschemes of the moduli
space M_A^{ss} corepresent $\mathcal{M}_X^s$ and $\mathcal{M}_X^{ss}$. The latter is projective.*

The article [267] also contains a discussion of the relative situation of a projective
morphism $X \to S$.

II. Moduli spaces in characteristic p.

Moduli spaces of semistable sheaves in positive and mixed characteristic exist as well.
The main missing ingredient was the boundedness of μ-semistable sheaves which has been
proved in Langer's seminal paper [356]. The paper [357] contains new estimates of Le
Potier-Simpson type. A survey of these new techniques is contained in [360]. A discussion
of the construction of moduli spaces via Simpson's techniques in positive and mixed characteristic under the assumption of boundedness can also be found in Maruyama's article
[388].

III. Moduli spaces of principal bundles.

Let G be a linear algebraic group. A G-principal bundle over a scheme X/k is a scheme
P with a right G-action and an invariant morphism $\pi : P \to X$ such that locally in the
étale topology on X there exist equivariant trivializations $P \cong G \times X$. The group G is
called the structure group of P. If F is any scheme on which G acts, we may form a new
bundle $P(F) := P \times^G F$ with fibre F, where $\times^G$ denotes the quotient for the equivalence
relation $(pg, f) = (p, gf)$ on $P \times F$. Of particular interest is the case, when F is a linear
representation of G. In this case $P(F)$ becomes a vector bundle.

Another important case is given by a group homomorphism $\alpha : G \to G'$ (e.g. the
inclusion of a subgroup). Then G acts on G' from the left via α, and $P \times^G G'$ becomes a
G'-principal bundle. Any G'-bundle of this form is said to admit a reduction of its structure
group to G.

If one is interested in stability notions for principal G-bundles, reductions of the structure group play a rôle similar to subsheaves in coherent sheaves. The relation can be understood as follows: If E is a vector bundle on X of rank r, we can naturally associate to it a principal $\mathrm{GL}(r)$-bundle $P := \mathbb{I}\mathrm{som}(k^r, E)$. Then E can be reconstructed from P by gluing in the standard representation of $\mathrm{GL}(r)$ on k^r, i.e. $E \cong P(k^r)$. Assume now that $F \subset E$ is a subbundle, say of rank r', such that E/F is again a vector bundle. Then we may form the bundle

$$P' := \{\varphi : k^r \xrightarrow{\cong} E(x) \mid x \in X, \varphi(k^{r'} \oplus 0) = F(x)\} \subset P.$$

This is a principal bundle for the parabolic structure group

$$B := \left\{ \left(\begin{array}{c|c} * & * \\ \hline 0 & * \end{array} \right) \right\} \subset \mathrm{GL}(r),$$

where the diagonal blocks in the block decomposition have the sizes r' and $r - r'$. The new B-principal bundle P' carries the information about E plus its subbundle F.

Let X be a smooth curve and G a reductive group. A principal G-bundle $P \to X$ is said to be (semi)stable if, for every reduction $P' \to X$ of the structure group G to a parabolic subgroup $B \subset G$ and any dominant character $\chi : B \to k^*$, the line bundle $P'(\chi)$ has degree $(\leq)0$. Moduli spaces for semistable principal bundles over a smooth curve were constructed by Ramanathan [226, 418]. The construction can be reduced to the construction of moduli for bundles by the choice of an embedding $G \to \mathrm{GL}(V)$.

Technical difficulties immediately arise, when the variety X is a surface or higher dimensional or becomes singular: whereas a torsion free sheaf on a smooth curve always is a vector bundle, one needs 'degenerated vector bundles', i.e. torsion free sheaves that are not necessarily locally free, in order to get compactified moduli spaces on higher dimensional or singular varieties. For principal bundles it is by no means clear, what a good notion of a 'degenerated principal bundle' should be. A solution for this problem on smooth projective varieties X has been worked out by Gómez and Sols [326]. They consider triples (P, E, ψ) where E is a torsion free sheaf, P is a principal G-bundle over the open subset $U \subset X$ where E is locally free and $\psi : E|_U \to P(\mathfrak{g})$ is an isomorphism to the adjoint bundle, associated to the adjoint action of G on its Lie algebra $\mathfrak{g}$. They can then define appropriate stability notions and construct projective moduli spaces. Quasi-projective moduli spaces of principal bundles had been constructed before in [342]. Alternative constructions have been given by Schmitt [425, 426]. Schmitt discusses moduli of principal bundles on a nodal curve in [427]. When G is semisimple, moduli spaces for semistable principal G-bundles were constructed in positive characteristic in a joint paper of Gómez, Langer, Schmitt and Sols [327]. See [328] for a short introduction to the subject and for restriction theorems for principal bundles in the sense of Chapter 7 see [282].

IV. Moduli spaces of decorated sheaves.

In Section 4.B we have discussed sheaves with one particular type of decoration, so-called framed modules. For framed modules on surfaces with a special framing along a curve Nevins proves in [403] the representability of the moduli functor by relating it to the Quot-scheme. No stability condition is needed, but the framing is supposed to be an isomorphism along the curve. In [404] another explicit example has been studied in detail. Nevins shows that for certain ruled surfaces over an elliptic curve and framing along a curve at infinity the cohomology of the moduli stack can be understood. Framed modules on the projective plane or a blow-up with a framing along a line, respectively the exceptional curve, play a prominent rôle in [396, 421].

Another type of decoration is provided by Le Potier's coherent systems of which pairs in the sense of Thaddeus [246] are a special case. A coherent system consists of a coherent sheaf E on a projective variety X together with a subspace $V \subset H^0(X, E)$. A stability condition can be written down, depending again on certain parameters, and moduli spaces have been constructed (see [147]). Coherent systems have been studied intensively in the case of curves (Brill-Noether theory). In particular a lot is known in this case about non-emptiness of the corresponding moduli spaces and the dependence on the parameters in the stability condition. See [290] and references therein. Coherent systems have also been used on surfaces, mostly as intermediate spaces between moduli spaces of sheaves of various ranks, but maybe the most spectacular appearance they have made recently is in the theory of curve counting invariants on Calabi-Yau threefolds. Curves on a Calabi-Yau threefold X can be counted in terms of certain intersection numbers of moduli spaces of maps from curves to X or of curves in X. These invariants run under the name of Gromov-Witten respectively Donaldson-Thomas. In [413] moduli spaces of pairs, i.e. coherent systems with a one-dimensional space of sections, are used to construct a moduli space of complexes $\mathcal{O}_X \to F$ (see below), where F is a pure sheaf supported in dimension one. The resulting new stable pairs invariants are closely related to Gromov-Witten, Donaldson-Thomas and physicists' BPS invariants. Presumably higher rank coherent systems on Calabi-Yau threefolds will play a rôle in the future.

Moduli spaces of triples parametrize two sheaves E, F together with a homomorphism $E \to F$. In contrast to framed modules and coherent systems both sheaves E and F are allowed to vary. Moduli spaces of triples have been studied intensively on curves, see [289] and references therein.

Clearly, many other types of interesting decorations can be considered and their moduli spaces can be studied. In a series of papers Schmitt has developed techniques that allow one, via GIT, to construct moduli spaces for very general decorations. Roughly, Schmitt considers a set of sheaves together with a set of morphisms corresponding to vertices respectively edges of a given quiver. See [428] and the monograph [429]. The latter also covers standard material from GIT.

Among the most interesting types of decorated sheaves are parabolic sheaves and Higgs bundles. They play a central rôle for quasi-projective varieties and variations of Hodge structures. Apart from the general existence results for moduli spaces of these sheaves (see [387, 443, 344] and [240]), most of the existing literature is restricted to the case of curves. For some results in higher dimensions and applications of parabolic sheaves and Higgs see [273, 280, 303, 324, 344, 349, 438].

V. Moduli spaces of complexes.

When the first edition of this book appeared there was considerable interest in constructing moduli spaces of sheaves with various kinds of additional structures. They were called decorated sheaves and a particular type, so-called framed modules, has been discussed in some detail in Section 4.B. Since then the focus has shifted more towards moduli spaces of complexes. A bounded complex of coherent sheaves $E^\bullet$ could be seen as a collection of sheaves E^i and homomorphisms $d^i : E^i \to E^{i+1}$ satisfying $d^{i+1} \circ d^i = 0$. It should be possible to write down stability conditions for such data and this has been done for complexes of length two. The new perspective however is that complexes should be considered as objects in the derived category. So a moduli space of complexes should parametrize complexes up to quasi-isomorphism. In particular, a deformation of a complex $E^\bullet$ does not necessarily induce deformations of the sheaves E^i themselves.

Without stability. There are general results for moduli spaces of complexes not satisfying any stability conditions. The resulting moduli spaces will only be algebraic spaces or stacks of a certain type.

For a flat projective morphism $X \to S$ of Noetherian schemes Inaba shows in [345]:

The étale sheafification of the functor that sends any locally Noetherian S-scheme T to the set of bounded complexes of coherent sheaves $E^\bullet \in D^b(T \times_S X)$ on $T \times_S X$ (up to tensor product with invertible sheaves coming from T) satisfying $\mathrm{Ext}_{X_t}^{-1}(E_t^\bullet, E_t^\bullet) = 0$ and $\mathrm{End}_{X_t}(E_t^\bullet) = k(t)$ for any point $t \in T$ is representable by a locally separated algebraic space.

The result was generalized, using completely different techniques, by Lieblich in [372]. He considers the case of a proper flat morphism of finite presentation between algebraic spaces fppf-locally presentable by schemes and shows:

The stack of relatively perfect complexes with vanishing $\mathrm{Ext}_{X_s}^{<0}(E_s^\bullet, E_s^\bullet)$ for every geometric point $s \in S$ is an algebraic stack of locally finite presentation over S.

The deformation theory of complexes (always as objects in the derived category, so only up to quasi-isomorphisms) on a fixed scheme X has been considered by Inaba in [348], where he proves the analogue of Theorem 4.5.3 for the moduli space of simple complexes. The deformation theory for objects of the derived category (with varying X) has been further studied in [372, 376, 341].

With stability. The concept of stability for complexes (as objects in the derived category) is problematic for various reasons. One being that the notion of subobjects does not make sense in a triangulated category.

In [348] Inaba suggests defining the notion of stability for complexes with respect to an ample sequence $\mathcal{L} = \{L_n\}_{n\geq}$ of objects in the derived category. The definition of an ample sequence is rather technical and goes back to Bondal and Orlov (see e.g. [338]), but the positive powers of an ample line bundle form such a sequence in the bounded derived category of coherent sheaves. Then a complex $E^\bullet$ is called (semi)stable if a certain inequality, generalizing the one imposed on semistable coherent sheaves in the obvious way, is satisfied for all morphisms $F^\bullet \to E^\bullet$ (in the derived category) that induce injective maps $\mathrm{Hom}(L_n, F^\bullet) \to \mathrm{Hom}(L_n, E^\bullet)$ for $n \gg 0$. So the latter condition replaces the notion of subobjects in an abelian category. Then for a flat projective morphism $X \to S$ of Noetherian schemes, an ample sequence $\mathcal{L} = \{L_n\}$, and a numerical polynomial P Inaba proves:

There exists a coarse moduli space for the functor that sends an S-scheme T to the set of $E^\bullet \in \mathrm{D}^b(T \times_S X)$ (up to tensor product with invertible sheaves coming from T) such that for every geometric point $s \in T$ the restriction E_s (semi)stable and $\mathrm{Ext}^i((L_n)_s, E_s^\bullet) = 0$ for $i \neq 0$ and $\dim \mathrm{Hom}((L_n)_s, E_s^\bullet) = P(n)$ for $n \gg 0$. Moreover, the moduli space of semistable complexes is projective over S.

In fact, in the proof the scheme X does not play a rôle and can be replaced by a *fibred triangulated category with base change property* over S. The construction is done, as in the case of coherent sheaves, via GIT and follows Maruyama's original construction quite closely.

A more conceptual notion of stability conditions on triangulated categories was worked out by Bridgeland in [294] following suggestions by M. Douglas. Very roughly, a stability condition refines the notion of a t-structure on a triangulated category $\mathcal{T}$ by decomposing $\mathcal{T}$ into abelian categories $\mathcal{P}(\phi) \subset \mathcal{T}$ indexed by real numbers $\phi \in \mathbb{R}$. The objects in $\mathcal{P}(\phi)$ are the semistable objects of *phase* ϕ and the objects in $\mathcal{P}(\phi)$ without proper subobjects are the stable objects. The existence of such stability conditions has been proved for the bounded derived category of coherent sheaves on special types of varieties, e.g. curves, abelian and K3 surfaces, projective spaces. It is generally believed that moduli spaces of (semi)stable objects should exist as (quasi)projective schemes at least when the category $\mathcal{T}$ is the bounded derived category of coherent sheaves on a projective variety.

For K3 surfaces Toda proves in [434] that the moduli stack of semistable objects with fixed phase and numerical class is an Artin stack of finite type. This has to be read as an assertion about the boundedness of semistable objects. For a related result see the appendix by Lieblich in [271].

Hein and Ploog define in [336] yet another notion of stability of complexes which is conceptually close to Inaba's notion and is inspired by Faltings' approach to semistability on curves and the higher-dimensional versions worked out by Álvarez-Cónsul and King [267]. An object A of a triangulated category $\mathcal{T}$ is called *Postnikov-stable* with respect to a finite Postnikov system $(C_\bullet, N)$, i.e. objects $C_j \in \mathcal{T}$, $j \in I$, and numbers N_j^i, $j \in I$, if $\dim \mathrm{Ext}^i(C_j, A) = N_j^i$ and A is right orthogonal to a certain convolution of $C_\bullet$. The notion can be compared to usual stability for coherent sheaves. The paper [336] also contains an explicit example where a moduli space of sheaves is compactified by adding Postnikov-stable complexes of length 2.

VI. Moduli of twisted sheaves.

Let X be a quasi-projective variety over a field k, let $\mathcal{A}$ be an Azumaya algebra and let $\alpha \in H^2_{\mathrm{et}}(X, \mathcal{O}_X^*)$ be its Brauer class. So $\mathcal{A}$ is an $\mathcal{O}_X$-algebra locally (in the étale topology) isomorphic to a matrix algebra. If $k = \mathbb{C}$, étale topology can be replaced by the analytic topology. For a locally free sheaf E on X, the two Azumaya algebras $\mathcal{A}$ and $\mathcal{A} \otimes \mathcal{E}nd(E)$ are Morita equivalent and define the same Brauer class α. Azumaya algebras of rank r^2 are parametrized by the non-abelian cohomology $H^1_{\mathrm{et}}(X, \mathrm{PGL}_r(\mathcal{O}_X))$ which also parametrizes Brauer-Severi varieties, i.e. locally trivial (in the étale topology) $\mathbb{P}^{r-1}$-bundles over X. Azumaya algebras and Brauer-Severi varieties are a classical subject of algebraic geometry (often with an arithmetical touch). But with Căldăraru's thesis [298, 299] Azumaya algebras and coherent sheaves over Azumaya algebras have entered mainstream moduli space theory. To fix notation, for an Azumaya algebra $\mathcal{A}$ on X with Brauer class α, we will call a coherent $\mathcal{A}$-module an α-twisted sheaf. In fact, when α is represented by a cocycle $\{\alpha_{ijk} \in H^0(U_{ijk}, \mathcal{O}_X^*)\}$ then there is an equivalence of the abelian category of coherent $\mathcal{A}$-modules with the abelian category of collections $\{E_i \in \mathrm{Coh}(U_i), \varphi_{ij} : E_i|_{U_{ij}} \cong E_j|_{U_{ij}}\}$ with $\varphi_{ij}\varphi_{jk}\varphi_{ki} = \alpha_{ijk} \cdot \mathrm{id}$. The latter are called, for obvious reasons, twisted sheaves.

Now, the main observation of Căldăraru is that although the universal family of stable sheaves over $M^s \times X$ (using the notation of Section 4.6) might not exist, it always exists as a twisted sheaf. To be more precise, by construction M^s is a universal geometric quotient of an open subset R^s of some Quot-scheme. In particular, the question whether a universal family on $M^s \times X$ exists is essentially equivalent to the question whether the universal quotient on $R^s \times X$ descends to $M^s \times X$. By Luna's Etale Slice Theorem 4.2.12 there exists an étale cover $\coprod U_i \to M^s$ such that a universal family $\coprod \mathcal{E}_i$ exists on $\coprod U_i \times X$. This then immediately leads to Căldăraru's observation:

Let M^s be a moduli space of stable sheaves on a projective scheme X. Then there exists a Brauer class α on M^s such that a universal family of stable sheaves $\mathcal{E}$ on $M^s \times X$ exists as an α-twisted sheaf on $M^s \times X$.

So in this context the Brauer class appears as an obstruction class to the existence of a universal family. In many interesting examples, the rôle of X and of the moduli space M^s is symmetric in the sense that X can also be viewed, via the universal family when it exists, as a moduli space of sheaves on M^s. If the universal family does not exist, i.e. the obstruction Brauer class α is non-trivial, then X parametrizes α-twisted sheaves on M^s. This point of view shows that moduli spaces of α-twisted sheaves are natural objects to study. There are other reasons for generalizing the existing theory to the twisted case, for which we refer to the literature.

Moduli spaces of twisted sheaves have been considered by Lieblich [373] in great detail. He first shows that the moduli stack of S-flat torsion free α-twisted sheaves on a scheme X, which is assumed smooth and projective over S, is an Artin stack locally of finite presentation. Moreover, Lieblich introduces the notion of (semi)stable twisted sheaves in terms of the Hilbert polynomial of the α-twisted sheaf viewed as a sheaf on the gerbe associated to α. He then shows:

If $X \to S$ is of relative dimension ≤ 2 and S is affine, then the moduli space of (semi)stable α-twisted sheaves exists. The moduli space of semistable sheaves is projective over S.

The definition of the moduli functor is similar to the one in the untwisted case. These results are applied in [374] to the so called period-index theorem of de Jong (see e.g. [302]).

A slightly different approach to semistability for twisted sheaves was proposed by Yoshioka in [455]. He interprets α-twisted sheaves on X as untwisted sheaves on the Brauer-Severi variety $\pi : \mathbb{P} \to X$ associated to α by pulling-back an $\{\alpha_{ijk}\}$-twisted sheaf F and tensoring it with the $\{\pi^{-1}\alpha_{ijk}^{-1}\}$-twisted sheaf $\mathcal{O}_\pi(1)$. The untwisted sheaves on $\mathbb{P}$ arising in this way can be described geometrically. Yoshioka introduces semistability of twisted sheaves F by testing subsheaves of the associated untwisted sheaf $\pi^* F \otimes \mathcal{O}_\pi(1)$. This reduces the construction of moduli spaces of these twisted sheaves on X to the usual one for sheaves on $\mathbb{P}$. In fact, the stability considered in [455] can be seen as a special case of Simpson's stability for sheaves over the Azumaya algebra $\mathcal{A}$ (see [240]).

In [337] Hoffmann and Stuhler consider simple twisted sheaves of rank one over the Azumaya algebra $\mathcal{A}$. Their moduli spaces are the twisted versions of the Picard scheme and considering not only locally free sheaves but also torsion free ones allows them to construct projective *twisted Picard schemes*. This was also studied by Lieblich in [375].

VII. Moduli spaces on non-reduced schemes.

Stable sheaves on non-reduced schemes always need special care. We have tried to cover this case by considering them as sheaves on some ambient projective space whose support happens to be non-reduced, but of course their moduli spaces have often very special properties.

In [347] moduli spaces on a non-reduced scheme X are stratified according to a stratification of X induced by a nilpotent sheaf of ideals $I \subset \mathcal{O}_X$ (see also [442]). This is applied to multiple curves and to the double plane viewed as a degeneration of a quadric surface. Earlier, in analogy to the curve case, the same author had studied moduli spaces of stable sheaves on reducible schemes consisting of two irreducible components intersecting in a divisor and had considered in particular moduli spaces on reducible quadrics, see [346].

Moduli spaces of sheaves on a smooth surface with support on multiple curves have been studied further and in great detail in [313, 314].

Part II

Sheaves on Surfaces

5

Construction Methods

The two most prominent methods for constructing vector bundles on surfaces are Serre's construction and elementary transformations. Both techniques will be used at several occasions in these notes. Section 5.3 contains examples of moduli spaces on K3 surfaces and fibred surfaces. In the latter case we discuss the relation between stability on the surface and stability on the fibres.

In order to motivate the first two sections let us recall some general facts about globally generated vector bundles.

Let $0 \to V \to H \to W \to 0$ be a short exact sequence of vector spaces and denote the dimension of V and W by v and r, respectively. Let s be an integer, $0 \leq s \leq \min\{v, r\}$, and let $M_s \subset \operatorname{Hom}(V, W)$ be the general determinantal variety of all homomorphisms of rank $\leq s$. Then M_s is a normal variety of codimension $(v - s)(r - s)$ and the singular part of M_s is precisely M_{s-1} (cf. [4] II §2). Let M'_s be the intersection of the pre-image of M_s under the surjection $\operatorname{Hom}(H, W) \to \operatorname{Hom}(V, W)$ and the open subset $U \subset \operatorname{Hom}(H, W)$ of surjective homomorphisms. Then M'_s is either empty or a normal subvariety of codimension $(v - s)(r - s)$ with $\operatorname{Sing}(M'_s) = M'_{s-1}$. Clearly, M'_s is invariant under the natural $\operatorname{GL}(W)$ action on U. Let M''_s be the image of M'_s under the bundle projection $U \to \operatorname{Grass}(H, r)$. Then M''_s has the analogous properties of M'_s.

Let X be a smooth variety. Suppose E is a locally free sheaf of rank r which is generated by its space of global sections $H := H^0(X, E)$. The evaluation homomorphism $H \otimes \mathcal{O}_X \to E$ induces a morphism $\varphi : X \to \operatorname{Grass}(H, r)$. If $V \subset H$ is a linear subspace of dimension v and if $M''_s \subset \operatorname{Grass}(H, r)$ is defined as above, then $X_s := \varphi^{-1}(M''_s) \subset X$ is by construction precisely the closed subscheme where the homomorphism $V \otimes \mathcal{O}_X \to E$ has rank less than or equal to s. Since $\operatorname{GL}(H)$ acts transitively on $\operatorname{Grass}(H, r)$, we may apply Kleiman's Transversality Theorem (cf. [98] III 10.8) and find that for generic choice of V the morphism $X \to \operatorname{Grass}(H, r)$ is transverse to any of the smooth subvarieties

$M_s'' \setminus M_{s-1}''$. It follows that X_s is either empty or a subvariety of codimension $(v-s)(r-s)$ with $\mathrm{Sing}(X_s) = X_{s-1}$.

Examples 5.0.1 — Let E be a globally generated rank r vector bundle and $H = H^0(X, E)$ the space of global sections as above.

1) Suppose $r > d := \dim(X)$ and let $v = r - d$, $s = r - d - 1$, so that X_s is precisely the locus where $V \otimes \mathcal{O}_X \to E$ is not fibrewise injective. If $V \subset H$ is general, X_s is either empty or has codimension $r - (r - d - 1) = d + 1$, hence is indeed empty. This means that there is a short exact sequence

$$0 \to \mathcal{O}_X^{\oplus r-d} \to E \to E' \to 0$$

for some locally free sheaf E' of rank d. In words, any globally generated vector bundle of rank bigger than the dimension of X is an extension of a vector bundle of smaller rank by a trivial bundle.

2) Let X be a surface and let $V \subset H$ be a general subspace with $v = r - 1$. Then X_{r-2} is empty or has codimension 2, and X_{r-3} is empty or has codimension 6 (hence is empty). Thus for a general choice of $r - 1$ global sections there is a short exact sequence

$$0 \to \mathcal{O}_X^{\oplus r-1} \to E \to F \to 0,$$

where F is of rank 1 almost everywhere, but has rank 2 precisely at a smooth scheme $Z = X_{r-2}$ of dimension 0, i.e. $F \cong \det(E) \otimes \mathcal{I}_Z$. For $r = 2$ this is part of the Serre correspondence between 0-cycles and rank two bundles which will be discussed in Section 5.1.

3) Again, let X be a surface and let $V \subset H$ be a general subspace with $v = r$. Then X_{r-1} is empty or has codimension 1 and X_{r-2} is empty or has codimension 4 (hence is empty). Thus for a general choice of r global sections there is a short exact sequence

$$0 \to \mathcal{O}_X^{\oplus r} \to E \to L \to 0,$$

where L is zero or a locally free sheaf of rank 1 on the smooth curve X_{r-1}. We say E is obtained by an elementary transformation of the trivial bundle along the smooth curve X_{r-1}. The details will be spelled out in Section 5.2.

Thus, the theory of globally generated bundles and determinantal varieties provides a uniform approach to the Serre correspondence and elementary transformations.

For the rest of this chapter we assume that X is a smooth projective surface over an algebraically closed field of characteristic 0 which, sometimes, will even be the field of complex numbers. By K_X we denote the canonical line bundle of X.

For the convenience of the reader we recall the following facts discussed in Chapter 1 and specify them for our situation. If F is a reflexive sheaf of dimension 2 on X and $q > 0$, then the sheaf $\mathcal{E}xt^q(F, \mathcal{O}_X)$ has codimension $\geq q + 2$ by Proposition 1.1.10 and must therefore vanish. Hence, F is locally free. If F is only torsion free then $\theta : F \to F^{\vee\vee}$ is a canonical embedding into a locally free sheaf. Again by Proposition 1.1.10 $\mathcal{E}xt^1(F, \mathcal{O}_X)$ has dimension 0 and $\mathcal{E}xt^2(F, \mathcal{O}_X) = 0$. Hence F is locally free outside a finite set of points in X and has homological dimension 1, i.e. if $\varphi : E \to F$ is any surjection with locally free E then $\ker(\varphi)$ is also locally free, or, still rephrasing the same fact, any saturated subsheaf of a locally free sheaf is again locally free. If $D \subset X$ is a divisor then clearly $\mathrm{dh}(\mathcal{O}_D) = 1$. Since locally any vector bundle G on D is isomorphic to $\mathcal{O}_D^r$, one gets $\mathrm{dh}(G) = 1$ as well. If $x \in X$ is a point, then $\mathrm{dh}(k(x)) = 2$. Finally, if $0 \to F' \to F \to F'' \to 0$ is a short exact sequence then $\mathrm{dh}(F') \leq \max\{\mathrm{dh}(F), \mathrm{dh}(F'') - 1\}$. If F is torsion free and $F' \subset F$ is locally free, then F/F' cannot contain 0-dimensional submodules.

5.1 The Serre Correspondence

The Serre correspondence relates rank two vector bundles on a surface X to subschemes of codimension 2. We begin with some easy observations.

If F is a torsion free sheaf of rank 1, then $F^{\vee\vee} =: L$ is a line bundle and $\mathcal{I} := F \otimes L^{\vee} \subset \mathcal{O}_X$ is the ideal sheaf of a subscheme Z of codimension at least 2, i.e. $F = L \otimes \mathcal{I}_Z$. Using the Hirzebruch-Riemann-Roch Theorem one gets $c_1(F) = c_1(L)$ and $c_2(F) = c_2(L \otimes \mathcal{I}_Z) = -c_2(\mathcal{O}_Z) = \ell(Z)$. Any torsion free sheaf F of arbitrary rank has a filtration with torsion free factors of rank 1: simply take any complete flag of linear subspaces of the stalk of F at the generic point of X and extend them to saturated subsheaves of F. For example, any torsion free sheaf of rank 2 admits an extension

$$0 \to L_1 \otimes \mathcal{I}_{Z_1} \to F \to L_2 \otimes \mathcal{I}_{Z_2} \to 0 \tag{5.1}$$

and the invariants of F are given by the product formula: $\det(F) = L_1 \otimes L_2$, $c_2(F) = c_1(L_1).c_1(L_2) + \ell(Z_1) + \ell(Z_2)$, and

$$\Delta(F) \;=\; 4c_2(F) - c_1^2(F) = 4\Big(\ell(Z_1) + \ell(Z_2)\Big) - \Big(c_1(L_1) - c_1(L_2)\Big)^2 \tag{5.2}$$

$$\geq \;-\Big(c_1(L_1) - c_1(L_2)\Big)^2 = -\Big(2c_1(L_1) - c_1(F)\Big)^2 \tag{5.3}$$

If F is locally free then Z_1 must be empty and if in addition Z_2 is not empty then the extension cannot split.

Theorem 5.1.1 — *Let $Z \subset X$ be a local complete intersection of codimension two, and let L and M be line bundles on X. Then there exists an extension*

$$0 \to L \to E \to M \otimes \mathcal{I}_Z \to 0$$

such that E is locally free if and only if the pair $(L^{\vee} \otimes M \otimes K_X, Z)$ has the Cayley-Bacharach property:

(CB) If $Z' \subset Z$ is a subscheme with $\ell(Z') = \ell(Z) - 1$ and $s \in H^0(X, L^{\vee} \otimes M \otimes K_X)$ with $s|_{Z'} = 0$, then $s|_Z = 0$.

Proof. Let us first show the 'only if' part. Assume the Cayley-Bacharach property does not hold, i.e. there exist a subscheme $Z' \subset Z$ and a section $s \in H^0(X, L^{\vee} \otimes M \otimes K_X)$ such that $\ell(Z') = \ell(Z) - 1$ and $s|_{Z'} = 0$ but $s|_Z \neq 0$. We have to show that given any extension $\xi : 0 \to L \to E \to M \otimes \mathcal{I}_Z \to 0$ the sheaf E is not locally free. Use the exact sequence $0 \to \mathcal{I}_Z \to \mathcal{I}_{Z'} \to k(x) \to 0$ induced by the inclusion $Z' \subset Z$ and the assumption to show that $H^1(X, L^{\vee} \otimes M \otimes K_X \otimes \mathcal{I}_Z) \to H^1(X, L^{\vee} \otimes M \otimes K_X \otimes \mathcal{I}_{Z'})$ is injective. The dual of this map is the natural homomorphism $\mathrm{Ext}^1(M \otimes \mathcal{I}_{Z'}, L) \to \mathrm{Ext}^1(M \otimes \mathcal{I}_Z, L)$ which is, therefore, surjective. Hence any extension ξ fits into a commutative diagram of the form

$$
\begin{array}{ccccccccc}
& & & & 0 & & 0 & & \\
& & & & \downarrow & & \downarrow & & \\
0 & \to & L & \to & E & \to & M \otimes \mathcal{I}_Z & \to & 0 \\
& & \| & & \downarrow & & \downarrow & & \\
0 & \to & L & \to & E' & \to & M \otimes \mathcal{I}_{Z'} & \to & 0 \\
& & & & \downarrow & & \downarrow & & \\
& & & & k(x) & = & k(x) & & \\
& & & & \downarrow & & \downarrow & & \\
& & & & 0 & & 0 & &
\end{array}
$$

Since L and $M \otimes \mathcal{I}_{Z'}$ are torsion free, E' is torsion free as well. Hence the sequence $0 \to E \to E' \to k(x) \to 0$ is non-split and E cannot be locally free.

For the other direction we use the assumption that Z is a local complete intersection. This implies that there are only finitely many subschemes $Z' \subset Z$ with $\ell(Z') = \ell(Z) - 1$. For let x be a closed point in the support of Z. Then there is presentation

$$
0 \longrightarrow \mathcal{O}_{X,x} \xrightarrow{\binom{f_2}{-f_1}} \mathcal{O}_{X,x}^{\oplus 2} \xrightarrow{(f_1\ f_2)} \mathcal{I}_{Z,x} \longrightarrow 0.
$$

Applying the functor $\mathcal{H}om(k(x), \cdot)$, we find $\mathcal{E}xt^1(k(x), \mathcal{I}_Z) \cong k(x)$, since $f_1, f_2 \in \mathfrak{m}_x$, so that there is precisely one subscheme $Z' \subset Z$ with $\mathcal{I}_{Z'}/\mathcal{I}_Z \cong k(x)$.

Suppose now that

$$
\xi : 0 \to L \to E \to M \otimes \mathcal{I}_Z \to 0
$$

is a non-locally free extension. Then there exists a non-split exact sequence $0 \to E \to E' \to k(x) \to 0$ where x is a singular point of E. The saturation of L in E' can differ from L only in the point x. Since L is locally free, it is saturated in E' as well. Thus we get a commutative diagram of the above form. Hence, the extension class ξ is contained in the image of the homomorphism $\mathrm{Ext}^1(M \otimes \mathcal{I}_{Z'}, L) \to \mathrm{Ext}^1(M \otimes \mathcal{I}_Z, L)$. Since the Cayley-Bacharach property ensures that the map $\mathrm{Ext}^1(M \otimes \mathcal{I}_{Z'}, L) \to \mathrm{Ext}^1(M \otimes \mathcal{I}_Z, L)$ is not surjective, we can choose ξ such that it is not contained in the image of this map for any of the finitely many Z' that could occur. The corresponding E will be locally free. $\square$

The Cayley-Bacharach property clearly holds for all Z if $H^0(X, L^\vee \otimes M \otimes K_X) = 0$.

Examples 5.1.2 — *i)* Let $X = \mathbb{P}^2$ and $x \in X$. Using Serre duality and the exact sequence $0 \to \mathcal{I}_x \to \mathcal{O}_X \to k(x) \to 0$, we find that $\mathrm{Ext}^1(\mathcal{I}_x, \mathcal{O}_X) \cong H^1(X, \mathcal{I}_x(-3))^\vee \cong H^0(X, k(x))^\vee \cong k$. Hence, up to scalars there is a unique non-split extension $0 \to \mathcal{O}_X \to E_x \to \mathcal{I}_x \to 0$. Since $H^0(X, K_X) = 0$, the Cayley-Bacharach Condition is satisfied and, therefore, this extension is locally free. Moreover, E_x is μ-semistable. Thus every point $x \in X$ corresponds to a μ-semistable vector bundle E_x.

ii) Let X be an arbitrary smooth surface and let $x \in X$ be a base point of the linear system $|L^\vee \otimes M \otimes K_X|$, i.e. all global sections of $L^\vee \otimes M \otimes K_X$ vanish in x. Then $(L^\vee \otimes M \otimes K_X, x)$ satisfies (CB). Hence there exists a locally free extension of the form $0 \to L \to E \to M \otimes \mathcal{I}_x \to 0$.

Analogously, if $x, y \in X$ are two points which *cannot* be separated by the linear system $|L^\vee \otimes M \otimes K_X|$, then there exists a locally free extension of the form $0 \to L \to E \to M \otimes \mathcal{I}_{\{x,y\}} \to 0$. These two examples are important for the study of surfaces of general type (cf. [229]).

Though the Serre correspondence works for higher dimensional varieties as well, it is in general not easy to produce vector bundles in this way. The reason being that codimension two subschemes of a variety of dimension > 2 are difficult to control.

On surfaces Serre's construction can be used to describe μ-stable rank two vector bundles with given determinant and large Chern number c_2.

Theorem 5.1.3 — *Let X be a smooth surface, H an ample divisor, and $\mathcal{Q} \in \mathrm{Pic}(X)$ a line bundle. Then there is a constant c_0 such that for all $c \geq c_0$ there exists a μ-stable rank two vector bundle E with $\det(E) \cong \mathcal{Q}$ and $c_2(E) = c$.*

Proof. First observe that it suffices to prove the theorem under the additional assumption that $\deg(\mathcal{Q})$ is sufficiently positive. For if the theorem holds for $\mathcal{Q}' = \mathcal{Q}(2nH)$, $n \gg 0$, and gives a μ-stable vector bundle E' with determinant $\mathcal{Q}'$ and second Chern class $c' \geq c_0'$ for some constant c_0', then $E = E' \otimes \mathcal{O}_X(-nH)$ is also μ-stable, has determinant $\mathcal{Q}$ and Chern class $c_2(E) = c' - nH.(c_1(\mathcal{Q}) + nH)$. Hence $c_0 = c_0' - nH.(c_1(\mathcal{Q}) + nH)$ will do.

Thus we may assume that $\deg(\mathcal{Q}) > 0$. The idea is to construct E as an extension of the form

$$0 \to \mathcal{O}_X \to E \to \mathcal{Q} \otimes \mathcal{I}_Z \to 0, \tag{5.4}$$

so that indeed $\det(E) = \mathcal{Q}$ and $c_2(E) = \ell(Z)$. Let $\ell_1 = h^0(K_X \otimes \mathcal{Q})$. Then for a generic 0-dimensional subscheme Z' of length $\ell(Z') \geq \ell_1$ the sheaf $K_X \otimes \mathcal{Q} \otimes \mathcal{I}_{Z'}$ has no non-trivial sections, so that for a generic subscheme Z of length $\ell(Z) > \ell_1$ the pair $(K_X \otimes \mathcal{Q}, Z)$ has the Cayley-Bacharach property (CB). Hence, under this hypothesis there exists an extension as above with locally free E. Suppose $M \subset E$ were a destabilizing line bundle. It follows from the inequality $\mu(M) \geq \mu(E) = \frac{1}{2}c_1(\mathcal{Q}).H > 0 = \mu(\mathcal{O}_X)$ that M cannot be contained in $\mathcal{O}_X$. Thus the composite homomorphism $M \to E \to \mathcal{Q} \otimes \mathcal{I}_Z$ is nonzero. It vanishes along a divisor D with $Z \subset D$ and $\deg(D) = \mu(\mathcal{Q}) - \mu(M) \leq \frac{1}{2}c_1(Q).H =: d$. The family of effective divisors of degree less than or equal to d is bounded. (This can be proved using the techniques developed in Chapter 3 or more easily using Chow points. For a proof see Lecture 16 in [192].) Let Y denote the Hilbert scheme that parametrizes effective divisors on X of degree $\leq d$, and let ℓ_2 be its dimension. For any integer $\ell > \max\{\ell_1, \ell_2\}$ let $\widetilde{Y}$ be the relative Hilbert scheme of pairs $[Z \subset D]$ where $[D] \in Y$ is an effective divisor and $Z \subset D$ is a tuple of ℓ distinct closed points on D. Then for each $[D] \in Y$ the fibre $\widetilde{Y}_{[D]}$ of the projection $\widetilde{Y} \to Y$ has dimension ℓ, so that $\dim(\widetilde{Y}) = \ell + \ell_2$. The image of $\widetilde{Y}$ in $\mathrm{Hilb}^\ell(X)$ under the projection $[Z \subset X] \mapsto Z$ has dimension $\leq \ell + \ell_2 < 2\ell = \dim(\mathrm{Hilb}^\ell(X))$. Hence, if Z is a generic ℓ-tuple of points, a divisor D containing Z and having degree $\leq d$ does not exist, which implies that the corresponding E is indeed μ-stable. $\square$

Remark 5.1.4 — The same method allows to construct μ-stable bundles E with vanishing obstruction space $\mathrm{Ext}^2(E, E)_0$. Such bundles correspond to smooth points in the moduli space $M(2, \mathcal{Q}, c)$. (Note that the vanishing condition is twist invariant, hence we may again assume that $\mathcal{Q}$ is as positive as we choose.)

Indeed, tensorizing the exact sequence (5.4) by $E^\vee \otimes K_X$, $\mathcal{Q}^\vee \otimes K_X$ and K_X, respectively, we get sequences

$$0 \to E^\vee \otimes K_X \to \mathcal{E}nd(E) \otimes K_X \to E \otimes \mathcal{I}_Z \otimes K_X \to 0 \tag{5.5}$$

$$0 \to \mathcal{Q}^\vee \otimes K_X \to E^\vee \otimes K_X \to \mathcal{I}_Z \otimes K_X \to 0 \tag{5.6}$$

$$0 \to K_X \to E \otimes K_X \to \mathcal{Q} \otimes \mathcal{I}_Z \otimes K_X \to 0 \tag{5.7}$$

From these we can read off that

$$
\begin{aligned}
\operatorname{ext}^2(E, E)_0 + h^2(\mathcal{O}_X) &= h^2(\mathcal{E}nd(E)) = h^0(\mathcal{E}nd(E) \otimes K_X) \\
&\leq h^0(E^\vee \otimes K_X) + h^0(E \otimes \mathcal{I}_Z \otimes K_X) \\
&\leq h^0(E^\vee \otimes K_X) + h^0(E \otimes K_X) \\
&\leq h^0(\mathcal{Q}^\vee \otimes K_X) + h^0(\mathcal{I}_Z \otimes K_X) \\
&\quad + h^0(K_X) + h^0(\mathcal{Q} \otimes K_X \otimes \mathcal{I}_Z),
\end{aligned}
$$

thus $\operatorname{ext}^2(E, E)_0 \leq h^0(\mathcal{Q}^\vee \otimes K_X) + h^0(\mathcal{I}_Z \otimes K_X) + h^0(\mathcal{Q} \otimes K_X \otimes \mathcal{I}_Z)$. The first term on the right hand side vanishes if $\deg(\mathcal{Q}) > \deg(K_X)$. The second and the third term vanish for generic Z of sufficiently great length. $\qquad\square$

The theorem above asserts the existence of μ-stable vector bundles for large second Chern numbers. It is not known if one can find stable bundles with given second Chern class $\tilde{c}_2 \in CH^2(X)$ and $c_2 \gg 0$. More precisely, one should ask if for a given line bundle $\mathcal{Q} \in \operatorname{Pic}(X)$ and a class $c \in CH^2(X)$ of degree zero one can construct a μ-stable rank two vector bundle E with $\tilde{c}(E) = c + c_2(E) \cdot x$, where $x \in X$ is a fixed base point and $c_2(E)$ is considered as an integer.

For the rest of Section 5.1 we assume for simplicity that our surface X is defined over the complex numbers. Since the Albanese variety $\operatorname{Alb}(X)$ is a first, though in general very rough, approximation of $CH^2(X)$, the following result can be regarded as a partial answer. Before stating the result, let us briefly recall the notion of the Albanese variety of a smooth variety X. By definition, $\operatorname{Alb}(X)$ is the abelian variety $H^0(X, \Omega_X)^\vee / H_1(X, \mathbb{Z})$ and, after having fixed a base point $x \in X$, the Albanese map is the morphism defined by

$$
A : X \to \operatorname{Alb}(X), \quad y \mapsto \int_x^y .
$$

The image of X generates $\operatorname{Alb}(X)$ as an abelian variety. In particular, the induced morphism $A : X^\ell \to \operatorname{Alb}(X)$, given by addition, is surjective for sufficiently large ℓ. Note that $A : X^\ell \to \operatorname{Alb}(X)$ is invariant with respect to the action of the symmetric group on X^ℓ. It therefore factors through the symmetric product and thus induces a morphism $A : \operatorname{Hilb}^\ell(X) \to \operatorname{Alb}(X)$. There is also a group homomorphism $\tilde{A} : CH^2(X) \to \operatorname{Alb}(X)$ which commutes with A and the map $X \to CH^2(X), x \mapsto [x]$. Both A and $\tilde{A}$ depend on the choice of the base point x. As a general reference for the Albanese map we recommend [254].

Proposition 5.1.5 — *For given $\mathcal{Q} \in \operatorname{Pic}(X)$, $x \in X$, $a \in \operatorname{Alb}(X)$, and a polarization H one can find an integer c_0 such that for $c \geq c_0$ there exists a μ-stable rank two vector bundle E with $\det(E) \cong \mathcal{Q}$, $c_2(E) = c$ and $\tilde{A}(\tilde{c}_2(E)) = a$.*

Proof. As above, we may assume that $\deg(\mathcal{Q}) > 0$. If Z is a codimension two subscheme and if E is a locally free sheaf fitting into a short exact sequence $0 \to \mathcal{O}_X \to E \to \mathcal{Q} \otimes \mathcal{I}_Z \to 0$, then $A(\tilde{c}_2(E)) = A(Z)$. Hence it is enough to show that the open subset $U \subset \mathrm{Hilb}^\ell(X)$ of those subschemes Z, for which a μ-stable locally free extension exists, maps surjectively to $\mathrm{Alb}(X)$. In the proof of 5.1.3 we have seen that U contains the set U' of all reduced Z which are not contained in any effective divisor D of degree $\leq d$ (notations as in 5.1.3) and satisfy $h^0(\mathcal{Q} \otimes K_X \otimes \mathcal{I}_{Z'}) = 0$ for all $Z' \subset Z$ with $\ell(Z') = \ell(Z) - 1$. We have also seen that the set of $Z \in \mathrm{Hilb}^\ell(X)$ that are contained in some divisor D as above has codimension $\geq \ell - \ell_2$. Choosing ℓ large enough we can make this codimension greater than $q = h^1(\mathcal{O}_X)$ which is the dimension of $\mathrm{Alb}(X)$ and hence an upper bound for the codimension of any fibre of the morphism $A : \mathrm{Hilb}^\ell(X) \to \mathrm{Alb}(X)$. Hence it suffices to show that $A : \mathrm{Hilb}^\ell(X)_0 \to \mathrm{Alb}(X)$ is surjective, where $\mathrm{Hilb}^\ell(X)_0 \subset \mathrm{Hilb}^\ell(X)$ is the open subscheme of all reduced $Z \in \mathrm{Hilb}^\ell(X)$ with $h^0(\mathcal{Q} \otimes K_X \otimes \mathcal{I}_{Z'}) = 0$ for all $Z' \subset Z$ of colength one.

Let $C \in |mH|$ be a smooth ample curve containing the fixed base point x. Since the group $H^1(X, \mathcal{O}_X(-C))$ vanishes, the restriction map $H^1(X, \mathcal{O}_X) \to H^1(C, \mathcal{O}_C)$ is injective. Using Hodge decomposition, this map is complex conjugate to the restriction map $H^0(X, \Omega_X) \to H^0(C, \Omega_C)$. It follows that the dual homomorphism $H^0(C, \Omega_C)^\vee \to H^0(X, \Omega_X)^\vee$ is surjective, and therefore the group homomorphism $\mathrm{Alb}(C) \to \mathrm{Alb}(X)$ is surjective as well. The Albanese map $S^\ell(C) \to \mathrm{Alb}(C) = \mathrm{Pic}^0(C)$ can also be described by $C \supset Z \mapsto \mathcal{O}_C(Z - \ell \cdot x) \in \mathrm{Pic}^0(C)$. Hence it suffices to find for any given line bundle $M \in \mathrm{Pic}^0(C)$ a reduced subscheme $Z \subset C$ such that the following conditions are satisfied: (1) $\mathcal{O}_C(Z - \ell \cdot x) \cong M$ and (2) $h^0(X, \mathcal{Q} \otimes K_X \otimes \mathcal{I}_{Z'}) = 0$ for every $Z' \subset Z$ of colength one. If $m \gg 0$ then $H^0(X, \mathcal{Q} \otimes K_X \otimes \mathcal{O}_X(-C)) = 0$, so that property (2) follows from the fact that $H^0(C, \mathcal{Q} \otimes K_X \otimes \mathcal{O}_C(-Z')) = 0$ for sufficiently large ℓ and any scheme $Z' \subset C$ of length $\ell - 1$. Finally, $M(\ell \cdot x)$ is very ample for $\ell \gg 0$ independently of M. Hence we easily find a reduced $Z \subset C$ with $\mathcal{O}_C(Z) \cong M(\ell \cdot x)$, i.e. satisfying condition (1). □

It is only natural to ask if Serre's construction can also be used to produce higher rank bundles. This is in fact possible as will be explained shortly. As a generalization of 5.1.3, 5.1.4, and 5.1.5 for the higher rank case one can prove

Theorem 5.1.6 — *For given $\mathcal{Q} \in \mathrm{Pic}(X)$, $r \geq 2$, $a \in \mathrm{Alb}(X)$, $x \in X$, and a polarization H one can find a constant c_0 such that for $c \geq c_0$ there exists a μ-stable vector bundle E with $\mathrm{rk}(E) = r$, $\det(E) \cong \mathcal{Q}$, and $c_2(E) = c$, such that $H^2(X, \mathcal{E}nd_0(E)) = 0$ and $\tilde{A}(\tilde{c}_2(E)) = a$.*

Proof. We only indicate the main idea of the proof. The details, though computationally more involved, are quite similar to the ones encountered before. First, one generalizes Serre's construction and considers extensions of the form

$$0 \to L \to E \to \bigoplus_{i=1}^{r-1} M_i \otimes \mathcal{I}_{Z_i} \to 0.$$

Assuming that all Z_i's are reduced and $Z_i \cap Z_j = \emptyset$ $(i \neq j)$, one can prove that a locally free extension exists if and only if $(L^\vee \otimes M_i \otimes K_X, Z_i)$ satisfies the Cayley-Bacharach property for all $i = 1, \dots, r - 1$. In order to construct vector bundles as asserted by the theorem one considers extensions of the form

$$0 \to \mathcal{Q}((1 - r)nH) \to E \to \bigoplus_{i=1}^{r-1} \mathcal{I}_{Z_i}(nH) \to 0,$$

for some sufficiently large integer n. Twisting with $\mathcal{Q}^\vee \otimes \mathcal{O}_X((r - 1)nH)$ yields the exact sequence

$$0 \to \mathcal{O}_X \to E' \to \bigoplus_{i=1}^{r-1} \mathcal{I}_{Z_i} \otimes \mathcal{Q}^\vee(rnH) \to 0,$$

where $E' := E \otimes \mathcal{Q}^\vee((r - 1)nH)$. Then the Cayley-Bacharach property holds for generic Z_i with $\ell(Z_i) \geq h^0(\mathcal{Q}^\vee(rnH) \otimes K_X) + 1$. Suppose now, that $F \subset E'$ is a destabilizing locally free subsheaf of rank $s < r$. If n was chosen large enough so that $\mu(\mathcal{Q}^\vee(rnH)) > 0$, then F must be contained in $\bigoplus_i \mathcal{I}_{Z_i} \otimes \mathcal{Q}^\vee(rnH)$, and passing to the exterior powers there is a nonzero and therefore injective homomorphism

$$\det(F) \otimes \mathcal{Q}^s(-rsnH) \longrightarrow \bigoplus_{1 \leq i_1 < \dots < i_s \leq r-1} \mathcal{I}_{Z_{i_1} \cup \dots \cup Z_{i_s}}.$$

(Note that the sheaf on the right hand side is the quotient of $\Lambda^s(\oplus_i \mathcal{I}_{Z_i})$ by its torsion submodule.) Thus there is an effective divisor D of degree

$$\deg(D) = s \cdot \mu(\mathcal{Q}^\vee(rnH)) - s \cdot \mu(F) \leq \mu(\mathcal{Q}^\vee(rnH))$$

which contains at least s of the $r - 1$ subschemes Z_i. As in the proof of 5.1.3 this is impossible if all Z_i are general and have sufficiently great length.

The vanishing of $H^2(X, \mathcal{E}nd(E)_0) = \mathrm{Ext}^2(E, E)_0$ is achieved as in 5.1.4. It is also not difficult to see that the proof of 5.1.5 goes through. $\qquad\square$

5.2 Elementary Transformations

Now, we come to the third example discussed in the introduction.

Definition 5.2.1 — *Let C be an effective divisor on the surface X. If F and G are vector bundles on X and C, respectively, then a vector bundle E on X is obtained by an elementary transformation of F along G if there exists an exact sequence*

$$0 \to E \to F \to i_*G \to 0,$$

where i denotes the embedding $C \subset X$.

If no confusion is likely, we just write G instead of i_*G, meaning G with its natural $\mathcal{O}_X$-structure.

Proposition 5.2.2 — *If F and G are locally free on X and C, respectively, then the kernel E of any surjection $\varphi : F \to i_*G$ is locally free. Moreover, if ρ denotes the rank of G, one has $\det(E) \cong \det(F) \otimes \mathcal{O}_X(-\rho \cdot C)$ and $c_2(E) = c_2(F) - \rho C.c_1(F) + \frac{1}{2}\rho C.(\rho C + K_X) + \chi(G)$.*

Proof. Since locally $G \cong \mathcal{O}_C^{\oplus \rho}$ and $0 \to \mathcal{O}_X(-C) \to \mathcal{O}_X \to \mathcal{O}_C \to 0$ is a locally free resolution on X, the sheaf i_*G is of homological dimension ≤ 1. This implies that $\mathrm{dh}(E) = 0$, i.e. E is locally free. The isomorphism $\det(E) \cong \det(F) \otimes \det(i_*G)^\vee \cong \det(F)(-\rho C)$ follows from the fact that G is trivial on the complement of finitely many points on C. Thus $\det(i_*G)$ and $\det(i_*\mathcal{O}_C^{\oplus \rho})$ are isomorphic on the complement of finitely many points, hence $\det(i_*G) \cong \det(i_*\mathcal{O}_C^{\oplus \rho}) \cong \mathcal{O}_X(\rho C)$. The formula for the second Chern class follows from $\chi(E) = \chi(F) - \chi(G)$ and the Hirzebruch-Riemann-Roch formula for E and F: $\chi(F) = \frac{1}{2}c_1(F).(c_1(F) - K_X) - c_2(F) + \mathrm{rk}(F)\chi(\mathcal{O}_X)$ and $\chi(E) = \frac{1}{2}c_1(E).(c_1(E) - K_X) - c_2(E) + \mathrm{rk}(F)\chi(\mathcal{O}_X)$. Inserting $c_1(E) = c_1(F) - \rho C$ gives the desired result. $\qquad\square$

Note that for a smooth (or at least reduced) curve C the characteristic $\chi(G)$ can be written as $\chi(G) = \deg(G) + \rho(1 - g(C)) = \deg(G) - \frac{\rho}{2}C.(K_X + C)$. Hence $c_2(E) = c_2(F) + (\deg(G) - \rho C.c_1(F)) + \frac{\rho(\rho-1)}{2}C^2$.

Example 5.2.3 — A trivial example is $\mathcal{O}_X(-C)$, which is the elementary transform of $\mathcal{O}_X$ along $\mathcal{O}_C$. Another example is provided by the sheaf $\Omega_X(\log C)$ of differentials with logarithmic poles along a smooth curve $C \subset X$. This is the locally free sheaf that is locally generated by dx_1/x_1 and dx_2, where (x_1, x_2) is a local chart and $x_1 = 0$ is the equation for C. The restriction map $\Omega_X \twoheadrightarrow \Omega_C$ twisted by $\mathcal{O}(C)$ yields an exact sequence

$$0 \to \Omega_X(\log C) \to \Omega_X(C) \to \Omega_C(C) \to 0.$$

Indeed, $f_1 dx_1/x_1 + f_2 dx_2/x_1$ is mapped to zero in $\Omega_C(C)$ if and only if $f_2 = g \cdot x_1$. Thus $\Omega_X(\log C)$ is the elementary transform of $\Omega_X(C)$ along $\Omega_C(C)$.

Let E be any vector bundle of rank r on a smooth projective surface X. For sufficiently large n the bundle $E^\vee(nH)$ is globally generated. The discussion in the introduction tells us that there is a short exact sequence

$$0 \to \mathcal{O}_X^{\oplus r} \to E^\vee(nH) \to M \to 0$$

for some line bundle M on a smooth curve $C \subset X$. Dualizing this sequence and twisting with $\mathcal{O}_X(nH)$ yields

$$0 \to E \to \mathcal{O}_X(nH)^{\oplus r} \to L \to 0,$$

with $L := \mathcal{E}xt_X^1(M, \mathcal{O}_X(nH))$. Note that L is a line bundle on C, as can easily be seen from the fact that, locally, $M \cong \mathcal{O}_C$ and $\mathcal{E}xt_X^1(\mathcal{O}_C, \mathcal{O}_X) = \mathcal{O}_C(C)$. In fact, $L \cong M^\vee \otimes \mathcal{O}_C(C + nH)$. Thus we have proved:

Proposition 5.2.4 — *Every vector bundle E of rank r can be obtained by an elementary transformation of $\mathcal{O}_X^{\oplus r}(nH)$, with $n \gg 0$, along a line bundle on a smooth curve $C \subset X$.* □

Similarly to Serre's construction, elementary transformations can be used to produce μ-stable vector bundles on X.

Theorem 5.2.5 — *For given $\mathcal{Q} \in \operatorname{Pic}(X)$, $r \geq 2$, ample divisor H and integer $c_0 \in \mathbb{Z}$, there exists a μ-stable vector bundle E with $\det(E) \cong \mathcal{Q}$, $\operatorname{rk}(E) = r$ and $c_2(E) \geq c_0$.*

Proof. Let C be a smooth curve. According to the Grothendieck Lemma 1.7.9, the torsion free quotients F of $\mathcal{O}_X^{\oplus r}$ with $\mu(F) \leq \frac{r-1}{r} C.H$ and $\operatorname{rk}(F) < r$ form a bounded family $\mathcal{C}$. Now $\hom(\mathcal{O}_X^{\oplus r}, \mathcal{O}_C(nH))$ grows much faster than $\hom(F, \mathcal{O}_C(nH))$ for any F in the family $\mathcal{C}$. Thus, if n is sufficiently large, a general homomorphism $\varphi : \mathcal{O}_X^{\oplus r} \to \mathcal{O}_C(nH)$ is surjective and does not factor through any $F \in \mathcal{C}$. Let E be the kernel of φ. Then E is locally free with $\det(E) = \mathcal{O}_X(-C)$ and $c_2(E) = nH.C \gg 0$. In order to see that E is μ-stable, let $E' \subset E$ be a saturated proper subsheaf, let F' be the saturation of E' in $\mathcal{O}_X^{\oplus r}$ and consider the subsheaf $F'/E' \subset \mathcal{O}_C(nH)$. If F'/E' is nonzero, then $\det(E') = \det(F') \otimes \mathcal{O}_X(-C)$, hence

$$\mu(E') = \mu(F') - \frac{C.H}{\operatorname{rk}(E')} < 0 - \frac{C.H}{\operatorname{rk}(E)} = \mu(E),$$

and we are done. If on the other hand $F'/E' = 0$ then $F := \mathcal{O}_X^{\oplus r}/E'$ is torsion free and φ factors through F. By construction F cannot be contained in $\mathcal{C}$, hence $\mu(F) > \frac{r-1}{r} C.H$. It follows that

$$\mu(E') = -\frac{r - \operatorname{rk}(E')}{\operatorname{rk}(E')} \mu(F) < -\frac{r - \operatorname{rk}(E')}{\operatorname{rk}(E')} \cdot \frac{r-1}{r} \cdot C.H < -\frac{C.H}{r} = \mu(E).$$

So E is indeed μ-stable.

If $\mathcal{Q}$ is an arbitrary line bundle, choose $m \gg 0$ in such a way that $\mathcal{Q}^{\vee}(rmH)$ is very ample, and pick a general curve $C \in |\mathcal{Q}^{\vee}(rmH)|$. If E is a μ-stable vector bundle constructed according to the recipe above with determinant $\det(E) \cong \mathcal{O}_X(-C) = \mathcal{Q}(-rmH)$, then $E(mH)$ is μ-stable with determinant $\mathcal{Q}$ and large second Chern class. $\square$

Remark 5.2.6 — One can in fact choose φ in the proof of the theorem in such a way that $\operatorname{Ext}^2(E, E)_0 \cong \operatorname{Hom}(E, E \otimes K_X)_0^{\vee}$ vanishes. First check that for n sufficiently large, any homomorphism $\psi_E : E \to E \otimes K_X$ can be extended to a homomorphism $\psi : \mathcal{O}_X^{\oplus r} \to \mathcal{O}_X^{\oplus r} \otimes K_X$. Conversely, such a homomorphism ψ leaves E invariant, if and only if there is a section $\psi' \in H^0(X, K_X)$ such that $\psi'\varphi = \varphi\psi$. It is easy to see that the condition on ψ to be traceless requires φ to factor through a quotient bundle $\mathcal{O}_X^{\oplus s}$, $0 < s < r$. As before, since the family of such quotients is obviously bounded, for sufficiently large n and general φ this will never be the case. $\square$

5.3 Examples of Moduli Spaces

Fibred surfaces. We first show that for certain polarizations on ruled surfaces the moduli space is empty. This will be a consequence of the relation between stability on the surface and stability on the fibres, which can be formulated for arbitrary fibred surfaces. The arguments may give a feeling for Bogomolov's restriction theorem proved in Chapter 7. For simplicity, we only deal with the rank two case, but see Remark 5.3.6.

Let X be a surface, let C be a smooth curve, and let $\pi : X \to C$ be a surjective morphism. Fix Chern classes c_1 and c_2. As usual, let $\Delta := 4c_2 - c_1^2$. By f we denote the homology class of the fibre of π.

Definition 5.3.1 — *A polarization H is called (c_1, c_2)-suitable if and only if for any line bundle $M \in \operatorname{Pic}(X)$ with $-\Delta \leq (2c_1(M) - c_1)^2$ either $f.(2c_1(M) - c_1) = 0$ or $f.(2c_1(M) - c_1)$ and $H.(2c_1(M) - c_1)$ have the same sign.*

Let $\eta \in C$ be the generic point of C and denote the generic fibre $X \times_C \operatorname{Spec}(k(\eta))$ by F_η. If E is a coherent sheaf on X, let E_η be the restriction of E to F_η.

Theorem 5.3.2 — *Let H be a (c_1, c_2)-suitable polarization and let E be a rank two vector bundle with $c_1(E) = c_1$ and $c_2(E) = c_2$. If E is μ-semistable (with respect to H), then E_η is semistable. If E_η is stable, then E is μ-stable.*

Proof. Let E be μ-semistable and let $M' \subset E_\eta$ be a rank one subbundle such that E_η/M' is locally free. Then there exists a unique saturated locally free subsheaf $M \subset E$ of rank one such that $M_\eta = M'$. By (5.3) we have $\Delta \geq -(2c_1(M) - c_1)^2$. If M_η is destabilizing, i.e. if $f.(2c_1(M) - c_1) > 0$ then, since H is (c_1, c_2)-suitable, also $H.(2c_1(M) - c_1) > 0$, contradicting the μ-semistability of E.

Conversely, assume that E_η is stable. If $M \subset E$ is a saturated subsheaf of rank one, then

$$f.c_1(M) = \deg(M_\eta) < \deg(E_\eta)/2 = (f.c_1)/2$$

Hence $f.(2c_1(M) - c_1) < 0$. Since H is (c_1, c_2)-suitable by assumption and since again $\Delta \geq -(2c_1(M) - c_1)^2$ by (5.3), we conclude that $H.(2c_1(M) - c_1) < 0$. Hence E is μ-stable with respect to H. □

Recall from Section 2.3 that E_η is semistable or geometrically stable if and only if the restriction of E to the fibre $\pi^{-1}(t)$ is semistable or geometrically stable, respectively, for all t in a dense open subset of C.

If $c_1(E).f \equiv 1(2)$, then, for obvious arithmetical reasons, E_η is geometrically stable if and only if E_η is semistable. Hence

Corollary 5.3.3 — *If $c_1.f \equiv 1(2)$ and if H is (c_1, c_2)-suitable, then a rank two vector bundle E with $c_1(E) = c_1$ and $c_2(E) = c_2$ is μ-stable if and only if E_η is stable. Moreover, E is μ-semistable if and only if E is μ-stable.* □

Corollary 5.3.4 — *If $X \to C$ is a ruled surface, then there exists no vector bundle E on X that is μ-semistable with respect to a $(c_1(E), c_2(E))$-suitable polarization and that satisfies $c_1(E).f \equiv 1(2)$.*

Proof. This is a consequence of the fact that there is no stable rank 2 bundle on $\mathbb{P}^1$. □

Remark 5.3.5 — The Hodge Index Theorem shows that for any choice of (c_1, c_2) there exists a suitable polarization. Indeed, let H be any polarization and define $H_n = H + nf$. Then H_n is ample for $n \geq 0$ and (c_1, c_2)-suitable for $n \geq \Delta \cdot (H.f)/2$. To see this let $\xi := 2c_1(M) - c_1$ for some line bundle M and assume that $\Delta \geq -\xi^2$. Since $f^2 = 0$ and $f.\left((f.\xi)H_n - (f.H_n)\xi\right) = 0$, the Hodge Index Theorem implies that

$$0 \geq \left((f.\xi)H_n - (f.H_n)\xi\right)^2 = (f.\xi)^2 H_n^2 - 2(f.\xi)(f.H_n)(H_n.\xi) + (f.H_n)^2\xi^2.$$

Dividing by $2(H_n.f)$ we obtain

$$(f.\xi)(H_n.\xi) \geq (f.\xi)^2 \left(\frac{H^2}{2H.f} + n\right) + \frac{H.f}{2}\xi^2.$$

Hence either $f.\xi = 0$, or $(f.\xi)^2 \geq 1$ and therefore

$$(f.\xi)(H_n.\xi) > n - \frac{H.f}{2}\Delta \geq 0$$

for sufficiently large n. The last inequality means that $H_n.\xi$ and $f.\xi$ have the same sign.

Remark 5.3.6 — Theorem 5.3.2 can be easily generalized to the higher rank case. In fact, for $r > 2$ one says that a polarization H is suitable if it is contained in the chamber close to the fibre class f. (For the concept of walls and chambers we refer to Appendix 4.C.) Numerically, this is described by the condition that for all $\xi \in \mathrm{Pic}(X)$ such that $-\frac{r^2}{4}\Delta \leq \xi^2 < 0$ either $\xi.f = 0$ or $\xi.f$ and $\xi.H$ have the same sign. The argument of the previous remark shows that if H is any polarization then $H + nf$ is suitable if $n \geq r^2(H.f)\Delta/8$ (see also Lemma 4.C.5).

K3 surfaces. In the second part of this section two examples of moduli spaces of sheaves on K3 surfaces are studied. We will see how the techniques introduced in the first two sections of this chapter can be applied to produce sheaves and to describe the global structure of the moduli spaces. We hope that studying the examples the reader may get a feeling for the geometry of these moduli spaces. They will also serve as an introduction for the general results on zero- and two-dimensional moduli spaces on K3 surfaces explained in Section 6.1. Both examples share a common feature. Namely, the canonical bundle of the moduli space of stable sheaves is trivial. This is a phenomenon which will be proved in broader generality in Chapter 8 and Chapter 10.

The canonical bundle of a K3 surface is trivial and the Euler characteristic of the structure sheaf is 2. Hence Serre duality takes the form $\mathrm{Ext}^i(A, B) = \mathrm{Ext}^{2-i}(B, A)^\vee$ for any two coherent sheaves A, B. Any stable sheaf E is simple, i.e. $\hom(E, E) = 1$, so that $\mathrm{ext}^2(E, E)_0 = \hom(E, E)_0 = \hom(E, E) - 1 = 0$. Thus any moduli space $M^s(2, \mathcal{Q}, \Delta)$ of stable rank 2 sheaves with determinant $\mathcal{Q}$ and discriminant Δ is empty or smooth of expected dimension

$$\dim M^s(2, \mathcal{Q}, \Delta) = \Delta - (r^2 - 1)\chi(\mathcal{O}) = \Delta - 6 = 4c_2 - c_1^2 - 6.$$

Example 5.3.7 — Let $X \subset \mathbb{P}^3$ be a general quartic hypersurface. By the adjunction formula X has trivial canonical bundle, and by the Lefschetz Theorem on hyperplane sections $\pi_1(X)$ is trivial ([179] Thm. 7.4). Hence X is a K3 surface. Moreover, by the Noether-Lefschetz Theorem (see [90] or [42]) its Picard group is generated by $\mathcal{O}_X(1)$, the restriction of the tautological line bundle on $\mathbb{P}^3$ to X. In particular, there is no doubt about the polarization of X which therefore will be omitted in the notation.

Consider the moduli spaces $M(2, \mathcal{O}_X(-1), c_2)$. For any rank two sheaf with determinant $\mathcal{O}_X(-1)$ μ-semistability implies μ-stability. Thus $M(2, \mathcal{O}_X(-1), c_2)$ is a smooth projective scheme. If $M(2, \mathcal{O}_X(-1), c_2)$ is not empty then its dimension is $4c_2 - 10$ (since $c_1(\mathcal{O}_X(-1))^2 = \deg(X) = 4$). In particular, if a stable sheaf with these invariants exists then $4c_2 \geq 10$. This is slightly stronger than the Bogomolov Inequality 3.4.1. The smallest possible moduli space is at least two-dimensional. In fact

Claim: $M := M(2, \mathcal{O}_X(-1), 3) \cong X$.

Proof. Since the reflexive hull of a μ-stable sheaf is again μ-stable, any $F \in M$ defines a point $F^{\vee\vee}$ in $M(2, \mathcal{O}_X(-1), c_2)$ with $c_2 \leq 3$. By the inequality above $c_2(F^{\vee\vee}) = 3$, i.e. $F \cong F^{\vee\vee}$ is locally free. For any $F \in M$ the Hirzebruch-Riemann-Roch formula gives $\chi(F) = 3$ and hence $\hom(F, \mathcal{O}_X) = h^2(F) \geq 3$ (we have $h^0(F) = 0$ because of the stability of F). Let $\varphi_i : F \to \mathcal{O}_X$, $i = 1, 2, 3$ be three linearly independent homomorphisms and let φ denote the sum $(\varphi_1, \varphi_2, \varphi_3) : F \to \mathcal{O}_X^3$. We claim that φ fits into a short exact sequence of the form

$$0 \longrightarrow F \overset{\varphi}{\longrightarrow} \mathcal{O}_X^3 \longrightarrow \mathcal{I}_x(1) \longrightarrow 0,$$

where $\mathcal{I}_x$ is the ideal sheaf of a point $x \in X$. If φ were not injective, then $\operatorname{im}(\varphi)$ would be of the form $\mathcal{I}_V(a)$ for some codimension two subscheme V. Since $\mathcal{I}_V(a) \subset \mathcal{O}_X^3$, one has $a \leq 0$. On the other hand, as a quotient of the stable sheaf F the rank one sheaf $\mathcal{I}_V(a)$ has non-negative degree. Therefore, $a = 0$. But then

$$\varphi : F \to \mathcal{I}_V \subset \mathcal{O}_X \subset \mathcal{O}_X^3$$

and hence the φ_i would only span a one-dimensional subspace of $\operatorname{Hom}(F, \mathcal{O}_X)$, which contradicts our choice. Therefore φ is injective. A Chern class calculation shows that its cokernel has determinant $\mathcal{O}_X(1)$ and second Chern class 1. Since $\operatorname{rk}(\operatorname{coker}(\varphi)) = 1$, it is enough to show that $\operatorname{coker}(\varphi)$ is torsion free. If not, let F' be the saturation of F in $\mathcal{O}_X^3$. Then F' is a rank two vector bundle as well and

$$\det(F) \subset \det(F') \cong \mathcal{O}_X(b) \subset \Lambda^2 \mathcal{O}_X^3$$

for some $-1 \leq b \leq 0$. Since both F and F' are locally free, $\det(F') \not\cong \det(F)$; hence $b = 0$. The quotient $\mathcal{O}_X^3/F'$ then is necessarily of the form $\mathcal{I}_V$ for a codimension two subscheme V. But $\operatorname{Hom}(\mathcal{O}_X, \mathcal{I}_V) = 0$ unless $V = \emptyset$, which then implies that $F' \cong \mathcal{O}_X^3$, contradicting again the linear independence of the φ_i. Eventually, we see that indeed any $F \in M$ is part of a short exact sequence of the form

$$0 \to F \to \mathcal{O}_X^3 \to \mathcal{I}_x(1) \to 0.$$

The stability of F implies $H^0(X, F) = 0$, so that the map $H^0(X, \mathcal{O}_X^3) \to H^0(X, \mathcal{I}_x(1)) \cong k^3$ is bijective. Hence $\operatorname{Ext}^1(F, \mathcal{O}_X) \cong H^1(X, F)^\vee = 0$. Inserting this bit of information into the Hirzebruch-Riemann-Roch formula above one concludes that $\hom(F, \mathcal{O}_X) = 3$.

This implies that φ (and hence the short exact sequence) is uniquely determined by F (up to the action of $GL(3)$).

On the other hand, if we start with a point $x \in X$ and denote the kernel of the evaluation map $H^0(X, \mathcal{I}_x(1)) \otimes \mathcal{O}_X \to \mathcal{I}_x(1)$ by F_x, then F_x is locally free and has no global section. Clearly, $h^0(X, F_x) = 0$ implies that F_x is stable; for any subsheaf possibly destabilizing F_x must be isomorphic to $\mathcal{O}_X$. In order to globalize this construction let $\Delta \subset X \times X$ denote the diagonal, $\mathcal{I}_\Delta$ its ideal sheaf, and let p and q be the two projections to X. Define a sheaf $\mathcal{F}$ by means of the exact sequence

$$0 \to \mathcal{F} \to p^*(p_*(\mathcal{I}_\Delta \otimes q^*\mathcal{O}_X(1))) \to \mathcal{I}_\Delta \otimes q^*\mathcal{O}_X(1) \to 0.$$

$\mathcal{F}$ is p-flat and $\mathcal{F}_x := \mathcal{F}|_{p^{-1}(x)} \cong F_x$. Thus $\mathcal{F}$ defines a morphism $X \to M$, $x \mapsto [F_x]$. The considerations above show that this map is surjective, because any F is part of an exact sequence of this form, and injective, because φ is uniquely determined by F. Since both spaces are smooth, $X \to M$ is an isomorphism.　　　　　　　　　　□

We will prove that 'good' two-dimensional components of the moduli space are always closely related to the K3 surface itself (6.1.14). In many instances the rôle of the two factors can be interchanged. Let us demonstrate this in our example. It is straightforward to complete the exact sequence

$$0 \to F_x \to H^0(X, \mathcal{I}_x(1)) \otimes \mathcal{O}_X \to \mathcal{I}_x(1) \to 0$$

to the following commutative diagram with exact rows

$$
\begin{array}{ccccccccc}
0 \to & F_x & \to & H^0(X, \mathcal{I}_x(1)) \otimes \mathcal{O}_X & \to & \mathcal{I}_x(1) & \to 0 \\
& \downarrow & & \downarrow & & \downarrow & \\
0 \to & \Omega_{\mathbb{P}^3}(1)|_X & \to & H^0(\mathbb{P}^3, \mathcal{O}(1)) \otimes \mathcal{O}_X & \to & \mathcal{O}_X(1) & \to 0 \\
& \downarrow & & \downarrow & & \downarrow & \\
0 \to & \mathcal{I}_x & \to & \mathcal{O}_X & \to & k(x) & \to 0.
\end{array}
$$

Both descriptions

$$0 \to F_x \to H^0(X, \mathcal{I}_x(1)) \otimes \mathcal{O}_X \to \mathcal{I}_x(1) \to 0 \qquad (5.8)$$

and

$$0 \to F_x \to \Omega_{\mathbb{P}^3}(1)|_X \to \mathcal{I}_x \to 0 \qquad (5.9)$$

are equivalent. Back to the proof, we had constructed the exact sequence

$$0 \to \mathcal{F} \to p^*(p_*(\mathcal{I}_\Delta \otimes q^*\mathcal{O}_X(1))) \to \mathcal{I}_\Delta \otimes q^*\mathcal{O}_X(1) \to 0$$

on $X \times X$. Restricting this sequence to $\{x\} \times X$ yields (5.8), and restricting it to the fibre $X \times \{x\}$ yields (5.9). (Use the exact sequence $0 \to \mathcal{I}_\Delta \otimes q^*\mathcal{O}(1) \to q^*\mathcal{O}(1) \to \mathcal{O}(1)_\Delta \to$

0 to see that $p_*(\mathcal{I}_\Delta \otimes q^*\mathcal{O}(1)) \cong \Omega_{\mathbb{P}^3}(1)|_X$.) Thus the vector bundle $\mathcal{F}$ on $X \times X$ identifies each factor as the moduli space of the other.

Example 5.3.8 — Let $\pi : X \to \mathbb{P}^1$ be an elliptic K3 surface with irreducible fibres. We furthermore assume that $X \to \mathbb{P}^1$ has a section $\sigma \subset X$. By the adjunction formula σ is a (-2)-curve. For the existence of such surfaces see [22]. If f denotes the class of a fibre, then $H = \sigma + 3f$, and more generally, $H_m := H + mf$ for $m \geq 0$, are ample divisors. This follows from the Nakai-Moishezon Criterion.

If $c_1.f \equiv 1(2)$, then for fixed c_2 and $m \gg 0$, the μ_{H_m}-semistability of a rank two vector bundle is equivalent to its μ_{H_m}-stability (cf. Corollary 5.3.3).

Claim: *If m is sufficiently large, then* $M := M_{H_m}(2, \mathcal{O}_X(\sigma - f), 1) \cong X$.

There are two ways to prove the claim. The first uses Serre's construction, and the second relies on the existence of a certain universal stable bundle on X, discovered by Friedman and Kametani-Sato, that restricts to a stable bundle on any fibre. For both approaches one needs Corollary 5.3.3 which says:
For $m \gg 0$ a bundle E with determinant Q such that $c_1(Q).f = 1$ is μ-stable (with respect to the polarization H_m) if and only if its restriction to the generic fibre is stable.
Note that $\Delta = 4c_2 - (\sigma - f)^2 = 4 - (-2 - 2) = 8$, so that $\dim(M) = 2$. As before this implies that any μ-stable sheaf in M is locally free.

Proof of the Claim via Serre's construction.
Let $[E] \in M$ be a closed point. By the Hirzebruch-Riemann-Roch formula $\chi(E) = \frac{1}{2}(\sigma - f)^2 - c_2 + 4 = 1$, and by stability $h^2(E) = \hom(E, \mathcal{O}_X) = 0$, so that $h^0(E) \geq 1$. Since the restriction of E to the generic fibre F_η is stable of degree 1, a global section $s \in H^0(X, E)$ can vanish in codimension two or along a divisor not intersecting the generic fibre, i.e. a union of fibres. Hence one always has an exact sequence of the form

$$0 \to \mathcal{O}_X(nf) \to E \to \mathcal{I}_Z \otimes \mathcal{O}_X(\sigma - (n+1)f) \to 0.$$

A comparison of the Chern classes yields the condition $1 = c_2(E) = n + \ell(Z)$. Hence either $n = 0$ and $Z = \{x\}$, i.e. s vanishes in codimension two at exactly one point $x \in X$, or $n = 1$ and $Z = \emptyset$, i.e. E is an extension of the line bundle $\mathcal{O}_X(\sigma - 2f)$ by $\mathcal{O}_X(f)$. The following calculations will be useful: essentially because of $\sigma^2 = -2$, there is no effective divisor D such that $\sigma \sim \nu f + D$ for any integer $\nu \geq 1$. This means that the groups $h^0(\mathcal{O}_X(\sigma - \nu f)) = h^2(\mathcal{O}_X(\nu f - \sigma))$ vanish. On the other hand, $h^2(\mathcal{O}_X(\sigma - \nu f)) = h^0(\mathcal{O}_X(\nu f - \sigma)) = 0$ because of stability: $\deg(\mathcal{O}_X(\nu f - \sigma)) = (\nu f - \sigma)(\sigma + (m+3)f) = \nu - m - 1 < 0$ for $m > \nu - 1$. It follows that $h^1(\mathcal{O}_X(\sigma - \nu f)) = h^1(\mathcal{O}(\nu f - \sigma)) = -\chi(\mathcal{O}_X(\sigma - \nu f)) = \nu - 1$.
Let us now take a closer look at the two cases $n = 0$ and $n = 1$:

i) An extension

$$0 \to \mathcal{O}_X \to E \to \mathcal{I}_x \otimes \mathcal{O}_X(\sigma - f) \to 0$$

is stable if and only if it is non-split and $x \notin \sigma$. *Moreover, for given* x *there is exactly one non-split extension.*

Proof. First check that indeed $\mathrm{ext}^1(\mathcal{I}_x \otimes \mathcal{O}_X(\sigma - f), \mathcal{O}_X) = 1$. Let E be the unique non-trivial extension. E is stable if and only if the restriction E_η to the generic fibre is stable. Let F be any smooth fibre that does not contain x. It suffices to show that the restriction map

$$\rho : \mathrm{Ext}^1_X(\mathcal{I}_x \otimes \mathcal{O}_X(\sigma - f), \mathcal{O}_X) \longrightarrow \mathrm{Ext}^1_F(\mathcal{O}_F(\sigma \cap F), \mathcal{O}_F)$$

is nonzero, for then E_F is stable and hence E_η is stable. Now ρ is dual to

$$\rho^\vee : H^0(F, \mathcal{O}_F(\sigma \cap F)) \longrightarrow H^1(X, \mathcal{I}_x(\sigma - f)).$$

Since $H^0(X, \mathcal{I}_x(\sigma - f)) = 0$, the kernel of ρ is precisely $H^0(X, \mathcal{I}_x(\sigma))$. Finally, since $h^0(\mathcal{O}_F(\sigma \cap F)) = 1$, the homomorphism $\rho^\vee$ is nonzero if and only if $h^0(\mathcal{I}_x(\sigma)) = 0$, i.e. if and only if $x \notin \sigma$. □

ii) An extension of the form

$$0 \to \mathcal{O}_X(f) \to E \to \mathcal{O}_X(\sigma - 2f) \to 0$$

is stable if and only if it does not split. The space of non-isomorphic non-trivial extensions is parametrized by $\mathbb{P}(H^1(\mathcal{O}_X(3f - \sigma))^\vee) \cong \mathbb{P}^1$.

Proof. Again, it suffices to check that a given non-trivial extension class is not mapped to zero by the restriction homomorphism

$$\begin{array}{ccc} \mathrm{Ext}^1_X(\mathcal{O}_X(\sigma - 2f), \mathcal{O}_X(f)) & \xrightarrow{\rho} & \mathrm{Ext}^1_F(\mathcal{O}_F(\sigma \cap F), \mathcal{O}_F) \\ \cong H^1(\mathcal{O}_X(3f - \sigma)) & & \cong H^1(\mathcal{O}_F(-\sigma \cap F)) \end{array}$$

for a general fibre F. Consider the exact sequence

$$0 \longrightarrow H^1(\mathcal{O}_X(2f - \sigma)) \xrightarrow{\cdot F} H^1(\mathcal{O}_X(3f - \sigma)) \longrightarrow H^1(\mathcal{O}_F(-\sigma \cap F)) \longrightarrow 0$$

of vector spaces of dimensions 1, 2 and 1, respectively. Using the Leray Spectral Sequence we can identify $H^1(\mathcal{O}_X(\nu f - \sigma)) \cong H^0(R^1\pi_*\mathcal{O}_X(-\sigma) \otimes \mathcal{O}_{\mathbb{P}^1}(\nu))$ for $\nu = 2, 3$, which implies that $R^1\pi_*\mathcal{O}_X(-\sigma) \cong \mathcal{O}_{\mathbb{P}^1}(-2)$. In this way the problem reduces to showing that varying the base point $\pi(F) \in \mathbb{P}^1$, which is the zero locus of a section $s \in H^0(\mathcal{O}_{\mathbb{P}^1}(1))$, leads to essentially different embeddings $H^0(\mathcal{O}_{\mathbb{P}^1}) \xrightarrow{\cdot s} H^0(\mathcal{O}_{\mathbb{P}^1}(1))$, which is obvious. □

iii) Let $\Delta \subset X \times X$ be the diagonal, $\mathcal{I}_\Delta$ its ideal sheaf and p and q the projections to the two factors. It follows from the Base Change Theorem and our computations of cohomology groups above that $B := R^1 p_*(\mathcal{I}_\Delta \otimes q^* \mathcal{O}_X(\sigma - f))$ is a line bundle. Similarly, one checks that $\mathrm{Ext}^1(\mathcal{I}_\Delta \otimes q^* \mathcal{O}_X(\sigma - f), p^*B) \cong \mathrm{Hom}(B, B) \cong \mathbb{C}$. Let

$$0 \to p^*B \to \mathcal{F} \to \mathcal{I}_\Delta \otimes q^* \mathcal{O}_X(\sigma - f) \to 0$$

be the unique non-trivial extension on $X \times X$. Then $\mathcal{F}$ is p-flat and $\mathcal{F}_{X \setminus \sigma}$ parametrizes stable sheaves. This produces an open embedding $X \setminus \sigma \to M$, whose complement is isomorphic to $\mathbb{P}^1$, by $i)$ and $ii)$. This proves the claim. □

Proof of the Claim via elementary transformations.

We have seen that $\mathrm{Ext}^1(\mathcal{O}_X(\sigma - f), \mathcal{O}_X(f))$ is one-dimensional. Hence there is a unique non-split extension

$$0 \to \mathcal{O}_X(f) \to G \to \mathcal{O}_X(\sigma - f) \to 0.$$

Obviously, $\det(G) \cong \mathcal{O}_X(\sigma)$ and $c_2(G) = 1$. By Remark 5.3.5 the polarization $H_2 = \sigma + 5f$ is $(\mathcal{O}(\sigma), 1)$-suitable. Since $f.H_2 = 1$ and $(\sigma - f).H_2 = 2$, the bundle G is μ-stable with respect to H_2.

Since $H^1(X, \mathcal{O}(f - \sigma)) = 0$, the map $H^1(X, \mathcal{O}(2f - \sigma)) \to H^1(F, \mathcal{O}(-(\sigma \cap F))$ is injective for any fibre F, i.e. the extension defining G is non-split on any fibre. Moreover, any stable sheaf E with the same invariants as G is isomorphic to G. Indeed, by the Hirzebruch-Riemann-Roch formula $\mathrm{Hom}(\mathcal{O}(f), E) \neq 0$ and by stability the cokernel of any homomorphism is torsion free, hence isomorphic to $\mathcal{O}(\sigma - f)$.

Let $x \in X$ be any closed point, $F := \pi^{-1}(\pi(x))$ the fibre through x and $\mathcal{I}_{F,x}$ the ideal sheaf of x in F. Since the extension

$$0 \to \mathcal{O}_F \to G_F \to \mathcal{O}_F(\sigma \cap F) \to 0$$

is non-split, $\mathrm{Hom}(\mathcal{I}_{F,x}(2\sigma), G_F) = 0$. Hence, by Serre duality $H^1(G_F^\vee \otimes \mathcal{I}_{F,x}(2\sigma)) = 0$. Since $\chi(G_F^\vee \otimes \mathcal{I}_{F,x}(2\sigma)) = 1$, there is a unique non-trivial homomorphism $\varphi : G \to \mathcal{I}_{F,x}(2\sigma)$ up to nonzero scalars. Again, since the extension defining G is non-split on F, the homomorphism φ must be surjective. Let E_x be the kernel of φ. Then $\det(E_x) \cong \mathcal{O}_X(\sigma - f)$ and $c_2(E_x) = 1$. Moreover, for the generic fibre F_η we have $E_x|_{F_\eta} \cong G|_{F_\eta}$, which implies that E_x is stable. In this way we get a stable bundle E_x for every point $x \in X$. To see that indeed $X \cong M$, it suffices to write down a universal family.

Let $\Delta \subset X \times_{\mathbb{P}^1} X \subset X \times X$ denote the diagonal and $\mathcal{I}$ the ideal sheaf of Δ as a subscheme in $X \times_{\mathbb{P}^1} X$. As before p and q denote the projections of $X \times X$ to the two factors. The Base Change Theorem and the dimension computations above imply that $L := p_*(\mathcal{I} \otimes q^*(G^\vee(2\sigma)))$ is a line bundle and that the natural homomorphism $q^*G \to p^*L^\vee \otimes \mathcal{I} \otimes q^* \mathcal{O}_X(2\sigma)$ is surjective. The kernel $\mathcal{E}$ defines a universal family. □

As in the previous example one might ask about the symmetry of the situation. Using

$$0 \to \mathcal{E} \to q^*G \to \mathcal{I} \otimes p^*L^{\vee} \otimes q^*\mathcal{O}_X(2\sigma) \to 0$$

one can compute the restriction of $\mathcal{E}$ to the fibres of q. We get

$$0 \to \mathcal{E}_{q^{-1}(x)} \to \mathcal{O}_X^2 \to \mathcal{I}_{F,x} \otimes L^{\vee} \to 0.$$

In particular, $c_1(\mathcal{E}_{q^{-1}(x)}) = \mathcal{O}_X(-f)$ and $c_2(\mathcal{E}_{q^{-1}(x)}) = \deg(\mathcal{I}_{F,x} \otimes L^{\vee}) = 2$. To see that $\deg(L^{\vee}|_F) = 3$ calculate as follows: Observe that the ideal sheaf of $X \times_{\mathbb{P}^1} X$ in $X \times X$ is given by $p^*\mathcal{O}_X(-f) \otimes q^*\mathcal{O}_X(-f)$ and that in $K(X \times X)$ we have the relation $[\mathcal{I}] = [\mathcal{O}_{X \times X}] - [p^*\mathcal{O}_X(-f) \otimes q^*\mathcal{O}_X(-f)] - [\mathcal{O}_\Delta]$. From this we deduce

$$\begin{aligned}
L &= p_*(\mathcal{I} \otimes q^*G^{\vee}(2\sigma)) = \det p_!(\mathcal{I} \otimes q^*G^{\vee}(2\sigma)) \\
&= \det p_!(q^*G^{\vee}(2\sigma)) \otimes (\det p_!(p^*\mathcal{O}_X(-f) \\
&\quad \otimes q^*G^{\vee}(2\sigma - f)))^{\vee} \otimes (\det p_!(q^*G^{\vee}(2\sigma)|_\Delta))^{\vee} \\
&= \mathcal{O}_X(\chi(G^{\vee}(2\sigma - f) \cdot f) \otimes (\det(G^{\vee}(2\sigma)))^{\vee} = \mathcal{O}_X(-3\sigma - 5f).
\end{aligned}$$

Hence $\deg(L^{\vee}|_F) = 3$.

The situation is not quite so symmetric as in 5.3.7, e.g. the determinant has even intersection with the class of the fibre. Nevertheless, the second factor is a moduli space of the first one. One can check that $\mathcal{E}_{q^{-1}(x)}$ is stable and that the dimension of its moduli space is $4c_2 - c_1^2 - 6 = 2$. In fact, $E = \mathcal{E}_{q^{-1}(x)}$ determines the point x uniquely by the condition that F is the only fibre where E is not semistable and that its destabilizing quotient is $\mathcal{I}_{F,x} \otimes L_F$. Details are left to the reader.

The reader may have noticed that in the second example we made little use of the fact that the elliptic surface is K3. Especially, the construction of the 'unique' bundle G goes through in broader generality.

Proposition 5.3.9 — *Let $X \to \mathbb{P}^1$ be a regular elliptic surface with a section $\sigma \subset X$. If H is the polarization $\sigma + (2\chi(\mathcal{O}_X) + 1)f$, then $M_H(2, K_X(-\sigma), 1)$ consists of a single reduced point which is given by the unique non-trivial extension $0 \to \mathcal{O}_X(f) \to G \to K_X(\sigma - f) \to 0$.* $\square$

Comments:

— Theorem 5.1.1 is standard by now (cf. [91], [132]).

— The existence of stable rank two bundles via Serre correspondence (5.1.3 and 5.1.4) was shown in [16].

— We would like to draw the attention of the reader to Gieseker's construction in [79]. Gieseker proved that for $c \geq 4([p_g/2] + 1)$ there exists a μ-stable rank two vector bundle E with $\det(E) \cong$

$\mathcal{O}_X$ and $c_2(E) = c$. Note that the bound is purely topological and does not depend on the polarization. As Corollary 5.3.4 shows, such a bound cannot be expected for $\det(E) \not\cong \mathcal{O}_X$.

— Proposition 5.1.5 is due to Ballico [12]. The statement about the existence and regularity of the bundle E in Theorem 5.1.6 was proved by W.-P. Li and Z. Qin [153]. The details of the proof can be found there. The assertion about the image under the Albanese map is a modification of Ballico's argument.

— Other existence results for higher rank are due to Sorger [241].

— Elementary transformations were intensively studied by Maruyama ([163, 167]). Proposition 5.2.4 and Theorem 5.2.5 are due to him.

— The notion of a suitable polarization was first introduced by Friedman in [67] for elliptic surfaces. He also proved Theorem 5.3.2. It seems Brosius observed Corollary 5.3.4 for the first time, though Takemoto in [244] already found that for $c_1.f \equiv 1(2)$ there exists no rank two vector bundle which is μ-stable with respect to every polarization. Suitable polarizations for higher rank vector bundles where considered by O'Grady [210]. He only discusses the case of an elliptic K3 surface, whose Picard group is spanned by the fibre class and the class of a section, but his arguments easily generalize.

— With the techniques of Example 5.3.7 one can attack a generic complete intersection of a quadric and a cubic hypersurface in $\mathbb{P}^4$. The moduli space $M(2, \mathcal{O}(-1), 3)$ is a reduced point. The locally free part of $M(2, \mathcal{O}(-1), 4)$ is isomorphic to the open subset of $Z \in \text{Hilb}^2(X)$, such that the line through Z meets X exactly in Z. This isomorphism was described in [190]. The birational correspondence between $M(2, \mathcal{O}(-1), 4)$ and $\text{Hilb}^2(X)$ reflects the projective geometry of X.

— Example 5.3.8 is entirely due to Friedman [68]. He treats it in the more general setting of elliptic surfaces which are not necessarily K3 surfaces. He also gives a complete description of the four-dimensional moduli space $M(2, \sigma, 2)$. It turns out that it is isomorphic to $\text{Hilb}^2(X)$, though the identification is fairly involved. The distinguished bundle G was also discovered by Qin [216] and in a broader context by Kametani and Sato [118]. We took the description as the unique extension from there. Friedman's point of view is that G is the unique bundle which restricts to a stable bundle on any fibre, even singular ones. For this purpose he generalizes results of Atiyah for vector bundles on singular nodal elliptic curves.

Appendix to Chapter 5

5.A Further comments (second edition)

I. General construction methods.

We are not aware of any genuinely new methods for constructing bundles on surfaces or higher dimensional varieties. So, Serre correspondence, elementary transformations and constructions that use a given fibre structure of the variety still seem the only techniques known.

The rôle of μ-stable vector bundles in the theory of coherent sheaves and Chow groups has been illuminated further in [392] where it is shown that the rational Chow ring is generated by the Chern characters of μ-stable vector bundles. This relates to Proposition 5.1.5, where given elements in the Albanese of the surface were realized by μ-stable bundles.

II. Moduli spaces on higher dimensional varieties.

Moduli spaces of sheaves and bundles on higher dimensional varieties are certainly interesting, but in general even the basic properties, e.g. smoothness, irreducibility, non-emptiness, are much harder to study. Without aiming at completeness, we mention a few references where moduli spaces of sheaves have been studied in geometrically concrete situations.

Fano varieties. We have avoided talking much about bundles on $\mathbb{P}^2$ or other rational surfaces. The standard reference for stable bundles on those surfaces and on projective spaces of higher dimension is still [212]. So we will not enter into the subject here, but only give pointers to a few papers we have become aware of since the first edition of this book.

The behaviour of moduli spaces of stable rank two bundles on $\mathbb{P}^d$ for growing second Chern class has been studied in [306]. This should be compared to work of O'Grady for surfaces that is discussed Chapter 9. Moduli spaces of stable rank two bundles on del Pezzo surface fibrations over $\mathbb{P}^1$ have been studied in [398, 399]. Moduli spaces of stable rank two bundles with small c_2 on cubic and quartic hypersurfaces $X \subset \mathbb{P}^4$ and their relations to the Fano variety of lines and the intermediate Jacobian of X have been studied in [277, 283, 315, 343].

Calabi-Yau threefolds. The main feature of sheaves and bundles on Calabi-Yau threefolds, i.e. smooth projective varieties of dimension three with $\omega_X \cong \mathcal{O}_X$, is that the expected dimension of their moduli spaces is zero. Indeed, by Serre duality $\mathrm{Ext}^1_X(E, E) \cong$

$\mathrm{Ext}_X^2(E, E)^\vee$, so there are as many first order deformations as there are potential obstructions. In this sense, each moduli space of stable sheaves on a Calabi-Yau threefold is expected to consist of a finite number of (very non-reduced) points which one might want to count. Very often however the moduli spaces are far from being zero-dimensional. Nevertheless a counting can be carried through by means of the *virtual fundamental class*, which has become an important tool in Gromov-Witten theory. The virtual fundamental class has been introduced by Li and Tian in [368] and by Behrend and Fantechi in [278]. The case that has been studied most is the case of sheaves concentrated on curves in the Calabi-Yau threefold. This leads to the so-called Donaldson-Thomas invariants which are closely related to Gromov-Witten invariants counting stable maps from curves to the variety (see e.g. [389] and comments in Section 4.D, IV).

It is probably fair to say that there are no general results concerning the geometry of moduli spaces of sheaves or bundles on Calabi-Yau threefolds. There are however many concrete examples that have been studied in detail, mostly in cases where the variety fibres over a surface or a curve, with generic fibres being elliptic curves respectively K3 surfaces. The most accessible ones are the relative moduli spaces, i.e. moduli spaces of sheaves concentrated on fibres of the given fibration. The reader may consult the articles [293, 371, 400, 401, 433, 437] and references therein.

6

Moduli Spaces on K3 Surfaces

By definition, K3 surfaces are surfaces with vanishing first Betti number and trivial canonical bundle. Examples of K3 surfaces are provided by smooth complete intersections of type $(a_1, \ldots, a_{n-2})$ in $\mathbb{P}_n$ with $\sum a_i = n + 1$, Kummer surfaces and certain elliptic surfaces. In the Enriques classification K3 surfaces occupy, together with abelian, Enriques and hyperelliptic surfaces, the distinguished position between ruled surfaces and surfaces of positive Kodaira dimension. The geometry of K3 surfaces and of their moduli space is one of the most fascinating topics in surface theory, bringing together complex algebraic geometry, differential geometry and arithmetic.

Following the general philosophy that moduli spaces of sheaves reflect the geometric structure of the surface it does not come as a surprise that studying moduli spaces of sheaves on K3 surfaces one encounters intriguing geometric structures. We will try to illuminate some aspects of the rich geometry of the situation.

We present the material at this early stage in the hope that having explicit examples with a rich geometry in mind will make the more abstract and general results, where the geometry has not yet fully unfolded, easier to access. At some points we make use of results presented later (Chapter 9, 10). In particular, a fundamental result in the theory, namely the existence of a symplectic structure on the moduli space of stable sheaves, will be discussed only in Chapter 10.

Section 6.1 gives an almost complete account of results due to Mukai describing zero- and two-dimensional moduli spaces. The result on the existence of a symplectic structure is in this section only used once (proof of 6.1.14) and there in the rather weak version that the canonical bundle of the moduli space of stable sheaves is trivial. In Section 6.2 we concentrate on moduli spaces of dimension ≥ 4. We prove that they provide examples of higher dimensional irreducible symplectic (or hyperkähler) manifolds. The presentation is based on the work of Beauville, Mukai, and O'Grady. Some of the arguments are only

166

sketched. Finally, the appendix contains a geometric proof of the irreducibility of the Quot-scheme $\text{Quot}(E, \ell)$ of zero-dimensional quotients of a locally free sheaf E.

6.1 Low-Dimensional ...

We begin this section with some technical remarks and the definition of the Mukai vector.

Definition 6.1.1 — *If E and F are coherent sheaves then the Euler characteristic of the pair (E, F) is*

$$\chi(E, F) := \sum (-1)^i \dim \text{Ext}^i(E, F).$$

$\chi(E, F)$ is bilinear in E and F and can be expressed in terms of their Chern characters. But before we can give the formula one more notation needs to be introduced.

Definition 6.1.2 — *If $v = \oplus v_i \in H^{ev}(X, \mathbb{Z}) = \bigoplus H^{2i}(X, \mathbb{Z})$ then $v^\vee := \oplus (-1)^i v_i$.*

The definition makes also perfect sense in the cohomology with rational or complex coefficients and in the Chow group. The notation is motivated by the fact that $ch^\vee(E) = ch(E^\vee)$ for any locally free sheaf E.

Lemma 6.1.3 — *If X is smooth and projective, then*

$$\chi(E, F) = \int_X ch^\vee(E).ch(F).td(X).$$

Proof. If E is locally free this follows directly from the Hirzebruch-Riemann-Roch formula and the multiplicativity of the Chern character:

$$\begin{aligned} \chi(E, F) = \chi(E^\vee \otimes F) &= \int_X ch(E^\vee \otimes F).td(X) \\ &= \int_X ch(E^\vee).ch(F).td(X) \\ &= \int_X ch^\vee(E).ch(F).td(X). \end{aligned}$$

If E is not locally free we consider a locally free resolution $E^\bullet \twoheadrightarrow E$ and use $ch(E^{\bullet\vee}) = ch^\vee(E^\bullet)$. $\qquad\square$

Definition 6.1.4 — *Let X be a smooth variety and let E be a coherent sheaf on X. Then the Mukai vector $v(E) \in H^{2*}(X, \mathbb{Q})$ of E is $ch(E).\sqrt{td(X)}$.*

Note that $td_0(X) = 1$ and hence the square root $\sqrt{td(X)}$ can be defined by a power series expansion.

Definition and Corollary 6.1.5 — *If X is smooth and projective, then*

$$(v, w) := -\int_X v^\vee . w$$

defines a bilinear form on $H^{2*}(X, \mathbb{Q})$. For any two coherent sheaves E and F one has $\chi(E, F) = -(v(E), v(F))$. □

Let now X be a K3 surface. If E is a coherent sheaf on X with $\text{rk}(E) = r$, $c_1(E) = c_1$, and $c_2(E) = c_2$, then $v(E) = (r, c_1, c_1^2/2 - c_2 + r)$. Clearly, we can recover r, c_1, and c_2 from $v(E)$.

Instead of $M(r, c_1, c_2)$ we will use the notation $M(v)$ for the moduli space of semistable sheaves, where $v = (r, c_1, c_1^2/2 - c_2 + r)$. If v is fixed we will also write M for $M(v)$. We denote the open subset parametrizing stable sheaves by M^s.

By 4.5.6 the expected dimension of M^s is $2rc_2 - (r-1)c_1^2 - 2(r^2-1) = (v, v) + 2$, which is always even, since the intersection form on X is even. The obstruction space (cf. Section 4.5) $\text{Ext}^2(E, E)_0$ vanishes, since by Serre duality $\text{Ext}^2(E, E)_0 \cong \text{Hom}(E, E)_0^\vee = 0$ for any $E \in M^s$. Hence by 4.5.4 the moduli space M^s is smooth. For the following we assume $r > 1$.

There are general results, mostly due to Mukai, which give a fairly complete description of moduli spaces of low dimensions, i.e. dimension ≤ 2. As M^s is even-dimensional, we are interested in zero- and two-dimensional examples.

Theorem 6.1.6 — *Suppose $(v, v) + 2 = 0$. If M^s is not empty, then M consists of a single reduced point which represents a stable locally free sheaf. In particular, $M^s = M$.*

Proof. Let E, F be semistable sheaves defining points $[E] \in M^s$ and $[F] \in M$. By 6.1.5 the Euler characteristic $\chi(F, E)$ depends only on the Chern classes of F and not on F itself. Since F and E have the same Chern classes, one has $\chi(F, E) = \chi(E, E) = -(v, v) = 2$. This implies that either $\text{Hom}(F, E) \neq 0$ or, by using Serre duality, that $\text{Hom}(E, F) \neq 0$. The stability of E and the semistability of F imply in both cases that $E \cong F$ (see 1.2.7).

It remains to show that E is locally free. For this purpose let G be the reflexive hull $E^{\vee\vee}$ of E and S the quotient of the natural embedding $E \subset G$. If the rank is two, one can argue as follows. G is still μ-semistable and hence satisfies the Bogomolov Inequality $4c_2(G) - c_1^2(G) \geq 0$. On the other hand, $4c_2(G) - c_1^2(G) = 4c_2(E) - 4 \cdot \ell(S) - c_1^2(E) = 6 - 4 \cdot \ell(S)$. Hence $\ell(S) \leq 1$, i.e. if E is not locally free, then $S \cong k(x)$ where x is a point in X. Denote by $\mathcal{E}$ the flat family on $\mathbb{P}(G) \times X$ defined by

$$0 \to \mathcal{E} \to q^*G \to (\pi \times 1)^* \mathcal{O}_\Delta \otimes p^* \mathcal{O}_\pi(1) \to 0,$$

where $\Delta \subset X \times X$ is the diagonal and $\pi : \mathbb{P}(G) \to X$ is the projection (for details see 8.1.7). Then $\text{Supp}(\mathcal{E}_t^{\vee\vee}/\mathcal{E}_t) = \pi(t)$ and for some $t_0 \in \pi^{-1}(x)$ the sheaf $\mathcal{E}_{t_0}$ is isomorphic to E. Hence for the generic $t \in \mathbb{P}(G)$ the sheaf $\mathcal{E}_t$ is stable but not isomorphic to E. Since the moduli space is zero-dimensional, this cannot happen. In fact a similar argument works in the higher rank case: Here one exploits the fact that $\mathrm{Quot}(G, \ell(S))$ is irreducible. This is proved in the appendix (Theorem 6.A.1). Thus $G \twoheadrightarrow S$ can be deformed to $G \twoheadrightarrow S_t$ with $\mathrm{Supp}(S) \neq \mathrm{Supp}(S_t)$. $\qquad\square$

Remark 6.1.7 — Note that the moduli space M might be non-empty even if the expected dimension $(v, v) + 2$ of the stable part M^s is negative. Indeed, $[\mathcal{O} \oplus \mathcal{O}]$ is a semistable sheaf with $(v(\mathcal{O} \oplus \mathcal{O}), v(\mathcal{O} \oplus \mathcal{O})) + 2 = -6$.

We now turn to moduli spaces of dimension two. In general there is no reason to expect that $M^s = M$ or that M is irreducible. But as above, whenever there exists a 'good' component of M^s, then both properties hold:

Theorem 6.1.8 — *Assume $(v, v) + 2 = 2$. If M^s has a complete irreducible component M_1, then $M_1 = M^s = M$, i.e. M is irreducible and all sheaves are stable. In particular, if $M^s = M$, then M is smooth and irreducible.*

Proof. The idea of the proof is a globalization of the proof of Theorem 6.1.6. The Hirzebruch-Riemann-Roch formula is replaced by Grothendieck's relative version.

Let us fix a quasi-universal family $\mathcal{E}$ over $M_1 \times X$ and denote the multiplicity $\mathrm{rk}(\mathcal{E})/r$ by s (cf. 4.6.2). Let $[F] \in M$ be an arbitrary point in the moduli space represented by a semistable sheaf F. For any $t \in M_1$ we have

$$\mathrm{Hom}(F, \mathcal{E}_t) = \begin{cases} 0 & \text{if } F^{\oplus s} \not\cong \mathcal{E}_t \\ k^{\oplus s} & \text{if } F^{\oplus s} \cong \mathcal{E}_t \end{cases}$$

and also

$$\mathrm{Ext}^2(F, \mathcal{E}_t) \cong \mathrm{Hom}(\mathcal{E}_t, F)^{\vee} = \begin{cases} 0 & \text{if } F^{\oplus s} \not\cong \mathcal{E}_t \\ k^{\oplus s} & \text{if } F^{\oplus s} \cong \mathcal{E}_t \end{cases}$$

Since $s \cdot \chi(F, \mathcal{E}_t) = \chi(\mathcal{E}_t, \mathcal{E}_t) = 0$ we also have

$$\mathrm{Ext}^1(F, \mathcal{E}_t) = \begin{cases} 0 & \text{if } F^{\oplus s} \not\cong \mathcal{E}_t \\ k^{\oplus 2s} & \text{if } F^{\oplus s} \cong \mathcal{E}_t. \end{cases}$$

Thus if $[F] \notin M_1$, then $\mathrm{Ext}^i(F, \mathcal{E}_t) = 0$ for all $t \in M_1$ and $i = 0, 1, 2$. Therefore we have $\mathcal{E}xt_p^i(q^*F, \mathcal{E}) = 0$. If $[F] \in M_1$, then $\mathcal{E}xt_p^i(q^*F, \mathcal{E}) = 0$ for $i = 0, 1$ and $\mathcal{E}xt_p^2(q^*F, \mathcal{E})(t_0) = k(t_0)^{\oplus s}$, where $t_0 = [F]$: this is an application of the Base Change Theorem. In our situation we can make it quite explicit. By [19] there exists a complex $\mathcal{P}^\bullet$ of locally free sheaves $\mathcal{P}^i$ of finite rank such that the i-th cohomology $\mathcal{H}^i(\mathcal{P}^\bullet)$ is isomorphic to $\mathcal{E}xt_p^i(q^*F, \mathcal{E})$ and $\mathcal{H}^i(\mathcal{P}^\bullet(t)) \cong \mathrm{Ext}^i(F, \mathcal{E}_t)$. This complex is bounded above, i.e. $\mathcal{P}^i = 0$ for $i \gg 0$. An argument of Mumford shows [194] that one can also assume that $\mathcal{P}^i = 0$ for $i < 0$. Since $\text{Ext}^i(F, \mathcal{E}_t) = 0$ for $i > 2$, the complex $\mathcal{P}^\bullet$ is exact at $\mathcal{P}^i$ for $i > 2$. The kernel $\ker(d_i)$ of the i-th differential is the kernel of a surjection of a locally free sheaf $\mathcal{P}^i$ to a torsion free sheaf $\text{im}(d_i)$. Hence $\ker(d_i)$ is locally free, since M_1 is a smooth surface. Replacing $\mathcal{P}^2$ by $\ker(d_2)$ we can assume that $\mathcal{P}^\bullet = \mathcal{P}^0 \xrightarrow{d_0} \mathcal{P}^1 \xrightarrow{d_1} \mathcal{P}^2$. We have seen that $\ker(d_0)$ is concentrated in t_0. At the same time, as a subsheaf of $\mathcal{P}^0$, it is torsion free, hence zero, i.e. d_0 is injective. Also $\ker(d_1)$ is locally free, contains the locally free sheaf $\mathcal{P}^0$ and actually equals it on the complement of the point t_0. Hence $\mathcal{P}_0 = \ker(d_1)$, i.e. $0 = \mathcal{H}^1(\mathcal{P}^\bullet) = \mathcal{E}xt_p^1(q^*F, \mathcal{E})$. For the last statement use $0 \to \text{im}(d_1) \to \mathcal{P}^2 \to \mathcal{H}^2(\mathcal{P}^\bullet) \to 0$ which shows that $\mathcal{H}^2(\mathcal{P}^\bullet)(t_0) = \mathcal{P}^2(t_0)/\text{im}(d_1)(t_0) \cong \mathcal{H}^2(\mathcal{P}^\bullet(t_0)) \cong \text{Ext}^2(F, \mathcal{E}_{t_0}) \cong k(t_0)^{\oplus s}$.

On the other hand, the Grothendieck-Riemann-Roch formula computes

$$a := ch([\mathcal{E}xt_p^0(q^*F, \mathcal{E})] - [\mathcal{E}xt_p^1(q^*F, \mathcal{E})] + [\mathcal{E}xt_p^2(q^*F, \mathcal{E})])$$

as an element of $H^*(M_1, \mathbb{Q})$ and shows that it only depends on $ch(q^*F)$ and $ch(\mathcal{E})$ as elements of $H^*(M_1 \times X, \mathbb{Q})$. Since $ch(F)$ is constant for all $[F] \in M$, in particular $ch(F) = s \cdot ch(\mathcal{E}_{t_0})$ even for $[F] \notin M_1$, one gets a contradiction by comparing $a = 0$ if $[F] \notin M_1$ and $0 \neq \chi(\mathcal{E}xt_p^2(q^*F, \mathcal{E})) = \langle a \cdot td(M), [M_1] \rangle$ if $[F] \in M_1$. $\quad\square$

Remark 6.1.9 — The assumption $M^s = M$ is satisfied frequently, e.g. if degree and rank are coprime any μ-semistable sheaf is μ-stable (see 1.2.14).

Note that under the assumption of the theorem any μ-stable $F \in M$ is locally free. Indeed, if $F \in M$ is μ-stable, then $G := F^{\vee\vee}$ is still μ-stable and thus defines a point in $M^s(r, c_1, c_2 - \ell)$. If F were not locally free, i.e. $\ell > 0$, the latter space would have negative dimension.

The following lemma will be needed in the proof of Proposition 6.1.13.

Lemma 6.1.10 — *Suppose that $(v, v) = 0$ and $M = M^s$. Moreover, assume that there is a universal family $\mathcal{E}$ on $M \times X$. Then*

$$\mathcal{E}xt_{p_{12}}^i(p_{13}^*\mathcal{E}, p_{23}^*\mathcal{E}) \cong \begin{cases} 0 & \text{if } i = 0, 1 \\ \mathcal{O}_\Delta & \text{if } i = 2, \end{cases}$$

where p_{ij} is the projection from $M \times M \times X$ to the product of the indicated factors, and $\Delta \subset M \times M$ is the diagonal.

Proof. **Step 1.** Let $t_0 \in M$ be a closed point representing a sheaf E. Then $\mathcal{E}xt_p^i (q^*E, \mathcal{E}) = 0$ for $i < 2$ and $\mathcal{E}xt_p^2(q^*E, \mathcal{E}) \cong k(t_0)$. The first statement was obtained in the proof of Theorem 6.1.8 together with a weaker form of the second statement, namely that the rank of $\mathcal{E}xt_p^2(q^*E, \mathcal{E})$ at t_0 is 1. It suffices therefore to show that for any tangent vector $S \cong \mathrm{Spec}(k[\varepsilon])$ at t_0 one has $k(t_0) \cong \mathcal{E}xt_p^2(q^*E, \mathcal{E}) \otimes \mathcal{O}_S(\cong \mathcal{E}xt_p^2(\mathcal{O}_S \otimes E, \mathcal{E}_S))$. Here $\mathcal{E}_S$ is the restriction of $\mathcal{E}$ to $S \times X$. This family fits into a short exact sequence $0 \to E \to \mathcal{E}_S \to E \to 0$ defining a class $\xi \in \mathrm{Ext}_X^1(E, E)$. Applying the functor $\mathrm{Hom}_{S \times X}(E[\varepsilon], \, . \,)$ to this sequence and using $\mathrm{Ext}_{S \times X}^i(E[\varepsilon], E) = \mathrm{Ext}_X^i(E, E)$ we get

$$\mathrm{Ext}_X^1(E, E) \xrightarrow{\xi} \mathrm{Ext}_X^2(E, E) \to \mathrm{Ext}_{S \times X}^2(E[\varepsilon], \mathcal{E}_S) \to \mathrm{Ext}_X^2(E, E) \to 0.$$

Since $\mathrm{Ext}_X^2(E, E) \cong k$ and since the cup product is non-degenerate by Serre duality, the map $\mathrm{Ext}_X^1(E, E) \xrightarrow{\xi} \mathrm{Ext}_X^2(E, E)$ is surjective. Hence the restriction homomorphism $\mathrm{Ext}_{S \times X}^2(E[\varepsilon], \mathcal{E}_S) \to \mathrm{Ext}_X^2(E, E)$ is an isomorphism.

Step 2. It follows from this and the spectral sequence

$$H^i(M, \mathcal{E}xt_p^j(q^*E, \mathcal{E})) \Longrightarrow \mathrm{Ext}_{M \times X}^{i+j}(q^*E, \mathcal{E})$$

that

$$\mathrm{Ext}_{M \times X}^n(q^*E, \mathcal{E}) \cong \begin{cases} k & \text{if } n = 2, \\ 0 & \text{else.} \end{cases}$$

Consider the Leray spectral sequence for the composition $\pi_1 = p_1 \circ p_{12} : M \times M \times X \to M \times M \to M$:

$$R^i \pi_{1*}\left(\mathcal{E}xt_{p_{12}}^j(p_{13}^*\mathcal{E}, p_{23}^*\mathcal{E})\right) \Longrightarrow \mathcal{E}xt_{\pi_1}^{i+j}(p_{13}^*\mathcal{E}, p_{23}^*\mathcal{E}).$$

As $\mathcal{E}xt_{p_{12}}^j(p_{13}^*\mathcal{E}, p_{23}^*\mathcal{E})$ is (set-theoretically) supported along the diagonal, the spectral sequence reduces to an isomorphism $\pi_{1*}\mathcal{E}xt_{p_{12}}^j(p_{13}^*\mathcal{E}, p_{23}^*\mathcal{E}) \cong \mathcal{E}xt_{\pi_1}^j(p_{13}^*\mathcal{E}, p_{23}^*\mathcal{E})$. It follows from the base change theorem and *Step 1*, that

$$\mathcal{E}xt_{\pi_1}^j(p_{13}^*\mathcal{E}, p_{23}^*\mathcal{E})(t) \cong \mathrm{Ext}_{M \times X}^j(\mathcal{E}_t \otimes \mathcal{O}_X, \mathcal{E}), \quad t \in M$$

for all j. This implies that $\mathcal{E}xt_{\pi_1}^2(p_{13}^*\mathcal{E}, p_{23}^*\mathcal{E})$ and hence $\mathcal{E}xt_{p_{12}}^2(p_{13}^*\mathcal{E}, p_{23}^*\mathcal{E})$ are line bundles on M and $\Delta \subset M \times M$, respectively. It remains to show that this line bundle is trivial. But as base change holds for $\mathcal{E}xt_{p_{12}}^2$ we have: $\mathcal{E}xt_{p_{12}}^2(p_{13}^*\mathcal{E}, p_{23}^*\mathcal{E})|_\Delta \cong \mathcal{E}xt_p^2(\mathcal{E}, \mathcal{E})$, and the trace map $tr_\mathcal{E} : \mathcal{E}xt_p^2(\mathcal{E}, \mathcal{E}) \to H^2(X, \mathcal{O}_X) \otimes \mathcal{O}_X$ (cf. 10.1.3) gives the desired isomorphism. $\qquad \square$

After having shown that in many cases the moduli space M is a smooth irreducible projective surface we go on and identify these surfaces in terms of their weight-two Hodge structures. Recall that a Hodge structure of weight n consists of a lattice $H_\mathbb{Z} \subset H_\mathbb{R}$ in a real vector space and a direct sum decomposition $H_\mathbb{C} := H_\mathbb{R} \otimes_\mathbb{R} \mathbb{C} \cong \bigoplus_{p+q=n} H^{p,q}$ such that $\overline{H^{p,q}} = H^{q,p}$. A $\mathbb{Q}$-Hodge structure is a $\mathbb{Q}$-vector space $H_\mathbb{Q} \subset H_\mathbb{R}$ in a real vector space of the same dimension and a decomposition of $H_\mathbb{C} = H_\mathbb{Q} \otimes_\mathbb{Q} \mathbb{C}$ as before.

Let Y be a compact Kähler manifold. Then there is a naturally defined weight n Hodge structure on $H^n(Y, \mathbb{Z})/\text{Torsion}$, which is given by $H^n(Y, \mathbb{C}) = \bigoplus_{p+q=n} H^{p,q}(Y)$. In particular, Y admits a natural weight-two Hodge structure on the second cohomology $H^2(Y, \mathbb{Z})$ defined by

$$H^2(Y, \mathbb{C}) = H^{2,0}(Y, \mathbb{C}) \oplus H^{1,1}(Y, \mathbb{C}) \oplus H^{0,2}(Y, \mathbb{C}).$$

If Y is a surface the intersection product defines a natural pairing on $H^2(Y, \mathbb{Z})$. The Global Torelli Theorem for K3 surfaces states that two K3 surfaces Y_1 and Y_2 are isomorphic if and only if there exists an isomorphism between their Hodge structures respecting the pairing, i.e. there exists an isomorphism $H^2(Y_1, \mathbb{Z}) \cong H^2(Y_2, \mathbb{Z})$ which maps $H^{2,0}(Y_1, \mathbb{C})$ to $H^{2,0}(Y_2, \mathbb{C})$ and which is compatible with the pairing. For details see [22], [26]. For surfaces one can also define a Hodge structure $\tilde{H}(Y, \mathbb{Z})$ on $H^{ev}(Y, \mathbb{Z}) = \bigoplus_i H^{2i}(Y, \mathbb{Z})$ as follows.

Definition 6.1.11 — *Let Y be a surface. $\tilde{H}(Y, \mathbb{Z})$ (or $\tilde{H}(Y, \mathbb{Q})$) is the natural weight-two Hodge structure on $\tilde{H}(Y, \mathbb{Z})$ given by $\tilde{H}^{2,0}(Y, \mathbb{C}) = H^{2,0}(Y, \mathbb{C})$, $\tilde{H}^{0,2}(Y, \mathbb{C}) = H^{0,2}(Y, \mathbb{C})$, and $\tilde{H}^{1,1}(Y, \mathbb{C}) = H^0(Y, \mathbb{C}) \oplus H^{1,1}(Y, \mathbb{C}) \oplus H^4(Y, \mathbb{C})$.*

Let $\tilde{H}(Y, \mathbb{Z})$ be endowed with the pairing $(., .)$ defined in 6.1.5. The restriction of $(., .)$ to $H^2(Y, \mathbb{Z})$ equals the intersection product. The inclusion $H^2(Y, \mathbb{Z}) \subset \tilde{H}(Y, \mathbb{Z})$ is compatible with the Hodge structure.

The Mukai vector v can be considered as an element of $\tilde{H}(Y, \mathbb{Z})$ of type $(1, 1)$. The expected dimension of M^s is two if and only if v is an isotropic vector.

Assume that v is an isotropic vector such that M^s has a complete component. By 6.1.8 the last condition is equivalent to $M^s = M$. Let $\mathcal{E}$ be a quasi-universal family over $M \times X$ of rank $s \cdot r$.

Definition 6.1.12 — *Let $f : H^*(X, \mathbb{Q}) \to H^*(M, \mathbb{Q})$ and $f' : H^*(M, \mathbb{Q}) \to H^*(X, \mathbb{Q})$ be the homomorphisms given by $f(c) = p_*(\mu.q^*(c))$ and $f'(c) = q_*(\nu.p^*(c))$, where $\mu := v^{\vee}(\mathcal{E})/s$ and $\nu := v(\mathcal{E})/s$.*

If we want to emphasize the dependence on $\mathcal{E}$ we write $f_{\mathcal{E}}$ and $f'_{\mathcal{E}}$. For any locally free sheaf W on M the family $\mathcal{E} \otimes p^*(W)$ is also quasi-universal. The corresponding homomorphisms are related as follows: $f_{\mathcal{E} \otimes p^*(W)}(c) = f_{\mathcal{E}}(c).(ch^{\vee}(W)/\text{rk}(W))$.

Proposition 6.1.13 — *Let v be an isotropic vector and assume $M^s = M$. Assume there exists a universal family $\mathcal{E}$ over $M \times X$. Then:*
i) M is a K3 surface.
ii) $f_{\mathcal{E}} \circ f'_{\mathcal{E}} = 1$.

iii) $f_\mathcal{E}$ defines an isomorphism of Hodge structures $\tilde{H}(X,\mathbb{Z}) \cong \tilde{H}(M,\mathbb{Z})$ which is compatible with the natural pairings.

Proof. Consider the diagram

$$
\begin{array}{ccccccc}
M \times M & & \xrightarrow{\quad\quad \pi_2 \quad\quad} & & & & M \\
\downarrow{\scriptstyle p_{12}} \searrow & & & & & & \| \\
& M \times M \times X & \xrightarrow{p_{23}} & M \times X & \xrightarrow{p} & M & \\
{\scriptstyle \pi_1}\downarrow & {\scriptstyle p_{13}}\downarrow & & {\scriptstyle q}\downarrow & & & \\
& M \times X & \xrightarrow{q} & X & & & \\
& {\scriptstyle p}\downarrow & & & & & \\
M & = & M & & & &
\end{array}
$$

Then, by the projection formula

$$
\begin{aligned}
f(f'(c)) &= p_*(\mu.q^*q_*(\nu.p^*c)) = p_*(\mu.p_{13*}p_{23}^*(\nu.p^*c)) \\
&= p_*p_{13*}(p_{13}^*\mu.p_{23}^*(\nu.p^*c)) = p_{1*}(p_{13}^*\mu.p_{23}^*\nu.p_{23}^*p^*c) \\
&= \pi_{1*}p_{12*}(p_{13}^*\mu.p_{23}^*\nu.p_{12}^*\pi_2^*c) = \pi_{1*}(p_{12*}(p_{13}^*\mu.p_{23}^*\nu).\pi_2^*c) \\
&= \pi_{1*}\left(p_{12*}\left(ch^\smile(p_{13}^*\mathcal{E}).ch(p_{23}^*\mathcal{E}).p_3^*td(X)\right).\pi_1^*\sqrt{td(M)}.\pi_2^*\sqrt{td(M)}.\pi_2^*c \right).
\end{aligned}
$$

It follows from Lemma 6.1.10 and the Grothendieck-Riemann-Roch formula for p_{12} that

$$
p_{12*}\left(ch^\smile(p_{13}^*\mathcal{E}).ch(p_{23}^*\mathcal{E}).p_3^*td(X)\right) = ch(\mathcal{E}xt_{p_{12}}^2(p_{13}^*\mathcal{E}, p_{23}^*\mathcal{E})) = ch(i_*\mathcal{O}_\Delta).
$$

Hence, $f(f'(c)) = \pi_{1*}(ch(i_*\mathcal{O}_\Delta).\pi_1^*\sqrt{td(M)}.\pi_2^*\sqrt{td(M)}.\pi_2^*c)$. Now the Grothendieck-Riemann-Roch formula applied to i says $ch(i_*\mathcal{O}_\Delta).td(M \times M) = i_*(ch(\mathcal{O}_\Delta).td(\Delta))$. Hence, $f(f'(c)) = \pi_{1*}(i_*ch(\mathcal{O}_\Delta).\pi_2^*c) = c$. Therefore, the homomorphisms f' and f are injective and surjective, respectively. Moreover, f preserves H^{odd} and H^{even}, because ν is an even class. Hence $H^1(M,\mathbb{Q}) = 0$. By 10.4.3 the moduli space admits a non-degenerate two-form and, therefore, the canonical bundle of M is trivial (This is the only place where we need a result of the later chapters). Using the Enriques-classification of algebraic surfaces one concludes that M is abelian or K3. Since $b_1(M) = 0$, it must be a K3 surface. In particular, $\dim H^*(M,\mathbb{Q}) = \dim H^*(X,\mathbb{Q})$. Hence, f and f' are isomorphisms.

The isomorphisms f and f' do respect the Hodge structure $\tilde{H}$. Indeed, f and f' are defined by the algebraic classes μ and ν, which are sums of classes of type (p,p). It is straightforward to check that this is enough to ensure that f and f' respect the Hodge type of an element $c \in \tilde{H}$. (Note that the compatibility with the Hodge structure is valid also for the case of a quasi-universal family.)

The compatibility with the pairing is shown by:

$$
\begin{aligned}
-(a, f(c)) &= \langle a^\vee . f(c), [M]\rangle = \langle a^\vee . p_*(\mu . q^*(c)), [M]\rangle \\
&= \langle p_*(p^*(a^\vee).\mu.q^*(c)), [M]\rangle = \langle p^*(a)^\vee.\mu.q^*(c), [M \times X]\rangle \\
&= \langle (p^*(a).\mu^\vee)^\vee.q^*(c), [M \times X]\rangle = \langle (p^*(a).\nu)^\vee.q^*(c), [M \times X]\rangle \\
&= \langle ((f'(a)^\vee.c), [X]\rangle = -(f'(a), c).
\end{aligned}
$$

If $c = f'(b)$ we conclude $(f'(a), f'(b)) = (a, f(f'(b))) = (a, b)$, i.e. f' is compatible with the pairing.

To conclude, we have to show that the isomorphisms f and f' are integral. Since $\sqrt{td(X)}$ and $\sqrt{td(M)}$ are integral, it is enough to show that $ch(\mathcal{E})$ is integral. This goes as follows. The first Chern class $c_1(\mathcal{E}) = ch_1(\mathcal{E})$ is certainly integral. Since $H^1(X, \mathbb{Z}) = 0$, it equals $p^*c_1(\mathcal{E}|_{M \times \{x\}}) + q^*c_1$. Since X and M are K3 surfaces, the intersection form is even. Hence $ch_2(\mathcal{E}) = c_1^2(\mathcal{E})/2 - c_2(\mathcal{E})$ is integral. Writing $ch(\mathcal{E}) = \sum_{p,q}^2 e^{p,q}$ with $e^{p,q} \in H^p(M, \mathbb{Q}) \otimes H^q(X, \mathbb{Q})$ this says that the classes $e^{2,0}, e^{0,2}, e^{4,0}, e^{2,2}$, and $e^{0,4}$ are all integral. Moreover, $ch(\mathcal{E}).p^*td(M) = \sum e^{p,q} + \sum e^{p,q}.p^* PD(pt)$, where $PD(pt)$ denotes the Poincaré dual of a point. Hence $q_*(ch(\mathcal{E}).p^*td(M)) = \sum (\int_M e^{4,q} + e^{0,q})$. On the other hand, $ch(q_!\mathcal{E})$ is integral and, by the Grothendieck-Riemann-Roch formula, equals $q_*(ch(\mathcal{E}).p^*td(M))$. Hence $e^{4,2}$ and $e^{4,4}$ are also integral. In particular, $ch_4(\mathcal{E}) = e^{4,4}$ is integral. The same argument applied to $p_*(ch(\mathcal{E}).q^*td(X))$ shows $e^{2,4}$ is integral. Hence, $ch_3(\mathcal{E}) = e^{4,2} + e^{2,4}$ is integral. Altogether this proves that $ch(\mathcal{E})$ is integral. $\qquad\square$

The orthogonal complement V of v in $\tilde{H}(X, \mathbb{Z})$ contains v. If we in addition assume that v is primitive, i.e. not divisible by any integer ≥ 2, then the quotient $V/\mathbb{Z}v$ is a free $\mathbb{Z}$-module. Since v is of pure type $(1, 1)$ and isotropic, the quotient $V/\mathbb{Z}v$ inherits the bilinear form and the Hodge structure of $\tilde{H}(X, \mathbb{Z})$.

Theorem 6.1.14 — *If v is isotropic and primitive and M^s has a complete component (i.e. $M^s = M$), then $f_\mathcal{E}$ defines an isomorphism of Hodge structures*

$$
H^2(M, \mathbb{Z}) \cong V/\mathbb{Z} \cdot v
$$

compatible with the natural pairing and independent of the quasi-universal family $\mathcal{E}$.

Proof. We first check that $f_\mathcal{E} : V \otimes \mathbb{Q} \subset \tilde{H}(X, \mathbb{Q}) \to \tilde{H}(M, \mathbb{Q})$ has no $H^0(M)$-component. Indeed, the $H^0(M) \otimes H^*(X)$ component of μ is $v^\vee$ and $v^\vee.c = -(v, c)$ and hence the $H^0(M)$ component of $f_\mathcal{E}(c)$ is $-(v, c)$, which vanishes for $c \in V$. Since

$$
f_{\mathcal{E} \otimes p^*W}(c) = f_\mathcal{E}(c).ch^\vee(W)/\mathrm{rk}(W),
$$

the H^2-component of $f_\mathcal{E}(c)$ for $c \in V$ is independent of $\mathcal{E}$. Thus we obtain a well-defined (i.e. independent of the quasi-universal family) map $f : V \to H^2(M, \mathbb{Q})$. The following computation shows that $f_{\mathcal{E}}(v)$ has trivial $H^2(M)$-component:

$$
\begin{aligned}
s \cdot f_{\mathcal{E}}(v) &= p_*(\mu.q^*(v)) \\
&= \sqrt{td(M)}.p_*(ch^{\vee}(\mathcal{E}).q^*\sqrt{td(X)}.q^*(ch(\mathcal{E}_{t_0}).\sqrt{td(X)})) \\
&= \sqrt{td(M)}.ch(\mathcal{E}xt_p^2(\mathcal{E}, q^*\mathcal{E}_{t_0})) \\
&= s^2 \cdot \sqrt{td(M)}.ch(k(t_0)) \\
&= s^2 \cdot \sqrt{td(M)}.PD(pt) \\
&= s^2 \cdot PD(pt),
\end{aligned}
$$

where $t_0 \in M$. Hence f defines a homomorphism $V/\mathbb{Z}v \to H^2(M, \mathbb{Q})$. If a universal family exists then this map takes values in the integral cohomology of M (Proposition 6.1.13). Hence $V/\mathbb{Z}v \cong H^2(M, \mathbb{Z})$. The general case is proved by deformation theory. The basic idea is to use the moduli space of polarized K3 surfaces and the relative moduli space of semistable sheaves. It is then not difficult to see that the moduli space M is a deformation of a fine moduli space on another nearby K3 surface. (For the complete argument see the proof of 6.2.5.) Since the map f is defined by means of the locally constant class μ, it is enough to prove the assertion for one fibre. $\square$

Corollary 6.1.15 — *Suppose that v is an isotropic vector and that $M^s = M$. Then there exists an isomorphism of rational Hodge structures $H^2(M, \mathbb{Q}) \cong H^2(X, \mathbb{Q})$ which is compatible with the intersection pairing.*

Proof. This follows from the theorem and the easy observation that

$$
\begin{aligned}
H^2(X, \mathbb{Q}) &\rightarrow \tilde{H}(X, \mathbb{Q}) \\
w &\mapsto (0, w, c_1.w/r)
\end{aligned}
$$

induces an isomorphism $H^2(X, \mathbb{Q}) \cong (V/\mathbb{Z}v) \otimes \mathbb{Q}$ of $\mathbb{Q}$-Hodge structures compatible with the pairing. The assumption that v be primitive is unnecessary, because we are only interested in $\mathbb{Q}$-Hodge structures. $\square$

6.2 ... and Higher-Dimensional Moduli Spaces

The aim of this section is to show that moduli spaces of sheaves on K3 surfaces have a very special geometric structure. They are Ricci flat and even hyperkähler. In fact, almost all known examples of hyperkähler manifolds are closely related to them. Thus the study of these moduli spaces sheds some light on the geometry of hyperkähler manifolds in general.

Let us begin with the definition of hyperkähler and irreducible symplectic manifolds.

Definition 6.2.1 — *A hyperkähler manifold is a Riemannian manifold (M, g) which admits two complex structures I and J such that $I \circ J = -J \circ I$ and such that g is a Kähler* *metric with respect to I and J. A complex manifold X is called irreducible symplectic if X is compact Kähler, simply connected and $H^{2,0}(X) = H^0(X, \Omega_X^2)$ is spanned by an everywhere non-degenerate two-form w.*

Recall, that a two-form is non-degenerate if the associated homomorphism $\mathcal{T}_X \to \Omega_X$ is an isomorphism. If (M, g) is hyperkähler and I and J are the two complex structures then $K := I \circ J$ is also a complex structure making g to a Kähler metric. If g is a hyperkähler metric then the holonomy of (M, g) is contained in $\mathrm{Sp}(m)$ where $\dim_{\mathbb{R}} M = 4m$. (M, g) is called *irreducible hyperkähler* if the holonomy equals $\mathrm{Sp}(m)$. If ω_I, ω_J, and ω_K denote the corresponding Kähler forms, then the linear combination $w = \omega_J + i \cdot \omega_K$ defines an element in $H^0(X, \Omega_X^2)$, where X is the complex manifold (M, I). Obviously, w is everywhere non-degenerate.

Theorem 6.2.2 — *If (M, g) is an irreducible compact hyperkähler manifold, then $X = (M, I)$ is irreducible symplectic. Conversely, if X is an irreducible symplectic manifold, then the underlying real manifold M admits a hyperkähler metric with prescribed Kähler class $[\omega_I]$.*

Proof. [25]   □

Even if one is primarily interested in hyperkähler metrics, this theorem allows one to work in the realm of complex geometry. In the sequel some examples of irreducible symplectic manifolds will be described, but the hyperkähler metric remains unknown, for Theorem 6.2.2 is a pure existence result based on Yau's solution of the Calabi conjecture.

Remark 6.2.3 — If X admits a non-degenerate two-form w, then $K_X \cong \mathcal{O}_X$. If X is compact, this implies that the Kodaira dimension of X is zero. In dimension two, according to the Enriques-Kodaira classification, a surface is irreducible symplectic if and only if it is a K3 surface. On the other hand, due to a result of Siu, any K3 surface admits a Kähler metric, hence is irreducible symplectic.

Theorem 6.2.4 — *Let X be an algebraic K3 surface. Then $\mathrm{Hilb}^n(X)$ is irreducible symplectic.*

Proof. $M := \mathrm{Hilb}^n(X)$ is smooth, projective and irreducible (cf. 4.5.10). Moreover, M can be identified with the moduli space of stable rank one sheaves with second Chern number n and therefore admits an everywhere non-degenerate two-form w (cf. 10.4.3). In order to prove that M is irreducible symplectic it therefore suffices to show that M is simply connected and that $\dim H^0(M, \Omega_M^2) = 1$.

For the second statement consider the complement $U \subset X^n := X \times \ldots \times X$ of the 'big diagonal' $\Delta := \{(x_1, \ldots, x_n) | x_i = x_j \text{ for some } i \neq j\}$. There is a natural morphism $\psi : U \rightarrow M$ mapping $(x_1, \ldots, x_n) \in U$ to $Z = \{x_1, \ldots, x_n\} \in M$. This morphism identifies the quotient of U by the action of the symmetric group S_n with an open subset V of M. Then $H^0(M, \Omega_M^2) \subset H^0(V, \Omega_V^2) = H^0(U, \Omega_U^2)^{S_n}$, where the latter is the space of S_n-invariant two-forms on U. But $H^0(U, \Omega_U^2) = H^0(X^n, \Omega_{X^n}^2)$ and $H^0(U, \Omega_U^2)^{S_n} = H^0(X^n, \Omega_{X^n}^2)^{S_n}$, since $\text{codim}(\Delta) = 2$. Since $H^0(X, \Omega_X^1) = 0$, we have $H^0(X^n, \Omega_{X^n}^2) \cong \bigoplus H^0(X, \Omega_X^2)$. Together with the canonical isomorphism $H^0(X^n, \Omega_{X^n}^2)^{S_n} \cong H^0(X, \Omega_X^2) \cong \mathbb{C}$ this yields $H^0(M, \Omega_M^2) = \mathbb{C}$. A similar argument shows $H^0(M, \Omega_M^1) \subset H^0(X, \Omega_X^1) = 0$, which immediately gives $b_1(M) = 0$.

In order to show $\pi_1(M) = \{1\}$ we argue as follows. The real codimension of Δ in X^n is 4, so that $\pi_1(U) \rightarrow \pi_1(X^n) = \{1\}$ is an isomorphism. And since $M \setminus V$ has real codimension 2 in M, the map $j : \pi_1(V) \rightarrow \pi_1(M)$ is surjective. The projection $pr : U \rightarrow V$ induces an isomorphism $S_n \rightarrow \pi_1(V)$ which is described as follows: Choose distinct points $x_1, \ldots, x_n \in X$ and take $(x_1, \ldots, x_n)$ and $\{x_1, \ldots, x_n\}$ as base points in U and V, respectively. For each $\pi \in S_n$ choose a path β_π in U connecting $(x_1, \ldots, x_n)$ and $(x_{\pi(1)}, \ldots, x_{\pi(n)})$. Then $\alpha_\pi = \psi \circ \beta_\pi$ is a closed path in V with base point $\{x_1, \ldots, x_n\}$. In order to prove that $\pi_1(M) = \{1\}$ it suffices to show that $j(\alpha_\pi)$ is null-homotopic in M. Since S_n is generated by transpositions, it suffices to consider the special case $\pi = (12)$. We may assume that x_1 and x_2 are contained some open set $W \subset X$ (in the classical topology) such that $W \cong B^4 \subset \mathbb{C}^2$ and $x_3, \ldots, x_n \in X \setminus W$. Then a path β_π can be described by rotating x_1 and x_2 in a complex line $\mathbb{C} \cap B^4$ around a point x_0. Now let x_1 and x_2 collide within this complex line to x_0, i.e. $\{x_1, x_2\}$ converges to $Z \subset X$ with $\text{Supp}(Z) = x_0$ and $(m_Z/m_Z^2)^* = T_{x_0}(\mathbb{C} \cap B)$. Then α_π is in M freely homotopic to the constant path $Z \cup \{x_2, \ldots, x_n\}$. Hence $j(\alpha_\pi) = 0$. $\qquad \square$

For the higher rank case we again use the Mukai vector

$$v = (v_0, v_1, v_2) = (r, c_1, c_1^2/2 - c_2 + r)$$

and denote the moduli space $M_H(r, c_1, c_2)$ by $M_H(v)$. Recall, $\dim M_H^s(v) = (v, v) + 2$, where $M_H^s(v)$ is the moduli space of stable sheaves. The component v_1 of the Mukai vector v is called *primitive* if it is indivisible as a cohomology class in $H^2(X, \mathbb{Z})$. Recall that from v one recovers the first Chern class c_1 and the discriminant Δ. We therefore have a chamber structure on the ample cone with respect to v (see 4.C).

Theorem 6.2.5 — *If v_1 is primitive and H is contained in an open chamber with respect to v, then $M_H(v)$ is an irreducible symplectic manifold.*

The proof of the theorem consists of two steps. We first prove it for a particular example (Corollary 6.2.7). The general result is obtained by a deformation argument.

We take up the setting of Example 5.3.8. Let $X \to \mathbb{P}_1$ be an elliptic K3 surface with fibre class $f \in H^2(X,\mathbb{Z})$ and a section $\sigma \subset X$ the class of which is also denoted by $\sigma \in H^2(X,\mathbb{Z})$. Assume $\mathrm{Pic}(X) = \mathbb{Z} \cdot \mathcal{O}(\sigma) \oplus \mathbb{Z} \cdot \mathcal{O}(f)$. Let v be a Mukai vector such that $v_1 = \sigma + \ell f$ and let H be suitable with respect to v (cf. 5.3.1). Note that $\sigma + mf$ is ample for $m \geq 3$ and suitable if $m - 3 \geq r^2 \Delta/8$ (cf. 5.3.6).

Proposition 6.2.6 — *Under the above assumptions the moduli space $M_H(v)$ is birational to* $\mathrm{Hilb}^n(X)$.

Proof. The proof is postponed until Chapter 11, (Theorem 11.3.2), where the proposition is treated as an example for the birational description of a moduli space.   □

Corollary 6.2.7 — *Under the assumptions of the proposition $M_H(v)$ is irreducible symplectic.*

Proof. We consider the following general situation. Let $f : X \to X'$ be a birational map between an irreducible symplectic manifold X and a compact manifold X' admitting a non-degenerate two-form w'. Let $U \subset X$ be the maximal open subset of f-regular points, i.e. $f|_U$ is a morphism. Then $\mathrm{codim}(X \setminus U) \geq 2$. Hence $\mathbb{C} \cdot w = H^0(X, \Omega_X^2) = H^0(U, \Omega_U^2)$. Moreover, $f^* : H^0(X', \Omega_{X'}^2) \to H^0(U, \Omega_U^2)$ is injective and thus $\mathbb{C} \cdot w' = H^0(X', \Omega_{X'}^2)$. We can write $f^* w'|_U = \lambda \cdot w|_U$ for some $\lambda \in \mathbb{C}^*$. Since w is non-degenerate everywhere, $f|_U$ is an embedding. The same arguments apply for the inverse birational map $f^{-1} :$ $X' \to X$. One concludes that there exists an open set $U' \subset X'$ such that $f^{-1}|_{U'}$ is regular, $\mathrm{codim}(X' \setminus U') \geq 2$, and $f : U \cong U'$. Moreover, this also implies $\pi_1(X') = \pi_1(U') = \pi_1(U) = \pi_1(X) = \{1\}$. Hence, if X' is Kähler then it is irreducible symplectic as well. Now, apply the argument to the birational correspondence between $\mathrm{Hilb}^n(X)$ and $M_H(v)$ postulated in 6.2.6.   □

The proof of Theorem 6.2.5 relies on the fact that any K3 surface can be deformed to an elliptic K3 surface. To make this rigorous one introduces the following functor:

Definition 6.2.8 — *Let d be a positive integer. Then $\underline{\mathcal{K}}_d$ is the functor $(Sch/\mathbb{C})^{\circ} \to (Sets)$ that maps a scheme Y to the set of all equivalence classes of pairs $(f : X \to Y, \mathcal{L})$ such that $f : X \to Y$ is a smooth family of K3 surfaces and for any $t \in Y$ the restriction $\mathcal{L}_t$ of $\mathcal{L}$ to the fibre $X_t = f^{-1}(t)$ is an ample primitive line bundle with $c_1^2(\mathcal{L}_t) = 2d$. Two pairs $(f : X \to Y, \mathcal{L})$, $(f' : X' \to Y, \mathcal{L}')$ are equivalent if there exists an Y-isomorphism $g : X \to X'$ and a line bundle $\mathcal{N}$ on Y such that $g^* \mathcal{L}' \cong \mathcal{L} \otimes f^* \mathcal{N}$.*

This is a very special case of the moduli functor of polarized varieties. The next theorem is an application of a more general result.

Theorem 6.2.9 — *The functor $\underline{\mathcal{K}}_d$ is corepresented by a coarse moduli space $\mathcal{K}_d$ which is a quasi-projective scheme.*

Proof. [260] □

Similar to moduli spaces of sheaves, the moduli space $\mathcal{K}_d$ is not fine, i.e. there is no universal family parametrized by $\mathcal{K}_d$. But, as for moduli space of sheaves, $\mathcal{K}_d$ is a PGL(N)-quotient $\pi : \mathcal{H}_d \to \mathcal{K}_d$ of an open subset $\mathcal{H}_d$ of a certain Hilbert scheme Hilb$(\mathbb{P}^N, P(n))$, where $P(n) = n^2 \cdot d + 2$. The universal family over the Hilbert scheme provides a smooth projective morphism $\psi : \mathcal{X} \to \mathcal{H}_d$ and a line bundle $\mathcal{L}$ on $\mathcal{X}$, such that $\pi(t) \in \mathcal{K}_d$ corresponds to the polarized K3 surface $(\mathcal{X}_t, \mathcal{L}_t)$.

An alternative construction of $\mathcal{K}_d$ can be given by using the Torelli Theorem for K3 surfaces. This approach immediately yields

Theorem 6.2.10 — *The moduli space $\mathcal{K}_d$ of primitively polarized K3 surfaces is an irreducible variety.*

Proof. [26] □

Using an irreducible component of $\mathcal{H}_d$, which dominates $\mathcal{K}_d$, the theorem shows that any two primitively polarized K3 surfaces (X, H) and (X', H') with $H^2 = H'^2$ are deformation equivalent. (In fact, $\mathcal{H}_d$ itself is irreducible.) More is known about the structure of $\mathcal{K}_d$ and the polarized K3 surfaces parametrized by it. We will need the following results: For the general polarized K3 surface $(X, H) \in \mathcal{K}_d$ one has Pic$(X) = \mathbb{Z} \cdot H$. 'General' here means for (X, H) in the complement of a countable union of closed subsets of $\mathcal{K}_d$. In fact the countable union of polarized K3 surfaces $(X, H) \in \mathcal{K}_d$ with $\rho(X) \geq 2$ is dense in $\mathcal{K}_d$. For the proof of these facts we refer to [26, 22].

It is the irreducibility of $\mathcal{K}_d$ which enables us to compare moduli spaces on different K3 surfaces:

Proposition 6.2.11 — *Let $v_0, v_2 \in \mathbb{Z}$ and $\varepsilon = \pm 1$. Then there exists a relative moduli space $\varphi : \mathcal{M} \to \mathcal{H}_d$ of semistable sheaves on the fibres of ψ such that: i) φ is projective, ii) for any $t \in \mathcal{H}_d$ the fibre $\varphi^{-1}(t)$ is canonically isomorphic to the moduli space $M_{\mathcal{L}_t}(v)$ of semistable sheaves on $\mathcal{X}_t$, where $v = (v_0, \varepsilon c_1(\mathcal{L}_t), v_2)$, and iii) φ is smooth at all points corresponding to stable sheaves.*

(Don't get confused by the extra sign ε. It is thrown in for purely technical reasons which will become clear later.)

Proof. i) and ii) follow from the general existence theorem for moduli spaces 4.3.7. Assertion iii) follows from the relative smoothness criterion 2.2.7: By Serre duality, we have $\text{Ext}^2_{\mathcal{X}_t}(E, E)_0 \cong \text{Hom}_{\mathcal{X}_t}(E, E)_0^{\vee} = 0$ for any stable sheaf E on the fibre $\mathcal{X}_t$. By Theorem 4.3.7 the relative moduli space $\mathcal{M} \to \mathcal{H}_d$ is a relative quotient of an open subset R of an appropriate Quot-scheme $\text{Quot}_{\mathcal{X}/\mathcal{H}_d}((\mathcal{L}^{\vee})^{\oplus P(m)}, P)$. Since $R \to \mathcal{M}$ is a fibre bundle over the stable sheaves, the morphism $\mathcal{M} \to \mathcal{H}_d$ is smooth at a point $[E] \in \mathcal{M}^s$ if and only if $R \to \mathcal{H}_d$ is smooth at $[q : (\mathcal{L}_t^{\vee})^{\oplus P(m)} \to E] \in R$ over it. Let K be the kernel of q. Since $\text{Ext}^1(K, E) \cap \text{Ext}^2(E, E)_0 = 0$, the tangent map $T_q R \to T_t \mathcal{H}_d$ is surjective by 2.2.7, 2.A.8 and 4.5.4. Hence $\mathcal{M} \to \mathcal{H}_d$ is smooth at all points corresponding to stable sheaves on the fibres. $\square$

Corollary 6.2.12 — *Let (X, H) and (X', H') be two polarized K3 surfaces with $H^2 = H'^2$ and let $v = (v_0, \varepsilon H, v_2)$, $v' = (v_0, \varepsilon H', v_2)$. Assume that every sheaf E in $M_H(v)$ and in $M_{H'}(v')$ is stable. Then $M_H(v)$ is irreducible symplectic if and only if $M_{H'}(v')$ is irreducible symplectic.*

Proof. Let $\mathcal{H}_d^0 \subset \mathcal{H}_d$ denote the dense open subset of regular values of φ. Then there exist points $t, t' \in \mathcal{H}_d^0$ such that $(\mathcal{X}_t, \mathcal{L}_t) = (X, H)$ and $(\mathcal{X}_{t'}, \mathcal{L}_{t'}) = (X', H')$. Hence the restriction of φ to $\mathcal{H}_d^0$ is a smooth projective family over a connected base with the two moduli spaces $M_H(v)$ and $M_{H'}(v')$ occurring as fibres over t and t', respectively. As for any smooth proper morphism over a connected base the fundamental groups and Betti numbers of all fibres of φ are equal. On the other hand, the Hodge numbers of the fibres are upper-semicontinuous. Since the Hodge spectral sequence degenerates on any fibre and hence the sum of the Hodge numbers equals the sum of the Betti numbers, the Hodge numbers of the fibres of φ stay also constant.

Therefore, if $M_H(v)$ is irreducible symplectic, then $M_{H'}(v')$ is simply connected and $h^{2,0}(M_{H'}(v')) = 1$. Since by 10.4.3 the moduli space $M_{H'}(v')$ admits a non-degenerate symplectic structure, $M_{H'}(v')$ is irreducible as well. $\square$

Proof of Theorem 6.2.5. Step 1. We first reduce to the case that $\rho(X) \geq 2$. If $\rho(X) = 1$, then $v_1 = \pm H$, where H is the ample generator of $\text{Pic}(X)$. As we have mentioned above, the set of polarized K3 surfaces $(X', H') \in \mathcal{K}_d$ with $\rho(X') \geq 2$ is a countable union of closed subsets which is dense in $\mathcal{K}_d$. On the other hand, the set of K3 surfaces $(X', H') \in \mathcal{K}_d$ such that $M_{H'}(v')$ is not smooth is a proper closed subset. Indeed, $M_{H'}(v')$ is smooth at stable points and the set of properly semistable sheaves is closed in $\mathcal{M}$ and does not dominate $\mathcal{H}_d$. Thus we can find $(X', H') \in \mathcal{K}_d$ such that $\rho(X') \geq 2$ and H' is generic with respect to v'. By 6.2.12 the moduli space $M_H(v)$ is irreducible symplectic if and only if $M_{H'}(v')$ is irreducible symplectic.

Step 2. We may assume $\rho(X) \geq 2$. Let us show that one can further reduce to the case that $H^2 > r^2 \Delta/8$. By assumption H is contained in an open chamber with respect to v. Hence there exists a polarization H' in the same chamber which is not linearly equivalent to v_1. Then we get $M_H(v) \cong M_{H'}(v) \cong M_{H'}(v(mH'))$ for any m. Here $v(mH')$ is the Mukai vector of $E \otimes \mathcal{O}(mH')$, where $E \in M_{H'}(v)$, and the second isomorphism is given by mapping E to $E \otimes \mathcal{O}(mH')$. For any m_0 there exists an integer $m \geq m_0$ such that $v_1(mH') = v_1 + rmH'$ is ample, contained in the chamber of H, and primitive. Hence $M_H(v) = M_{v_1+rmH'}(v(mH'))$. Clearly, $(v_1 + rmH')^2$ can be made arbitrarily large for $m_0 \gg 0$.

Step 3. Assume now that (X, H) is a polarized K3 surface with $H^2 > r^2\Delta/8$. Let X' be an elliptic K3 surface with $\text{Pic}(X') = \mathbb{Z} \cdot f \oplus \mathbb{Z} \cdot \sigma$ as in Proposition 6.2.6. Then $H' := \sigma + (H^2 + 2)f$ is ample and suitable with respect to $v' = (v_0, H', v_2)$ by 5.3.6. Moreover, $H'^2 = H^2 =: 2d$. Thus $(X, H), (X', H') \in \mathcal{K}_d$. Hence $M_H(v)$ is irreducible symplectic if and only if $M_{H'}(v')$ is irreducible symplectic which was the content of Proposition 6.2.6. $\qquad\square$

In Section 6.1 we first established that any two-dimensional moduli space is a K3 surface, i.e. irreducible symplectic, and then determined its Hodge structure. Here we proceed along the same line. After having achieved the first half we now go on and study the Hodge structure of the moduli space.

A weight-two Hodge structure on any compact Kähler manifold is given by the Hodge decomposition $H^2(X, \mathbb{C}) = H^{2,0}(X) \oplus H^{1,1}(X) \oplus H^{0,2}(X)$. For an irreducible symplectic manifold the full information about the decomposition is encoded in the inclusion of the one-dimensional space $H^{2,0}(X) \subset H^2(X, \mathbb{C})$, i.e. a point in $\mathbb{P}(H^2(X, \mathbb{C})^\vee)$. This point is called the *period point* of X. Next we introduce an auxiliary quadratic form, by means of which one can recover the whole weight-two Hodge structure from the period point in $\mathbb{P}(H^2(X, \mathbb{C})^\vee)$.

Definition and Theorem 6.2.13 — *If X is irreducible symplectic of dimension $2n$, then there exists a canonical integral form q of index $(3, b_2(X) - 3)$ on $H^2(X, \mathbb{Z})$ given by*

$$q(aw + \alpha + b\bar{w}) = \lambda \cdot \left(ab + (n/2) \int_X (w\bar{w})^{n-1}\alpha^2 \right),$$

where $\alpha \in H^{1,1}(X)$, $\mathbb{C} \cdot w = H^0(X, \Omega_X^2)$ with $\int (w\bar{w})^n = 1$ and λ is a positive scalar. Moreover,

$$\binom{2n}{n} \cdot q(\beta)^n = \lambda^n \cdot \int_X \beta^{2n}.$$

Proof. [25], [73] $\qquad\qquad\qquad\qquad\qquad\qquad\qquad\qquad\qquad\qquad\qquad\square$

Note that for K3 surfaces this is just the intersection pairing.

For the higher dimensional examples constructed above one can identify the weight-two Hodge structure endowed with this pairing. We begin with the Hilbert scheme. The higher rank case is based on this computation.

Theorem 6.2.14 — *Let X be a K3 surface and $n > 1$. Then there exists an isomorphism of weight-two Hodge structures compatible with the canonical integral forms*

$$H^2(\mathrm{Hilb}^n(X), \mathbb{Z}) \cong H^2(X, \mathbb{Z}) \oplus \mathbb{Z} \cdot \delta,$$

where on the right hand side δ is a class of type $(1,1)$ and the integral form is the direct sum of the intersection pairing on X and the integral form given by $\delta^2 = -2(n-1)$. The constant λ in 6.2.13 is $1/2$.

Proof. [25]   □

For the higher rank case, we recall and slightly modify the definition of the map

$$f : H^*(X, \mathbb{Q}) \to H^2(M_H(v), \mathbb{Q})$$

introduced in the proof of 6.1.14. If $\mathcal{E}$ is a quasi-universal family over $M_H(v) \times X$ of rank $m \cdot r$ and $c \in H^*(X, \mathbb{Z})$, then $f(c) := p_*\{\mu.q^*(c)\}_2/m$, where this time $\mu := ch^\vee(\mathcal{E}).q^*\sqrt{td(X)}$. This differs from the original definition by the factor $\sqrt{td(M)}$. As before, denote by V the orthogonal complement of $v \in \tilde{H}(X, \mathbb{Q})$ endowed with the quadratic form and the induced Hodge structure. Note that under our assumption $\dim(M_H(v)) > 2$ the vector v is no longer isotropic, i.e. $v \notin V$. A priori, f need not be integral even if $M_H(v)$ admits a universal family.

Theorem 6.2.15 — *Under the assumptions of 6.2.5 the homomorphism f defines an isomorphism of integral Hodge structures $V \cong H^2(M_H(v), \mathbb{Z})$ compatible with the quadratic forms.*

Proof. [210]   □

This theorem due to O'Grady nicely generalizes Beauville's result for the Hilbert scheme and Mukai's computations in the two-dimensional case. Indeed, if $v = (1, 0, 1 - n)$, then $M_H(v) = \mathrm{Hilb}^n(X)$ and $V = \{(a, b, a(n-1))|a \in \mathbb{Z}, b \in H^2(X, \mathbb{Z})\}$. And with its induced Hodge structure V is isomorphic to the direct sum of $H^2(X, \mathbb{Z})$ and $\mathbb{Z}$ where for $a \in \mathbb{Z}$ one has $q(a) = ((a, 0, a(n-1)), (a, 0, a(n-1))) = -2a^2(n-1)$.

We conclude this section by stating a result which indicates that although the moduli spaces $M_H(v)$ provide examples of higher dimensional compact hyperkähler manifolds, they do not furnish completely new examples.

Theorem 6.2.16 — *Let v_1 be primitive and H contained in an open chamber with respect to v. Then $M_H(v)$ and $\mathrm{Hilb}^n(X)$ are deformation equivalent, where $n = (v, v)/2 + 1$*

Proof. [113] □

Appendix to Chapter 6

6.A The Irreducibility of the Quot-scheme

In the appendix we prove that the Quot-scheme $\mathrm{Quot}(E, \ell)$ of zero-dimensional quotients of length ℓ of a fixed locally free sheaf E is irreducible. This result was used on several occasions in this chapter but it is also interesting for its own sake.

In the following the *socle* of a zero-dimensional sheaf T at a point x is the $k(x)$-vector space of all elements $t \in T_x$ which are annihilated by the maximal ideal of $\mathcal{O}_{X,x}$. This usage of the word 'socle' differs from that in Section 1.5. Exercise: In what sense are they related?

If $\mathrm{rk}(E) = 1$, then $\mathrm{Quot}(E, \ell)$ is isomorphic to the Hilbert scheme $\mathrm{Hilb}(X, \ell)$. The latter was shown to be smooth and irreducible (4.5.10).

Theorem 6.A.1 — *Let X be a smooth surface, E a locally free sheaf and $\ell > 0$ an integer. Then $\mathrm{Quot}(E, \ell)$ is an irreducible variety of dimension $\ell(\mathrm{rk}(E) + 1)$.*

Proof. The assertion is proved by induction over ℓ. If $\ell = 1$, then $\mathrm{Quot}(E, 1) = \mathbb{P}(E)$, which is clearly irreducible.

Let $0 \to N \to \mathcal{O}_{\mathrm{Quot}} \otimes E \to T \to 0$ be the universal quotient family over $Y_\ell :=$ $\mathrm{Quot}(E, \ell) \times X$. For any point $(s, x) \in Y_\ell$ and a non-trivial homomorphism $\lambda : N_s(x) \to k(x)$ we can form the push-out diagram

$$
\begin{array}{ccccccccc}
0 & \longrightarrow & k(x) & \longrightarrow & T'_{s'} & \longrightarrow & T_s & \longrightarrow & 0 \\
  &   & \lambda \uparrow &   & s' \uparrow &   & \| &   &   \\
0 & \longrightarrow & N_s & \longrightarrow & E & \xrightarrow{s} & T_s & \longrightarrow & 0.
\end{array}
$$

Thus sending $(s, x, \langle \lambda \rangle)$ to (s', x) defines a morphism $\psi : \mathbb{P}(N) \to Y_{\ell+1}$. We want to use the diagram

$$ Y_\ell \xleftarrow{\varphi} \mathbb{P}(N) \xrightarrow{\psi} Y_{\ell+1} $$

for an induction argument. For each $(s : E \to T_s, x) \in Y_\ell$ let $i(s, x) := \hom(k(x), T_s)$ denote the dimension of the socle of T_s at x. Then i is an upper semicontinuous function on Y_ℓ: let $Y_{\ell,i}$ denote the stratum of points of socle dimension i. It is not difficult to see that if $y = \varphi(\langle \lambda \rangle)$ and $y' = \psi(\langle \lambda \rangle)$ for some $\langle \lambda \rangle \in \mathbb{P}(N)$, then $|i(y) - i(y')| \leq 1$. This shows that

$$ \psi^{-1}(Y_{\ell+1,j}) \subset \bigcup_{|i-j| \leq 1} \varphi^{-1}(Y_{\ell,i}). \tag{6.1} $$

Check that $\psi^{-1}(s',x) = \mathbb{P}(\mathrm{Socle}(T'_x)^{\vee}) \cong \mathbb{P}^{i(s',x)-1}$, and $\varphi^{-1}(s,x) = \mathbb{P}(N_s(x)) \cong$
$\mathbb{P}^{\dim N_s(x)-1}$. Moreover, using a minimal projective resolution of T_s over the local ring
$\mathcal{O}_{X,x}$, one shows: $\dim N_s(x) = \mathrm{rk}(E)+i(s,x)$. Using this relation, the information about
the fibre dimension, and the relation (6.1) one proves by induction that $\mathrm{codim}(Y_{\ell,i}, Y_\ell) \geq$
$2i$ for all $i \geq 0$ and $\ell > 0$. Let $0 \to \mathcal{A} \to \mathcal{B} \to N \to 0$ be a locally free resolution of N.
Then $\mathrm{rk}(\mathcal{B}) = \mathrm{rk}(\mathcal{A}) + \mathrm{rk}(E)$, and $\mathbb{P}(N) \subset \mathbb{P}(\mathcal{B})$ is the vanishing locus of the homomorphism $\pi^*\mathcal{A} \to \pi^*\mathcal{B} \to \mathcal{O}_{\mathbb{P}(\mathcal{B})}(1)$, $\pi : \mathbb{P}(\mathcal{B}) \to Y_\ell$ denoting the projection. In particular, as
Y_ℓ is irreducible, and $\mathbb{P}(N)$ is locally cut out by $\mathrm{rk}(\mathcal{A}) = \mathrm{rk}(\mathcal{B}) - \mathrm{rk}(E)$ equations in $\mathbb{P}(\mathcal{B})$,
every irreducible component of $\mathbb{P}(N)$ has dimension $\geq \dim(Y_\ell) + \mathrm{rk}(E) - 1$. Now it is
easy to see that $\varphi^{-1}(Y_{\ell,0})$ is irreducible and has the expected dimension and that $\varphi^{-1}(Y_{\ell,i})$
is too small for all $i \geq 1$ to contribute other components. Hence $\mathbb{P}(N)$ is irreducible, and
as the composition $\mathbb{P}(N) \to Y_{\ell+1} \to \mathrm{Quot}(E, \ell + 1)$ is surjective, $Y_{\ell+1}$ is irreducible as
well.   □

Comments:

— 6.1.6, 6.1.8, 6.1.14 are contained in Mukai's impressive article [188]. He also applies the results
to show the algebraicity of certain cycles in the cohomology of the product of two K3 surfaces.

— For a more detailed study of rigid bundles, i.e. zero-dimensional moduli spaces, see Kuleshov's
article [133].

— The relation between irreducible symplectic and hyperkähler manifolds was made explicit in
Beauville's paper [25]. He also proved Theorem 6.2.4 for all K3 surfaces provided the Hilbert scheme
is Kähler. That this holds in general follows from a result of Varouchas [258]. Furthermore, Beauville
described another series of examples of irreducible symplectic manifolds, so called generalized Kummer varieties, starting with a torus.

— The main ingredient for 6.2.5, namely the existence of the symplectic structure is due to Mukai.
His result will be discussed in detail in Chapter 10. The irreducibility and 1-connectedness was first
shown in the rank two case by Göttsche and Huybrechts in [87] and for arbitrary rank by O'Grady in
[210]. The proof we presented follows [210].

— The calculation of the Hodge structure of the Hilbert scheme (Theorem 6.2.14) is due to Beauville [25]. Note that the Hodge structures (without metric) of any weight of the Hilbert scheme of an
arbitrary surface can be computed. This was done by Göttsche and Soergel [86].

— The description in the higher rank case (Theorem 6.2.15) is due to O'Grady. The proof relies
on the proof of 6.2.5, but a more careful description of the birational correspondence between moduli
space and Hilbert scheme on an elliptic surface is needed. Once the assertion is settled in this case,
the general case follows immediately by using the irreducibility of the moduli space $\mathcal{K}_d$. This part is
analogous to an argument of Mukai's.

— In [87] Göttsche and Huybrechts computed all the Hodge numbers of the moduli space of rank
two sheaves. They coincide with the Hodge numbers of the Hilbert scheme of the same dimension.
But this is not surprising after having established 6.2.16.

— Theorem 6.2.16 was proved by Huybrechts [113]. The proof is based on the fact that any two
birational symplectic manifolds are deformation equivalent. Note that in [113] the proof is given only
for the rank two case, but the techniques can now be extended to cover the general case as well.

— The results of [113] (and their generalizations) show that all known examples of irreducible
symplectic manifolds, i.e. compact irreducible hyperkähler manifolds, are deformation equivalent

either to the Hilbert scheme of a K3 surface or to a generalized Kummer variety. In particular, all examples have second Betti number either 7, 22, or 23.

— The irreducibility of $\mathrm{Quot}(E, \ell)$ (Theorem 6.A.1) was obtained by J. Li [148] for $\mathrm{rk}(E) = 2$ and by Gieseker and Li [82] for $\mathrm{rk}(E) \geq 2$. The proof sketched in the appendix is due to Ellingsrud and Lehn [58]. They also show that the fibres of the natural morphism $\mathrm{Quot}(E, \ell) \to S^\ell(X)$ are irreducible. Note that in the case $\mathrm{rk}(E) = 1$ Theorem 6.A.1 reduces to the theorem of Fogarty that the Hilbert scheme of points on an irreducible smooth surface is again irreducible (cf. 4.5.10).

6.B Further comments (second edition)

I. Derived categories.

Bounded derived categories of coherent sheaves have always been a fundamental concept and an important technical tool in algebraic geometry, but over the last decade they have been studied quite intensively also in their own right. This is partially motivated by Kontsevich's homological mirror symmetry. We just mention a few results that are closely related to moduli spaces of sheaves. For further information and general results see [338]. To set up notation, let $D^b(X)$ be the bounded derived category of the abelian category $\mathrm{Coh}(X)$ of coherent sheaves on a smooth projective variety X.

Derived categories of K3 surfaces. Already Mukai proved in [188] that a two-dimensional moduli space M of sheaves on a K3 surface X satisfying the conditions of Proposition 6.1.13 is derived equivalent to the surface X itself. More precisely, the Fourier-Mukai functor $\Phi_{\mathcal{E}} : \mathcal{F}^\bullet \mapsto Rp_*(q^*\mathcal{F}^\bullet \otimes^L \mathcal{E})$ describes an exact linear equivalence

$$D^b(X) \xrightarrow{\sim} D^b(M).$$

Later Orlov proved in [411] that every Fourier-Mukai partner of a K3 surface X is of this form. More precisely, he proves the equivalence of the following three statements:

i) Two K3 surfaces X and Y are derived equivalent, i.e. there exists an exact linear equivalence $\Phi : D^b(X) \xrightarrow{\sim} D^b(Y)$.

ii) The K3 surface Y is isomorphic to X or Y is isomorphic to a moduli space of stable sheaves on X satisfying the conditions of Proposition 6.1.13.

iii) There exists an isomorphism of Hodge structures $\widetilde{H}(X, \mathbb{Z}) \cong \widetilde{H}(Y, \mathbb{Z})$ respecting the Mukai pairing.

Note that it is not difficult to show that any smooth projective variety Y derived equivalent to a K3 surface X is itself a K3 surface.

As was shown in [339] these statements are also equivalent to the following one, which a priori seems stronger than *ii)*:

iv) The K3 surface Y is isomorphic to X or Y is isomorphic to a fine moduli space of μ-stable vector bundles on X.

In [339] one finds yet another characterization of derived equivalent K3 surfaces in terms of equivalences between certain abelian categories associated naturally to polarized K3 surfaces.

Twisted derived categories. As was explained in Section 4.D, a substitute for the missing universal sheaf in the case of a moduli space which is only coarse is provided by a twisted universal sheaf. This was first observed in [298]. Căldăraru used this observation to relate the derived category $D^b(X)$ of a K3 surface X to the bounded derived category

$D^b(\text{Coh}(M, \alpha))$ of a coarse projective moduli space of stable sheaves M endowed with a Brauer class α such that the universal sheaf exists as a $1 \boxtimes \alpha$-twisted sheaf on $X \times M$. The proof is identical to the one in the case of a fine moduli space.

In [340] the equivalence of the statements *i)-iii)* was proved in the twisted case. The Mukai lattice with the natural weight two Hodge structure has to be modified in order to take the twist by a Brauer class into account.

Derived categories of abelian varieties. The theory for abelian varieties and for abelian surfaces in particular is even better understood (but not quite so interesting) as the theory for K3 surfaces. Roughly, two abelian varieties A and B are derived equivalent if and only if there exists a particular kind of isomorphism $A \times \hat{A} \cong B \times \hat{B}$. Derived equivalence can also be explained in terms of moduli spaces of semi-homogeneous vector bundles. These results are mainly due to Orlov [412] and Polishchuk [414], see [338], Ch. 10 or [415].

General classification. It is not difficult to show that two derived equivalent smooth projective varieties are of the same dimension. So, in particular, a Fourier-Mukai partner of a surface is again a surface. Results of Bridgeland, Maciocia and Kawamata combined prove the following fundamental result (see e.g. [338], Ch. 11):

If X is neither an elliptic nor an abelian nor a K3 surface, then any smooth projective surface derived equivalent to X is already isomorphic to X.

In fact, if an elliptic surface $X \to C$ admits a non-isomorphic Fourier-Mukai partner, then by a result of Kawamata the elliptic fibration is relatively minimal. The most interesting elliptic surfaces are of Kodaira dimension one. In this case, Bridgeland and Maciocia [292] prove:

Any Fourier-Mukai partner of a relatively minimal elliptic surface $X \to C$ of Kodaira dimension $\text{kod}(X) = 1$ is isomorphic to a moduli space $M(0, rf, c)$ of semistable sheaves of rank zero and first Chern class a multiple of the fibre class f.

II. Hilbert schemes of K3 surfaces.

The cohomology of the Hilbert schemes $\text{Hilb}^n(X)$ of any surface X is a much studied object. It is best understood when X is the affine plane, a K3 or abelian surface. For an abelian surface X the Hilbert scheme $\text{Hilb}^{n+1}(X)$ splits after a finite étale cover into X and the generalized Kummer variety $K^n(X)$. The latter behaves in many respects like the Hilbert scheme of a K3 surface. For instance, it is irreducible symplectic.

Cohomology ring. A first numerical approach via the Weil conjectures goes back to Göttsche's thesis [83]. Already then it was made clear that for a complete understanding, instead of looking at $H^*(\mathrm{Hilb}^n(X))$ one better considers the cohomology of all Hilbert schemes at the same time, i.e. $H^*(\coprod_n \mathrm{Hilb}^n(X))$. The latter was later reinterpreted by Nakajima (see e.g. [395]) and independently by Grojnowski [329] as a Fock space with an action of a Heisenberg Lie algebra modelled on the cohomology of X.

However, the ring structure of the cohomology was much harder to get. For surfaces with trivial canonical bundle this is now fully understood due to work of Lehn and Sorger [363, 364, 365], Li, Qin, and Wang [370], and Vasserot [436]. For a survey of these results see [319, 362]. The ring structure of the cohomology for generalized Kummer varieties was described by Britze [295].

Topological invariants. The topological invariants like Betti numbers and the Euler characteristic of the Hilbert scheme $\mathrm{Hilb}^n(X)$ of a K3 surface or of the generalized Kummer variety $K^n(A)$ of an abelian surface A are known due to work of Göttsche and Soergel [83, 86]. The Euler characteristic for the generalized Kummer variety was later computed again by Debarre in [311] (using ideas of [276]) and once more in [331].

Chern numbers. Certain Chern numbers of Hilbert schemes of K3 surfaces and generalized Kummer varieties have been computed. See [318, 405] or the more recent [284] and the references therein. Chern classes of the tangent bundle or of certain tautological bundles can be considered as elements in the rational cohomology or in the Chow ring. An intriguing conjecture concerning the latter case has been put forward by Beauville and verified for many Hilbert schemes by Voisin in [439]. In [385] Markman proved that in general the Chern classes of $\mathrm{Hilb}^n(X)$ of a K3 surface X are not contained in the subring generated by $H^2(X)$.

Lagrangian fibrations. If $X \to \mathbb{P}^1$ is an elliptic K3 surface, its Hilbert scheme $\mathrm{Hilb}^n(X)$ comes with a morphism $\mathrm{Hilb}^n(X) \to \mathbb{P}^n$ which is a Lagrangian fibration, i.e. the restriction of the non-degenerate two-form $w \in H^0(\mathrm{Hilb}^n(X), \Omega^2)$ to the general (and in fact to any) fibre is trivial. Moreover, the general fibre is an abelian variety, which in this case can explicitly be described as the product of n generic fibres of $X \to \mathbb{P}^1$.

A general conjecture predicts that an irreducible symplectic manifold Y admits a rational Lagrangian fibration if there exists a class $\alpha \in H^2(Y, \mathbb{Z}) \cap H^{1,1}(Y)$ with $q(\alpha) = 0$. Here q denotes the quadratic form in Theorem 6.2.13. If α is a nef class, then the fibration is conjectured to be regular. For more details and a survey of the theory of compact hyperkähler manifolds see Part III in [330].

The conjecture has been proved for Hilbert schemes of K3 surfaces X with Picard rank one independently by Markushevich in [386] and by Sawon in [422]. Note that in this case for a class α to exist on $\mathrm{Hilb}^n(X)$ one needs $2(g-1)n^2 = H^2$, where H is the generator of $\mathrm{Pic}(X)$. In both papers the fibration was first shown to be rational and using an argument of

Yoshioka regularity was then shown in [422]. It is maybe noteworthy that both approaches use twisted sheaves and the corresponding derived categories.

The analogous result for generalized Kummer varieties has been established around the same time by Gulbrandsen in [332]. Derived categories make their appearance again, but the proof is simpler and does not use twisted sheaves.

Tautological bundles on the Hilbert scheme. To any line bundle L on the surface X one can naturally associate the tautological line bundle L_n on $\mathrm{Hilb}^n(X)$, which by definition is $p_*(q^*L \otimes \mathcal{O}_{\mathcal{Z}})$, where $\mathcal{Z} \subset \mathrm{Hilb}^n(X) \times X$ is the universal subscheme. The Euler-Poincaré characteristic of L_n is given by $\chi(L_n) = \binom{\chi(L)+n-1}{n}$ (see [318, 330]). It is more complicated to understand the single cohomology groups and the ring structure of L_n and its tensor products. This problem has been treated in [310, 424]. The stability of L_2 with respect to an appropriate polarization has been proven in [430], where one can also find an explicit computation of its Chern character which seems unknown in general. The total Chern class of L_n was computed in terms of a Nakajima basis in [363].

III. Moduli spaces of sheaves on K3 surfaces.

Moduli spaces of semistable sheaves on K3 and abelian surfaces are still studied intensively. They provide the only known examples of irreducible symplectic manifolds and their geometry is rich and intricate for various reasons. Hilbert schemes, as the moduli spaces of rank one sheaves, form the most accessible class among them.

Deformation equivalence and birational correspondences of moduli spaces. Theorem 6.2.16 was subsequently generalized by O'Grady and Yoshioka. In [451] Yoshioka treats the case $v(kr, k\ell, s)$ with $r + \ell$ and v primitive and $(v, v)/2 + 1 \geq k^2 - 1$ or $k = 1$. The paper also generalizes O'Grady's description of the Mukai pairing of the moduli space (see Theorem 6.2.15) to the case of non-primitive v_1.

We also wish to mention the paper [325] where the fundamental fact that curves in K3 surfaces tend to be (Brill-Noether) generic (cf. [138]) was used to give a new proof for the irreducibility of some of the moduli spaces of stable sheaves on K3 surfaces.

Since complete moduli spaces of stable sheaves on K3 surfaces are deformation equivalent to Hilbert schemes, their cohomology is essentially known. In this context we mention Markman's result in [384] saying that the Künneth components of the universal family of a fine moduli space generate the cohomology of the moduli space. This observation for moduli spaces on rational surfaces and on curves goes back to Ellingsrud and Strømme [317] respectively Beauville [275].

Birational correspondences between moduli spaces of stable sheaves and Hilbert schemes have been observed in many instances. This has been crucial for e.g. Theorem 6.2.15.

A thorough study of birational correspondences between various moduli spaces of stable sheaves on a fixed K3 surface has been undertaken by Markman in [383].

Singular moduli spaces. In the papers [406, 407] O'Grady studies two singular moduli spaces of semistable sheaves and describes explicitly resolutions which turn out to be irreducible symplectic. These two examples are referred to in the literature as *O'Grady's sporadic examples* which together with the two series provided by the Hilbert schemes of points on a K3 surface and the generalized Kummer varieties are the only known examples of deformation classes of irreducible symplectic manifolds.

In [406] O'Grady studies the 10-dimensional moduli space $M(v)$ of semistable sheaves with Mukai vector $v = (2, 0, 6)$. Due to the existence of properly semistable sheaves, $M(v)$ is not smooth. But it is shown to be birational to an irreducible symplectic manifold $\tilde{M}(v)$ with second Betti number $b_2(\tilde{M}(v)) \geq 24$, which excludes $\tilde{M}(v)$ from being deformation equivalent to $\mathrm{Hilb}^5(X)$, the Hilbert scheme of a K3 surface, or $K^5(A)$ the generalized Kummer variety associated to an abelian surface, which have $b_2 = 23$ respectively $b_2 = 7$. Later Rapagnetta showed in [420] that $b_2(\tilde{M}(v)) = 24$. He also computed the natural quadratic form on $H^2(\tilde{M}(v), \mathbb{Z})$. The Euler number of $\tilde{M}(v)$ was computed by Mozgovoy in his thesis [393].

The second example of O'Grady [407] produces a new six-dimensional irreducible symplectic manifold. The construction starts with a singular moduli space of rank two sheaves on an abelian surface. Since the second Betti number in this example turns out to be $b_2 = 8$, this new manifold is not deformation equivalent to any of the other known examples. In [419] Rapagnetta computes the second integral cohomology with its quadratic form and the Euler number.

Maybe the most remarkable feature of O'Grady's examples is that they cannot be generalized. One would think that the same idea of resolving singular moduli spaces of semistable sheaves on K3 or abelian surfaces could lead to many more examples or at least put the original two in a series comparable to that of Hilbert scheme and generalized Kummer varieties. However, as has been shown by Kaledin, Lehn, and Sorger in [351, 366] by a careful analysis of the singularities no other moduli space of semistable sheaves admits a projective symplectic resolution. That O'Grady's moduli spaces $M(2, 0, 2n)$ do not admit a symplectic resolution for $n \geq 3$ had before been deduced by Namikawa [402] from O'Grady's calculation, by Kaledin and Lehn [350] from an analysis of the singularities, and by Choy and Kiem in [300, 301] via a completely different method relying on properties of the virtual Hodge polynomial for singular varieties admitting crepant resolutions.

Two-dimensional moduli spaces. Complete moduli spaces of stable sheaves on K3 surfaces of dimension two have been shown to be K3 surfaces themselves (see Proposition

6.1.13). In a series of papers Madonna and Nikulin have tried to understand when exactly the moduli space is isomorphic to the original K3 surface. See [378] and references therein. In [379] the analogous question for moduli spaces of higher dimensions and their relation to the Hilbert scheme of the original K3 surfaces is studied.

What about cases when properly semistable sheaves exist? This situation has been studied in [266] and [410]. In the first paper Abe shows that the moduli space is still birational to a smooth K3 surface and studies its singularities. In the second paper this is studied from the view point of moduli spaces of *twisted* sheaves. Whenever the chosen polarization is non-generic, so allowing for semistable but not stable sheaves, one introduces an extra parameter (cf. [57]) and considers moduli spaces of twisted stable sheaves. The two-dimensional moduli spaces are shown to be normal for certain choices of the twisting parameter and the singularities are studied.

7

Restriction of Sheaves to Curves

In this chapter we take up a problem already discussed in Section 3.1. We try to understand how μ-(semi)stable sheaves behave under restriction to hypersurfaces. At present, there are three quite different approaches to this question, and we will treat them in separate sections. None of these methods covers the results of the others completely.

The theorems of Mehta and Ramanathan 7.2.1 and 7.2.8 show that the restriction of a μ-stable or μ-semistable sheaf to a *general* hypersurface of *sufficiently high degree* is again μ-stable or μ-semistable, respectively. It has the disadvantage that it is not effective, i.e. there is no control of the degree of the hypersurface, which could, a priori, depend on the sheaf itself. However, such a bound, depending only on the rank of the sheaf and the degree of the variety, is provided by Flenner's Theorem 7.1.1. Since it is based on a careful exploitation of the Grauert-Mülich Theorem in the refined form 3.1.5, it works only in characteristic zero and for μ-semistable sheaves. In that respect, Bogomolov's Theorem 7.3.5 is the strongest, though one has to restrict to the case of smooth surfaces. It says that the restriction of a μ-stable vector bundle on a surface to *any* curve of sufficiently high degree is again μ-stable, whereas the theorems mentioned before provide information for *general* hypersurfaces only. Moreover, the bound in Bogomolov's theorem depends on the invariants of the bundle only. This result provides an important tool for the investigation of the geometry of moduli spaces in the following chapters.

7.1 Flenner's Theorem

Let X be a normal projective variety of dimension n over an algebraically closed field of characteristic zero and let $\mathcal{O}(1)$ be a very ample line bundle on X. Furthermore, let $Z \subset \Pi \times X = \prod_{i=1}^{\ell} |\mathcal{O}_X(a)| \times X$ be the incidence variety of complete intersections $D_1 \cap \ldots \cap D_\ell$ with $D_i \in |\mathcal{O}_X(a)|$. (For the notation compare Section 3.1.) Recall that $q : Z \to X$ is a product of projective bundles over X (cf. Section 2.1) and therefore

193

$\mathrm{Pic}(Z) \cong q^*\mathrm{Pic}(X) \oplus p^*\mathrm{Pic}(\Pi)$. The same holds true for any open subset of Z containing all points of codimension one.

$$
\begin{array}{ccccc}
Z_s & \subset & Z & \xrightarrow{q} & X \\
\downarrow & & \downarrow p & & \\
s & \in & \prod_{i=1}^{\ell} |\mathcal{O}(a)| & =: \Pi &
\end{array}
$$

If E is μ-semistable, then for a general complete intersection $Z_s = p^{-1}(s)$ one has (Theorem 3.1.5):

$$
\delta\mu(E|_{Z_s}) := \max\{\mu_i(E|_{Z_s}) - \mu_{i+1}(E|_{Z_s})\} < -\mu_{\min}(T_{Z/X}|_{Z_s}).
$$

Roughly, the proof of the Grauert-Mülich Theorem was based on this inequality and the upper bound $-\mu_{\min}(T_{Z/X}|_{Z_s}) \le a^{\ell+1} \cdot \deg(X)$. The Theorem of Flenner combines the inequality for $\delta\mu$ with a better bound for $-\mu_{\min}(T_{Z/X}|_{Z_s})$. The new bound allows one to conclude that $\delta\mu = 0$, i.e. $E|_{Z_s}$ μ-semistable, for $a \gg 0$. Note that in the following theorem only the rank of E enters the condition on a.

Theorem 7.1.1 — *Assume*

$$
\frac{\binom{a+n}{a} - \ell \cdot a - 1}{a} > \deg(X) \cdot \max\left\{\frac{r^2-1}{4}, 1\right\}.
$$

If E is a μ-semistable sheaf of rank r, then the restriction $E|_{D_1 \cap ... \cap D_\ell}$ to a general complete intersection with $D_i \in |\mathcal{O}(a)|$ is μ-semistable.

Proof. The proof is divided into several steps. We eventually reduce the assertion to the case that X is a projective space. *Step 1.* We claim that it suffices to show

$$
-\mu_{\min}(T_{Z/X}|_{Z_s}) \le \frac{a^{\ell+1}}{\binom{n+a}{a} - a \cdot \ell - 1} \cdot \deg(X). \tag{7.1}
$$

Assume E is of rank r and the restriction $E|_{Z_s}$ is not μ-semistable for general $s \in \Pi$. Let $0 \subset F_0 \subset F_1 \subset ... \subset F_j = q^*E|_Z$ be the relative Harder-Narasimhan filtration with respect to the family $p : Z \to \Pi$. Then for some i

$$
\delta\mu(E|_{Z_s}) = \mu_i - \mu_{i+1} = \frac{\deg((F_{i+1}/F_i)|_{Z_s})}{\mathrm{rk}(F_{i+1}/F_i)} - \frac{\deg((F_{i+2}/F_{i+1})|_{Z_s})}{\mathrm{rk}(F_{i+2}/F_{i+1})}
$$

$$
\ge \frac{a^{\ell}}{\mathrm{l.c.m.}(\mathrm{rk}(F_{i+1}/F_i), \mathrm{rk}(F_{i+2}/F_{i+1}))}.
$$

Indeed, since $\det(F_{i+1}/F_i) \cong q^*\mathcal{Q} \otimes p^*\mathcal{M}$, where $\mathcal{Q} \in \mathrm{Pic}(X)$ and $\mathcal{M} \in \mathrm{Pic}(\Pi)$, one has $\deg((F_{i+1}/F_i)|_{Z_s}) = \deg(\mathcal{Q}|_{Z_s}) = \deg(\mathcal{Q}) \cdot a^{\ell}$. Using the inequality

$$\text{l.c.m.}(\text{rk}(F_{i+1}/F_i), \text{rk}(F_{i+2}/F_{i+1})) \le \max\left\{1, \frac{r^2-1}{4}\right\}$$

and (7.1) we obtain

$$\frac{a^\ell}{\max\{1, \frac{r^2-1}{4}\}} \le \delta\mu(E|_{z_s}) \le -\mu_{\min}(T_{Z/X}|_{z_s}) \le \frac{a^{\ell+1}}{\binom{n+a}{a} - a \cdot \ell - 1} \cdot \deg(X)$$

which immediately contradicts the assumption

$$\deg(X) \cdot \max\left\{1, \frac{r^2-1}{4}\right\} < \frac{\binom{n+a}{a} - a \cdot \ell - 1}{a}.$$

Step 2. Since $\mathcal{O}(1)$ is very ample, one finds a linear system $\mathbb{P} := \mathbb{P}(V) \subset |\mathcal{O}(1)|$ such that $\phi_V : X \to \mathbb{P}$ is a finite surjective morphism. Since $\mu_{\min}(T_{Z/X}|_{z_s})$ can only decrease when specializing Z_s, it is enough to show (7.1) for a general complete intersection $D_1 \cap \ldots \cap D_\ell$ with $D_i \in \mathbb{P}(S^a V^{\vee}) \subset |\mathcal{O}(a)|$. Moreover, we may replace the incidence variety $Z \subset \Pi \times X$ by $\tilde{Z} := \Pi_V \times_\Pi Z$, where $\Pi_V = \prod_{i=1}^{\ell} \mathbb{P}(S^a V^{\vee})$: Since $T_{Z/X}|_{z_s}$ and $T_{\tilde{Z}/X}|_{z_s}$ are related via the exact sequence

$$0 \longrightarrow T_{\tilde{Z}/X}|_{z_s} \longrightarrow T_{Z/X}|_{z_s} \longrightarrow \mathcal{O}_{Z_s}^m \longrightarrow 0$$

$(m = h^0(X, \mathcal{O}(a)) - \dim S^a V)$, one has $-\mu_{\min}(T_{Z/X}|_{z_s}) \le -\mu_{\min}(T_{\tilde{Z}/X}|_{z_s})$. Thus it suffices to show

$$-\mu_{\min}(T_{\tilde{Z}/X}|_{z_s}) \le \frac{a^{\ell+1}}{\binom{n+a}{a} - a \cdot \ell - 1} \cdot \deg(X)$$

with $s \in \Pi_V$.

Step 3. If $Z' \subset \prod_{i=1}^{\ell} |\mathcal{O}_{\mathbb{P}}(a)| \times \mathbb{P} = \Pi_V \times \mathbb{P}$ denotes the incidence variety on $\mathbb{P}$, then $\tilde{Z} = Z' \times_{\mathbb{P}} X$. Hence $(1 \times \phi)^* T_{Z'/\mathbb{P}} \cong T_{\tilde{Z}/X}$. Using the above exact sequence this yields

$$\begin{aligned}
\mu_{\min}(T_{\tilde{Z}/X}|_{z_s}) &= \mu_{\min}((1 \times \phi)^*(T_{Z'/\mathbb{P}})|_{z_s}) \\
&= \mu_{\min}(T_{Z'/\mathbb{P}}|_{z'_s}) \cdot \deg(\phi) \\
&= \mu_{\min}(T_{Z'/X}|_{z'_s}) \cdot \deg(X).
\end{aligned}$$

This completes the reduction to the case $X = \mathbb{P}$.

Step 4. We now prove $-\mu_{\min}(T_{Z/X}|_{z_s}) \le \frac{a^{\ell+1}}{\binom{n+a}{a} - a \cdot \ell - 1}$ for the case $X = \mathbb{P} = \mathbb{P}(V)$. To shorten notation we introduce $A := S^a(V)$ and $N := \binom{n+a}{a}$.

Let $Z \subset \Pi_V \times \mathbb{P}$ be the incidence variety and let $v : \mathbb{P} \to \mathbb{P}(A)$ be the Veronese embedding. Then Z is the pull-back of the incidence variety

$$\left\{(H_1, \ldots, H_\ell, x) \in \prod_{i=1}^{\ell} \mathbb{P}(A^{\vee}) \times \mathbb{P}(A) | x \in H_i\right\},$$

which is canonically isomorphic to $\mathbb{P}(\mathcal{T}_{\mathbb{P}(A)}(-1)) \times_{\mathbb{P}(A)} \ldots \times_{\mathbb{P}(A)} \mathbb{P}(\mathcal{T}_{\mathbb{P}(A)}(-1))$. Hence Z is as a $\mathbb{P}$-scheme isomorphic to $\mathbb{P}(v^*(\mathcal{T}_{\mathbb{P}(A)}(-1))) \times_{\mathbb{P}} \ldots \times_{\mathbb{P}} \mathbb{P}(v^*(\mathcal{T}_{\mathbb{P}(A)}(-1)))$, and the relative Euler sequence takes the form

$$0 \longrightarrow \mathcal{O}^\ell \longrightarrow p^*\mathcal{O}(1)^\ell \otimes q^*v^*(\Omega_{\mathbb{P}(A)}(1)) \longrightarrow \mathcal{T}_{Z/\mathbb{P}} \longrightarrow 0.$$

Therefore, $-\mu_{\min}(\mathcal{T}_{Z/\mathbb{P}}|_{Z_s}) \leq -\mu_{\min}(v^*(\Omega_{\mathbb{P}(A^\vee)}(1))|_{Z_s})$. Using the Euler sequence on $\mathbb{P}(A)$:

$$0 \longrightarrow \Omega_{\mathbb{P}(A)}(1) \longrightarrow A \otimes \mathcal{O}_{\mathbb{P}(A)} \longrightarrow \mathcal{O}(1) \longrightarrow 0$$

the pull-back $v^*(\Omega_{\mathbb{P}(A)}(1))$ can naturally be identified with the kernel of $A \otimes \mathcal{O}_{\mathbb{P}} \to \mathcal{O}(a)$. According to the notation of Section 1.4 this is $\mathcal{K}_a^a$, which will be abbreviated by $\mathcal{K}$. We conclude the proof by showing

$$-\mu_{\min}(\mathcal{K}|_{Z_s}) \leq \frac{a^{\ell+1}}{\binom{n+a}{a} - a \cdot \ell - 1}.$$

Recall from 1.4.5 that $\mathcal{K}$ is semistable and by the exact sequence

$$0 \longrightarrow \mathcal{K} \longrightarrow A \otimes \mathcal{O}_{\mathbb{P}} \longrightarrow \mathcal{O}(a) \longrightarrow 0$$

one has $\mu(\mathcal{K}) = -\frac{a}{N-1}$.

If $Y = Z_s$ is a general complete intersection $D_1 \cap \ldots \cap D_\ell$ with $D_i \in \mathbb{P}(A^\vee)$, then the Koszul complex takes form

$$0 \to \Lambda^\ell B(-\ell a) \to \ldots \to \Lambda^2 B(-2a) \to B(-a) \to \mathcal{O}_{\mathbb{P}} \to \mathcal{O}_Y \to 0,$$

where $B \subset A$ is the subspace spanned by the sections cutting out Y. Splitting the Koszul complex into short exact sequences we obtain

$$0 \to E_{j+1} \to \Lambda^j B(-ja) \to E_j \to 0 \tag{7.2}$$

with $E_{\ell+1} = 0$ and $E_0 = \mathcal{O}_Y$. From the dual of the short exact sequence defining $\mathcal{K}$,

$$0 \to \mathcal{O}(-a) \to A^\vee \otimes \mathcal{O}_{\mathbb{P}} \to \mathcal{K}^\vee \to 0,$$

one gets short exact sequences for the exterior powers of $\mathcal{K}^\vee$:

$$0 \to \Lambda^{q-1}\mathcal{K}^\vee(-a) \to \Lambda^q A^\vee \otimes \mathcal{O} \to \Lambda^q \mathcal{K}^\vee \to 0$$

For $b < 0$ and $\nu < n$ the cohomology groups $H^\nu(\mathbb{P}, \Lambda^q A^\vee(b))$ vanish. This gives isomorphisms

$$H^0(\Lambda^q \mathcal{K}^\vee(b)) = H^1(\Lambda^{q-1}\mathcal{K}^\vee(b-a)) = \ldots = H^\nu(\Lambda^{q-\nu}\mathcal{K}^\vee(b-\nu a))$$

for all $b < 0$ and $\nu < n$. By Lemma 1.4.5 and Corollary 3.2.10 the sheaf $\Lambda^q \mathcal{K}^\vee$ is μ-semistable, so that $H^0(\mathbb{P}, \Lambda^q \mathcal{K}^\vee(b)) = 0$ as soon as $0 > b + \mu(\Lambda^q \mathcal{K}^\vee) = b - q \cdot \mu(\mathcal{K})$, which is equivalent to $b < -\frac{q \cdot a}{N-1}$. By tensorizing the sequences (7.2) with $\Lambda^p \mathcal{K}^{\vee}$ and passing to cohomology we get the exact sequences

$$\Lambda^j B \otimes H^j(\Lambda^p \mathcal{K}^{\vee}(b - ja)) \to H^j(E_j \otimes \Lambda^p \mathcal{K}^{\vee}(b)) \to H^{j+1}(E_{j+1} \otimes \Lambda^p \mathcal{K}^{\vee}(b))$$

The term on the left vanishes for all $j = 0, \ldots, \ell$ as soon as $b < -(p+\ell) \cdot a/(N-1)$. For such b one gets

$$H^0(Y, \Lambda^p \mathcal{K}^{\vee} \otimes \mathcal{O}_Y(b)) = H^0(E_0 \otimes \Lambda^p \mathcal{K}^{\vee}(b)) \subset \ldots \subset H^{\ell+1}(E_{\ell+1} \otimes \Lambda^p \mathcal{K}^{\vee}(b)) = 0.$$

Hence, $H^0(Y, \Lambda^p \mathcal{K}^{\vee} \otimes \mathcal{O}_Y(b)) = 0$ for $b < -\frac{(p+\ell) \cdot a}{N-1}$.

If $p^* \mathcal{K} \to F$ denotes the minimal destabilizing quotient with respect to the family $p :$ $Z \to \Pi_V$, then $\det(F) \cong q^* \mathcal{O}(b) \otimes p^* \mathcal{M}$ with $\mathcal{M} \in \mathrm{Pic}(\Pi_V)$. Hence the surjection $\mathcal{K}|_Y \to F|_Y$ defines a non-trivial element in $H^0(Y, \Lambda^q \mathcal{K}^{\vee} \otimes \mathcal{O}(b))$, where $q = \mathrm{rk}(F)$ and therefore

$$b \geq -\frac{(q+\ell) \cdot a}{N-1}.$$

On the other hand, $-1 \geq b$, for b is a negative integer, and thus

$$q \geq \frac{N - 1 - a \cdot \ell}{a}.$$

Both inequalities together imply

$$\mu_{\min}(\mathcal{K}|_Y) = \frac{b \cdot a^\ell}{q} \geq -\frac{(q+\ell) \cdot a}{(N-1) \cdot q} \cdot a^\ell \geq -\frac{a^{\ell+1}}{N - a \cdot \ell - 1}.$$

$\square$

7.2 The Theorems of Mehta and Ramanathan

In this section we work over an algebraically closed field of arbitrary characteristic.

Theorem 7.2.1 — *Let X be a smooth projective variety of dimension $n \geq 2$ and let $\mathcal{O}(1)$ be a very ample line bundle. Let E be a μ-semistable sheaf. Then there is an integer a_0 such that for all $a \geq a_0$ there is a dense open subset $U_a \subset |\mathcal{O}(a)|$ such that for all $D \in U_a$ the divisor D is smooth and $E|_D$ is again μ-semistable.*

Proof. Let a be a positive integer and let as before

$$Z_a \xrightarrow{q} X$$
$$\downarrow p$$
$$\Pi_a := |\mathcal{O}(a)|$$

be the universal family of hypersurface sections.

The μ-semistable sheaf E is torsion free and for any a and general $D \in \Pi_a$ the restriction $E|_D$ is again torsion free (Lemma 1.1.13). Moreover, q^*E is flat over Π_a, since, independently of $D \in \Pi_a$, the restriction $E|_D$ has the same Hilbert polynomial $P(E|_D, m) = P(E, m) - P(E, m - a)$. According to the theorem on the relative Harder-Narasimhan filtration (cf. 2.3.2). there is a dense open subset $V_a \subset \Pi_a$ and a quotient $q^*E|_{Z_{V_a}} \to F_a$ that restricts to the minimal destabilizing quotient of $E|_D$ for all $D \in V_a$. Let $\mathcal{Q}$ be an extension of $\det(F_a)$ to some line bundle on all of Z_a. Then $\mathcal{Q}$ can be uniquely decomposed as

$$\mathcal{Q} = q^*\mathcal{L}_a \otimes p^*\mathcal{M}$$

with $\mathcal{L}_a \in \text{Pic}(X)$ and $\mathcal{M} \in \text{Pic}(\Pi_a)$.

Lemma 7.2.2 — *Let $a \geq 3$. If $\mathcal{L}'$ and $\mathcal{L}''$ are line bundles on X such that $\mathcal{L}'|_D \cong \mathcal{L}''|_D$ for all D in a dense subset of Π_a, then $\mathcal{L}' \cong \mathcal{L}''$.*

Proof. Let $\mathcal{L} = (\mathcal{L}')^{-1} \otimes \mathcal{L}''$. Then $h^0(\mathcal{L}|_D) = 1 = h^0(\mathcal{L}^\vee|_D)$ for all D in a dense subset of Π_a. By semicontinuity, $h^0(\mathcal{L}|_D), h^0(\mathcal{L}^\vee|_D) \geq 1$ for all $D \in \Pi_a$. Thus $\mathcal{L}|_D \cong \mathcal{O}_D$ if D is integral. Now the set B_a of all integral divisors in Π_a is open and its complement has codimension at least 2 (Use Bertini's theorem and the assumption $a \geq 3$). Therefore, there is an isomorphism $\mathcal{N} \to p_*q^*\mathcal{L}|_{B_a}$ for some line bundle $\mathcal{N} \in \text{Pic}(B_a) = \text{Pic}(\Pi_a)$ and an isomorphism $p^*\mathcal{N} \to q^*\mathcal{L}$ on $p^{-1}(B_a)$, hence on the whole of Z_a. This implies $\mathcal{L} = \mathcal{O}_X$ and $\mathcal{N} = \mathcal{O}_{\Pi_a}$. $\square$

Let $U_a \subset V_a$ denote the dense open set of points $D \in V_a$ such that D is smooth and $E|_D$ torsion free (cf. 1.1.13).

Lemma 7.2.3 — *Let $a_1, \ldots, a_\ell$ be positive integers, $a = \sum a_i$, and let $D_i \in U_{a_i}$ be divisors such that $D = \sum D_i$ is a divisor with normal crossings. Then there is a smooth locally closed curve $C \subset \Pi_a$ containing the point $D \in \Pi_a$ such that $C \setminus \{D\} \subset U_a$ and such that $Z_C = C \times_{\Pi_a} Z_a$ is smooth in codimension 2.*

Remark 7.2.4 — If $D_1 \in U_{a_1}$ is given, one can always find $D_i \in U_{a_i}$ for $i \geq 2$ such that $D = \sum D_i$ is a divisor with normal crossings.

Proof of the lemma. A general line $L \subset \Pi_a$ through the closed point D will not be contained in the complement of U_a. Then $L \setminus U_a$ is a finite set containing D. Let $C = L \cap U_a \cup \{D\}$. The curve C is completely determined by the choice of a hyperplane H in the cotangent space $\Omega_{\Pi_a}([D])$. We must choose H in such a way that Z_C is smooth in codimension 2. Let $z \in D = \bigcup D_i$ be a closed point in the fibre over $D \in \Pi_a$. The homomorphism

$$p_z^* : \Omega_{\Pi_a}([D]) \to \Omega_{Z_a}(z)$$

is injective if and only if z is not contained in any of the intersections $D_i \cap D_j$, and the kernel is 1-dimensional otherwise. Choose H such that the corresponding projective subspace does not contain any of the images of the maps

$$D_i \cap D_j \to \mathbb{P}(\Omega_{\Pi_a}([D])^\vee), z \mapsto \ker(p_z^*).$$

Then $H \to \Omega_{Z_a}(z)$ is injective for all points $z \in D$ outside a closed subset of codimension 2, and Z_C is smooth in these points. This means that the set of points where Z_C fails to be smooth has codimension at least three in Z_C. $\qquad \square$

Let $\mu(a)$ and $r(a)$ denote the slope and the rank of the minimal destabilizing quotient of $E|_D$ for a general point $D \in \Pi_a$. Then $1 \leq r(a) \leq \mathrm{rk}(E)$ and

$$\frac{\mu(a)}{a} = \frac{\deg(\mathcal{L}_a)}{r(a)} \in \frac{\mathbb{Z}}{\mathrm{rk}(E)!} \subset \mathbb{Q}.$$

Lemma 7.2.5 — *Let $a_1, \ldots, a_j$ be positive integers, $a = \sum a_i$. Then $\mu(a) \geq \sum \mu(a_i)$, and in case of equality $r(a) \leq \min\{r(a_i)\}$.*

Corollary 7.2.6 — *$r(a)$ and $\frac{\mu(a)}{a}$ are constant for $a \gg 0$.*

Proof. The function $a \mapsto \frac{\mu(a)}{a}$ takes values in a discrete subset of $\mathbb{Q}$ and is bounded from above by $\mu(E)$. Therefore it attains its maximum value on any subset of $\mathbb{N}$. Suppose the maximum on the set of integers ≥ 2 is attained at b_0 and the maximum on all positive integers ≥ 2 coprime to b_0 is attained at b_1. If β_0 and β_1 are any positive integers and $b = \beta_0 b_0 + \beta_1 b_1$, then the lemma says

$$\frac{\mu(b)}{b} \geq \frac{\beta_0 b_0}{b} \frac{\mu(b_0)}{b_0} + \frac{\beta_1 b_1}{b} \frac{\mu(b_1)}{b_1}$$

$$\geq \frac{\beta_0 b_0}{b} \frac{\mu(b_1)}{b_1} + \frac{\beta_1 b_1}{b} \frac{\mu(b_1)}{b_1}$$

$$= \frac{\mu(b_1)}{b_1}.$$

Hence $\frac{\mu(b)}{b} = \frac{\mu(b_1)}{b_1}$ and also $\frac{\mu(b)}{b} = \frac{\mu(b_0)}{b_0}$ for all b that can be written as a positive linear combination of b_0 and b_1, hence in particular for all $b > b_0 b_1$. A similar argument shows that $r(b)$ is eventually constant. $\qquad \square$

Proof of the lemma. Let D_i be divisors satisfying the requirements of Lemma 7.2.3 and let C be a curve with the properties of 7.2.3. Over V_a there exists the minimal destabilizing quotient $q^*E|_{Z_{V_a}} \to F$. Its restriction to $V_a \cap C$ can uniquely be extended to a C flat quotient $q^*E|_{Z_C} \to F_C$. The flatness of F_C implies that $P(F_C|_D) = P(F_{C,c})$ for all $c \in C \setminus \{D\}$. Hence $\mathrm{rk}(F_C|_D) = r(a)$ and $\mu(F_C|_D) = \mu(a)$.

Let $\bar{F} = F_C|_D/T(F_C|_D)$. Then $\mathrm{rk}(\bar{F}|_{D_i}) = \mathrm{rk}(\bar{F}) = \mathrm{rk}(F_C|_D) = r(a)$ and $\mu(a) = \mu(F_C|_D) \geq \mu(\bar{F})$. Moreover, since $\bar{F}$ is pure, the sequence

$$0 \to \bar{F} \to \bigoplus_i \bar{F}|_{D_i} \to \bigoplus_{i<j} \bar{F}|_{D_i \cap D_j} \to 0$$

is exact modulo sheaves of dimension $n-3$. Computing the coefficients of degree $n-2$ in the Hilbert polynomials of these sheaves (use the Hirzebruch-Riemann-Roch formula), we get the equation:

$$r(a)\left(\mu(\bar{F}) - \frac{1}{2}D(D + K_X).H^{n-2}\right)$$

$$= \sum_i r(a)\left(\mu(\bar{F}|_{D_i}).H^n - \frac{1}{2}D_i(D_i + K_X).H^{n-2}\right)$$

$$- \sum_{i<j} \mathrm{rk}(\bar{F}|_{D_i \cap D_j})D_i.D_j.H^{n-2}.$$

Cancelling superfluous terms one gets

$$\mu(\bar{F}) = \sum_i \left(\mu(\bar{F}|_{D_i}) - \frac{1}{2}\sum_{j \neq i}\left(\frac{\mathrm{rk}(\bar{F}|_{D_i \cap D_j})}{r(a)} - 1\right)a_i a_j\right).$$

Let $F_i = \bar{F}|_{D_i}/T(\bar{F}|_{D_i})$. Then

$$\mu(F_i) \leq \mu(\bar{F}|_{D_i}) - \sum_{j \neq i}\left(\frac{\mathrm{rk}(\bar{F}|_{D_i \cap D_j})}{r(a)} - 1\right)a_i a_j$$

$$\leq \mu(\bar{F}|_{D_i}) - \frac{1}{2}\sum_{j \neq i}\left(\frac{\mathrm{rk}(\bar{F}|_{D_i \cap D_j})}{r(a)} - 1\right)a_i a_j.$$

It follows that $\mu(a) \geq \mu(\bar{F}) \geq \sum_i \mu(F_i) \geq \sum_i \mu(a_i)$. Moreover, if $\mu(a) = \sum_i \mu(a_i)$ we must have equality everywhere. In particular, $\mathrm{rk}(\bar{F}|_{D_i \cap D_j}) = r(a)$ and $\mu(F_i) = \mu(a_i)$. Since F_i has the minimal possible slope, $r(a) = \mathrm{rk}(F_i) \leq r(a_i)$. $\square$

Supplement to the proof: if $\mu(a)/a = \mu(a_i)/a_i$ and $r(a) = r(a_i)$ for all i, then $F_C|_{D_i}$ differs from the minimal destabilizing quotient of $E|_{D_i}$ only in dimension $n-3$, in particular their determinant line bundles as sheaves on D_i are equal. This can be used to prove:

Lemma 7.2.7 — *There is a line bundle $\mathcal{L} \in \mathrm{Pic}(X)$ such that $\mathcal{L}_a \cong \mathcal{L}$ for all $a \gg 0$.*

Proof. Let $d_0 \geq 3$ be an integer such that $r(a)$ and $\frac{\mu(a)}{a}$ are constant for all $a \geq d_0$. Let $a \geq 2d_0 + 1$ and let $d_1 = a - d_0$. Choose $D_0 \in U_{d_0}$ arbitrary and let $D_1 \in U_{d_1}$ such that

$D = D_0 + D_1$ is a normal crossing divisor. Let C be a curve as in the previous lemma and consider the quotient $q^*E|_{Z_C} \to F_C$ as above. Extend $\det(F_C|_{Z_C^{reg}})$ to a C-flat sheaf $\mathcal{A}$ on Z_C. Then $\mathcal{A}|_{D'} \cong \mathcal{L}_a|_{D'}$ for all $D' \in C \setminus \{D\}$ and, since $F_C|_{D_i}$ differs from the minimal destabilizing quotient only in dimension $n-3$, $\mathcal{A}|_{D_i} \cong \mathcal{L}_{d_i}$ outside a set of codimension 2 in D_i for $i = 0, 1$. By semicontinuity there exist non-trivial homomorphisms $\mathcal{L}_a|_D \to \mathcal{A}|_D$ and $\mathcal{A}|_D \to \mathcal{L}_a|_D$. Hence there exists a non-trivial homomorphism $\mathcal{L}_a|_{D_i} \to \mathcal{L}_{d_i}|_{D_i}$ for some i. Since both line bundles are of the same degree, it is an isomorphism. Which in turn implies that also on the other component there is such an isomorphism. Hence $\mathcal{L}_a|_{D_i} = \mathcal{L}_{d_i}|_{D_i}$ for $i = 0, 1$. Since D_0 was arbitrary, Lemma 7.2.2 implies that $\mathcal{L}_a \cong \mathcal{L}_{d_0}$ for all $a \geq 2d_0 + 1$. $\qquad\square$

We can now finish the proof of Theorem 7.2.1: suppose the theorem were false, i.e. we had $\deg(\mathcal{L})/r < \mu(E)$ and $r < \mathrm{rk}(E)$, where $r = r(a)$ for $a \gg 0$. Let a be sufficiently large, let $D \in U_a$, and let $E|_D \to F_D$ be the minimal destabilizing quotient. There is a large open subscheme $D' \subset D$ such that $F_D|_{D'}$ is locally free of rank r. This induces a homomorphism $\sigma_D : \Lambda^r E|_D \to \mathcal{L}|_D$ which is surjective over D' and morphisms

$$D' \to \mathrm{Grass}(E, r) \to \mathbb{P}(\Lambda^r E).$$

Consider the exact sequence

$$\mathrm{Hom}(\Lambda^r E, \mathcal{L}(-a)) \to \mathrm{Hom}(\Lambda^r E, \mathcal{L}) \to \mathrm{Hom}(\Lambda^r E|_D, \mathcal{L}|_D) \to \mathrm{Ext}^1(\Lambda^r E, \mathcal{L}(-a)).$$

By Serre's theorem and Serre duality, one has for $i = 0, 1$

$$\mathrm{Ext}^i(\Lambda^r E, \mathcal{L}(-a))^{\vee} = H^{n-i}(X, \Lambda^r E \otimes \mathcal{L}^{\vee} \otimes \omega_X(a)) = 0$$

for all $a \gg 0$, since by assumption $n \geq 2$. Hence if a is sufficiently large, σ_D extends uniquely to a homomorphism $\sigma : \Lambda^r E \to \mathcal{L}$. The support of the cokernel of this homomorphism meets the ample divisor D in a subset of codimension 2. Hence σ is surjective on a large open subset $X' \subset X$ with $D' = X' \cap D$. We want the induced morphism $i : X' \to \mathbb{P}(\Lambda^r E)$ to factorize through $\mathrm{Grass}(E, r)$. The ideal sheaf of $\mathrm{Grass}(E, r)$ in $\mathbb{P}(\Lambda^r E)$ is generated by finitely many sheaves $\mathcal{I}_\nu \subset S^\nu(\Lambda^r E)$, $\nu \leq \nu_0$. The morphism i factors through $\mathrm{Grass}(E, r)$ if and only if the composite maps

$$\psi_\nu : \mathcal{I}_\nu \longrightarrow S^\nu(\Lambda^r E) \longrightarrow \mathcal{L}^\nu$$

vanish. But we know already that the restriction of ψ_ν to D vanishes, so that we can consider ψ_ν as elements in $\mathrm{Hom}(\mathcal{I}_\nu, \mathcal{L}^\nu(-a))$. Clearly, these groups vanish for $a \gg 0$. This proves that F_D extends to a quotient $F_{X'}$ of $E|_{X'}$ which is locally free of rank r with $\det(F_{X'}) = \mathcal{L}|_{X'}$. Hence

$$\mu(F_{X'}) = \frac{\deg(\mathcal{L})}{r} < \mu(E).$$

This contradicts the assumption that E is μ-semistable and, thus, concludes the proof of Theorem 7.2.1   □

We now turn to the restriction of μ-stable sheaves.

Theorem 7.2.8 — *Let X be a smooth projective variety of dimension $n \geq 2$ and let $\mathcal{O}(1)$ be a very ample line bundle. Let E be a μ-stable sheaf. Then there is an integer a_0 such that for all $a \geq a_0$ there is a dense open subset $W_a \subset |\mathcal{O}(a)|$ such that for all $D \in W_a$ the divisor D is smooth and $E|_D$ is μ-stable.*

The techniques to prove the theorem are quite similar to the ones encountered before. The main difficulty is the fact that a destabilizing subsheaf of a μ-semistable sheaf is not unique. By 1.5.9 a μ-semistable sheaf which is simple but not μ-stable has a proper extended socle. Thus we first show that the restriction is simple and then use the extended socle (rather its quotient) as a replacement for the minimal destabilizing quotient.

Lemma 7.2.9 — *For $a \gg 0$ and general $D \in |\mathcal{O}(a)|$ the restriction $E|_D$ is simple.*

Proof. Let F be the double dual of E. Then for arbitrary a and general $D \in |\mathcal{O}(a)|$, $F|_D$ is the double dual of $E|_D$ (cf. Section 1.1). Since E and $E|_D$ are torsion free and F and $F|_D$ are reflexive (cf. 1.1.13), there are injective homomorphisms

$$\mathrm{End}(E) \to \mathrm{End}(F) \qquad \text{and} \qquad \mathrm{End}(E|_D) \to \mathrm{End}(F|_D).$$

Therefore, it suffices to show that $F|_D$ is simple for $a \gg 0$ and D general. But if E is μ-stable, then so is F. In particular, F is simple. Consider the exact sequence

$$\mathrm{Hom}(F, F(-a)) \to \mathrm{End}(F) \to \mathrm{End}(F|_D) \to \mathrm{Ext}^1(F, F(-a)).$$

Recall the spectral sequence $H^i(X, \mathcal{E}xt^j(F, F \otimes \omega_X(a))) \Rightarrow \mathrm{Ext}^{i+j}(F, F \otimes \omega_X(a))$. For sufficiently large $a \gg 0$ we get

$$\mathrm{Ext}^1(F, F(-a))^\vee \cong \mathrm{Ext}^{n-1}(F, F \otimes \omega_X(a)) \cong H^0(X, \mathcal{E}xt^{n-1}(F, F) \otimes \omega_X(a)).$$

But $\mathcal{E}xt^{n-1}(F, F) = 0$, since F is reflexive and thus $\mathrm{dh}(F) \leq n - 2$ (cf. Section 1.1). Hence for a sufficiently large, $\mathrm{End}(F) \to \mathrm{End}(F|_D)$ is surjective.   □

Let $a_0 \geq 3$ be an integer such that for all $a \geq a_0$ and general $D \in |\mathcal{O}(a)|$ the restriction $E|_D$ is μ-semistable and simple. Suppose $E|_D$ is not μ-stable for a general D. Then $E|_{D_\eta}$ is geometrically μ-unstable for the generic point $\eta \in |\mathcal{O}(a)|$, i.e. the pull-back to some extension of $k(\eta)$ is not μ-stable. This follows from the openness of stability (cf. 2.3.1). Since $E|_{D_\eta}$ is simple, the sheaf $E|_{D_\eta}$ is stable if and only if it is geometrically stable (Lemma 1.5.10). Hence $E|_{D_\eta}$ is not μ-stable. In fact, the extended socle of $E|_{D_\eta}$

is a proper destabilizing subsheaf (1.5.9). Extend the corresponding quotient sheaf F_η to a coherent quotient $q^*E \to F_a$ over all of Z_a. Let W_a denote the dense open subset of points $D \in |\mathcal{O}(a)|$ such that D is smooth and F_a is flat over W_a. Then $E|_D \to F|_D$ is a destabilizing quotient for all $D \in W_a$.

Lemma 7.2.10 — *If $E|_{D_0}$ is μ-stable for some $D_0 \in W_a$, $a \geq a_0$, then $E|_{D'}$ is μ-stable for all $D' \in W_{a'}$ and all $a' \geq 2a$.*

Proof. Choose $D_1 \in W_{a'-a}$ such that $D = D_0 + D_1$ is a normal crossing divisor, and let $C \subset |\mathcal{O}(a')|$ be a curve as in the proof of Lemma 7.2.3. Then the destabilizing quotient $F_a|_{Z_C \setminus \{D\}}$ can be extended to a flat quotient F_C of $q^*E|_{Z_C}$. Then $F_C|_{D_i}$ destabilizes $E|_{D_i}$ in contradiction to the assumptions. □

Assume now the theorem is false. Then $E|_D$ is unstable for all $a \geq a_0$ and general $D \in W_a$. As before there are line bundles $\mathcal{L}_a \in \mathrm{Pic}(X)$ such that $\det(F_a|_D) = \mathcal{L}_a|_D$ for all $D \in W_a$ and all $a \geq a_0$. The same argument as in Lemma 7.2.7 shows: if $a_1, \ldots, a_j$ are integers $\geq a_0$ and $a = \sum a_i$, and if $D_i \in W_{a_i}$ are points such that $D = \sum D_i$ is a normal crossing divisor, then $\mathcal{L}_a|_{D_i}$ is the determinant line bundle of some destabilizing quotient of $E|_{D_i}$.

Lemma 7.2.11 — *If D is a smooth projective variety, and if E_D is a μ-semistable sheaf, then the set T_D of determinant bundles of destabilizing quotients of E_D is finite and its cardinality is bounded by $2^{\mathrm{rk}(E_D)}$.*

Proof. Let $\mathcal{L}_1, \ldots, \mathcal{L}_\rho$, $\rho \leq \mathrm{rk}(E_D)$, be the determinant bundles of the factors of some Jordan Hölder filtration of E_D. Then T_D is contained in the set of line bundles of the form $\bigotimes_{i \in I} \mathcal{L}_i$, where $I \subset \{1, \ldots, \rho\}$. □

Let $a \geq 2a_0$, and let $D \in W_{a_0}$ be an arbitrary point. We saw that $\mathcal{L}_a|_D \in T_D$. In fact, we get a function

$$\varphi : \mathbb{N}_{\geq 2a_0} \longrightarrow \prod_{D \in W_{a_0}} T_D.$$

Let $\sim$ be the equivalence relation on $\mathbb{N}_{\geq 2a_0}$ generated by: $a \sim a'$ if the set of points $s \in W_a$ with $\varphi(a)(s) = \varphi(a')(s)$ is dense in W_{a_0}. Then there are at most $2^{\mathrm{rk}(E)}$ distinct equivalence classes, and in particular, there is at least one infinite class N. For assume that there are distinct classes $N_1, \ldots, N_\ell$, $\ell > 2^{\mathrm{rk}(E)}$. Choose representatives $a_i \in N_i$. For fixed $D \in W_{a_0}$, we have

$$\varphi(a_1)([D]), \ldots, \varphi(a_\ell)([D]) \in T_D.$$

Since $\ell > |T_D|$, at least two of these elements must be equal. In this way we can pick for any $D \in W_{a_0}$ a pair of indices i, j. But the set of all these pairs is finite. Hence their is at least one pair i, j which is associated to all points in a dense subset of W_{a_0}. But by definition this means $a_i \sim a_j$, hence $N_i = N_j$, a contradiction.

Lemma 7.2.12 — *There is a line bundle $\mathcal{L} \in \mathrm{Pic}(X)$ such that $\mathcal{L} \cong \mathcal{L}_a$ for all $a \in N$.*

Proof. If $\varphi(a)$ equals $\varphi(a')$ on a dense subset of W_{a_0} then $\mathcal{L}_a|_D \cong \mathcal{L}_{a'}|_D$ for all D in a dense subset of $|\mathcal{O}(a)|$, so that Lemma 7.2.2 implies $\mathcal{L}_a \cong \mathcal{L}_{a'}$. $\qquad\square$

Finally, let $N' \subset N$ be an infinite subset such that F_a has the same rank, say r, for all $a \in N'$. Summing up, we have: there is a line bundle $\mathcal{L}$ on X and an integer $0 < r < \mathrm{rk}(E)$ such that $\deg(\mathcal{L}) = r\mu(E)$ and such that for all $a \in N'$ and general $D \in W_{a'}$ there is a destabilizing quotient $E|_D \to F_D$ with $\mathrm{rk}(F_D) = r$ and $\det(F_D) = \mathcal{L}|_D$. But the arguments at the end of the proof of the previous theorem show that this suffices to construct a destabilizing quotient $E \to F_X$ for sufficiently large a. This contradicts the assumptions of the theorem. $\qquad\square$

7.3 Bogomolov's Theorems

This section is devoted to a number of results due to Bogomolov. The original references are [28, 29, 31]. In our presentation we use the fact that tensor powers of μ-semistable sheaves are again μ-semistable (in characteristic zero), i.e. we build on the Grauert-Mülich Theorem and Maruyama's results, discussed in Chapter 3. In this we deviate from Bogomolov's line of argument, which is independent of these theorems. However, the essential ideas are all due to Bogomolov. In the following let X be a smooth projective surface over an algebraically closed field of characteristic zero.

Recall that Bogomolov's inequality 3.4.1 states that the discriminant of any μ-semistable torsion free sheaf is nonnegative. Before we begin to improve upon this result, we give a short elegant variant of the proof of 3.4.1, say as a warm-up for calculations with discriminants, following an argument of Le Potier. Using one of the restriction theorems of the previous sections one can generalize the inequality to sheaves on higher dimensional varieties.

Theorem 7.3.1 — *Let X be a smooth projective variety of dimension n and H an ample divisor on X. If F is a μ-semistable torsion free sheaf, then*

$$\Delta(F).H^{n-2} \geq 0.$$

Proof. By 7.1.1 or 7.2.1 the restriction of F to a general complete intersection $X' := D_1 \cap \ldots \cap D_{n-2}$ with $D_i \in |aH|$ and $a \gg 0$ is again μ-semistable and torsion free.

Since $a^{n-2}\Delta(F).H^{n-2} = \Delta(F|_{X'})$, we may reduce to the case of a μ-semistable sheaf on a surface. Thus, let H be an ample divisor on a surface X and let F be a torsion free μ-semistable sheaf. As in the earlier proof we may assume that F is locally free and has trivial determinant. By Theorem 3.1.4, the vector bundles $F^{\otimes n}$ are all μ-semistable. They have trivial determinant and their ranks and discriminants are given by $r_n = r^n$ and $\Delta_n = nr^{2(n-1)}\Delta(F)$. Replacing H by some large multiple, it follows from the restriction theorem of Flenner or Mehta-Ramanathan, that $F|_C$ – and hence also $F^{\otimes n}|_C$ — is semistable for a general curve $C \in |H|$. In particular, it follows from Lemma 3.3.2 that there is a positive constant γ, depending only on X, such that $h^0(F^{\otimes n}) \leq \gamma \cdot r_n$. By Serre duality, and enlarging γ if necessary, we also get $h^2(F^{\otimes n}) \leq \gamma \cdot r_n$, and therefore $\chi(F^{\otimes n}) \leq 2\gamma \cdot r^n$. On the other hand, the Hirzebruch-Riemann-Roch formula for bundles with vanishing first Chern class says:

$$\chi(F^{\otimes n}) = r_n\chi(\mathcal{O}_X) - \frac{\Delta_n}{2r_n} = r^n\chi(\mathcal{O}_X) - \frac{n}{2}r^{n-2}\Delta(F).$$

If n goes to infinity, this contradicts $\chi(F^{\otimes n}) \leq 2\gamma \cdot r^n$, unless $\Delta(F) \geq 0$. $\qquad\square$

Corollary 7.3.2 — *Let F be a torsion free sheaf. If F is μ-semistable with respect to an ample divisor H, then the discriminants of the μ-Jordan-Hölder factors of F satisfy the inequality $\Delta(gr_i^{JH}(F)) \leq \Delta(F)$ for all i.*

Proof. Assume first that an arbitrary filtration of F with torsion free factors F_i of rank r_i and first Chern classes γ_i is given. Let $r := \sum_i r_i = \mathrm{rk}(F)$ and $\gamma := \sum_i \gamma_i = c_1(F)$. Recall that the Chern character and the discriminant are related by $2r \cdot ch_2 = c_1^2 - \Delta$. The additivity of the Chern character in short exact sequences therefore provides the first equality in the following identity and a direct calculation gives the second:

$$\sum_i \frac{\Delta(F_i)}{r_i} - \frac{\Delta(F)}{r} = \sum_i \frac{\gamma_i^2}{r_i} - \frac{\gamma^2}{r} = \frac{1}{r}\sum_{i<j} r_i r_j \left(\frac{\gamma_i}{r_i} - \frac{\gamma_j}{r_j}\right)^2. \qquad (7.3)$$

Now if the factors F_i arise from a Jordan-Hölder filtration of F, then $(\gamma_i/r_i - \gamma_j/r_j).H = 0$, and therefore $(\gamma_i/r_i - \gamma_j/r_j)^2 \leq 0$ for all i, j, by the Hodge Index Theorem. Since $\Delta(F_i) \geq 0$ by Bogomolov's Inequality, we get

$$\frac{\Delta(F_i)}{r_i} - \frac{\Delta(F)}{r} \leq -\sum_{j\neq i} \frac{\Delta(F_j)}{r_j} \leq 0,$$

which is even stronger than the assertion of the corollary. $\qquad\square$

We can rephrase the Bogomolov Inequality as follows: if $\Delta(F) < 0$ for some torsion free sheaf F, then F must be μ-unstable with respect to all *all* polarizations H on X.

Indeed, the next theorem implies that one can find a single subsheaf which is destabilizing for all polarizations. Before stating the theorem, we introduce some notations: let Num denote the free $\mathbb{Z}$-module $\mathrm{Pic}(X)/\equiv$, where $\equiv$ means numerical equivalence. Its rank ρ is called the Picard number of X. The intersection product defines an integral quadratic form on Num, whose real extension to $\mathrm{Num}_{\mathbb{R}}$ is of type $(1, \rho - 1)$ by the Hodge Index Theorem. Let K^+ denote the open cone

$$K^+ = \{D \in \mathrm{Num}_{\mathbb{R}} | D^2 > 0, \, D.H > 0 \text{ for all ample divisors } H\}.$$

Note that the second condition is added only to pick one of the two connected components of the set of all D with $D^2 > 0$. This cone contains the cone of ample divisors and in turn is contained in the cone of effective divisors. K^+ satisfies the following property:

$$D \in K^+ \Leftrightarrow D.L > 0 \text{ for all } L \in \overline{K^+} \setminus \{0\}. \tag{7.4}$$

For any pair of sheaves G, G' with nonzero ranks let

$$\xi_{G',G} := c_1(G')/\mathrm{rk}(G') - c_1(G)/\mathrm{rk}(G) \in \mathrm{Num}_{\mathbb{R}}.$$

Theorem 7.3.3 — *Let F be a torsion free coherent sheaf with $\Delta < 0$. Then there is a non-trivial saturated subsheaf F' with $\xi_{F',F} \in K^+$. In particular, if F is a torsion free sheaf which is μ-semistable with respect to a divisor in K^+, then $\Delta(F) \geq 0$.*

Before we prove the theorem the reader may check the following identities: let $0 \to F' \to F \to F'' \to 0$ be a short exact sequence of non-trivial torsion free coherent sheaves. If $G \subset F'$ is a non-trivial subsheaf, then

$$\xi_{G,F} = \xi_{F',F} + \xi_{G,F'}. \tag{7.5}$$

And if $G'' \subset F''$ is a proper subsheaf of rank s and G the kernel of the surjection $F \to F''/G''$, then

$$\xi_{G,F} = \frac{r'(r'' - s)}{(r' + s)r''} \cdot \xi_{F',F} + \frac{s}{r' + s} \cdot \xi_{G'',F''}, \tag{7.6}$$

where, of course, r, r' and r'' are the ranks of F, F' and F'', respectively. Note that in both cases the coefficients in the linear combinations are positive numbers.

Proof of the theorem. If $\rho = 1$, the claim follows directly from the previous theorem: any saturated destabilizing subsheaf suffices. So assume that $\rho \geq 2$. For any nonzero $\xi \in \mathrm{Num}_{\mathbb{R}}$ let $\mathcal{C}(\xi)$ denote the open subcone $\{D \in \overline{K^+} | D.\xi > 0\}$. Property (7.4) says that ξ is in K^+ if and only if $\mathcal{C}(\xi) = \overline{K^+} \setminus \{0\}$.

If $r = 1$, then $F \cong L \otimes \mathcal{I}_Z$, where L is a line bundle and $\mathcal{I}_Z$ the ideal sheaf of a zero-dimensional subscheme $Z \subset X$, and $\Delta(F) = 2\ell(Z) \geq 0$. Now assume that $\Delta(F) < 0$.

Let F' be a saturated destabilizing subsheaf with respect to some polarization H, and let F'' be the quotient F/F'. Then writing the identity (7.3) in the form

$$\frac{\Delta'}{r'} + \frac{\Delta''}{r''} = \frac{\Delta}{r} + \frac{r'r''}{r}\xi^2_{F',F},$$

we see that either $\xi^2_{F',F} > 0$, and we are done, or that Δ' or Δ'' is negative. In this case we can assume by induction that there is either a saturated subsheaf $G \subset F'$ with $\xi_{G,F} \in K^+$ or a saturated subsheaf $G'' \subset F''$ with $\xi_{G'',F''} \in K^+$. In the latter case, let G be the kernel of $F \to F''/G''$. In any case, $\xi_{G,F}$ is a positive linear combination of $\xi_{F',F}$ and some element $\xi \in K^+$ by (7.5) and (7.6). Now by assumption, $\xi_{F',F}$ is not in K^+ and therefore $\mathcal{C}(\xi_{F',F})$ is a proper subcone of $\overline{K^+} \setminus \{0\}$. But ξ is strictly positive on the closure of $\mathcal{C}(\xi_{F',F})$ in $\overline{K^+} \setminus \{0\}$. Thus $\mathcal{C}(\xi_{G,F})$ contains this closure and a fortiori $\mathcal{C}(\xi_{F',F})$ as proper subcones. Hence replacing F' by G strictly enlarges the cone $\mathcal{C}(\xi_{F',F})$. Repeating this process we get a sequence of strictly increasing subcones of $\overline{K^+} \setminus \{0\}$ until at some point $\xi^2_{F',F} > 0$. All we are left with is to prove that this process must terminate: let $H_1, \dots, H_\rho$ be ample divisors whose classes in $\mathrm{Num}_{\mathbb{R}}$ form an $\mathbb{R}$-basis and are contained in $\mathcal{C}(\xi_{F',F})$. Let G be any subsheaf of F with $\mathcal{C}(\xi_{G,F}) \supset \mathcal{C}(\xi_{F',F})$. Then $\xi_{G,F}$ is contained in the lattice $\frac{1}{r!}\mathrm{Num}$ and satisfies the relations

$$0 < \xi_{G,F}.H_j < \mu^{H_j}_{\max}(F) - \mu^{H_j}(F)$$

for all j. That is, ξ is contained in a bounded discrete and hence finite subset of $\mathrm{Num}_{\mathbb{R}}$. $\square$

Having found a subsheaf $F' \subset F$ with $\xi_{F',F} \in K^+$ the next step is to improve the theorem in a quantitative direction by giving a lower bound for the positive square $\xi^2_{F',F}$:

Theorem 7.3.4 — *Let F be a torsion free coherent sheaf with $\Delta < 0$. Then there is a saturated subsheaf F' with $\xi_{F',F} \in K^+$ satisfying the inequality*

$$\xi^2_{F',F} \geq -\frac{\Delta}{r^2(r-1)}.$$

Proof. If $F' \subset F$ is a saturated subsheaf with $\xi_{F',F} \in K^+$, then the Hodge Index Theorem implies

$$\xi^2_{F',F} \leq (\xi_{F',F}.H)^2/H^2 \leq \left(\mu^H_{\max}(F) - \mu^H(F)\right)^2/H^2$$

for any ample divisor H. In particular, the numbers $\xi^2_{F',F}$ for varying F' are bounded from above. Let F' be such that $\xi^2_{F',F}$ attains its maximum value. As before, let F'' be the quotient F/F'. Suppose now that $\Delta' < 0$ and let $G \subset F'$ be a saturated subsheaf with $\xi_{G,F'} \in K^+$. Since $\xi_{G,F'}$ and $\xi_{F',F}$ are both elements in the positive cone K^+, the Hodge Index Theorem shows that

$$|\xi_{G,F}| = |\xi_{G,F'} + \xi_{F',F}| \geq |\xi_{G,F'}| + |\xi_{F',F}| > |\xi_{F',F}|,$$

contradicting the maximality of F'. Here we have used the notation $|\xi| = (\xi^2)^{1/2}$. Hence $\Delta' \geq 0$. Assume now that

$$\frac{\Delta}{r} < -r(r-1)\xi_{F',F}^2.$$

Using the additivity relation (7.3) again, we get

$$\frac{\Delta''}{r''} \leq \frac{\Delta}{r} + \frac{rr'}{r''}\xi_{F',F}^2 < -\frac{r''r(r-1) - rr'}{r''}\xi_{F',F}^2 = -r^2\frac{r''-1}{r''}\xi_{F',F}^2 < 0. \tag{7.7}$$

Arguing by induction on the rank, we can now apply the theorem to F''. As before let $G'' \subset F''$ be a destabilizing subsheaf of rank s satisfying the relation

$$\xi_{G'',F''}^2 \geq -\frac{\Delta''}{r''^2(r''-1)} > \frac{r^2}{r''^2}\xi_{F',F}^2.$$

For the last inequality use (7.7). Let G denote the kernel of $F \to F''/G''$. Using (7.6) we have

$$|\xi_{G,F}| \geq \frac{r'(r''-s)}{(r'+s)r''} \cdot |\xi_{F',F}| + \frac{s}{r'+s} \cdot |\xi_{G'',F''}|$$

$$> \frac{r'(r''-s)}{(r'+s)r''} \cdot |\xi_{F',F}| + \frac{s}{r'+s} \cdot \frac{r}{r''} \cdot |\xi_{F',F}| = |\xi_{F',F}|.$$

Again this contradicts the maximality of F', and therefore proves the theorem. $\qquad\square$

We are now prepared to prove Bogomolov's effective restriction theorem. Let r be an integer greater than 1, and let R be the maximum of the numbers $\binom{r}{\ell}\binom{r-2}{\ell-1}$ for all $1 \leq \ell < r$. (Certainly the maximum is attained for $\ell = \lfloor\frac{r}{2}\rfloor$.)

Theorem 7.3.5 — *Let F be a locally free sheaf of rank $r \geq 2$. Assume F is μ-stable with respect to an ample class $H \in K^+ \cap Num$. Let $C \subset X$ be a smooth curve with $[C] = nH$. If $2n > \frac{R}{r}\Delta(F) + 1$, then $F|_C$ is a stable sheaf.*

Proof. Suppose C satisfies the conditions of the theorem and $F|_C$ has a destabilizing quotient E of rank s. By taking exterior powers we wish to reduce the proof to the case $s = 1$. Then $\Lambda^s F|_C \to \Lambda^s E$ is still destabilizing and $\Lambda^s E$ is a line bundle, but $\Lambda^s F$ need not be μ-stable. At this point we evoke the Theorem 3.2.11, that powers of μ-stable bundles are μ-polystable. Recall, that the proof, which we only sketched, relies on the Kobayashi-Hitchin correspondence (or the interpretation of stable bundles on a curve in terms of unitary representations). Thus, there is a decomposition $\Lambda^s F = \bigoplus F_i$, where the bundles F_i are μ-stable with slope $\mu(F)$. We replace the numerical assumption $2n > \frac{R}{r}\Delta(F) + 1$ by the two inequalities

$$2n > \Delta(\Lambda^s F) + 1 \tag{7.8}$$

and

$$n^2 H^2 = C^2 > \Delta(\Lambda^s F). \tag{7.9}$$

Indeed, (7.8) follows from $\mathrm{rk}(\Lambda^s F) = \binom{r}{s}$ and $\Delta(\Lambda^s F) = \binom{r-1}{s-1}\binom{r}{s}\frac{\Delta(F)}{r}$. The second inequality is a consequence of the first and $\Delta(\Lambda^s F) \geq 0$: slightly improving (7.8) by using integrality we get $n \geq \Delta(\Lambda^s F)/2+1$. Hence $n^2 H^2 \geq n^2 \geq \Delta(\Lambda^s F)^2/4+1+\Delta(\Lambda^s F) > \Delta(\Lambda^s F)$.

Next consider the exterior power $\Lambda^s F \to \Lambda^s F|_C \to \Lambda^s E =: L$, where L is a line bundle with $\mu(L) = \mu(E) \leq \mu(F|_C) = \mu(\Lambda^s F|_C)$. We may assume that the induced homomorphism $F_0 \to F|_C \to L$ is not trivial. Replacing L by the image of $F_0 \to L$, which has even smaller degree, and using $\Delta(F_0) \leq \Delta(\Lambda^s F)$ by 7.3.2, we obtain a μ-stable bundle F_0 with a destabilizing line bundle $F_0|_C \to L$ such that $2n > \Delta(F_0) + 1$ and $C^2 > \Delta(F_0)$. The case $\mathrm{rk}(F_0) = 1$ can be excluded by a lemma stated after the proof. If $\mathrm{rk}(F_0) > 1$ we have concluded our reduction to the case of a rank one destabilizing line bundle, i.e. we may assume that F is μ-stable of rank $r \geq 2$ with

$$2n > \Delta(F) + 1 \tag{7.10}$$

and

$$C^2 > \Delta(F) \geq \Delta(F)/(r - 1) \tag{7.11}$$

and that $F|_C \to E$ is a destabilizing quotient of rank one.

Let G be the kernel of the composite homomorphism $F \to F|_C \to E$. Then $c_1(G) = c_1(F) - C$ and $\chi(G) = \chi(F) - (\deg(E) + 1 - g(C))$. Expressing the Euler characteristic of F and G in terms of their discriminants we get (cf. 5.2.2)

$$\Delta(G) = \Delta(F) - 2(\deg(F|C) - r \deg(E)) - (r - 1)C^2.$$

Since E is destabilizing, $\deg(F|C) - r \deg(E) \geq 0$.

$$\Delta(G) \leq \Delta(F) - (r - 1)C^2 < 0$$

because of (7.11). By the previous theorem there is a saturated subsheaf $G' \subset G$ of rank, say, t with

$$\xi_{G',G} \in K^+ \quad \text{and} \quad \xi_{G',G}^2 \geq -\frac{\Delta(G)}{r^2(r - 1)}.$$

Then $\xi_{G',G} = \xi_{G',F} + \frac{1}{r}C$. The stability of F implies that $\xi_{G',F}.C < 0$, and since the intersection product on Num takes integral values,

$$0 < \xi_{G',G}.C \leq -\frac{n}{rt} + \frac{n^2}{r}H^2.$$

For any two divisors D and D' in K^+ the inequality $(DD')^2 \geq D^2 D'^2$ holds. Apply this to C and $\xi_{G',G}$ and get

$$-\frac{\Delta(G)}{r^2(r-1)}n^2 H^2 \leq \xi_{G',G}^2 C^2 \leq \left(\frac{n^2}{r}H^2 - \frac{n}{rt}\right)^2.$$

Using the estimate $\Delta(G) \leq \Delta(F) - (r-1)n^2 H^2$ and cancelling common factors we get

$$-\frac{\Delta(F)}{r-1}H^2 \leq -\frac{2n}{t}H^2 + \frac{1}{t^2},$$

hence

$$2n \leq \frac{t}{r-1}\Delta(F) + \frac{1}{tH^2} \leq \Delta(F) + 1,$$

which contradicts (7.10). □

In the proof we made use of the following lemma:

Lemma 7.3.6 — *Let F be a μ-semistable vector bundle and $\Lambda^s F \to M$ be a rank one torsion free quotient with $\mu(\Lambda^s F) = \mu(M)$. If the restriction $\Lambda^s F|_C \to M_C$ to a curve C is the s-th exterior power of a locally free quotient $F|_C \to E$ of rank s, then $\Lambda^s F \to M$ is induced by a torsion free quotient $F \to \tilde{E}$ of rank s. In particular, if F is μ-stable, then $s = \mathrm{rk}(F)$.*

Proof. The technique to prove this was already used twice in Section 7.2. Let

$$\mathrm{Grass}(F, s) \subset \mathbb{P}(\Lambda^s F)$$

be the Plücker embedding of the relative Grassmannian. Its ideal sheaf is generated by $\mathcal{I}_\nu \subset S^\nu(\Lambda^s F)$. In fact, it is generated by the Plücker relations which are the image of a homomorphism $\Lambda^{s+1}F \otimes \Lambda^{s-1}F \to S^2(\Lambda^s F)$. The quotient $\Lambda^s F \to M$ corresponds to a section of $\mathbb{P}(\Lambda^s F|_U) \to U$, where $U = X \setminus \mathrm{Supp}(M^{\sim\sim}/M)$. The image of this section is contained in $\mathrm{Grass}(F, s)$ if and only if the composite maps

$$\mathcal{I}_\nu \to S^\nu(\Lambda^s F) \to S^\nu(M)$$

vanish or, equivalently, if the composition

$$\Phi : \Lambda^{s-1}F \otimes \Lambda^{s-1}F \to S^2\Lambda^s F \to S^2(M)$$

vanishes. Standard calculations show $\mu(\Lambda^{s-1}F \otimes \Lambda^{s-1}F) = 2s\mu(F)$ and $\mu(S^2 M) = 2s\mu(F)$. The existence of $F|_C \to E$ implies that the curve $C \subset U$ is mapped to $\mathrm{Grass}(F, s)$ by the cross section that corresponds to the homomorphism $\Lambda^s F|_U \to M|_U$. Hence Φ is an element in $\mathrm{Hom}(\Lambda^{s+1}F \otimes \Lambda^{s-1}F, S^2(M)(-C))$. Using the μ-semistability of $\Lambda^{s-1}F \otimes \Lambda^{s-1}F$ (cf. 3.2.10) and $\mu(\Lambda^{s-1}F \otimes \Lambda^{s-1}F) > \mu(S^2 M(-C))$, this yields $\Phi = 0$, i.e. U maps to $\mathrm{Grass}(F, s)$. □

Remark 7.3.7 — Of course, the theorem remains valid if F is only torsion free but the curve C avoids the singularities of F. One can also weaken the assumption on the class H: let H be an arbitrary class in $K^+ \cap \text{Num}$ and let F be a μ-stable vector bundle with respect to H. If we furthermore assume that also all exterior powers of F are μ-stable with respect to H, which is automatically satisfied if $\text{rk}(F) \leq 3$, then the conclusion of the theorem holds true. In Chapter 11 this will be applied to minimal surfaces of general type and $H = K_X$ which is only big and nef.

Comments:
— We wish to emphasize that the results in Section 7.1 and 7.3 assume that the characteristic of our base field is zero. The restriction theorems of Mehta and Ramanathan are valid in positive characteristic as well. Unfortunately, it is not effective. In fact, an *effective* restriction theorem would settle the open question whether families of semistable sheaves with fixed topological data are bounded in positive characteristic. (Added in second edition: The boundedness has been proved now by Langer [356], see Appendix 7.A.)

— The proof of Flenner's Theorem 7.1.1 follows quite closely the original presentation in [63], though we avoided the use of spectral sequences. Since its proof relies on the Grauert-Mülich Theorem, and hence the Harder-Narasimhan filtration, it does not generalize to the case of μ-stable sheaves.

— The references for the theorems of Mehta and Ramanathan (7.2.1, 7.2.8) are of course [175] and [176]. Also see [174]. The complete argument for the fact, used in Lemma 7.2.2, that the complement of the integral divisors has codimension at least two can be found in [175].

— Tyurin generalized their arguments (cf. [252] and for a more detailed proof [111]) and showed that a family of μ-stable rank two bundles on a surface restricts stably to a general ample curve of high degree.

— One should also be aware of the following result due to Maruyama ([164], also [174]):
If X is smooth and projective and $\mathcal{O}(1)$ is very ample, then the restriction of a μ-semistable sheaf of rank $< \dim(X)$ to the generic hypersurface is again μ-semistable.
The proof of it is rather easy, but as it has obviously no application to sheaves on surfaces, we omitted the proof.

— Proofs of Theorem 7.3.5 for rank two bundles for special cases can be found at various places. O'Grady in [206] treats the case $\text{Pic}(X) = \mathbb{Z}$ and Friedman and Morgan give a proof for the case $c_1 = 0$ [71]. The complete proof is in Bogomolov's papers [29] and [31].

— In special cases one can improve the results. Hein [101] and Anghel [1] deal with the case of rank two bundles on K3 surfaces and on abelian surfaces, respectively.

Appendix to Chapter 7

7.A Further comments (second edition)

I. Restriction theorems in positive characteristic.

In [335] Hein considers μ-semistable vector bundles of rank two on surfaces over an algebraically closed field of arbitrary characteristic. He proves an effective restriction theorem for general curves in $|\mathcal{O}(a)|$ with the lower bound for a only depending on the degree of the surface and on the discriminant. He deduces from it a boundedness result for μ-semistable vector bundles of rank two and a weak version of Bogomolov's inequality, both thus valid in arbitrary characteristic. The paper also contains a result similar to Bogomolov's restriction theorem, showing that the restriction of a μ-stable vector bundle of rank two to any smooth curve in $|\mathcal{O}(a)|$ for a large enough is stable.

An effective restriction result in arbitrary characteristic for μ-stable vector bundles of rank two or three on a smooth projective variety of arbitrary dimension had earlier been proved in [397]. Nakashima also proves that the restriction to any smooth $D \in |\mathcal{O}(a)|$ is μ-stable for a large, but the bound for a depends on the Harder-Narasimhan filtration of the tangent bundle.

For *strongly semistable* vector bundles on a surface in positive characteristic, i.e. bundles that remain semistable even after Frobenius pull-back, Brenner shows in [291] that the restriction to generic curves of high degree need not be strongly semistable again. This has to be compared with Langer's result (Thm. 3.1 in [356]) which can be read as a result bounding the minimal and maximal slope of the Harder-Narasimhan filtration of the restriction of a strongly semistable torsion free sheaf.

The most general restriction theorem in positive characteristic was proved by Langer (cf. Thm. 5.2 in [356]). We only give a special case of it:

Let E be a μ-semistable torsion free sheaf of rank r and

$$a > \left\lfloor \frac{r}{r-1} \Delta(E) H^{n-2} + \frac{1}{dr(r-1)} + \frac{(r-1)\beta_r}{dr} \right\rfloor.$$

Then the restriction $E|_D$ is μ-semistable for any normal $D \in |\mathcal{O}(aH)|$ with $E|_D$ torsion free.

Here, $d = \deg(X) = H^n$ and $\beta_r = (k\frac{r(r-1)}{p-1}d)^2$ with $k \gg 0$ such that $\mathcal{T}_X(k)$ is globally generated.

8

Line Bundles on the Moduli Space

This chapter is devoted to the study of line bundles on the moduli space. In Sections 8.1 and 8.2 we first discuss a general method for associating to a flat family of coherent sheaves a determinant line bundle on the base of this family. The next step is to construct such determinant bundles on the moduli space of semistable sheaves even if there is no universal family. Having done this we study the properties of two particular line bundles $\mathcal{L}_0$ and $\mathcal{L}_1$ on the moduli space of semistable torsion free sheaves on a smooth surface. Whereas $\mathcal{L}_0 \otimes \mathcal{L}_1^m$ is ample relative to $\mathrm{Pic}(X)$ for sufficiently large m, the linear system $|\mathcal{L}_1^m|$ contracts certain parts of the moduli space and in fact defines a morphism from the Gieseker-Maruyama moduli space of semistable sheaves to the Donaldson-Uhlenbeck compactification of the moduli space of μ-stable vector bundles. The presentation of the material is based on the work of J. Le Potier and J. Li.

In the final section we compare the canonical bundle of the good part of the moduli space with the line bundle $\mathcal{L}_1$. This is an application of the Grothendieck-Riemann-Roch formula.

8.1 Construction of Determinant Line Bundles

Let X be a smooth projective variety of dimension n. The Grothendieck group $K(X)$ of coherent sheaves on X becomes a commutative ring with $1 = [\mathcal{O}_X]$ by putting $[F_1] \cdot [F_2] := [F_1 \otimes F_2]$ for locally free sheaves F_1 and F_2. Two classes u and u' in $K(X)$ are said to be numerically equivalent: $u \equiv u'$, if their difference is contained in the radical of the quadratic form $(a, b) \mapsto \chi(a \cdot b)$. Let $K(X)_{\mathrm{num}} = K(X)/\equiv$. If $A \subset K(X)$ is any subset, let $A^{\perp} \subset K(X)$ be the subset of all elements orthogonal to A with respect to this quadratic form. By the Hirzebruch-Riemann-Roch formula we have

$$\chi(a \cdot b) = \int_X ch(a)ch(b)td(X)$$

213

Thus the numerical behaviour of $a \in K(X)_{\mathrm{num}}$ is determined by its associated rank $\mathrm{rk}(a)$ and Chern classes $c_i(a)$.

A flat family $\mathcal{E}$ of coherent sheaves on X parametrized by S defines an element $[\mathcal{E}] \in K^0(S \times X)$, and as the projection $p : S \times X \to S$ is a smooth morphism, there is a well defined homomorphism $p_! : K^0(S \times X) \to K^0(S)$ (cf. 2.1.11).

Definition 8.1.1 — *Let $\lambda_\mathcal{E} : K(X) \to \mathrm{Pic}(S)$ be the composition of the homomorphisms:*

$$K(X) \xrightarrow{q^*} K^0(S \times X) \xrightarrow{\cdot [\mathcal{E}]} K^0(S \times X) \xrightarrow{p_!} K^0(S) \xrightarrow{\det} \mathrm{Pic}(S).$$

Here is a list of some easily verified properties of this construction:

Lemma 8.1.2 — *i) If $0 \to \mathcal{E}' \to \mathcal{E} \to \mathcal{E}'' \to 0$ is a short exact sequence of S-flat families of coherent sheaves then $\lambda_\mathcal{E} \cong \lambda_{\mathcal{E}'} \otimes \lambda_{\mathcal{E}''}$.*
ii) If $\mathcal{E}$ is an S-flat family and $f : S' \to S$ a morphism then for any $u \in K(X)$ one has $\lambda_{f_X^* \mathcal{E}}(u) = f^* \lambda_\mathcal{E}(u)$.
iii) If G is an algebraic group, S a scheme with a G-action and $\mathcal{E}$ a G-linearized S-flat family of coherent sheaves on X, then $\lambda_\mathcal{E}$ factors through the group $\mathrm{Pic}^G(S)$ of isomorphism classes of G-linearized line bundles on S.
iv) Let $\mathcal{E}$ be an S-flat family of coherent sheaves of class $c \in K(X)_{\mathrm{num}}$ and let $\mathcal{N}$ be a locally free $\mathcal{O}_S$-sheaf. Then $\lambda_{\mathcal{E} \otimes p^* \mathcal{N}}(u) \cong \lambda_\mathcal{E}(u)^{\mathrm{rk}(\mathcal{N})} \otimes \det(\mathcal{N})^{\chi(c \otimes u)}$.

Proof. The last assertion follows from the projection formula for direct image sheaves: $R^i p_*(\mathcal{E} \otimes p^* \mathcal{N}) = R^i p_*(\mathcal{E}) \otimes \mathcal{N}$, and the general isomorphism $\det(A \otimes B) \cong \det(A)^{\mathrm{rk}(B)} \otimes \det(B)^{\mathrm{rk}(A)}$ for arbitrary locally free sheaves. $\square$

Examples 8.1.3 — *i)* Let $x \in X$ be a smooth point and $u = [\mathcal{O}_x]$ the class of the structure sheaf of x. Let E be an S-flat family of sheaves on X and $E_\bullet \to E$ a finite locally free resolution. Then by 8.1.2 *i)*

$$\lambda_E(u) = \bigotimes_i \lambda_{E_i}(u)^{(-1)^i} = \bigotimes_i \det(R^\bullet p_*(E_i \otimes \mathcal{O}_x))^{(-1)^i}.$$

Now $\det(R^\bullet p_*(E_i \otimes \mathcal{O}_x)) = \det(p_* E_i|_{S \times \{x\}}) = p_*(\det(E_i)|_{S \times \{x\}})$. Hence

$$\lambda_E(u) = p_* \bigotimes_i \det(E_i)^{(-1)^i}|_{S \times \{x\}} = p_*(\det(E)|_{S \times \{x\}}).$$

ii) Let $H \subset X$ be a very ample divisor and let $h = [\mathcal{O}_H]$ be its class in $K(X)$. Then $[\mathcal{O}_X(\ell)] = (1 - h)^{-\ell} = 1 + \ell h + \binom{\ell+1}{2} h^2 + \ldots$. In Section 4.3 we used the line bundles $\det(p_!(\widetilde{F} \otimes q^* \mathcal{O}_X(\ell)))$ on the quotient scheme $\mathrm{Quot}(\mathcal{H}, P)$ in the construction of the moduli spaces. These bundles are very ample for $\ell \gg 0$. Using the λ-formalism above we can express them as follows:

$$\det(p_!(\widetilde{F} \otimes q^* \mathcal{O}_X(\ell))) = \lambda_{\widetilde{F}}([\mathcal{O}_X(\ell)])$$
$$= \lambda_{\widetilde{F}}(1) \otimes \lambda_{\widetilde{F}}(h)^\ell \otimes \ldots \otimes \lambda_{\widetilde{F}}(h^n)^{\binom{\ell+n-1}{n}}.$$

In particular, $\det(p_!(\widetilde{F} \otimes q^* \mathcal{O}_X(\ell)))$ does not, in general, depend linearly on ℓ and projective embeddings given by multiples of this line bundle might be quite different for different ℓ.

iii) Let $\mathcal{E}$ be a universal family parametrized by the moduli space M^s. As above we find that the dominant term in the ℓ-expansion of $\lambda_{\mathcal{E}}([\mathcal{O}_X(\ell)])$ is $\lambda_{\mathcal{E}}(h^n)^{\binom{\ell+n-1}{n}}$. Now $h^n = \sum_{j=1}^{\deg(X)}[\mathcal{O}_{x_j}]$ where $x_1, \ldots, x_{\deg(X)}$ are the intersection points of n general hyperplanes. According to the example in *i)* we can write

$$\lambda_{\mathcal{E}}(h^n)^{\binom{\ell+n-1}{n}} = \bigotimes_{j=1}^{\deg(X)} \det(\mathcal{E})^{\binom{\ell+n-1}{n}}|_{M^s \times \{x_j\}}.$$

If $\mathcal{E}$ is replaced by $\mathcal{E} \otimes p^* L$ for some line bundle L on M^s, the expression on the right hand side changes by $L^{\mathrm{rk}(\mathcal{E}) \deg(X)\binom{\ell+n-1}{n}}$. Thus, if L is very negative, $\lambda_{\mathcal{E} \otimes L}([\mathcal{O}_X(\ell)])$ becomes very negative for $\ell \gg 0$. □

For any class c in $K(X)_{\mathrm{num}}$, we write $c(m) := c \cdot [\mathcal{O}_X(m)]$ and denote by $P(c)$ the associated Hilbert polynomial $P(c, m) = \chi(c(m))$. If F is an S-flat family of coherent sheaves with Hilbert polynomial $P(c)$ the points $s \in S$ such that F_s is of class c form an open and closed subscheme of S. This follows from the fact that for a flat family F the Euler characteristic $s \mapsto \chi(F_s)$ is a locally constant function. As a consequence the moduli space $M(P)$ decomposes into finitely many open and closed subschemes $M(c_i)$, where c_i runs through the set of classes with $P(c_i) = P$. A universal family $\mathcal{E}$ on $M^s(c) \times X$ is well-defined only up to tensorizing with a line bundle on $M^s(c)$. Part *iv)* of the lemma shows that $\lambda_{\mathcal{E}}(u)$ is independent of this ambiguity, if $\chi(c \otimes u) = 0$, i.e. if u is orthogonal to c. We therefore define:

Definition 8.1.4 — *For a given class $c \in K(X)_{\mathrm{num}}$ let*

$$K_c = c^\perp \qquad and \qquad K_{c,H} = c^\perp \cap \{1, h, h^2, \ldots, h^n\}^{\perp\perp}.$$

The following theorem says that the condition $c \perp u$ is also sufficient to get a well-defined determinant line bundle on $M^s(c)$ by means of u. More precisely:

Theorem 8.1.5 — *Let c be a class in $K(X)_{\mathrm{num}}$. Then there are group homomorphisms $\lambda^s : K_c \to \mathrm{Pic}(M^s(c))$ and $\lambda : K_{c,H} \to \mathrm{Pic}(M(c))$ with the following properties:*

1. λ and λ^s commute with the inclusion $K_{c,H} \subset K_c$ and the restriction homomorphism $\mathrm{Pic}(M(c)) \to \mathrm{Pic}(M^s(c))$.

2. If $\mathcal{E}$ is a flat family of semistable sheaves of class c on X parametrized by S, and if $\phi_\mathcal{E} : S \to M(c)$ is the classifying morphism, then λ and $\lambda_\mathcal{E} : K(X) \to \mathrm{Pic}(S)$ commute with the inclusion $K_{c,H} \subset K(X)$ and the homomorphism $\phi_\mathcal{E}^* : \mathrm{Pic}(M(c)) \to \mathrm{Pic}(S)$.

3. If $\mathcal{E}$ is a flat family of stable sheaves of class c on X parametrized by S, then λ^s and $\lambda_\mathcal{E} : K(X) \to \mathrm{Pic}(S)$ commute with the inclusion $K_c \subset K(X)$ and the homomorphism $\phi_\mathcal{E}^* : \mathrm{Pic}(M^s(c)) \to \mathrm{Pic}(S)$.

In order to prove the theorem we have to recall the set-up of the construction of $M(c)$ in Section 4.3: we choose a very large integer m, fix a vector space V of dimension $P(c, m)$ and let $\mathcal{H} := V \otimes \mathcal{O}_X(-m)$. Let $R(c) \subset \mathrm{Quot}(\mathcal{H}, P)$ denote the open subscheme of those quotients $[q : \mathcal{H} \to F]$ for which F is a semistable sheaf of class c and q induces an isomorphism $V \to H^0(F(m))$. There is a universal family $\mathcal{O}_{R(c)} \otimes \mathcal{H} \to \widetilde{F}$. If m was chosen large enough and $\ell \gg 0$, $R(c)$ is the set of semistable points in $\overline{R(c)}$ with respect to the action of $\mathrm{SL}(V)$ and the canonical linearization of $\lambda_{\widetilde{F}}([\mathcal{O}_X(\ell)])$. Moreover, $M(c) = R(c)/\!\!/\mathrm{SL}(V)$. The determinant bundle $\det(\widetilde{F})$ of the universal family induces a morphism $\det : R(c) \to \mathrm{Pic}(X)$ such that $\det(\widetilde{F}) = \det_X^*(\mathcal{P}) \otimes p^*\mathcal{A}$ where $\mathcal{P}$ is the Poincaré line bundle on $\mathrm{Pic}(X) \times X$ and $\mathcal{A}$ some line bundle on $R(c)$. (Of course, $\det : R(c) \to \mathrm{Pic}(X)$ can be the constant morphism, for example if $\dim(c) = \deg(P(c)) \leq \dim(X) - 2$.) We fix these notations for the rest of this section.

Proof of the theorem. Let $u \in K(X)_{\mathrm{num}}$ be an arbitrary class and consider the line bundle $L := \lambda_{\widetilde{F}}(u)$ on $R(c)$. L inherits a $\mathrm{GL}(V)$-linearization from $\widetilde{F}$. We want to know whether L descends to a line bundle on $M(c)$ or $M^s(c)$.

According to the criterion of Theorem 4.2.15 we must control the action of the stabilizer subgroup in $\mathrm{GL}(V)$ of points in closed orbits. The orbit of a point $[q : \mathcal{H} \to F] \in R(c)$ is closed if and only if F is a polystable sheaf, i.e. if it is isomorphic to a direct sum $\bigoplus_i F_i \otimes_k W_i$ with distinct stable sheaves F_i and k-vector spaces W_i. The stabilizer of $[q]$ then is isomorphic to $\mathrm{Aut}(F) \cong \prod \mathrm{GL}(W_i)$, and an element $(A_1, \ldots, A_\ell)$, $A_i \in \mathrm{GL}(W_i)$, acts on the fibre

$$L([q]) \cong \bigotimes_i \left(\det(H^\bullet([F_i] \cdot u))^{\dim(W_i)} \otimes (\det(W_i))^{\chi([F_i] \cdot u)} \right)$$

via multiplication with the number $\prod_i \det(A_i)^{\chi(u \cdot [F_i])}$ (cf. the remarks following 2.1.11). Let $c_i = [F_i]$, and let r and r_i be the multiplicities of F and F_i, respectively. By construction, we have for all ℓ:

$$r_i\chi(c \cdot [\mathcal{O}_X(\ell)]) = r_i P(F(\ell)) = r P(F_i(\ell)) = r\chi(c_i \cdot [\mathcal{O}_X(\ell)]).$$

This is equivalent to: $\chi((rc_i - r_ic) \cdot h^\ell) = 0$ for all ℓ, i.e. $(rc_i - r_ic) \in \{1, h, \ldots, h^n\}^\perp$.

Now distinguish two cases: if F is in fact stable, so that $\text{Aut}(F) \cong \mathbb{G}_m$, and if $u \in K_c$, then $A \in \mathbb{G}_m(k)$ acts by $A^{\chi(u \cdot c)} = A^0 = 1$. If on the other hand F is not stable but $u \in K_{c,H}$, then we have $\chi((rc_i - r_ic) \cdot u) = 0$, since $(rc_i - r_ic) \in \{1, h, \dots, h^n\}^{\perp}$. Therefore $\chi(c_i \cdot u) = \frac{r_i}{r}\chi(c \cdot u) = 0$. Thus, again, any element in the stabilizer subgroup acts trivially. It follows that $u \in K_c$ or $u \in K_{c,H}$ are sufficient conditions on u to let the line bundle L descend to bundles $\lambda^s(u)$ on $M^s(c)$, or $\lambda(u)$ on $M(c)$, respectively.

It remains to check the commutativity relations. Part 1 of the theorem is trivial. To get the universal properties 2 and 3 proceed as follows: suppose $\mathcal{E}$ is an S-flat family of semistable sheaves of class c. Let $\pi : \widetilde{S} = \mathbb{I}\text{som}(V, p_*(\mathcal{E} \otimes \mathcal{O}_X(m))) \to S$ be the frame bundle (cf. 4.2.3) associated to the locally free sheaf $p_*(\mathcal{E} \otimes \mathcal{O}_X(m))$, and let $\widetilde{\phi}_{\mathcal{E}} : \widetilde{S} \to R(c)$ be the classifying morphism for the quotient $V \otimes \mathcal{O}_{\widetilde{S} \times X} \to \pi^*\mathcal{E}$ which is the composition of the tautological trivialization (4.2.6) and the evaluation map. $\widetilde{\phi}_{\mathcal{E}}$ is a $\text{GL}(V)$-equivariant morphism, and $\pi \circ \widetilde{\phi}_{\mathcal{E}} = \phi_{\mathcal{E}} \circ \pi$, where $\phi_{\mathcal{E}} : S \to M$ is the classifying morphism for the family $\mathcal{E}$.

$$
\begin{array}{ccc}
\widetilde{S} & \xrightarrow{\widetilde{\phi}_{\mathcal{E}}} & R(c) \\
\pi \downarrow & & \downarrow \pi \\
S & \xrightarrow{\phi_{\mathcal{E}}} & M
\end{array}
$$

We obtain the following sequence of $\text{GL}(V)$-equivariant isomorphisms

$$\pi^*\phi_{\mathcal{E}}^*\lambda(u) = \widetilde{\phi}_{\mathcal{E}}^*\pi^*\lambda(u) = \widetilde{\phi}_{\mathcal{E}}^*\lambda_{\widetilde{F}}(u) = \lambda_{\widetilde{\phi}_{\mathcal{E}}^*\widetilde{F}}(u) = \lambda_{\pi^*\mathcal{E}}(u) = \pi^*\lambda_{\mathcal{E}}(u).$$

Assertions 2 and 3 follow from this and the fact that $\pi^* : \text{Pic}(S) \to \text{Pic}^{\text{GL}(V)}(\widetilde{S})$ is injective. (cf. 4.2.16). $\qquad\square$

Before we describe natural line bundles in the image of λ, we want to raise the question of how many line bundles one can construct this way. The best result in this direction is due to J. Li. Unfortunately, the techniques developed here are not sufficient to cover his result. In particular, we have not explained the relation to gauge theory essential for its proof. We only state the following special case of the result in [151] which can be conveniently formulated in the language introduced above.

Theorem 8.1.6 — *If X is a regular surface, i.e. $q(X) = 0$, then*

$$\lambda : K(X)_c \otimes \mathbb{Q} \to \text{Pic}(M^s(2, \mathcal{Q}, c_2)) \otimes \mathbb{Q}$$

is surjective for $c_2 \gg 0$. $\qquad\square$

Let X be a surface. We will see in Example 8.1.8 *ii)* below that only the degree of classes u in $CH^2(X)$ matters for the restriction of $\lambda(u)$ to the moduli space $M(r, \mathcal{Q}, c_2)$

of semistable sheaves with fixed determinant $\mathcal{Q}$, i.e. we can reduce from $K(X)$ to the group $\mathbb{Z} \oplus \mathrm{Pic}(X) \oplus \mathbb{Z}$, sending u to the triple $(\mathrm{rk}(u), \det(u), \chi(u))$. Moreover, the condition to be orthogonal to c imposes a linear condition on u.

The above theorem gives reason to expect that for large second Chern number c_2 the Picard group of the moduli space $M^s(r, \mathcal{Q}, c_2)$ contains a subgroup which is (roughly) of the form $\mathrm{Pic}(X) \oplus \mathbb{Z}$. More evidence is given by the following example:

Example 8.1.7 — Let E be a μ-stable locally free sheaf of rank r, determinant $\mathcal{Q}$ and second Chern class $c_2(E) = c_2 - 1$, and let $\pi : S := \mathbb{P}(E) \to X$ be its projectivization. Let $\gamma : S \to S \times X$ be the graph of π and let F be the kernel of the surjective homomorphism

$$q^*E \longrightarrow q^*E|_{\gamma(S)} = \gamma_*\pi^*E \longrightarrow \gamma_*\mathcal{O}_\pi(1).$$

Then F is an S-flat family with fibres

$$F_s = \ker(E \to E(x) \xrightarrow{s} k(x))$$

for $x = \pi(s)$ and $s \in \mathbb{P}(E(x)) = \pi^{-1}(x) \subset S$. As E is μ-stable, the same is true for F_s. Moreover, there is a unique way of embedding F_s into E. Hence F_s and $F_{s'}$ are non-isomorphic for all $s \neq s'$ in S, and F induces an injective morphism $S \to M^s(r, \mathcal{Q}, c_2)$. Applying $\lambda(u)$ to the short exact sequence

$$0 \to F \to q^*E \to \gamma_*\mathcal{O}_\pi(1) \to 0$$

we get according to 8.1.2:

$$\lambda_F(u) = \lambda_{q^*E}(u) \otimes \lambda_{\gamma_*\mathcal{O}_\pi(1)}(u)^{\vee}.$$

As γ is the graph of π,

$$\lambda_{\gamma_*\mathcal{O}_\pi(1)}(u) = \det(\mathcal{O}_\pi(1) \otimes \pi^*u) = \mathcal{O}_\pi(\mathrm{rk}(u)) \otimes \pi^*(\det(u)).$$

Hence $\lambda_F(u) \cong \mathcal{O}_\pi(-\mathrm{rk}(u)) \otimes \pi^*(\det(u))^{\vee}$. We will see in the next chapter that sheaves E as above always exist for large c_2. The calculations then show that $\mathrm{Pic}(M^s(r, \mathcal{Q}, c_s))$ contains $\mathbb{Z} \oplus \mathrm{Pic}(X)$. $\square$

In the following we investigate some particular classes in $K(X)_{c,H}$ and their associated line bundles.

Examples 8.1.8 — Let X be a smooth variety of dimension n, H a very ample divisor and c a class in $K(X)_{\mathrm{num}}$.
i) For any pair of integers $0 \leq i < j \leq n$, the class $v_{ij}(c) := -\chi(c \cdot h^j) \cdot h^i + \chi(c \cdot h^i) \cdot h^j$ is an element in $K_{c,H}$, as is rather obvious.

ii) Let $D_0, D_1 \in K(X)$ be the classes of zero-dimensional sheaves of the same length, and let $D = D_0 - D_1$. Then $D \equiv 0$, so that D is in particular an element in $K_{c,H}(X)$. Moreover, $\lambda(D) \cong \det^*(\mathcal{M})$ for some line bundle $\mathcal{M}$ on $\mathrm{Pic}(X)$, where $\det : M(c) \to \mathrm{Pic}(X)$ is the determinant morphism. It clearly suffices to prove this assertion for the special case that D_i is the structure sheaf of a closed point x_i. Then we have the following isomorphisms of line bundles on $R(c)$:

$$\lambda_{\widetilde{F}}(D_i) \cong p_*(\det(\widetilde{F})|_{R \times \{x_i\}}) \cong \mathcal{A} \otimes \det^*(\mathcal{P}|_{\mathrm{Pic}(X) \times \{x_i\}}),$$

where as before $\mathcal{P}$ is the Poincaré line bundle on $\mathrm{Pic}(X) \times X$. Thus

$$\lambda_{\widetilde{F}}(D) = \det^*(\mathcal{P}|_{\mathrm{Pic}(X) \times \{x_0\}} \otimes \mathcal{P}^{\vee}|_{\mathrm{Pic}(X) \times \{x_1\}}).$$

iii) Observe that in the expression $v_i(c) := v_{in}(c) = -\chi(c \cdot h^n) \cdot h^i + \chi(c \cdot h^i) \cdot h^n$ the first coefficient $\chi(c \cdot h^n)$ equals $\mathrm{rk}(c) \deg(X)$, whereas h^n is represented by $\deg(X)$ points on X. Choose a fixed base point $x \in X$ and define

$$u_i(c) := -r \cdot h^i + \chi(c \cdot h^i) \cdot [\mathcal{O}_x].$$

Then $v_i(c) = \deg(X) \cdot u_i(c) + \chi(c \cdot h^i) \cdot (h^n - \deg(X) \cdot [\mathcal{O}_x]))$. It follows from part *ii)* that

$$\lambda(v_i(c)) \cong \lambda(u_i(c))^{\deg(X)} \otimes \det^*(\mathcal{M})$$

for some line bundle $\mathcal{M}$ on $\mathrm{Pic}(X)$. $\qquad\square$

The line bundles $\lambda(u_i(c))$ play an important rôle in the geometry of the moduli spaces. We therefore define:

Definition 8.1.9 — *Let $x \in X$ be a closed point, let $u_i(c) = -r \cdot h^i + \chi(c \cdot h^i) \cdot [\mathcal{O}_x]$, $i \geq 0$, and let $\mathcal{L}_i \in \mathrm{Pic}(M(c))$ be the line bundle*

$$\mathcal{L}_i := \lambda(u_i(c))$$

for $i \geq 0$. The restriction of the line bundles $\mathcal{L}_i$ to the fibres $\det^{-1}(\mathcal{Q})$ of the determinant $\det : M(c) \to \mathrm{Pic}(X)$ is independent of the choice of x.

Proposition 8.1.10 — *Let $\tau_m : M(c) \to M(c(m))$ be the isomorphism which is induced by $[F] \mapsto [F \otimes \mathcal{O}_X(m)]$. Then*

$$\tau_m^* \mathcal{L}_i \cong \bigotimes_{\nu \geq 0} \mathcal{L}_{i+\nu}^{\binom{m+\nu-1}{\nu}}.$$

Proof. Recall that $[\mathcal{O}_X(m)] = \sum_{\nu \geq 0} \binom{m+\nu-1}{\nu} h^\nu \in K(X)$, and of course $[\mathcal{O}_x] \cdot h^i = 0$ for $i > 0$. Hence

$$u_i(c(m)) \cdot [\mathcal{O}_X(mH)]$$
$$= \left(-r \cdot h^i + \sum_{\nu \geq 0} \binom{m+\nu-1}{\nu} \chi(c \cdot h^{i+\nu}) \cdot [\mathcal{O}_x]\right) \cdot \sum_{j \geq 0} \binom{m+j-1}{j} h^j$$
$$= \sum_{\nu \geq 0} \binom{m+\nu-1}{\nu} \left(-r \cdot h^{i+\nu} + \chi(c \cdot h^{i+\nu}) \cdot [\mathcal{O}_x]\right)$$
$$= \sum_{\nu \geq 0} \binom{m+\nu-1}{\nu} u_{i+\nu}(c).$$

Applying λ we get the isomorphism of the proposition. $\square$

Theorem 8.1.11 — *Let (X, H) be a smooth polarized projective variety, and let c be the class of a torsion free sheaf of rank $r > 0$. For $m \gg 0$ the line bundle $\mathcal{L}_0$ on $M(c(m))$ is relatively ample with respect to the determinant morphism $\det : M(c(m)) \to \mathrm{Pic}(X)$.*

Proof. Recall that in the general set-up explained above for all points $[q : \mathcal{H} \to F] \in R(c)$ the sheaf $F(m)$ is regular and $V \to H^0(F(m))$ is an isomorphism. Hence the universal family yields isomorphisms $V \otimes \mathcal{O}_{R(c)} \cong p_* \widetilde{F}(m)$ and $\det(V) \otimes \mathcal{O}_{R(c)} \cong \det p_!(\widetilde{F}(m))$. Since $u_0(c(m)) = -r \cdot [\mathcal{O}_X] + P(m) \cdot [\mathcal{O}_x]$, we get

$$\lambda_{\widetilde{F}(m)}(u_0(c(m))) \cong \det(V)^{-r} \otimes (\det \widetilde{F}|_{R(c) \times \{x\}})^{P(m)}$$
$$\cong \det(V)^{-r} \otimes (\mathcal{A} \otimes \det{}^* \mathcal{P}|_{\mathrm{Pic}(X) \times \{x\}})^{P(m)}$$

Theorem 4.A.1 says that $\mathcal{A}$ is ample relative to $\mathrm{Pic}(X)$ and that some tensor power of $\mathcal{A}$ descends to a line bundle on $M(c)$ which is again ample relative to $\mathrm{Pic}(X)$. This shows that the line bundle $\mathcal{L}_0$ on $M(c(m))$ is ample relative to $\mathrm{Pic}(X)$. $\square$

Remark 8.1.12 — *i)* If X is of dimension one, only $\mathcal{L}_0$ is non-trivial and the theorem says that $\mathcal{L}_0$ is ample relative to $\mathrm{Pic}(X)$. For the case of a surface only $\mathcal{L}_0$ and $\mathcal{L}_1$ are non-trivial and for $m \gg 0$ the line bundle $\mathcal{L}_0 \otimes \mathcal{L}_1^m$ on the moduli space $M(c)$ is ample on the fibres of $\det : M(c) \to \mathrm{Pic}(X)$.

ii) For later use we point out that the argument above shows that on any fibre of the morphism 'det' the line bundles $\mathcal{A}$ and $\lambda_{\widetilde{F}(m)}(u_0(c(m)))$ are isomorphic as $\mathrm{SL}(V)$-linearized line bundles.

8.2 A Moduli Space for μ-Semistable Sheaves

Let X be a smooth projective surface with an ample divisor H. Fix a class $c \in K(X)_{\mathrm{num}}$ with rank r and Chern classes c_1 and c_2, and a line bundle $\mathcal{Q}$ with $c_1(\mathcal{Q}) = c_1$. Proposition 8.1.10 and Theorem 8.1.11 show that the line bundle $\mathcal{L}_0 \otimes \mathcal{L}_1^m$ is ample on $M(r, \mathcal{Q}, c_2)$ for sufficiently large m. What can be said about $\mathcal{L}_1$ itself? It is clear that the class of $\mathcal{L}_1$ must be contained in the closure of the ample cone. It will be shown that for sufficiently large m the linear system $|\mathcal{L}_1^m|$ is base point free and leads to a morphism from $M(r, \mathcal{Q}, c_2)$ to the Donaldson-Uhlenbeck compactification of the moduli space of μ-stable vector bundles as defined in gauge theory. In fact the main purpose of this section is to construct a moduli space $M^{\mu ss} = M^{\mu ss}(r, \mathcal{Q}, c_2)$ for μ-semistable sheaves. The assertions about the linear system $|\mathcal{L}_1^m|$ on $M(r, \mathcal{Q}, c_2)$ will follow from this.

In order to demonstrate some properties of the linear system $|\mathcal{L}_1^m|$, we study the line bundle $\lambda(u_1)$ in the following examples for two particular families. These provide strong hints which sheaves in M cannot possibly be separated and which on the contrary should be expected to be separable.

Example 8.2.1 — Let E be a torsion free sheaf of rank r on X. For $\ell \geq 0$ consider the scheme $\mathrm{Quot}(E, \ell)$ that parametrizes zero-dimensional quotients of E of length ℓ. There is a universal exact sequence

$$0 \to F \to \mathcal{O}_{\mathrm{Quot}} \otimes E \to T \to 0$$

of families on X parametrized by $\mathrm{Quot}(E, \ell)$. Let c be the class of F_s for some $s \in S$ and let $u_1 = u_1(c)$. From the short exact sequence one gets an isomorphism $\lambda_F(u_1) = \lambda_{q^* E}(u_1) \otimes \lambda_T(u_1)^\vee \cong \lambda_T(u_1)^\vee$. Recall that any zero-dimensional sheaf is semistable, so that T induces a morphism Φ_T from $\mathrm{Quot}(E, \ell)$ to the moduli space $M(\ell) \cong S^\ell(X)$, cf. 4.3.6. Since u_1 is orthogonal to any zero-dimensional sheaf we can apply Theorem 8.1.5 and conclude that $\lambda_F(u_1) = \Phi_T^* \lambda(u_1)^\vee$. We claim that $\lambda(u_1)^\vee$ is an ample line bundle on $S^\ell(X)$. To see this consider the quotient map $\pi : X^\ell \to S^\ell(X)$ for the action of the symmetric group. Let $pr_i : X^\ell \to X$ denote the projection to the i-th factor and $\mathcal{O}_\Delta$ the structure sheaf of the diagonal $\Delta \subset X \times X$. Then $\mathcal{E} := \bigoplus_{i=1}^\ell pr_{i,X}^* \mathcal{O}_\Delta$ is an equivariant flat family of sheaves on X of length ℓ and π is the classifying morphism for $\mathcal{E}$. Clearly

$$\lambda_{\mathcal{E}}(u_1)^\vee \cong \bigotimes_i pr_i^* \det(u_1)^\vee \cong \bigotimes_i pr_i^* \mathcal{O}_X(r \cdot H)$$

is an ample line bundle on X^ℓ. On the other hand $\lambda_{\mathcal{E}}(u_1)^\vee \cong \pi^* \lambda(u_1)^\vee$. Since π is finite, $\lambda(u_1)^\vee$ is ample as well. Since the fibres of Φ_T are connected, we conclude that for sufficiently large n the complete linear system $|\lambda_F(u_1)^n|$ separates points s and s' in $\mathrm{Quot}(E, \ell)$ if and only if $\pi(T_s) \neq \pi(T_{s'})$. Note that if E is μ-semistable or μ-stable then the same holds for all F_s, $s \in \mathrm{Quot}(E, \ell)$. ☐

Example 8.2.2 — Let F' and F'' be coherent sheaves on X of rank r' and r'', respectively. The projective space $\mathbb{P} := \mathbb{P}(\mathrm{Ext}^1(F'', F')^\vee \oplus k)$, parametrizes all extensions of F'' by

F', including the trivial one, $F' \oplus F''$, and there is a tautological family

$$0 \to q^* F' \otimes p^* \mathcal{O}_{\mathbb{P}}(1) \to \mathcal{F} \to q^* F'' \to 0$$

on $\mathbb{P} \times X$. Let $u \in K(X)$ be orthogonal to F'. Then

$$\lambda_{\mathcal{F}}(u) \cong \lambda_{q^* F' \otimes p^* \mathcal{O}_{\mathbb{P}}(1)}(u) \otimes \lambda_{q_* F''}(u) \cong \mathcal{O}_{\mathbb{P}}(1)^{\chi(F' \otimes u)} \cong \mathcal{O}_{\mathbb{P}},$$

since $\chi([F'] \cdot u) = 0$ by assumption. This applies in particular to the following situation: Let F be a μ-semistable sheaf of class c and let $u = u_1(c)$. If $F' \subset F$ is μ-destabilizing, then $[F'] \perp u_1(c)$. The argument above shows that no power of $\lambda_{\mathcal{F}}(u_1)$ can separate F and $F' \oplus F/F'$.   □

We begin with the construction of $M^{\mu ss}$: the family of μ-semistable sheaves of class c is bounded (cf. 3.3.7), so that for sufficiently large m all of them are m-regular. Let $R^{\mu ss} \subset \mathrm{Quot}(\mathcal{H}, P)$ be the locally closed subscheme of all quotients $[q : \mathcal{H} \to F]$ such that F is μ-semistable of rank r, determinant $\mathcal{Q}$ and second Chern class c_2 and such that q induces an isomorphism $V \to H^0(F(m))$. The group $\mathrm{SL}(V)$ acts on $R^{\mu ss}$ by composition. The universal quotient $\tilde{q} : \mathcal{O}_{R^{\mu ss}} \otimes \mathcal{H} \to \widetilde{F}$ allows to construct a line bundle

$$\mathcal{N} := \lambda_{\widetilde{F}}(u_1(c)).$$

on $R^{\mu ss}$.

Proposition 8.2.3 — *There is an integer $\nu > 0$ such that the line bundle $\mathcal{N}^\nu$ is generated by $\mathrm{SL}(V)$-invariant global sections.*

The main technique to prove the proposition consist in the following: if S parametrizes a family F of μ-semistable sheaves, and if $C \in |aH|$ is a general smooth curve and $a \gg 0$, then restricting F to $S \times C$ produces a family of generically μ-semistable sheaves on C (cf. Chapter 7) and therefore a rational map $S \to M_C$ from S to the moduli space M_C of semistable sheaves on the curve C. The ample line bundle $\mathcal{L}_0$ on M_C pulls back to a power of $\lambda_F(u_1(c))$, and in this manner we can produce sections in the latter line bundle. In detail:

Let $i : C \to X$ be a smooth curve in the linear system $|aH|$. For any class $w \in K(X)$, let $w|_C := i^* w$ be the induced class in $K(C)$. In particular, $c|_C$ is completely determined by its rank r and the restriction $\mathcal{Q}|_C$. Clearly, $P' = P(c|_C)$ is also given by $P'(n) = P(c, n) - P(c, n - a)$. Let m' be a large positive integer, $\mathcal{H}' = \mathcal{O}_C(-m')^{P'(m')}$, and let $Q_C \subset \mathrm{Quot}_C(\mathcal{H}', P')$ be the closed subset of quotients with determinant $\mathcal{Q}|_C$. Moreover, let $\mathcal{O}_{Q_C} \otimes \mathcal{H}' \to \widetilde{F}'$ be the universal quotient and consider the line bundle $\mathcal{L}'_0 = \lambda_{\widetilde{F}'}(u_0(c|_C))$ on Q_C. If m' is sufficiently large the following hold:

1. Given a point $[q : \mathcal{H}' \to E] \in Q_C$, the following assertions are equivalent:

    1.1. E is a (semi)stable sheaf and $V \to H^0(E(m'))$ is an isomorphism.

    1.2. $[q]$ is a (semi)stable point in Q_C for the action of $\mathrm{SL}(P'(m'))$ with respect to the canonical linearization of $\mathcal{L}'_0$.

    1.3. There is an integer ν and a $\mathrm{SL}(P'(m'))$-invariant section σ in $(\mathcal{L}'_0)^\nu$ such that $\sigma([q]) \neq 0$,

2. Two points $[q_i : \mathcal{H}' \to E_i]$, $i = 1, 2$ are separated by invariant sections in some tensor power of $\mathcal{L}'_0$, if and only if either both are semistable points but E_1 and E_2 are not S-equivalent or one of them is semistable but the other is not.

Suppose now that F is an S-flat family of μ-semistable torsion free sheaves on X. The assumption that F_s is torsion free for all $s \in S$ implies that the restriction $\mathcal{F} := F|_{S \times C}$ is still S-flat (Lemma 2.1.4) and that there is an exact sequence

$$0 \to F \otimes \mathcal{O}_X(-a) \to F \to \mathcal{F} \to 0 \qquad (8.1)$$

Increasing m' if necessary we can assume that in addition to the assertions 1 and 2 above we also have:

3. $\mathcal{F}_s$ is m'-regular for all $s \in S$.

Then $p_*(\mathcal{F}(m'))$ is a locally free $\mathcal{O}_S$-sheaf of rank $P'(m')$. Let $\pi : \widetilde{S} \to S$ be the associated projective frame bundle. It parametrizes a quotient $\mathcal{O}_{\widetilde{S}} \otimes \mathcal{H}' \to \pi^* \mathcal{F} \otimes \mathcal{O}_\pi(1)$ which in turn induces a $\mathrm{SL}(P'(m'))$-invariant morphism $\Phi_{\mathcal{F}} : \widetilde{S} \to Q_C$. If G is an algebraic group acting on S and if F carries a linearization with respect to this action, then $\widetilde{S}$ inherits a G-action which commutes with the SL-action such that π and $\Phi_{\mathcal{F}}$ are both equivariant for $G \times \mathrm{SL}$.

Before we go on, we need to compare certain determinant line bundles. Consider the following element in $K(X)_{\mathrm{num}}$:

$$w := -\chi(c \cdot h[\mathcal{O}_C]) \cdot 1 + \chi(c \cdot [\mathcal{O}_C]) \cdot h$$

As $[\mathcal{O}_C] = ah - \binom{a}{2}h^2 \in K(X)$ we have

$$
\begin{aligned}
w - w(-a) = w \cdot [\mathcal{O}_C] &= -\chi(c \cdot h[\mathcal{O}_C]) \cdot [\mathcal{O}_C] + \chi(c \cdot [\mathcal{O}_C]) \cdot h[\mathcal{O}_C] \\
&= a^2 \cdot (-\chi(ch^2) \cdot h + \chi(c \cdot h) \cdot h^2) \\
&= a^2 \cdot v_1(c) \equiv a^2 \deg(X) \cdot u_1(c).
\end{aligned}
$$

and

$$w|_C = -\chi(c|_C \cdot h|_C) \cdot 1 + \chi(c|_C) \cdot h|_C = v_0(c|_C) \equiv a \deg(X) \cdot u_0(c|_C)$$

From the short exact sequence (8.1) we get

$$\lambda_F(w(-a))^{\vee} \otimes \lambda_F(w) = \lambda_{\mathcal{F}}(w)$$

and

$$\lambda_{\mathcal{F}}(u_0(c|_C))^{a\,\deg(X)} \cong \lambda_F(w - w(-a)) \cong \lambda_F(u_1(c))^{a^2\,\deg(X)}. \qquad (8.2)$$

Returning to the situation

$$\widetilde{S} \xrightarrow{\Phi_{\mathcal{F}}} Q_C$$
$$\pi \downarrow$$
$$S$$

above we get:

$$
\begin{aligned}
\Phi_{\mathcal{F}}^*(\mathcal{L}_0')^{\deg(C)} &\cong \Phi_{\mathcal{F}}^*(\lambda_{\widetilde{F}'}(v_0(c|_C))) \cong \lambda_{\pi^*\mathcal{F}\otimes\mathcal{O}_{\pi}(1)}(v_0(c|_C)) && \text{by 8.1.2 ii)}\\
&\cong \lambda_{\pi^*\mathcal{F}}(v_0(c|_C)) && \text{by 8.1.2 iv)}\\
&\cong \pi^*\lambda_{\mathcal{F}}(v_0(c|_C)) && \text{by 8.1.2 ii)}\\
&\cong \pi^*\lambda_F(u_1(c))^{a^2\,\deg(X)} && \text{by (8.2)}
\end{aligned}
$$

Assume now that σ is an SL-invariant section in $(\mathcal{L}_0')^{\nu\,\deg(C)}$. Then $\Phi_{\mathcal{F}}^*(\sigma)$ is a $G \times$ SL-invariant section and therefore descends to a G-invariant section in $\lambda_F(u_1(c))^{\nu a^2\,\deg(X)}$. In this way we get a linear map

$$s_F : H^0\left(Q_C, (\mathcal{L}_0')^{\nu\,\deg(C)}\right)^{\mathrm{SL}} \longrightarrow H^0\left(S, \lambda_F(u_1(c))^{\nu a^2\,\deg(X)}\right)^G.$$

We conclude (cf. Theorem 4.3.3 and Definition 4.2.9):

Lemma 8.2.4 —

1. *If $s \in S$ is a point such that $F_s|_C$ is semistable then there is an integer $\nu > 0$ and a G-invariant section $\bar{\sigma}$ in $\lambda_F(u_1(c))^{\nu}$ such that $\bar{\sigma}(s) \neq 0$.*
2. *If s_1 and s_2 are two points in S such that either $F_{s_1}|_C$ and $F_{s_2}|_C$ are both semistable but not S-equivalent or one of them is semistable and the other is not, then there are G-equivariant sections in some tensor power of $\lambda_F(v_1(c))$ that separate s_1 and s_2.* □

Proposition 8.2.3 now follows trivially from the first part of the lemma: Just apply it to the case $S = R^{\mu ss}$ and $G = \mathrm{SL}(V)$. □

If $\mathcal{N}^{\nu}$ is generated by invariant sections, we can also find a finite dimensional subspace $W \subset W_{\nu} := H^0(R^{\mu ss}, \mathcal{N}^{\nu})^{\mathrm{SL}(V)}$ that generates $\mathcal{N}^{\nu}$. Let $\varphi_W : R^{\mu ss} \to \mathbb{P}(W)$ be the induced $\mathrm{SL}(P(m))$-invariant morphism. We claim that

$$M_W := \varphi_W(R^{\mu ss})$$

is a projective scheme. In fact, one has the following general result:

Proposition 8.2.5 — *If T is a separated scheme of finite type over k, and if $\varphi : R^{\mu ss} \to T$ is any invariant morphism, then the image of φ is proper.*

Proof. This is a direct consequence of Langton's Theorem: let $t_0 \in \overline{\varphi(R^{\mu ss})}$ be a closed point. Then there is a discrete valuation ring A with quotient field K and a morphism $f : \text{Spec}(A) \to T$ that maps the closed point ξ_0 to t_0 and the generic point ξ_1 to a point t_1 in the image of φ. Let $y_1 \in \varphi^{-1}(t_1)$ be a closed point in the fibre, then $k(t_1) \subset k(y_1)$ is a finite extension, and there is a finite extension field K' of K and a homomorphism $k(y_1) \to K'$ such that

$$
\begin{array}{ccc}
K' & \leftarrow & k(y_1) \\
\uparrow & & \uparrow \\
K & \leftarrow & k(t_1)
\end{array}
$$

commutes. Let $A' \subset K'$ be a discrete valuation ring that dominates A. Geometrically, $k(y_1) \to K'$ corresponds to a morphism $g' : \text{Spec}(K') \to \text{Spec}(k(y_1)) \to R^{\mu ss}$ and thus to a quotient $[q_{K'} : K' \otimes \mathcal{H} \to F_{K'}]$.

$$
\begin{array}{ccccc}
\text{Spec}(K') & \longrightarrow & & & R^{\mu ss} \\
\downarrow & \searrow & & & \downarrow \\
& & \text{Spec}(A') & & \\
& & \downarrow & & \\
\text{Spec}(K) & \to & \text{Spec}(A) & \to & T
\end{array}
$$

According to Langton's Theorem 2.B.1, the family $F_{K'}$ extends to an A'-flat family $F_{A'}$ of μ-semistable sheaves. Since A' is a local ring and therefore $p_*(F_{A'}(m))$ a free A'-module of rank $P(m)$, there is a quotient $[q_{A'} : A' \otimes \mathcal{H} \to F_{A'}]$. Let $f' : \text{Spec}(A') \to R^{\mu ss}$ be the induced morphism. Since $K' \otimes F_{A'} \cong F_{K'}$, the quotients $K' \otimes q_{A'}$ and $q_{K'}$ differ by an element in $\underline{\text{SL}(V)}(K')$. But φ is an invariant morphism, so that $\varphi \circ f'|_{\text{Spec}(K')} = \varphi \circ g' = f \circ \pi|_{\text{Spec}(K')}$, where $\pi : \text{Spec}(A') \to \text{Spec}(A)$ is the natural projection. Since T is separated, we have $\varphi \circ f' = f \circ \pi$. Thus if ξ_0' is the closed point in $\text{Spec}(A')$, we see that

$$
t_0 = f(\xi_0) = f(\pi(\xi_0')) = \varphi(f'(\xi_0'))
$$

and is therefore contained in the image of φ. $\qquad\square$

Proposition 8.2.6 — *There is an integer $N > 0$ such that $\bigoplus_{\ell \geq 0} W_{\ell N}$ is a finitely generated graded ring.*

Proof. Let $\nu \geq 0$ be an integer such that $\mathcal{N}^\nu$ is generated by a finite dimensional subspace $W \subset W_\nu$. For $d \geq 1$ let W^d be the image of the multiplication map $W \otimes \ldots \otimes W \to W_{d\nu}$, and let $W' \subset W_{d\nu}$ be a finite dimensional space containing W^d. Then W^d and

W' generate $\mathcal{N}^{d\nu}$ and there is a finite morphism $\pi_{W'/W} : M_{W'} \to M_W$ such that $\varphi_W = \pi_{W'/W} \circ \varphi_{W'}$ and $\pi^*_{W'/W} \mathcal{O}_{M_W}(d) \cong \mathcal{O}_{M_{W'}}(1)$. Moreover, there are inclusions

$$
\begin{array}{ccccc}
W^d & \subset & W' & & \\
\cap & & \cap & & \\
H^0(M_W, \mathcal{O}(d)) & \subset & H^0(M_{W'}, \mathcal{O}(1)) & \subset & W_{d\nu}
\end{array}
$$

and $\pi_{W'/W}$ is an isomorphism, if and only if $H^0(M_W, \mathcal{O}(d)) = H^0(M_{W'}\mathcal{O}(1))$. Clearly, the projective system $\{M_W, \pi_{W'/W}\}$ has a limit since it is dominated by $R^{\mu ss}$. If the limit is isomorphic to, say, M_W with $W \subset W_N$, then $H^0(M_W, \mathcal{O}(k)) = W_{kN}$ for all $k \geq 0$. $\square$

Definition 8.2.7 — *Suppose that N is a positive integer as in the proposition above. Let $M^{\mu ss} = M^{\mu ss}(c)$ be the projective scheme*

$$
\mathrm{Proj} \bigoplus_{k \geq 0} H^0(R^{\mu ss}, \mathcal{N}^{kN})^{\mathrm{SL}(P(m))},
$$

and let $\pi : R^{\mu ss} \to M^{\mu ss}$ be the canonically induced morphism.

This resembles very much the GIT construction of Chapter 4. The main difference is that $\mathcal{N}$ is not ample. And indeed, $M^{\mu ss}$ will in general not be a categorical quotient of $R^{\mu ss}$. Still, $M^{\mu ss}$ has a certain universal property. Namely, let $\mathcal{M}^{\mu ss}$ denote the functor which associates to S the set of isomorphism classes of S-flat families of torsion free μ-semistable sheaves of class c on X. It is easy to construct a natural transformation $\mathcal{M}^{\mu ss} \to \underline{M}^{\mu ss}$ with the property that for any S-flat family F of μ-semistable sheaves and classifying morphism $\Phi_F : S \to M^{\mu ss}$ the pull-back of $\mathcal{O}_{M^{\mu ss}}(1)$ via Φ_F is isomorphic to $\lambda_F(u_1(c))^N$. Furthermore, the triple $(M^{\mu ss}, \mathcal{O}(1), N)$ is uniquely characterized by this property up to unique isomorphism and replacing $(\mathcal{O}(1), N)$ by some multiple $(\mathcal{O}(d), dN)$. In particular, the construction of $M^{\mu ss}$ does not depend on the choice of the integer m. We omit the details.

Definition and Theorem 8.2.8 — *Because of the universal property of M the functor morphism $\mathcal{M} \to \mathcal{M}^{\mu ss}$ induces a morphism $\gamma : M \longrightarrow M^{\mu ss}$ such that $\gamma^* \mathcal{O}(1) \cong \mathcal{L}_1^N$.*$\square$

In order to understand the geometry of $M^{\mu ss}$ and the morphism γ better, we need to study the morphism $\pi : R^{\mu ss} \to M^{\mu ss}$ in greater detail and see which points in $R^{\mu ss}$ are separated by π and which are not. The ultimate aim of this section is to show that at least pointwise $M^{\mu ss}$ can be identified with the Donaldson-Uhlenbeck compactification. See also Remark 8.2.17.

Example 8.2.9 — Recall that $M(1, \mathcal{O}_X, \ell) \cong \mathrm{Hilb}^\ell(X)$. According to the calculations in Example 8.2.1, there is an isomorphism $\mathcal{L}_1 \cong g^*\mathcal{L}$, where $g : \mathrm{Hilb}^\ell(X) \to S^\ell(X)$ is the morphism constructed in 4.3.6 and $\mathcal{L}$ is an ample line bundle on $S^\ell(X)$. It follows from Zariski's Main Theorem that $H^0(\text{Hilb}^\ell(X), \mathcal{L}_1^\nu) = H^0(S^\ell(X), \mathcal{L}^\nu)$. This leads to a complete description of the morphism γ in this particular case: $M(1, \mathcal{O}_X, \ell)^{\mu ss} = S^\ell(X)$ and $\gamma = g$.

Definition 8.2.10 — *Let F be a μ-semistable torsion free sheaf on X. Let $gr^\mu F$ be the graded object associated to a μ-Jordan-Hölder filtration of F with torsion free factors. Then $gr^\mu F$ is torsion free. Let F^{**} denote the double dual of $(gr^\mu F)$: it is a μ-polystable locally free sheaf, and let $l_F : X \to \mathbb{N}_0$ be the function $x \mapsto \ell((F^{**}/gr^\mu F)_x)$, which can be considered as an element in the symmetric product $S^l X$ with $l = c_2(F) - c_2(F^{**})$. Both F^{**} and l_F are well-defined invariants of F, i.e. do not depend on the choice of the μ-Jordan-Hölder filtration (cf. 1.6.10).*

Theorem 8.2.11 — *Let F_1 and F_2 be two μ-semistable sheaves of rank r and fixed Chern classes $c_1, c_2 \in H^*(X)$. Then F_1 and F_2 define the same closed point in $M^{\mu ss}$ if and only if $F_1^{**} \cong F_2^{**}$ and $l_{F_1} = l_{F_2}$.*

Proof. One direction is easy to prove: if F is μ-semistable, and if $gr^\mu(F)$ is the torsion free graded object associated to an appropriate μ-Jordan Hölder filtration of F, then we can construct a flat family $\mathcal{F}$ parametrized by $\mathbb{P}^1$ such that $\mathcal{F}_\infty \cong gr^\mu(F)$ and $\mathcal{F}_t \cong F$ for all $t \neq \infty$. Hence the induced classifying morphism $\Phi_\mathcal{F} : \mathbb{P}^1 \to M^{\mu ss}$ maps $\mathbb{P}^1$ to a single point. This means that $[F] = [gr^\mu(F)]$ in $M^{\mu ss}$. We may therefore restrict ourselves to μ-polystable sheaves: let F be μ-polystable torsion free, and let $E = F^{**}$ be its double dual. Then F is (non-uniquely) represented by a closed point y in $\text{Quot}(E, \ell)$, where $\ell = c_2(F) - c_2(E)$. Any other μ-polystable torsion free sheaf F' satisfies the conditions $(F')^{**} = F^{**}$ and $l_F = l_{F'}$ if and only if F' is represented by a closed point y' in $\text{Quot}(E, \ell)$, such that y and y' lie in the same fibre of the morphism $\psi : \text{Quot}(E, \ell) \to S^\ell(X)$. But any such fibre is connected, and as we saw in Example 8.2.1, the restriction of $\mathcal{N}$ to a fibre is trivial. This means that any fibre of ψ is contracted to a single point by the morphism $j : \text{Quot}(E, \ell) \to M^{\mu ss}$ associated to the family F. This proves the 'if'– direction of the theorem.

The 'only if'– direction is done in two steps:

Lemma 8.2.12 — *Let F_1 and F_2 be μ-semistable sheaves on X. If a is a sufficiently large integer and $C \in |aH|$ a general smooth curve, then $F_1|_C$ and $F_2|_C$ are S-equivalent if and only if $F_1^{**} \cong F_2^{**}$.*

Proof. Let $gr^\mu(F_1)$ be the graded object of a μ-Jordan-Hölder filtration of F_1 with torsion free factors. Using the theorems of Mehta-Ramanathan 7.2.8 or Bogomolov 7.3.5

we can choose a so large that the restriction of any summand of F_1^{**} to any smooth curve in $|aH|$ is stable again. Now choose C in such a way that it avoids the finite set of all singular points of $gr^\mu(F_1)$. Then $gr^\mu(F_1)|_C \cong F_1^{**}|_C$ is the graded object of a Jordan-Hölder filtration of $F_1|_C$. This shows that for a general curve C of sufficiently high degree $F_1|_C$ and $F_2|_C$ are S-equivalent if and only if $F_1^{**}|_C \cong F_2^{**}|_C$. For $a \gg 0$ and $i = 0, 1$ we have $\text{Ext}^i(F_1^{**}, F_2^{**}(-C)) = 0$ (and the same with the rôles of F_1^{**} and F_2^{**} exchanged), so that $\text{Hom}_X(F_1^{**}, F_2^{**}) \cong \text{Hom}_C(F_1^{**}|_C, F_2^{**}|_C)$. This means that $F_1^{**}|_C = F_2^{**}|_C$ if and only if $F_1^{**} = F_2^{**}$. $\qquad\square$

In particular, if $F_1^{**} \ncong F_2^{**}$ then any two points in $R^{\mu ss}$ representing F_1 and F_2 can be separated by invariant sections in some tensor power of $\mathcal{N}$ by the second part of Lemma 8.2.4. The most difficult case therefore is that of two sheaves F_1 and F_2 with $F_1^{**} \cong F_2^{**} =: E$ but $l_{F_1} \neq l_{F_2}$. Let $\ell = c_2(F_i) - c_2(E) = \sum_{x \in X} l_{F_i}(x)$. We have already seen that the fibres of the morphism $\text{Quot}(E, \ell) \to S^\ell(X)$ are contracted to points by $j : \text{Quot}(E, \ell) \to M^{\mu ss}$. As $S^\ell(X)$ is normal, $j|_{\text{Quot}(E,\ell)_{red}}$ factors through a morphism $\hat{\jmath} : S^\ell(X) \to M^{\mu ss}$. Clearly, the proof of the theorem is complete if we can show the following proposition:

Proposition 8.2.13 — *The morphism $\hat{\jmath} : S^\ell(X) \to M^{\mu ss}$ is a closed immersion.*

Without further effort, just using what we have proved so far, we can at least state the following: as $\hat{\jmath}^*(\mathcal{O}_{M^{\mu ss}}(1))$ is ample by Example 8.2.2, $\hat{\jmath}$ must be finite. Moreover, using Lemma 8.2.4 and Bogomolov's Restriction Theorem 7.3.5, one can show that j separates points $s, s' \in \text{Quot}(E, \ell)$ if the corresponding zero-dimensional sheaves T, T' have set-theoretically distinct support. Hence $\hat{\jmath}$ is, generically, an embedding. This does not quite suffice to prove the proposition. The path to the proof begins with a detour:

Let $pr_i : X^\ell \to X$ be the projection onto the i-th factor. If $\mathcal{L}$ is an arbitrary line bundle on X, then $\otimes_i pr_i^* \mathcal{L}$ has a natural linearization for the action of the symmetric group S_ℓ and descends to a line bundle $\tilde{\mathcal{L}}$ on $S^\ell(X)$. If $\gamma_1, \dots, \gamma_\ell$ are ℓ global sections, we can form the symmetrized tensor

$$\frac{1}{\ell!} \sum_{\pi \in S_\ell} \gamma_{\pi(1)} \otimes \dots \otimes \gamma_{\pi(\ell)}$$

which descends to a section $\gamma_1 \cdot \dots \cdot \gamma_\ell$ of $\tilde{\mathcal{L}}$. If C is a curve defined by a section γ in $\mathcal{L}$, let $\tilde{C}$ denote the Cartier divisor on $S^\ell(X)$ given by $\gamma \cdot \dots \cdot \gamma$. It is easy to see that if γ runs through an open subset of section in $\mathcal{L}$ then the corresponding sections $\gamma \cdot \dots \cdot \gamma$ span $H^0(S^\ell(X), \tilde{\mathcal{L}})$. Furthermore, if $\mathcal{L}$ is ample, then $\tilde{\mathcal{L}}$ is ample as well.

Lemma 8.2.14 — *i) Let T be an S-flat family of zero-dimensional sheaves on X of length ℓ, inducing a classifying morphism $\Phi_T : S \to S^\ell(X)$. Let $C \subset X$ be a smooth curve and* let $p : S \times X \to S$ be the projection. The exact sequence $T \otimes \mathcal{O}(-C) \to T \to T|_{S \times C}$ induces a homomorphism $\psi : p_* T(-C) \to p_* T$ between locally free sheaves of rank ℓ. Then

$$\{\det(\psi) = 0\} = \Phi_T^{-1}(\tilde{C}) \tag{8.3}$$

ii) Moreover, if S is integral, and if $T_s \cap C = \emptyset$ for some (and hence general) $s \in S$, then $p_*(T|_{S \times C})$ is a torsion sheaf on S of projective dimension 1. If

$$0 \to \mathcal{A} \xrightarrow{\psi} \mathcal{B} \to p_*(T|_{S \times C}) \to 0$$

is any resolution by locally free sheaves $\mathcal{A}$ and $\mathcal{B}$ of necessarily the same rank, then (8.3) holds for ψ.

Proof. i) Let $\pi : \mathrm{Drap} \to S$ denote the relative flag scheme (cf. 2.A.1) of all full flags

$$0 \subset F_1 T \subset \ldots \subset F_\ell T = T_s, \quad s \in S.$$

The factors of the universal flag parametrized by Drap have length one and induce a morphism $\tilde{\Phi}_T : \mathrm{Drap} \to X^\ell$ so that the diagram

$$
\begin{array}{ccc}
\mathrm{Drap} & \xrightarrow{\tilde{\Phi}_T} & X^\ell \\
\pi \downarrow & & \pi \downarrow \\
S & \xrightarrow{\Phi} & S^\ell(X)
\end{array}
$$

commutes. As S is the scheme-theoretic image of $\pi : \mathrm{Drap} \to S$, it suffices to prove (8.3) for $\pi^{-1}(\psi)$ instead of ψ. Now $\pi^{-1}(\psi)$ has diagonal form with respect to the filtrations $p_* F_\bullet T(-C)$ and $p_* F_\bullet T$ of $\pi^* p_* T(-C)$ and $\pi^* p_* T$, respectively. Hence if ψ_i, $i = 1, \ldots, \ell$, are the induced maps on the factors, we have $\pi^{-1}(\det(\psi)) = \det(\pi^{-1}(\psi)) = \prod_i \det(\psi_i)$. As $\pi^{-1}(\tilde{C}) = \sum_i pr_i^{-1}(C)$, it suffices to show (8.3) for each ψ_i instead of ψ, i.e. for the case $\ell = 1$. But this case can immediately be reduced to the case $S = X$, $T = \mathcal{O}_\Delta$, when the assertion is obvious.

ii) It is clear that under the given assumptions $p_* T|_{S \times C}$ is a torsion sheaf. Hence, the homomorphism $p_* T(-C) \to p_* T$ is generically isomorphic and therefore injective everywhere, so that indeed $p_* T|_{S \times C}$ has projective dimension 1. It is a matter of local commutative algebra to see that the Cartier divisor defined by $\det(\psi)$ is independent of the resolution. $\qquad \square$

Approaching our original goal, let E be a locally free sheaf and consider the variety $S = \mathrm{Quot}(E, \ell)$ parametrizing a tautological families F and T that fit into an exact sequence

$$0 \to F \to \mathcal{O}_S \otimes E \to T \to 0.$$

Let $\Phi_T : \mathrm{Quot}(E,\ell) \to S^\ell(X)$ be the morphism associated to T. Let $C \in |aH|$ be an arbitrary smooth curve, and let G be a locally free sheaf on X with the property that $H^1(F_s \otimes G|_C) = 0 = H^1(E \otimes G|_C)$ for all $s \in S$. Then there is a short exact sequence

$$0 \longrightarrow p_*(F \otimes G|_C) \overset{\psi}{\longrightarrow} p_*(\mathcal{O}_S \otimes (E \otimes G)|_C) \longrightarrow p_*(T \otimes G|_C) \longrightarrow 0.$$

As the conditions of part *ii)* of the lemma are satisfied, we get

$$\mathrm{div}(\det(\psi)) = \Phi_{G \otimes T}^{-1}(\tilde{C}) = \mathrm{rk}(G) \cdot \Phi_T^{-1}(\tilde{C}).$$

Now locally free sheaves G of the type above span $K(X)$. Hence by linearity we get the following result:

Lemma 8.2.15 — *For any $w \in K(X)$ the following holds: the homomorphism $F \to \mathcal{O}_S \otimes E$ induces a rational homomorphism*

$$\bar{\psi} : \lambda_{F|_{S \times C}}(w) \longrightarrow \lambda_{\mathcal{O}_S \otimes E|_C}(w) \cong \mathcal{O}_S$$

with $\mathrm{div}(\bar{\psi}) = \mathrm{rk}(w) \cdot \Phi_T^{-1}(\tilde{C})$, i.e. $\bar{\psi}$ has zeros or poles depending on the sign of $\mathrm{rk}(w)$. $\square$

Proof of Proposition 8.2.13. Let now E be a μ-polystable locally free sheaf of rank r and determinant $\mathcal{Q}$, and $\ell = c_2(c) - c_2(E) \geq 0$. If a is sufficiently large, and if $C \in |aH|$ is an *arbitrary* smooth curve, then $E|_C$ is again polystable by Bogomolov's Restriction Theorem 7.3.5. The two families F and $E_S = \mathcal{O}_S \otimes E$ on $S = \mathrm{Quot}(E,\ell)$ induce homomorphisms

$$s_{E_S} : H^0\left(Q_C, (\mathcal{L}_0')^{\nu a \deg(X)}\right)^{\mathrm{SL}} \to H^0\left(S, \lambda_{E_S}(u_1(E))^{\nu a^2 \deg(X)}\right) = H^0(S, \mathcal{O}_S)$$

and

$$s_F : H^0\left(Q_C, (\mathcal{L}_0')^{\nu a \deg(X)}\right)^{\mathrm{SL}} \to H^0\left(S, \lambda_F(u_1(c))^{\nu a^2 \deg(X)}\right).$$

On the complement U of $\Phi_T^{-1}(\tilde{C})$ in S the two line bundles on the right hand side are isomorphic and $s_{E_S}(\sigma)|_U = s_F(\sigma)|_U$ for any invariant section σ. Moreover, the rational homomorphism $\bar{\psi}$ maps $s_F(\sigma)$ to $s_{E_S}(\sigma)$. Since $E|_C$ is polystable, there is an integer ν and a section $\sigma_0 \in H^0(Q_C, (\mathcal{L}_0')^{\nu a \deg(X)})^{\mathrm{SL}}$ such that $s_{E_S}(\sigma_0) \neq 0$. Therefore, $s_F(\sigma_0)$ must have zeros of precisely the same order as the poles of $\bar{\psi}$ up to an additional factor $n := \nu a^2 \deg(X)$. Hence, Lemma 8.2.15 says that the vanishing divisor of $s_F(\sigma_0)$ equals $n \cdot r \cdot \Phi_T^{-1}(\tilde{C})$. We finally conclude: σ_0 induces a section σ_0' in some tensor power of $\mathcal{O}_{M^{\mu ss}}(1)$ such that the vanishing divisor associated to $\hat{\jmath}^{-1}(\sigma_0')$ on $S^\ell(X)$ is a multiple of $\tilde{C}$. But we have seen before that these divisors span a very ample linear system as C runs through all smooth curves in the linear system $|aH|$ for sufficiently large a. Hence $\hat{\jmath}$ is an embedding. $\square$

Corollary 8.2.16 — $\gamma : M \to M^{\mu ss}$ *embeds the open subscheme* $M^{\mu,lf} \subset M$ *of μ-stable locally free sheaves. In particular,* $d := \dim \gamma(M) \geq \dim(M^{\mu,lf})$ *and*

$$h^0(M, \mathcal{L}_1^\ell) \sim \ell^d.$$

Proof. Theorem 8.2.11 implies that $\gamma|_{M^{\mu,lf}}$ is injective. But in fact, the proof of Lemma 8.2.12 shows that $M^{\mu,lf}$ embeds into the moduli space of stable sheaves on C, where C is any smooth curve in $|aH|$ for sufficiently large C, which implies that $\gamma|_{M^{\mu,lf}}$ is an embedding. The second assertion is clear, as $\mathcal{O}_{M^{\mu ss}}(1)$ is ample. $\qquad\square$

Remark 8.2.17 — Let $M^{\mu-\text{poly}}(r, \mathcal{Q}, c_2) \subset M(r, \mathcal{Q}, c_2)$ denote the subset representing μ-polystable locally free sheaves. The previous results can be interpreted as follows: set-theoretically, there is a stratification

$$M^{\mu ss}(r, \mathcal{Q}, c_2) = \coprod_{\ell \geq 0} M^{\mu-\text{poly}}(r, \mathcal{Q}, c_2 - \ell) \times S^\ell(X).$$

We will briefly indicate how this is related to gauge theory: In order to study differentiable structures on a simply connected real 4-dimensional smooth manifold N, Donaldson introduced moduli spaces $M_N^{asd}(2, 0, c_2)$ of irreducible antiselfdual SU(2)-connections in a C^∞-complex vector bundle with second Chern class c_2 on N, equipped with a Riemannian metric. He proved that if N is the underlying C^∞-manifold of a smooth complex projective surface X with the Hodge metric, then there is an analytic isomorphism between $M_N^{asd}(2, 0, c_2)$ and the moduli space $M_X^{\mu,lf}(2, 0, c_2)$ of μ-stable locally free sheaves on X of rank 2 and the given Chern classes. In general, the space M^{asd} is not compact. As Donaldson pointed out, results of Uhlenbeck can be interpreted as follows: the disjoint union

$$\coprod_{\ell \geq 0} M_N^{asd}(2, 0, c_2 - \ell) \times S^\ell(N)$$

can be given a natural topology which makes the disjoint union a compact space and induces the given topology on each stratum. The closure of M^{asd} in this union is called the Donaldson-Uhlenbeck compactification. Li [148] and Morgan [181] show that there is a homeomorphism $\gamma(M) \longrightarrow \overline{M^{asd}}$ extending the analytic isomorphism $M^{\mu,lf} \to M^{asd}$ constructed by Donaldson. For more information on the relation to gauge theory, see the books of Donaldson and Kronheimer [46] and Friedman and Morgan [71] and the references given there.

8.3 The Canonical Class of the Moduli Space

Let $M_0 \subset M(r, \mathcal{Q}, c_2)$ be the open subscheme of stable sheaves F with rank r, determinant $\mathcal{Q}$, second Chern class c_2 and $\mathrm{Ext}^2(F, F)_0 = 0$. According to Theorem 4.5.4, M_0 is smooth of expected dimension $\Delta - (r^2 - 1) \cdot \chi(\mathcal{O}_X)$. We will later see that M_0 is dense in $M(r, \mathcal{Q}, c_2)$ for sufficiently large discriminant Δ and that the complement has large codimension. The purpose of this section is to relate the canonical bundle of M_0 to the line bundle $\mathcal{L}_1$ studied in the last section. The main technical tool here is the Grothendieck-Riemann-Roch formula. It states that for any class $\beta \in K^0(X \times S)$ one has

$$ch(p_! \beta) = p_*(ch(\beta).q^* td(X)) \text{ in } CH^*(S)_{\mathbb{Q}}.$$

(Recall that $p : S \times X \to S$ and $q : S \times X \to X$ are the projections.)

Let F be an S-flat family of sheaves on X. Then there is a bounded complex $F^\bullet$ of locally free sheaves which is quasi-isomorphic to F, and

$$[\mathcal{E}xt_p(F, F)] = \sum_{i \geq 0} (-1)^i [\mathcal{E}xt_p^i(F, F)] = p_!(F^{\bullet\vee} \otimes F^\bullet)$$

is an element in $K^0(S)$. If $\gamma \in CH^*(S)_{\mathbb{Q}}$, let γ_i denote the homogeneous component of γ of degree i.

Proposition 8.3.1 — *Let F be an S-flat family of sheaves on X of rank r, determinant $\mathcal{Q}$ and Chern classes c_1 and c_2. Let*

$$\Delta(F) = 2rc_2(F) - (r-1)c_1(F)^2 \in CH^*(S \times X)$$

denote the discriminant of the family F. Then the following equations hold in $CH^*(S)_{\mathbb{Q}}$.
i) $c_1([\mathcal{E}xt_p(F, F)]) = \frac{1}{2} \{p_*(\Delta(F).q^* K_X)\}_1.$
ii) $c_1(\lambda_F(u_1)) = \frac{1}{2} \{p_*(\Delta(F).q^* H)\}_1.$

Proof. Both results are direct applications of the Grothendieck-Riemann-Roch formula.
i) By Grothendieck-Riemann-Roch we have

$$c_1([\mathcal{E}xt_p(F, F)]) = c_1(p_!(F^{\bullet\vee} \otimes F^\bullet)) = \left\{ p_*(ch(F^{\bullet\vee}).ch(F^\bullet).q^* td(X)) \right\}_1.$$

As these Chern class calculations are purely formal, we can use the identity (3.4) on page 79 and write

$$ch(F^{\bullet\vee}).ch(F^\bullet) = r^2 - c_2(F^{\bullet\vee} \otimes F^\bullet) + \ldots = r^2 - \Delta(F) + \ldots,$$

where the dots $\ldots$ indicate terms of degree ≥ 4. On the other hand

$$td(X) = 1 - \frac{1}{2} K_X + \frac{1}{12}(c_1^2(X) + c_2(X)).$$

Hence the only term of degree 3 in $ch(F^{\bullet\vee}).ch(F^\bullet).q^*td(X)$ is $\frac{1}{2}\Delta(F).q^*(K_X)$, and terms of other degrees do not contribute to the left hand side of the equation in *i*).

ii) By definition, $u_1 = -r \cdot h + \chi(c \cdot h) \cdot [\mathcal{O}_x]$. Since $\chi(c \cdot h) = c_1.H - \frac{r}{2}H^2 - \frac{r}{2}H.K_X$ one gets

$$
\begin{aligned}
ch(u_1) &= -r \cdot ch(h) + \chi(c \cdot h) \cdot ch(\mathcal{O}_x) \\
&= -r\left(H - \frac{1}{2}H^2\right) + \left(c_1.H - \frac{r}{2}H^2 - \frac{r}{2}H.K_X\right) \\
&= -rH + c_1.H - \frac{r}{2}H.K_X
\end{aligned}
$$

and therefore $ch(u_1).td(X) = -rH + c_1.H$. The assumption that the family F has fibrewise determinant $\mathcal{Q}$ implies that $\det(F) = p^*\mathcal{S} \otimes q^*\mathcal{Q}$ for some line bundle $\mathcal{S}$ on S, so that $c_1(F) = p^*c_1(\mathcal{S}) + q^*c_1(\mathcal{Q}) =: p^*s + q^*c_1$. Now

$$
c_1(\lambda_F(u_1)) = c_1(p_!(F \cdot q^*u_1)) = \{p_*(ch(F).q^*(ch(u_1).td(X)))\}_1 .
$$

After expansion of $\{ch(F).q^*(ch(u_1).td(X)) - \frac{1}{2}\Delta(F).q^*H\}_3$ and cancellation of most terms the only thing left is

$$
\frac{1}{2}(c_1(F)^2 - 2c_1(F).q^*c_1).q^*H = \frac{1}{2}(p^*s^2 - q^*c_1^2).q^*H = \frac{1}{2}p^*s^2.q^*H.
$$

Integration of this term along the fibres of p gives 0, as asserted. $\qquad\square$

As an immediate consequence of the proposition we see that if K_X and H are linearly dependent over $\mathbb{Q}$, i.e. if $K_X = \varepsilon \cdot H \in \mathrm{Pic}(X) \otimes \mathbb{Q}$, then, under the hypotheses of the proposition, one also has $c_1([\mathcal{E}xt_p(F, F)]) = \varepsilon \cdot c_1(\lambda_F(u_1))$. We can reduce to the following cases:

1. $\varepsilon = -1 \Leftrightarrow -K_X$ is ample, i.e. X is a Del Pezzo surface.

2. $\varepsilon = 0 \Leftrightarrow X$ is a minimal surface of Kodaira dimension 0.

3. $\varepsilon = 1 \Leftrightarrow K_X$ is ample, i.e. X is a minimal surface of general type without (-2)-curves.

Let $M_0 \subset M = M(r, \mathcal{Q}, c_2)$ be the open subset of points F where F is a stable sheaf with $\mathrm{Ext}^2(F, F)_0 = 0$, and let R_0 be the pre-image of M_0 under the quotient morphism $\pi : R \to M$. Moreover, let $\widetilde{F}$ denote the universal family on $R_0 \times X$. Then

Theorem 8.3.2 — $\pi^*K_{M_0} \cong \det[\mathcal{E}xt_p(\widetilde{F}, \widetilde{F})]$.

This is a direct consequence of Theorem 10.2.1. Here, we must appeal to the patience of the reader. $\qquad\square$

Theorem 8.3.3 — *Let* (X, H) *be a polarized projective surface with* $K_X = \varepsilon \cdot H$, $\varepsilon = -1, 0$ *or* 1. *Then* $K_{M_0} \cong \mathcal{L}_1^\varepsilon$ *modulo torsion line bundles.*

Proof. It follows from the discussion above and the theorem, that

$$\pi^* K_{M_0} = \det([\mathcal{E}xt_p(\widetilde{F}, \widetilde{F})]) \cong \lambda_{\widetilde{F}}(u_1)^\varepsilon = \pi^* \lambda(u_1)^\varepsilon = \pi^* \mathcal{L}_1^\varepsilon$$

modulo torsion line bundles on R_0. As π^* is injective (4.2.16), the assertion of the theorem follows. $\square$

Note that we can state the isomorphism of the theorem only up to torsion line bundles, because the Chern class computations above were carried out in $CH^*(R_0)_\mathbb{Q}$.

Combining Theorem 8.3.3 and Corollary 8.2.16 we see that the canonical bundle on the moduli space of μ-stable vector bundles is anti-ample for Del Pezzo surfaces, is torsion for minimal surfaces of Kodaira dimension zero, and is ample for surfaces with ample canonical bundle. This gives strong evidence that the moduli spaces of higher rank sheaves detect the place of the surface in the Enriques classification.

Comments:
— The homomorphism λ was introduced by Le Potier [144]. Theorem 8.1.5 is taken from that paper. Le Potier also shows that $\mathcal{L}_1^N$ is globally generated for sufficient divisible N, and that the induced morphism $\varphi_{|\mathcal{L}_1^N|}$ separates the open part of μ-stable locally free sheaves from its complement in the moduli space. His approach is a generalization of [52]. The comparison with the Donaldson-Uhlenbeck compactification in the case of rank two sheaves with trivial determinant was done by J. Li [148]. The line bundle used by Li can be compared to $\mathcal{L}_1$ by the following lemma:

Lemma 8.3.4 — *If* $c_1 = 0$ *and* $r = 2$, *then for any smooth curve* $C \in |kH|$ *and* $\theta_C \in \mathrm{Pic}^{g(c)-1}(C)$ *one has* $[\theta_C] = -\frac{k}{2} u_1$ *as classes in* $K(X)$. *In particular, if a universal family* $\mathcal{E}$ *exists, then* $\mathcal{L}_1^k = \det p_!(\mathcal{E} \otimes q^* \theta_C)^{-2}$.

— The construction of the 'moduli space' of μ-semistable sheaves is essentially contained in J. Li's paper [148]. The proof of 8.2.11 is a mixture of methods from [144] and [148], though in order to prove an equivalent of 8.2.13, Li varies the curve $C \in |aH|$ and uses relative moduli spaces for one-dimensional families of curves, instead of varying $[F]$ in $\mathrm{Quot}(E, \ell)$ as we did in the proof presented in these notes. One should also mention that both approaches of Le Potier and Li were motivated by Donaldson's non-vanishing result [47]. Li also shows in [148] that the image of $\gamma : M \to M^{\mu ss}$ is homeomorphic to the Donaldson-Uhlenbeck compactification in gauge theory. For this see also the work of Morgan [181].
— The surjectivity of the map λ (Theorem 8.1.6) for $q(X) = 0$ can be deduced from Li's results in [151] in the case of rank two sheaves. He developes a more general technique to produce line bundles by starting with the K-group of the product $X \times X$. His proof relies on the computation of the second cohomology of the moduli space via gauge theory [150]. It would be nice to have an algebraic argument of this part. In the description of the Picard group of the moduli space of curves the same problem arises. In order to show the surjectivity of a natural map to the Picard group one uses transcendental information about the second cohomology.

— For information related to Lemma 8.2.14 see the paper of Knudson and Mumford [126].
— For details about the Grothendieck-Riemann-Roch formula see the book of Fulton [73].
— The identification (8.3.1, 8.3.3) of the canonical class of the 'good' part of the moduli space in the rank two case was done in [149] and [112].

8.4 Further comments (second edition)

I. Theta functions.

In the comments to the first edition of the book we have mentioned that Faltings had given in [61] a GIT-free approach to the construction of moduli spaces of bundles on curves via generalized theta-functions. See [320] for an update.

In [267] it was shown by Álvarez-Cónsul and King that also in higher dimension theta functions exist in abundance and can be used to embed the moduli space of semistable sheaves explicitly. To be more precise let us consider the moduli space $M(c)$ as in Section 8.1 and consider the class $u_{AK} := ([U_0 \otimes \mathcal{O}_X(-n)] - [U_1 \otimes \mathcal{O}_X(-m)])^\vee$. Here U_0 and U_1 are vector spaces of dimension $P(n)$ respectively $P(m)$, where $P(n) = \chi(c(n))$ is the Hilbert polynomial of semistable sheaves of class c. Tacitly we assume that $m \gg n \gg 0$ as in Section 4.D. Then $u_{AK} \in K_{c,H}$ in the sense of Definition 8.1.4. Thus by Theorem 8.1.5 we know that $\lambda(u_{AK})$ defines a line bundle on $M(c)$. The first remark is that by the construction method used in [267] one has:

The line bundle $\lambda(u_{AK}) \in \mathrm{Pic}(M(c))$ is ample.

The even more interesting aspect of this line bundle is that global sections can be produced rather explicitly as follows. Consider homomorphisms $\delta : U_1 \otimes \mathcal{O}_X(-m) \to U_0 \otimes \mathcal{O}_X(-n)$.

To every such δ one can associate naturally a global section $\theta_\delta \in H^0(M(c), \lambda(u_{AK}))$. Moreover, θ_δ vanishes in $[F] \in M(c)$ if and only if the induced map $\mathrm{Hom}(U_0, F(n)) \to \mathrm{Hom}(U_1, F(m))$ is not bijective.

The idea of constructing sections of determinant bundles in this way is classical (see e.g. [146]), but what is less clear and had been carried through only for curves by Faltings is that these sections yield a projective embedding of the moduli space. This is the main result of [267], Sect. 7, which is based on the construction of the moduli space via the embedding into the moduli space of representations of a certain Kronecker quiver (see Section 4.D). We restrict to the case of characteristic zero.

There exist finitely many homomorphisms $\delta_0, \dots, \delta_N : U_1 \otimes \mathcal{O}_X(-m) \to U_0 \otimes \mathcal{O}_X(-n)$ such that the induced sections yield a projective embedding:

$$(\theta_{\delta_0}, \dots, \theta_{\delta_N}) : M(c) \to \mathbb{P}^N.$$

The problem of extending Faltings' construction to higher dimensions has also been studied by Hein in [333, 334]. It is intimately related to the so-called strange duality conjecture, which shall be discussed next.

II. Strange duality.

Strange duality is a phenomenon that has been intensively studied for moduli spaces of bundles on curves. We will only mention a few aspects of the corresponding theory for surfaces which is less well developed. For the case of curves see [279] and the references in [380]. The theory for surfaces was initiated by J. Le Potier who studied the case of the projective plane. We recommend [423] for a survey of the subject and comments on its history.

For our discussion let X be an algebraic surface and let $c, c' \in K(X)_{\text{num}}$ with $\chi(c \cdot c') = 0$. In other words, $c \in K_{c'}$ and $c' \in K_c$. Consider the two moduli spaces $M(c)$ and $M(c')$ of semistable sheaves with Chern character c respectively c'. We suppress the choice of the ample divisors H and H' with respect to which stability is considered. To simplify even further we shall tacitly assume $c \in K_{c',H'}$ and $c' \in K_{c,H}$. Then by Theorem 8.1.5 the two line bundles $\lambda(c') \in \operatorname{Pic}(M(c))$ and $\lambda(c) \in \operatorname{Pic}(M(c'))$ are well defined. Strange duality is a conjectural relation between the space of global sections $H^0(M(c), \lambda(c')^{\vee})$ and $H^0(M(c'), \lambda(c)^{\vee})$.

To establish any relation between these two spaces of global sections on a priori completely unrelated moduli spaces one could try to compute their dimensions. Another approach would be to construct a natural map between these two spaces. In an ideal world one could apply the following construction. Suppose that for all $E \in M(c)$ and for all $F \in M(c')$ one has $H^2(E \otimes F) = 0$. Then consider the closed subvariety D in the product $M(c) \times M(c')$ defined as $D = \{(E, F) \mid H^0(E \otimes F) \neq 0\}$. Due to the assumption $\chi(c \cdot c') = 0$, we expect D to be proper and, in fact, to be of codimension one. The associated line bundle can be obtained by a universal construction similar to the one for λ. Consider the product $M(c) \times X \times M(c')$ with the projections p_i and p_{ij}. Suppose both moduli spaces are fine with universal sheaves $\mathcal{E}$ on $M(c) \times X$ and $\mathcal{E}'$ on $X \times M(c')$ and consider $\mathcal{F} := p_{12}^* \mathcal{E} \otimes p_{23}^* \mathcal{E}'$. For simplicity we will assume that all higher Tor-groups for this tensor product are trivial. Then let $\Theta \in \operatorname{Pic}(M(c) \times M(c'))$ be the dual of $\det Rp_{13*}(\mathcal{F})$. Under all our (over)simplifying assumptions one has $\mathcal{O}(D) = \Theta$. Moreover, if X is regular, then $\Theta = \lambda(c)^{\vee} \boxtimes \lambda(c')^{\vee}$. (The argument behind this fact is that $\lambda(c)$ only depends on the numerical class of c and is similar to the computation in Example 8.1.8.) Thus $D \in H^0(M(c), \lambda(c)^{\vee}) \otimes H^0(M(c'), \lambda(c)^{\vee})$ which can also be interpreted as a homomorphism

$$D : H^0(M(c), \lambda(c')^{\vee})^{\vee} \to H^0(M(c'), \lambda(c)^{\vee}). \tag{8.4}$$

Strange duality now asks

Under which conditions is (8.4) bijective?

Not surprisingly, the answer depends very much on the type of surfaces. It has been studied in quite some detail for the projective plane, abelian surfaces and (elliptic) K3 surfaces. E.g. in [309] the bijectivity of (8.4) has been shown in many case for $X = \mathbb{P}^2$. In [382] the bijectivity of (8.4) has been shown for elliptic K3 surfaces under additional assumptions, e.g. that $c_1(c).f = c_1(c').f = 1$ where f is the class of the elliptic fibre. In [380] numerical evidence for the bijectivity of (8.4) for general K3 surfaces has been deduced from [318]. Similar considerations for abelian surfaces can be found in [381].

9

Irreducibility and Smoothness

For small discriminants, moduli spaces of semistable sheaves can look rather wild: their dimension need not be the expected one, they need not be irreducible nor need they be reduced let alone non-singular. This changes if the discriminant increases: the moduli spaces become irreducible, if we fix the determinant, normal, of expected dimension, and the codimension of the locus of points which are singular or represent μ-unstable sheaves increases. This behaviour is the subject of the present chapter. The results in this chapter are due to Gieseker, Li and O'Grady. Our main source for the presentation is O'Grady's article [209].

Let X be a smooth projective surface, H a very ample divisor on X, and K a canonical divisor. We write $\mathcal{O}_X(1) = \mathcal{O}_X(H)$ for the corresponding line bundle.

9.1 Preparations

Fix a rank $r \geq 2$, a line bundle $\mathcal{Q} \in \text{Pic}(X)$ and Chern classes $c_1 = c_1(\mathcal{Q}), c_2$. Let $\Delta = 2rc_2 - (r-1)c_1^2$ and P be the associated discriminant and Hilbert polynomial, respectively. Let $M = M(\Delta)$ be short for the moduli space $M_X(r, \mathcal{Q}, c_2)$. By the Bogomolov Inequality 3.4.1 $M(\Delta)$ is empty, unless $\Delta \geq 0$, as we will assume from now on. Recall some elements of the construction of $M(\Delta)$ in Section 4.3: there is an integer $m \gg 0$ such that the following holds: Let $\mathcal{H} = k^{P(m)} \otimes \mathcal{O}_X(-m)$ and let $R \subset \text{Quot}_X(\mathcal{H}, P)$ be the locally closed subscheme consisting of those quotients $q : \mathcal{H} \to F$ where F is semistable, $H^0(\mathcal{H}(m)) \to H^0(F(m))$ is an isomorphism, and $\det(F) \cong \mathcal{Q}$. Then there is a morphism $\pi : R \to M$ such that M is a good quotient for the $\text{SL}(P(m))$-action on R. (These notations differ slightly from those in Chapter 4 as we have fixed the determinant!)

Let e be a nonnegative real number. Let $R(e)$ be the closed subset in R of quotients $\mathcal{H} \to F$, where F is e-unstable. (For e-stability see 3.A). This set is certainly invariant under the group action, so that $M(e) := \pi(R(e))$ is closed as well.

239

Theorem 9.1.1 — *There is a constant $B = B(r, H, X)$ such that*

$$\dim R(e) \leq d(e) + \text{end}(\mathcal{H}) - 1$$
$$\dim M(e) \leq d(e) + r^2 - 1$$

with $d(e) = (1 - \frac{1}{2r})\Delta + (3r - 1)e^2 + \frac{r[KH]_+}{2|H|}e + B$.

Proof. Let F be a semistable sheaf with $[F] \in R(e)$. Then there is a filtration $0 \subset \mathcal{H}_0 \subset \mathcal{H}_1 \subset \ldots \subset \mathcal{H}_\ell = \mathcal{H}$ such that $F_1 := \mathcal{H}_1/\mathcal{H}_0$ is an e-destabilizing submodule of $F = \mathcal{H}/\mathcal{H}_0$, i.e. $\mu(F_1) \geq \mu(F) - e|H|/\text{rk}(F_1)$, and such that $\mathcal{H}_2/\mathcal{H}_1 \subset \ldots \subset \mathcal{H}/\mathcal{H}_1$ is the Harder-Narasimhan filtration of F/F_1. This filtration defines a point y in the flag-scheme $Y = \text{Drap}(\mathcal{H}, P_\bullet)$, with $P_i = P(\mathcal{H}_i/\mathcal{H}_{i-1})$ for $i = 0, \ldots, \ell$. There is a natural morphism $f : Y \to R$ given by forgetting all of the flag except $\mathcal{H}_0$, and $R(e)$ is the union of the images of all Y appearing in this way. By Grothendieck's Lemma 1.7.9 the number of such flag-schemes is finite. In order to bound the dimension of $R(e)$ it is therefore enough to bound the dimension of Y. By Proposition 2.A.12 and the definition of the groups $\text{Ext}_\pm$ in 2.A.3 one has

$$\dim(Y) \leq \text{ext}^0_+(\mathcal{H}, \mathcal{H}) \leq \text{end}(\mathcal{H}) - 1 + \text{ext}^1_-(F, F).$$

The estimate for $\dim R(e)$ follows from this and Proposition 3.A.2. Any fibre of $\pi : R \to M$ contains a closed orbit whose dimension is given by the difference of $\text{end}(\mathcal{H})$ and the dimension of the stabilizer of a polystable sheaf of rank r. The dimension of this stabilizer is bounded by r^2. Hence for any point $[F] \in M$ one has $\dim \pi^{-1}([F]) \geq \text{end}(\mathcal{H}) - r^2$, and therefore $\dim M(e) \leq \dim R(e) - (\text{end}(\mathcal{H}) - r^2)$. This proves the second claim. $\square$

Recall (cf. 4.5.8) that there is a number β_∞ such that for any point $[F] \in M^s(\Delta)$ one has dimension bounds

$$\Delta - (r^2 - 1)\chi(\mathcal{O}_X) \leq \dim_{[F]} M \leq \Delta - (r^2 - 1)\chi(\mathcal{O}_X) + \beta_\infty.$$

Using the theorem above we can, at least for sufficiently large discriminant Δ, exclude the possibility of irreducible components in M which parametrize semistable sheaves which are not μ-stable. Let R^μ and M^μ denote the open subschemes of μ-stable sheaves in R and M, respectively.

Theorem 9.1.2 — *If $\Delta - (r^2 - 1)\chi(\mathcal{O}_X) > (1 - \frac{1}{2r})\Delta + B$, then R^μ and M^μ are dense in R and M, respectively. In particular, $\dim Z \geq \Delta - (r^2 - 1)\chi(\mathcal{O}_X)$ for all irreducible components Z of $M(\Delta)$. Moreover, $\text{codim}(M \setminus M^\mu, M) \geq \frac{1}{2r}\Delta - (r^2 - 1)(\chi(\mathcal{O}_X) + 1) - B$.*

Proof. By definition, $R - R^\mu = R(0)$. The assumption of the theorem and the dimension bound for $R(0)$ of Theorem 9.1.1 give:

$$\dim R(0) \ \leq \ d(0) + \text{end}(\mathcal{H}) - 1 = \left(1 - \frac{1}{2r}\right)\Delta + \text{end}(\mathcal{H}) - 1 + B$$
$$< \ \Delta - (r^2 - 1)\chi(\mathcal{O}_X) + \text{end}(\mathcal{H}) - 1$$

By Proposition 4.5.9 for any point $[\varphi] \in R$ one has

$$\Delta - (r^2 - 1)\chi(\mathcal{O}_X) + \text{end}(\mathcal{H}) - 1 \leq \dim_{[\varphi]} R.$$

Therefore the μ-unstable locus in R is of smaller dimension than any component of R, which means that R^μ is dense in R. Hence M^μ is dense in M, too, and the remaining two estimates of the theorem follow from $\dim M^s \geq \exp \dim M(\Delta) = \Delta - (r^2 - 1)\chi(\mathcal{O}_X)$ and $\dim M(0) \leq (1 - \frac{1}{2r})\Delta + B + r^2 - 1$. $\qquad\square$

9.2 The Boundary

Let F be a flat family of torsion free sheaves of rank r on X parametrized by a scheme S. The boundary of S by definition is the set

$$\partial S = \{s \in S | F_s \text{ is not locally free}\}$$

Lemma 9.2.1 — ∂S *is a closed subset of* S, *and if* $\partial S \neq \emptyset$, *then* $\text{codim}(\partial S, S) \leq r - 1$.

Proof. Choose an epimorphism $L_0 \to F$ with L_0 a locally free sheaf on $S \times X$ of constant rank ℓ_0. For example, $L_0 = p^* p_*(F \otimes q^* \mathcal{O}(n)) \otimes q^* \mathcal{O}(-n)$ for $n \gg 0$ would do. Then the kernel L_1 is S-flat and fibrewise locally free, hence locally free on $S \times X$ of rank $\ell_1 = \ell_0 - r$. If ϕ denotes the homomorphism $L_1 \to L_0$, then: F_s is locally free at $x \in X \Leftrightarrow F$ is locally free at $(s, x) \Leftrightarrow \phi(s, x)$ has rank ℓ_1. Hence the set Y of points (s, x) where F is not locally free can be endowed with a closed subscheme structure given by the $\ell_1 \times \ell_1$-minors of ϕ, and by the dimension bounds for determinantal varieties one gets $\text{codim}(Y, S \times X) \leq r + 1$ (cf. [4] Ch. II). Since fibrewise F is torsion free and therefore locally free outside a zero-dimensional subscheme, the projection $Y \to S$ is finite with set-theoretic image ∂S. This proves the lemma. $\qquad\square$

We want to extend the definition of the boundary to subsets of M, though in general there is no universal family which could be used. Consider the good quotient morphism $\pi : R \to M = M(\Delta)$. If $Z \subset M$ is a locally closed subset, say $Z = \overline{Z} \cap U$ for some open $U \subset M$, then $\pi^{-1}(Z)$ is closed in $\pi^{-1}(U)$, and $\partial \pi^{-1}(Z)$ is an invariant closed subset in $\pi^{-1}(Z)$ and $\pi^{-1}(U)$. This implies that $\partial Z := \pi(\partial \pi^{-1}(Z))$ is a closed subset in Z. Consider the boundary of the open subset $Z^\mu := \{[F] \in Z | F \text{ is } \mu\text{-stable}\} \subset Z$. If

$\partial Z^\mu \neq \emptyset$, then $\mathrm{codim}(\partial Z^\mu, Z^\mu) \leq r - 1$. For $\pi : R \to M$ is a principal bundle at stable points, so that the estimate of the lemma carries over to Z^μ.

In the following we will need the polynomial $\vartheta(r) = (6r^3 - \frac{55}{4}r^2 + 11r - 3)r$. The particular values of its coefficients are not really interesting unless one wants to do specific calculations in which case they could most likely be improved.

Theorem 9.2.2 — *There are constants A_1, C_1, C_2 depending on $r, \chi(\mathcal{O}_X), H^2, HK$, and K^2 such that if $\Delta \geq A_1$ and if $Z \subset M$ is an irreducible closed subset with*

$$\dim Z \geq \left(1 - \frac{r-1}{2\vartheta(r)}\right)\Delta + C_1\sqrt{\Delta} + C_2$$

then $\partial Z^\mu \neq \emptyset$.

The proof of this theorem will be given in Section 9.5.

9.3 Generic Smoothness

For any coherent sheaf F let

$$\beta(F) := \mathrm{ext}^2(F, F)_0 = \hom(F, F \otimes K)_0,$$

where the subscript 0 indicates the subspace of traceless extension classes and homomorphisms, respectively. If Z parametrizes a family F, let $\beta(Z) := \min\{\beta(F_s)|s \in Z\}$, which is the generic value of β on Z if Z is irreducible. If F is μ-semistable torsion free then we have the uniform bound $\beta(F) \leq \beta_\infty$ (cf. 4.5.8).

Definition 9.3.1 — *A sheaf F is good, if F is μ-stable and $\beta(F) = 0$.*

It is clear from Corollary 4.5.4 that at good points the moduli space $M(\Delta)$ is smooth of the expected dimension. If we want to bound the dimension of the locus of sheaves which are not good, then half of the problem is solved by Theorem 9.1.1. For the other half consider the closed set

$$W = \{[F] \in M(\Delta)|\beta(F) > 0\}$$

(As before one ought to define W as the image of the corresponding closed subset in R).

Theorem 9.3.2 — *There is a constant $C_3 \geq C_2$ depending on r, X, H, such that for all $\Delta \geq A_1$*

$$\dim W \leq \left(1 - \frac{r-1}{2\vartheta(r)}\right)\Delta + C_1\sqrt{\Delta} + C_3\,.$$

Again we postpone the proof to a later section (see 9.6) and derive some consequences first: Suppose that Δ satisfies the following conditions:

1. $\Delta > A_1$

2. $\Delta - (r^2 - 1)\chi(\mathcal{O}_X) \geq (1 - \frac{1}{2r})\Delta + B + r^2 + 1$

3. $\Delta - (r^2 - 1)\chi(\mathcal{O}_X) \geq (1 - \frac{r-1}{2\vartheta(r)})\Delta + C_1\sqrt{\Delta} + C_3 + 2.$

Then we can apply Theorems 9.1.2 and 9.3.2 and conclude that the points in $M(\Delta)$ which are not good form a closed subset of codimension at least 2. This leads to the following result:

Theorem 9.3.3 — *There is a constant A_2 depending on r, X, H such that if $\Delta \geq A_2$ then*

1. *Every irreducible component of $M(\Delta)$ contains good points. In particular, it is generically smooth and has the expected dimension.*
2. *$M(\Delta)$ is normal and $M^s(\Delta)$ is a local complete intersection.*

Proof. Choose A_2 such that for $\Delta \geq A_2$ the conditions (1)–(3) are simultaneously satisfied. Then the good points are dense in R. Hence R is generically smooth and has the expected dimension. By Proposition 2.2.8 R is a local complete intersection. Moreover, the singular points have large codimension. Hence R satisfies the condition S_2 and is normal by the Serre-Criterion ([98] II 8.23). As a GIT-quotient of R, M is normal, too. It follows from Luna's Etale Slice Theorem 4.2.12, that $M^s(\Delta)$ is a local complete intersection if this holds for R. $\qquad\square$

9.4 Irreducibility

Assume now that $\Delta \geq A_2$, and let $[F] \in M(\Delta)$ be a good point. Let F' be the kernel of any surjection $F \to k(p)$, where $p \in X$ is a point at which F is locally free and $k(p)$ is the structure sheaf of p. Then F' is μ-stable and $\text{Hom}(F', F' \otimes K) \subset \text{Hom}(F, F \otimes K)$, implying that F' is again good. In particular, F' is contained in a single irreducible component of $M(\Delta')$, $\Delta' := \Delta + 2r$, and this component does not change if $[F]$ or the morphism $F \to k(p)$ vary in connected families. This proves the following lemma:

Lemma 9.4.1 — *If Λ_Δ denotes the set of irreducible components of $M(\Delta)$, then sending $[F]$ to $[F']$ induces a well-defined map $\phi : \Lambda_\Delta \to \Lambda_{\Delta+2r}$.* $\qquad\square$

Our aim is to show that for sufficiently large Δ the map ϕ is surjective and that this implies that Λ_Δ consists of a single point.

Theorem 9.4.2 — *There is a constant A_3 such that for all $\Delta \geq A_3$ the following holds:*

1. *Every irreducible component of M contains a point $[F]$ which represents a good locally free sheaf F.*

2. *Every irreducible component of M contains a point $[F]$ such that both F and $F^{\vee\vee}$ are good and $\ell(F^{\vee\vee}/F) = 1$.*

This is a refinement of Theorem 9.3.3. Its proof uses the same techniques as the proof of Theorem 9.3.2 and will also be given in Section 9.6.

Now we have collected enough machinery to prove

Theorem 9.4.3 — *There is a constant A_4 such that for all $\Delta \geq A_4$ the moduli space $M(\Delta)$ is irreducible.*

Proof. Clearly part 2 of Theorem 9.4.2 implies that the map $\phi : \Lambda_{\Delta-2r} \to \Lambda_\Delta$ is surjective for $\Delta \geq A_3$. For if $Z \in \Lambda_\Delta$, pick a good point $[F] \in Z$ with $\ell(F^{\vee\vee}/F) = 1$ such that $F^{\vee\vee}$ is good. Then the component containing $F^{\vee\vee}$ is mapped to Z.

Hence it suffices to show, that if $[E_1], [E_2]$ are any two good locally free sheaves, then some power ϕ^ℓ will map their components to the same point in $\Lambda_{\Delta+2r\ell}$. This, together with the surjectivity of ϕ and the finiteness of Λ_Δ, implies that Λ_Δ contains only one point for sufficiently large Δ. Now $E_1(m)$ and $E_2(m)$ are globally generated for sufficiently large m. Choosing $r - 1$ generic global sections one finds exact sequences (cf. 5.0.1)

$$0 \to \mathcal{O}(-m)^{r-1} \to E_i \to \hat{\mathcal{Q}} \otimes \mathcal{I}_{Z_i} \to 0$$

with $\hat{\mathcal{Q}} = \mathcal{Q} \otimes \mathcal{O}((r - 1)m)$ and zero-dimensional subschemes $Z_i \subset X$. Let $\mathcal{I}_Z = \mathcal{I}_{Z_1} \cap \mathcal{I}_{Z_2}$ and define sheaves $F_i \subset E_i$ by

$$
\begin{array}{ccccccccc}
0 & \to & \mathcal{O}(-m)^{r-1} & \to & E_i & \to & \hat{\mathcal{Q}} \otimes \mathcal{I}_{Z_i} & \to & 0 \\
& & \| & & \uparrow & & \uparrow & & \\
0 & \to & \mathcal{O}(-m)^{r-1} & \to & F_i & \to & \hat{\mathcal{Q}} \otimes \mathcal{I}_Z & \to & 0.
\end{array}
$$

F_1 and F_2 are good points in $M(\Delta + 2r\ell)$ for $\ell = \ell(Z) - \ell(Z_i)$ and determine the images of the components of E_1 and E_2 under the map ϕ^ℓ. The open subset in

$$\mathbb{P}(\mathrm{Ext}^1(\hat{\mathcal{Q}} \otimes \mathcal{I}_Z, \mathcal{O}(-m)^{r-1})^\vee)$$

that parametrizes good points is nonempty, for it contains the extensions defining F_1 and F_2, and is certainly irreducible. This forces F_1 and F_2 to lie in the same component of $M(\Delta + 2r\ell)$. $\square$

9.5 Proof of Theorem 9.2.2

Proposition 9.5.1 — *Let $C \in |nH|$ be a smooth curve and let M_C be the moduli space of semistable sheaves on C of rank r and determinant $\mathcal{Q}|_C$. Let $Z \subset M$ be a closed irreducible subvariety with $\partial Z = \emptyset$. If $\dim Z > \dim M_C$, then there is a point $[F] \in Z$ such that $F|_C$ is not stable.*

Proof. Assume to the contrary that $F|_C$ is stable for all $[F] \in Z$. Then the restriction $F \mapsto F|_C$ defines a morphism $\varphi : Z \to M_C$. By equation 8.2 in Section 8.2 we know $\varphi^*(\mathcal{L}_0')^{n \deg(X)} \cong \mathcal{L}_1^{n^2 \deg(X)}|_Z$, where $\mathcal{L}_0'$ is an ample line bundle on M_C (cf. 8.1.12). Moreover, sections in some high power of $\mathcal{L}_1$ define an embedding of $M^\mu \setminus \partial M$ (cf. 8.2.16). By assumption $Z \subset M^\mu \setminus \partial M$. Hence the line bundle $\mathcal{L}_1|_Z$ is ample. Thus φ is finite and $\dim(Z) \leq \dim(M_C)$ $\qquad\square$

Proposition 9.5.2 — *Let C be a smooth connected curve of genus $g \geq 2$, and let $\mathcal{F}$ be a flat family of locally free sheaves of rank r on C parametrized by a k-scheme S of finite type. Then the closed set*

$$S^{us} = \{s \in S | \mathcal{F}_s \text{ is not geometrically stable}\}$$

is empty, or has codimension $\leq \frac{r^2}{4} g$ in S.

As the moduli space need not be fine, the proposition cannot be applied to the moduli space M_C itself. Indeed, the codimension of the subset parametrizing properly semistable sheaves can be larger than predicted by the proposition.

Proof. Suppose $F = \mathcal{F}_s$ is not stable for some closed point $s \in S$. Let $m = \operatorname{reg}(F)$ be the regularity of F and $\phi : G := \mathcal{O}_C(-m) \otimes H^0(F(m)) \to F$ the evaluation map. There is an open neighbourhood U of s in S and a morphism $U \to \operatorname{Quot}_C(G; P(F))$ mapping s to $[\phi]$ such that $\mathcal{F}|_U$ is the pullback of the universal quotient. Hence, it suffices to prove the proposition for the following 'universal example': $\mathcal{F}$ is the universal family parametrized by the open subset $S \subset \operatorname{Quot}(G; P(F))$ corresponding to all points $\phi : G \to F$ such that F is an m-regular locally free sheaf and ϕ induces an isomorphism $H^0(G(m)) \to H^0(F(m))$.

Let d be the degree of F. For any pair (d_1, r_1) of integers with $0 < r_1 < r$ let $r_2 = r - r_1$, $d_2 = d - d_2$ and let $P_i(m) = r_i m + (d_i + r_i(1 - g))$ denote the corresponding Hilbert polynomial. Finally, let $P_0 = P(G) - P(F)$. The relative Quot-scheme $D(d_1, r_1) := \operatorname{Quot}(\mathcal{F}, P_2)$ is an open subset of the flag-scheme $\operatorname{Drap}(G; P_0, P_1, P_2)$. Consider the canonical projection $\pi : D(d_1, r_1) \to S$. The image of π is precisely the closed subset of points s in S such that $\mathcal{F}_s$ has a submodule of degree d_1 and rank r_1. A point $y \in D(d_1, r_1)$ corresponds to a filtration $0 \subset G_0 \subset G_1 \subset G_2 = G$, and $s := \pi(y)$ then corresponds to the quotient $G \to G/G_0 =: F$. Let $F_i = G_i/G_0$ be the induced filtration of F. The smoothness obstruction for S is contained in $\text{Ext}^1(G_0, F)$. As $\text{Hom}(G, G) \cong \text{Hom}(G, F)$ and $\text{Ext}^i(G, F) = 0$ for $i \geq 1$ by definition of S, we have $\text{Ext}^1(G_0, F) \cong \text{Ext}^2(F, F) = 0$, since C is a curve. Thus S is smooth. Moreover, there is an exact sequence

$$\ldots \to \text{Ext}^i_-(F, F) \to \text{Ext}^i(G, F) \to \text{Ext}^i_+(G, G) \to \text{Ext}^{i+1}_-(F, F) \to \ldots \quad (9.1)$$

(We leave it as an exercise to the reader to establish this sequence. Recall the definition of the groups $\text{Ext}_\pm$ in Appendix 2.A and write down an appropriate short exact sequence which leads to the desired sequence. Cf. [51]). Because of $\text{Ext}^i(G, F) = 0$ for $i \geq 1$, we get $\text{Ext}^i_+(G, G) \cong \text{Ext}^{i+1}_-(F, F) = 0$ for $i \geq 1$ (Use the spectral sequences 2.A.4 and $\dim(C) = 1$). Now $\text{Ext}^1_+(G, G)$ is the obstruction space for the smoothness of $D(d_1, r_1)$ (cf. Proposition 2.A.12). Hence $D(d_1, r_1)$ is smooth as well. By Proposition 2.2.7 there is an exact sequence

$$0 \to \text{Hom}(F_1, F/F_1) \to T_y D(d_1, r_1) \xrightarrow{T\pi} T_s S \to \text{Ext}^1(F_1, F/F_1) \quad (9.2)$$

In fact, it follows from $\text{Ext}^i(F_1, F/F_1) = \text{Ext}^i_+(F, F)$, the long exact sequence

$$\ldots \to \text{Ext}^i_+(F, F) \to \text{Ext}^i_+(G, G) \to \text{Ext}^i(G_0, F) \to \text{Ext}^{i+1}_+(F, F) \to \ldots$$

(again, we leave it to the reader to establish this sequence) and the vanishing results listed above, that the last homomorphism in (9.2) is surjective. Using Riemann-Roch, this implies that

$$\text{codim}(\pi(D(d_1, r_1)), S) \leq \text{ext}^1(F_1, F/F_1) = r_1 r_2 \left(g - 1 + \frac{d_1}{r_1} - \frac{d_2}{r_2} \right) + \text{hom}(F_1, F_2).$$

Now S^{us} is the union of all $\pi(D(d_1, r_1))$, where (d_1, r_1) satisfies $d_1/r_1 \geq d/r$. Note that by the Grothendieck Lemma 1.7.9 there are only finitely many such flag varieties which are nonempty.

Let V be an irreducible component of S^{us}. Assume first that a general point of V corresponds to a semistable sheaf F. Then V is the image of $D(d_1, r_1)$ for a pair (d_1, r_1) with $d_1/r_1 = d/r = d_2/r_2$. Hence

$$\text{codim}(V) \leq r_1 r_2 (g - 1) + \text{hom}(F_1, F/F_1) \leq r_1 r_2 g \leq \frac{r^2}{4} g.$$

Here we used that F/F_1 is also semistable and therefore $\text{hom}(F_1, F/F_1) \leq r_1 r_2$.

Assume now that a general point of V corresponds to a sheaf F which is not semistable and let (d_1, r_1) denote degree and rank of the maximal destabilizing subsheaf of F. Then $\text{Hom}(F_1, F/F_1) = 0$. Therefore, $D(d_1, r_1) \to S$ is, generically, a closed immersion with image V of codimension $\text{codim}(V, S) \leq r_1 r_2(g - 1 + \frac{d_1}{r_1} - \frac{d_2}{r_2})$. In case

$$\frac{d_1}{r_1} - \frac{d_2}{r_2} \leq \frac{r_1 + r_2 - 1}{r_1 r_2},$$

we get $\text{codim}(V, B) \leq r_1 r_2 g - (r_1 - 1)(r_2 - 1) \leq \frac{r^2}{4}g$, and we are done. Hence it suffices to show that the alternative relation

$$\frac{d_1}{r_1} - \frac{d_2}{r_2} > \frac{r_1 + r_2 - 1}{r_1 r_2}$$

is impossible. Otherwise, the slightly stronger inequality $(d_1 - 1)/r_1 \geq (d_2 + 1)/r_2$ must hold, since the involved degrees and ranks are integers.

This means that the kernel F_1' of any surjection $F_1 \to k(P)$, $P \in C$, is still destabilizing. Hence there is a component $D' \subset D(d_1 - 1, r_1)$ which surjects onto V. The fibre dimension of this morphism is greater than or equal to $\dim(D') - \dim(V) = \chi(F_1', F/F_1') - \chi(F_1, F/F_1) = r_1 + r_2$ by the Riemann-Roch formula. But the tangent space to the fibre of $D' \to V$ at a point $[F \to F/F_1'] \in D'$ is given by

$$\text{Hom}(F_1', F/F_1') \cong \text{Hom}(F_1', F/F_1) \oplus \text{Hom}(F_1', k(P)).$$

In order to get a contradiction it suffices to show $\hom(F_1', F/F_1) < r_2$. But this is equivalent to the claim that $\text{Hom}(F_1', F_2) \to \text{Ext}^1(\mathcal{O}_P, F/F_1)$ is not surjective, and, by Serre duality, that $\text{Hom}(F/F_1, F_1 \otimes \omega_C) \to \text{Hom}(F/F_1, \mathcal{O}_P \otimes \omega_C)$ is not trivial. But this is certainly true for an appropriate choice of $F_1 \to k(P)$. $\qquad \square$

Proposition 9.5.3 — *Let $C \in |nH|$ be a smooth curve, let $e > 0$ be a rational number and let $d(e)$ be the quantity defined in Theorem 9.1.1. Let $Z \subset M$ be a closed irreducible subset with $\partial Z = \emptyset$, and suppose that $\dim Z > \dim M_C$ and $\dim Z > \frac{r^2}{4}g(C) + d(e) + r^2$. Then there is a point $[F] \in Z$ such that F is e-stable and $F|_C$ is unstable.*

Proof. Since $\dim Z > \dim M_C$, by Proposition 9.5.1 there is a point $[F]$ in Z such that $F|_C$ is unstable. Let $Z' \subset R$ be an irreducible component of the pre-image $\pi^{-1}(Z)$ under the morphism $\pi : R \to M$ which maps onto Z. Then

$$\dim Z' > d(e) + \text{end}(\mathcal{H}) + \frac{r^2}{4}g(C).$$

Let $(Z')^{us}$ denote the (nonempty!) closed subset corresponding to sheaves whose restriction to C is unstable. Then by Proposition 9.5.2 and 9.1.1

$$\dim(Z')^{us} > d(e) + \text{end}(\mathcal{H}) > \dim R(e).$$

This implies that there is a point $[\mathcal{H} \to F] \in Z'$ such that $F|_C$ is unstable and F is e-stable. $\qquad \square$

Proposition 9.5.4 — *Let $Z \subset M$ be closed and irreducible. Let $C \in |nH|$ be a smooth curve and $e = (r - 1)\frac{CH}{|H|}$. Suppose that Z contains a point $[F]$ such that F is e-stable but $F|_C$ is unstable. If $\dim Z > \exp \dim M + \beta_\infty + \frac{r^2}{4} - \frac{r-1}{2}C(C - K)$, then $\partial Z \neq \emptyset$.*

Proof. Assume to the contrary that $\partial Z = \emptyset$ so that all sheaves corresponding to points in Z are locally free. By assumption $F|_C$ is unstable, i.e. there is an exact sequence

$$0 \to F' \to F|_C \to F'' \to 0$$

with locally free $\mathcal{O}_C$-modules F' and F'' with $\mu(F') \geq \mu(F'')$. Let E be the kernel of the composite homomorphism

$$F \to F|_C \to F''.$$

Since F and F'' are locally free on X and C, respectively, E is locally free, too. F can be recovered from E and the homomorphism $q_0 : E(C)|_C \to F' \otimes \mathcal{O}_C(C)$:

$$0 \to F \to E(C) \to F'(C) \to 0.$$

q_0 corresponds to a closed point in the quotient scheme $\Sigma := \mathrm{Quot}_C(E(C)|_C, P(F'(C)))$. Using Corollary 2.2.9, together with the notations introduced there, we can give a lower bound for the dimension of Σ:

$$\begin{aligned}
\dim \Sigma \;\geq\;& \chi(F'', F'(C)) = r'r''(\mu' + C^2 - \mu'' + 1 - g(C)) \\
\geq\;& r'r''(C^2 - \tfrac{1}{2}C(C+K)) \geq \frac{r-1}{2}C(C-K).
\end{aligned}$$

(Recall that $\mu' = \mu(F') \geq \mu'' = \mu(F'')$!) Let $q^*E(C)|_C \to G$ be the universal quotient on $\Sigma \times C$ and define a family $\mathcal{F}$ of sheaves on X by the exact sequence

$$0 \to \mathcal{F} \to q^*E(C) \to i_*G \to 0,$$

where $i : \Sigma \times C \to \Sigma \times X$ is the inclusion. Then $\mathcal{F}$ is Σ-flat, $\mathcal{F}_{q_0} = F$ by construction and $\mathcal{F}_\sigma$ is μ-stable for all points $\sigma \in \Sigma$. To see this let $A \subset \mathcal{F}_\sigma$ be a subsheaf of rank $\mathrm{rk}(A) < r$. There are inclusions

$$A \subset \mathcal{F}_\sigma \subset E(C) \subset F(C).$$

Hence the e-stability of F implies:

$$\mu(A) < \mu(F) + CH - \frac{e|H|}{\mathrm{rk}(A)} = \mu(F) - \frac{r - \mathrm{rk}(A) - 1}{\mathrm{rk}(A)} CH \leq \mu(F) = \mu(\mathcal{F}_\sigma).$$

The family $\mathcal{F}$ induces a morphism $\Sigma \to M^\mu$. Let Σ' be an irreducible component of the fibre product $\Sigma \times_{M^\mu} Z^\mu$. Then

$$\begin{aligned}
\dim \Sigma' \;\geq\;& \dim Z + \dim \Sigma - \dim T_{[F]}M \\
>\;& (\exp \dim M + \beta_\infty - \dim T_{[F]}M) + \left(\dim \Sigma - \frac{r-1}{2}C(C-K)\right) + \frac{r^2}{4} \\
\geq\;& \frac{r^2}{4}.
\end{aligned}$$

Since $\partial Z = \emptyset$, $\partial \Sigma'$ must also be empty. But for any point $\sigma \in \Sigma$ corresponding to a short exact sequence

$$0 \to \mathcal{F}_\sigma \to E(C) \to G_\sigma \to 0$$

the sheaf $\mathcal{F}_\sigma$ is locally free on X if and only if G_σ is locally free on C. Hence, Σ' parametrizes locally free sheaves of rank r' on C and thus induces a C-morphism

$$\varphi : \Sigma' \times C \to \mathrm{Grass}(E(C)|_C, r'),$$

where $r' = \mathrm{rk}(F')$ as above. $\mathrm{Grass}(E(C)|_C, r')$ is a locally trivial fibre bundle over C with fibres isomorphic to $\mathrm{Grass}(k^r, r')$ and of dimension $r'(r - r') \le \frac{r^2}{4}$. Since $\dim \Sigma' > \frac{r^2}{4}$, for a fixed point $c \in C$ the morphism

$$\varphi(c) : \Sigma' \to \mathrm{Grass}(E(C)(c), r')$$

cannot be finite. Let Σ'' be a component of a fibre of $\varphi(c)$ of dimension ≥ 1. Then

$$\varphi'' = \varphi|_{\Sigma'' \times C} : \Sigma'' \times C \to \mathrm{Grass}(E(C)|_C, r')$$

contracts the fibre $\Sigma'' \times \{c\}$. The Rigidity Lemma (cf. [195] Prop. 6.1, p. 115) then forces φ'' to contract all fibres and to factorize through the projection onto C. But this would mean that all points in Σ'' parametrize the same quotient, which is absurd. □

We summarize the results: suppose C is a smooth curve in the linear system $|nH|, n \ge 1$. Then

$$g(C) - 1 = \frac{1}{2}C(C + K) = \frac{1}{2}n^2 H^2 + \frac{1}{2}nKH$$

and, by Corollary 4.5.5,

$$\dim M_C = (r^2 - 1)(g(C) - 1) = \frac{r^2 - 1}{2}(n^2 H^2 + nKH).$$

What we have proved so far is the following: suppose $Z \subset M$ is an irreducible closed subset such that

$$\dim Z > \frac{r^2 - 1}{2}(n^2 H^2 + nKH) =: \phi_0(n)$$

$$\dim Z > \left(1 - \frac{1}{2r}\right)\Delta + \left(3r^3 - \frac{55}{8}r^2 + 5r - 1\right)H^2 n^2$$
$$+ \left(\frac{5}{8}r^2 - \frac{1}{2}r\right)[KH]_+ n + \frac{5}{4}r^2 + B =: \phi_1(n)$$

$$\dim Z > \Delta - \frac{r - 1}{2}H^2 n^2 + \frac{r - 1}{2}[KH]_+ n$$
$$+ \beta_\infty + \frac{r^2}{4} - (r^2 - 1)\chi(\mathcal{O}) =: \phi_2(n)$$

(here ϕ_0, ϕ_1 and ϕ_2 are the constants of Propositions 9.5.3 and 9.5.4 for $e = (r-1)\frac{C.H}{|H|}$, expressed as functions of n). Then 9.5.3 and 9.5.4 together imply that $\partial Z \neq \emptyset$. We need to analyze the growth relations between ϕ_0, ϕ_1 and ϕ_2. First observe that $\phi_0(n) \leq \phi_1(n)$ for all $n \geq 0$. Next, consider the 'leading terms' of ϕ_1 and ϕ_2:

$$\tilde{\phi}_1(n) \ := \ \left(1 - \frac{1}{2r}\right)\Delta + \left(3r^3 - \frac{55}{8}r^2 + 5r - 1\right)H^2 n^2$$

$$\tilde{\phi}_2(n) \ := \ \Delta - \frac{r-1}{2}H^2 n^2.$$

Then the equation $\tilde{\phi}_1(x) = \tilde{\phi}_2(x)$ has the positive solution

$$x_0 = \sqrt{\frac{\Delta}{\vartheta(r)H^2}},$$

where $\vartheta(r) = 6r^4 - \frac{55}{4}r^3 + 11r^2 - 3r$. The quadratic polynomials $\phi_1(x)$ and $\phi_2(x)$ attain their minimum and maximum value at

$$x_1 = -\frac{(5r^2 - 4r)[KH]_+}{(48r^3 - 110r^2 + 80r - 16)H^2} < 0$$

and $x_2 = \frac{[KH]_+}{2H^2}$, respectively. Hence, if $\Delta \geq A_1 := \vartheta(r)H^2 \cdot (2 + \frac{[HK]_+}{2H^2})^2$, then $x_0 - 2 \geq x_2 \geq 0 \geq x_1$. Let $n_0 = \lfloor x_0 \rfloor$. Then $n_0 \geq 1$ and

$$\phi_1(n_0) \leq \phi_1(x_0), \qquad \phi_2(n_0) \leq \phi_2(x_0 - 1),$$

as n_0 is in the range where ϕ_1 is increasing and ϕ_2 is decreasing. We conclude: if $\Delta \geq A_1$ and $\dim Z \geq \max\{\phi_1(x_0), \phi_2(x_0 - 1)\}$ then $\partial Z \neq \emptyset$. Now express $\phi_1(x_0)$ and $\phi_2(x_0-1)$ in terms of Δ, using the definition of x_0, and check that their maximum is not greater than the constant given in Theorem 9.2.2. Therefore, if Z satisfies the assumptions of Theorem 9.2.2, then $\partial Z \neq \emptyset$. Under the same assumptions let $Z' \subset \pi^{-1}(Z) \subset R$ be an irreducible component that dominates Z. Then $\partial Z' \neq \emptyset$ and

$$
\begin{aligned}
\dim \partial Z' \ &\geq \ \dim Z' - (r-1) && \text{by 9.2.1} \\
&\geq \ \dim Z + (\mathrm{end}(\mathcal{H}) - r^2) - (r-1) \\
&\geq \ \phi_1(0) + \mathrm{end}(\mathcal{H}) - (r^2 + r - 1) \\
&= \ \left(1 - \frac{1}{2r}\right)\Delta + B + \mathrm{end}(\mathcal{H}) + \left(\frac{1}{4}r^2 - r + 1\right) \\
&> \ \left(1 - \frac{1}{2r}\right)\Delta + B + \mathrm{end}(\mathcal{H}) - 1 \\
&\geq \ \dim R(0) && \text{by 9.1.1.}
\end{aligned}
$$

Hence, $\partial(Z')^\mu \neq \emptyset$, and thus $\partial Z^\mu \neq \emptyset$. This finishes the proof of the theorem. $\square$

9.6 Proof of Theorem 9.3.2

Let F be a torsion free sheaf of rank r on X and let $T = F^{\vee\vee}/F$. Since T is zero-dimensional, F and $F^{\vee\vee}$ have the same rank and slope, $F^{\vee\vee}$ is μ-stable if and only if F is μ-stable; and $\Delta(F) = \Delta(F^{\vee\vee}) + 2r\ell(T)$. Note that $\mathcal{E}xt^1(F, \mathcal{O}) \cong \mathcal{E}xt^2(T, \mathcal{O})$ is zero-dimensional of length $\ell(T)$. Consider now a flat family F of torsion free sheaves parametrized by S and let $T_s = (F_s)^{\vee\vee}/F_s$.

Lemma 9.6.1 — *The function $s \mapsto \ell(T_s)$ is semicontinuous. If S is reduced and $\ell(T_s)$ is constant then forming the double dual commutes with base change and $F^{\vee\vee}$ is locally free.*

Proof. Choose a locally free resolution $0 \to L_1 \to L_0 \to F \to 0$. Dualizing yields an exact sequence

$$0 \to \mathcal{H}om(F, \mathcal{O}) \to L_0^{\vee} \to L_1^{\vee} \to \mathcal{E}xt^1(F, \mathcal{O}) \to 0.$$

This shows that $F \mapsto \mathcal{E}xt^1(F, \mathcal{O})$ commutes with base change and proves the semicontinuity. If S is reduced and $\ell(T_s)$ is constant then $\mathcal{E}xt^1(F, \mathcal{O})$ is S-flat. But then $F^{\vee} = \mathcal{H}om(F, \mathcal{O})$ is also S-flat and forming the dual commutes with base change. □

The *double-dual stratification* of S by definition is given by the subsets

$$S_\nu = \{s \in S | \ell(T_s) \geq \nu\}.$$

These are closed according to the lemma.

Let $Z \subset M(\Delta)$ be a closed irreducible subset and assume that ∂Z^μ is nonempty and $\beta(Z) > 0$. Define a sequence of triples

$$Y_i \subset Z_i \subset M_i = M(\Delta_i), i = 1, \ldots, n,$$

by the following procedure: $\Delta_0 = \Delta$, $Z_0 = Z$, $Y_i \subset \partial Z_i^\mu$ is an irreducible component of the maximal open stratum of the double-dual stratification of ∂Z_i^μ. Let ℓ_i be the constant value of ℓ on Y_i. Then sending $[F]$ to $[F^{\vee\vee}]$ defines a morphism $Y_i \to M_{i+1} = M(\Delta_{i+1})$, where $\Delta_{i+1} = \Delta_i - 2r\ell_i$. Finally, let Z_{i+1} be the closure of the image of this morphism. This process breaks off, say at the index n, when $Y_n = \emptyset$. (It must certainly come to an end, as $\Delta_i \geq 0$ for all i by the Bogomolov Inequality).

Remark 9.6.2 — Strictly speaking we have defined the double dual stratification only for schemes which parametrize flat families, i.e. on $\partial Z^\mu \times_M R$ rather than on ∂Z itself. But obviously the stratification is invariant under the group action on R and therefore projects to a stratification on ∂Z^μ. Similarly, the morphism to M_1 etc. is defined first on $Y \times_M R$ but factors naturally through Y. □

How do $\dim Z_i$ and $\beta(Z_i)$ change? Let $[E] \in Z_i$ be a general point. Then by construction E is a μ-stable locally free sheaf. There is a classifying morphism

$$\mathrm{Quot}(E, \ell_i) \longrightarrow \partial M_{i-1}^{\mu}$$

sending $[\phi : E \to T]$ to $\ker(\phi)$. This is easily seen to be an injective morphism. If $[F] \in Y_{i-1}$, then the fibre of $Y_{i-1} \to Z_i$ over $[F^{\vee\vee}]$ is contained in the image of $\mathrm{Quot}(F^{\vee\vee}, \ell_i)$. But by Theorem 6.A.1 $\mathrm{Quot}(E, \ell_i)$ is irreducible of dimension $\ell_i(r+1)$. In particular,

$$\begin{aligned}
\dim Z_i \;&\geq\; \dim Y_{i-1} - \ell_i(r+1) \\
&\geq\; \dim Z_{i-1} - (r-1) - \ell_i(r+1) \qquad \text{by 9.2.1} \\
&\geq\; \dim Z_{i-1} - (2r-1)\ell_i - 1,
\end{aligned}$$

and summing up:

$$\dim Z_n \geq \dim Z - (2r-1)\sum_{i=1}^{n} \ell_i - N, \tag{9.3}$$

where N is the number of times that equality holds in

$$\dim Z_i \geq \dim Z_{i-1} - (2r-1)\ell_i - 1. \tag{9.4}$$

To get a bound on N, consider a general sheaf $[E] \in Z_i$ and a sheaf $[F] \in Y_{i-1}$ with $F \subset E$. Then $\mathrm{Hom}(F, F \otimes K) \subset \mathrm{Hom}(E, E \otimes K)$ so that

$$\beta(Z_i) = \beta(E) \geq \beta(F) \geq \beta(Z_{i-1}) > 0 \tag{9.5}$$

What happens if equality holds in (9.4)? In this case $\mathrm{Quot}(E, \ell_i)_{\mathrm{red}}$ must be contained in Y_{i-1}, since it is an irreducible scheme. We claim that in this situation the strict inequality $\beta(F) < \beta(E)$ holds for $F = \ker(\phi)$, when $[\phi : E \to T] \in \mathrm{Quot}(E, \ell_i)$ is a general point: namely, let $\phi : E \to E \otimes K$ be a non-trivial traceless homomorphism. Then $\phi(x)$ cannot be a multiple of the identity on $E(x)$ for all $x \in X$. Thus for a general $\phi : E \to T$ the kernel F is not preserved by ϕ. Thus $\phi \notin \mathrm{Hom}(F, F \otimes K)$ and $\beta(F) < \beta(E)$. This argument shows that we can sharpen (9.5) each time that equality holds in (9.4): we get $\beta(Z_n) \geq \beta(Z_0) + N$. This can be used to give a bound for N:

$$N \leq N + \beta(Z_0) \leq \beta(Z_n) \leq \beta_\infty \tag{9.6}$$

Recall that $\Delta_n = \Delta - 2r\sum_{i=1}^{n} \ell_i$. Using this, and the inequalities (9.3) and (9.6) we get:

$$\dim Z_n - \left(1 - \frac{1}{2r}\right)\Delta_n \geq \dim Z - \left(1 - \frac{1}{2r}\right)\Delta - \beta_\infty. \tag{9.7}$$

We are now ready to prove Theorem 9.3.2: define

$$C_3 := \max\left\{ C_2 + \beta_\infty, \frac{A_1}{2r} + 2\beta_\infty - (r^2 - 1)\chi(\mathcal{O}_X) \right\}.$$

If Theorem 9.3.2 were false, let $Z \subset W$ be an irreducible component of W with

$$\dim Z > \left(1 - \frac{r-1}{2\vartheta}\right)\Delta + C_1\sqrt{\Delta} + C_3.$$

By the definition of W we also have $\beta(Z) > 0$. Since $C_3 \geq C_2$, Theorem 9.2.2 can be applied to Z, so that $\partial Z^\mu \neq \emptyset$. The procedure above leads to the construction of a closed irreducible subset $Z_n \subset M(\Delta_n)$ such that $\partial Z_n^\mu = \emptyset$ and such that estimate (9.7) holds. It suffices to show that C_3 was chosen large enough so that Z_n still satisfies the conditions of Theorem 9.2.2 and therefore provides the contradiction $\partial Z_n^\mu \neq \emptyset$. Firstly, Z_n parametrizes, generically, μ-stable sheaves. Hence

$$\dim Z_n \leq \exp \dim M(\Delta_n) + \beta_\infty = \Delta_n + \beta_\infty - (r^2 - 1)\chi(\mathcal{O}_X).$$

This and (9.7) give:

$$
\begin{aligned}
\Delta_n/2r &\geq \left(\dim Z_n - (1 - \tfrac{1}{2r})\Delta_n\right) + \left((r^2 - 1)\chi(\mathcal{O}_X) - \beta_\infty\right) \\
&\geq \left(\dim Z_n - (1 - \tfrac{1}{2r})\Delta\right) + \left((r^2 - 1)\chi(\mathcal{O}_X) - 2\beta_\infty\right) \\
&\geq \left(\tfrac{1}{2r} - \tfrac{r-1}{2\vartheta}\right)\Delta + C_1\sqrt{\Delta} + \left(C_3 + (r^2 - 1)\chi(\mathcal{O}_X) - 2\beta_\infty\right) \\
&\geq A_1/2r,
\end{aligned}
$$

since $(\tfrac{1}{2r} - \tfrac{r-1}{2\vartheta}) > 0$. Secondly, again using the estimate (9.7):

$$
\dim Z_n - \left(\left(1 - \frac{r-1}{2\vartheta}\right)\Delta_n + C_1\sqrt{\Delta_n} + C_2\right)
$$

$$
\geq \dim Z - \left(\left(1 - \frac{r-1}{2\vartheta}\right)\Delta + C_1\sqrt{\Delta} + C_3\right)
$$

$$
+ \left(\frac{1}{2r} - \frac{r-1}{2\vartheta}\right)(\Delta - \Delta_n) + C_1(\sqrt{\Delta} - \sqrt{\Delta_n})
$$

$$
+ (C_3 - C_2 - \beta_\infty)
$$

$$
\geq 0,
$$

where all the terms on the right hand side are nonnegative by the assumption on $\dim Z$ and the definition of C_3. $\qquad\square$

Proof of Theorem 9.4.2

Let $A_3 = A_2 + 2r(\frac{\beta_\infty}{r-1} + 1)$, and let Z be an irreducible component of $M(\Delta)$ for $\Delta \geq A_3$. By the choice of the constants A_i we have $A_3 \geq A_2 \geq A_1$. Thus, for $\Delta \geq A_3$ Theorem 9.3.3 applies and says that Z has the expected dimension. Moreover, the conditions 1. – 3. on page 243 are satisfied. Hence, Theorem 9.2.2 applies, and we can conclude that $\partial Z^\mu \neq \emptyset$. Let Y be an irreducible component of the maximal open stratum of the double-dual stratification of ∂Z^μ, ℓ the constant value $\ell(T_s)$, $s \in Y$, and let Z' be the closure of the image of the morphism $Y \to M' = M(\Delta')$, $\Delta' = \Delta - 2r\ell$, as above. Then $\dim Z' \geq \dim Y - (r+1)\ell$. Distinguish the following two cases:

Case 1. Suppose Z contains no points corresponding to good locally free sheaves. Then Y is an open dense subset of Z and it follows:

$$\dim Z' \quad \geq \quad \exp \dim M - (r+1)\ell \tag{9.8}$$

$$= \quad \exp \dim M' + (r-1)\ell. \tag{9.9}$$

Either $\Delta' = \Delta - 2r\ell \geq A_2$, then M' is generically good by Theorem 9.3.3, hence $\dim Z' \leq \exp \dim M'$, a contradiction; or $\Delta' = \Delta - 2r\ell < A_2$, then $2r\ell > \Delta - A_2 \geq A_3 - A_2 = 2r(\frac{\beta_\infty}{r-1} + 1)$ so that $(r-1)\ell > \beta_\infty$ and

$$\dim Z' > \exp \dim M' + \beta_\infty,$$

again a contradiction. This proves part 1 of the theorem.

Case 2. Suppose ∂Z^μ is a proper subset of Z^μ. We must show that $\ell = 1$. In this case $\dim Y \geq \dim Z - (r-1)$ by 9.2.1, so that instead of (9.9) we get

$$\dim Z' \geq \exp \dim M' + (r-1)\ell - (r-1) = \exp \dim M' + (r-1)(\ell-1).$$

The very same arguments as in Case 1 lead to a contradiction, unless $\ell = 1$ as asserted. $\square$

Comments:

— The main references for this chapter are articles of O'Grady, Li and Gieseker-Li. The general outline of our presentation follows the article [209] of O'Grady. The bounds O'Grady gives for Δ are all explicit. Moreover, he can further improve these bounds in the rank 2 case. Our presentation is less ambitious: even though all bounds could easily be made explicit we tried to keep the arguments as simple as possible. As a result, some of the coefficients in the statements are worse than those in the O'Grady's paper.

— Generic smoothness was first proved by Donaldson [47] for sheaves of rank 2 and trivial determinant, and by Zuo [263] and Friedman for general determinants. Their methods did not give effective bounds.

— Asymptotic irreducibility was obtained by a very different and also very interesting method by Gieseker and Li for rank 2 sheaves in [81] and for arbitrary rank in [82].

— Asymptotic normality was proved by J. Li [149].

10

Symplectic Structures

A *symplectic structure* on a non-singular variety M is by definition a non-trivial regular two-form, i.e. a global section $0 \neq w \in H^0(M, \Omega_M^2)$. Any such two-form defines a homomorphism $T_M \cong \Omega_M^\vee \to \Omega_M$ which we will also denote by w. This homomorphism satisfies $w^* = -w$, i.e. w is alternating. Conversely, any such alternating map defines a symplectic structure.

The symplectic structure w is called (generically) *non-degenerate* if $w : T_M \to \Omega_M$ is (generically) bijective. The symplectic structure is *closed* if $dw = 0$. Sometimes we will also call a regular two-form on a singular variety a symplectic structure. Note that our definition of a symplectic structure is rather weak. Usually one requires a symplectic structure to be closed and non-degenerate.

Any non-degenerate symplectic structure defines an isomorphism $\Lambda^n w : K_M^\vee \cong K_M$, where $n = \dim M$. In particular, $K_M^2 \cong \mathcal{O}_M$. Using the Pfaffian one can in fact show that $K_M \cong \mathcal{O}_M$. Any generically non-degenerate symplectic structure is non-degenerate on the complement of the divisor defined by $\Lambda^n w \in H^0(M, K_M^2)$.

By definition a compact surface admits a symplectic structure if and only if $p_g > 0$. Going through the classification one checks that the only surfaces with a non-degenerate symplectic structure are K3 and abelian surfaces.

The general philosophy that moduli spaces of sheaves on a surface inherit properties from the surface suggests that on a symplectic surface the moduli space should carry a similar structure. That this is indeed the case will be shown in this lecture. We will also discuss how holomorphic one-forms on the surface give rise to one-forms on the moduli space.

In Section 10.2 we give a description of the tangent bundle of the good part of the moduli space in terms of the Kodaira-Spencer map. In Section 10.3 one- and two-forms on the moduli space are constructed using the Atiyah class of a (quasi)-universal family. The question under which hypotheses these forms are non-degenerate is studied in the final

Section 10.4. We begin with a discussion of the technical tools for the investigations in this chapter.

10.1 Trace Map, Atiyah Class and Kodaira-Spencer Map

In this section we recall the definition of the cup product (or Yoneda pairing) for Ext-groups of sheaves and complexes of sheaves, the trace map and the Atiyah class of a complex. These are the technical ingredients for the geometric results of the following sections.

In the following let Y be a k-scheme of finite type.

10.1.1 The cup product — Let $E^\bullet$ and $F^\bullet$ be finite complexes of locally free sheaves. $\mathcal{H}om^\bullet(E^\bullet, F^\bullet)$ is the complex with

$$\mathcal{H}om^n(E^\bullet, F^\bullet) = \bigoplus_i \mathcal{H}om(E^i, F^{i+n})$$

and differential

$$d(\varphi) = d_F \circ \varphi - (-1)^{\deg \varphi} \cdot \varphi \circ d_E. \tag{10.1}$$

If $G^\bullet$ is another finite complex of locally free sheaves, composition yields a homomorphism

$$\mathcal{H}om^\bullet(F^\bullet, G^\bullet) \otimes \mathcal{H}om^\bullet(E^\bullet, F^\bullet) \xrightarrow{\circ} \mathcal{H}om^\bullet(E^\bullet, G^\bullet) \tag{10.2}$$

such that $d(\psi \circ \varphi) = d(\psi) \circ \varphi + (-1)^{\deg \psi} \psi \circ d(\varphi)$ for homogeneous elements φ and ψ. For any two finite complexes $A^\bullet$ and $B^\bullet$ of coherent sheaves on X there is a cup product

$$\mathbb{H}^i(A^\bullet) \otimes \mathbb{H}^j(B^\bullet) \xrightarrow{\circ} \mathbb{H}^{i+j}(A^\bullet \otimes B^\bullet),$$

most conveniently defined via Čech cohomology: let $\mathcal{U} = \{U_i\}_{i \in I}$ be an open affine covering of Y, indexed by a well ordered set I. The intersection $U_{i_0 \ldots i_p} = \bigcap_{j=0}^p U_{i_j}$ is again affine for any finite (ordered) subset $\{i_0 < \ldots < i_p\} \subset I$. For any sheaf F consider the complex $C^\bullet(F, \mathcal{U})$ of k-vector spaces with homogeneous components

$$C^p(F, \mathcal{U}) = \prod_{i_0 < \ldots < i_p} \Gamma(F, U_{i_0 \ldots i_p})$$

and differential

$$(\check{d}\alpha)_{i_0 \ldots i_{p+1}} = \sum_{j=1}^{p+1} (-1)^j \alpha_{i_0 \ldots \hat{i}_j \ldots i_{p+1}}|_{U_{i_0 \ldots i_{p+1}}}.$$

If $F^\bullet$ is a finite complex, we can form the double complex $C^\bullet(F^\bullet, \mathcal{U})$ with anticommuting differentials $d' = \check{d} : C^p(F^q, \mathcal{U}) \to C^{p+1}(F^q, \mathcal{U})$ and $d'' = (-1)^p \cdot d_F : C^p(F^q, \mathcal{U}) \to$

$C^p(F^{q+1}, \mathcal{U})$. The cohomology of the total complex associated to $C^\bullet(F^\bullet, \mathcal{U})$ computes $\mathbb{H}^\bullet(F^\bullet)$. Now define a cup product

$$C^p(A^q, \mathcal{U}) \otimes C^{p'}(B^{q'}, \mathcal{U}) \longrightarrow C^{p+p'}((A \otimes B)^{q+q'}, \mathcal{U})$$

by

$$(\alpha \otimes \beta)_{i_0 \ldots i_{p+p'}} = (-1)^{qp'} \cdot \alpha_{i_0 \ldots i_p}|_{U_{i_0 \ldots i_{p+p'}}} \otimes \beta_{i_p \ldots i_{p+p'}}|_{U_{i_0 \ldots i_{p+p'}}}.$$

Thus composition induces a product

$$\mathrm{Ext}^i(F^\bullet, G^\bullet) \otimes \mathrm{Ext}^j(E^\bullet, F^\bullet) \longrightarrow \mathrm{Ext}^{i+j}(E^\bullet, G^\bullet).$$

In particular, $\mathcal{H}om^\bullet(E^\bullet, E^\bullet)$ has the structure of a sheaf of differential graded algebras and its cohomology $\mathrm{Ext}^\bullet(E^\bullet, E^\bullet)$ inherits a k-algebra structure. If we interpret $\mathrm{Ext}^i(E^\bullet, F^\bullet)$ as $\mathrm{Hom}_\mathcal{D}(E^\bullet, F^\bullet[i])$, where $\mathcal{D}$ is the derived category of quasi-coherent sheaves, then the cup product for Ext-groups is simply given by composition.

10.1.2 The trace map — For any locally free sheaf E let $tr_E : \mathcal{E}nd(E) \to \mathcal{O}_Y$ denote the trace map, which can be defined locally after trivializing E. More generally, if $E^\bullet$ is a finite complex of locally free sheaves, define a trace

$$tr_{E^\bullet} : \mathcal{H}om^\bullet(E^\bullet, E^\bullet) \longrightarrow \mathcal{O}_Y$$

by setting $tr_{E^\bullet}|_{\mathcal{H}om(E^i, E^j)} = 0$, except in the case $i = j$, when we put $tr_{E^\bullet}|_{\mathcal{E}nd(E^i)} = (-1)^i tr_{E^i}$. Also let

$$i_{E^\bullet} : \mathcal{O}_Y \longrightarrow \mathcal{H}om^0(E^\bullet, E^\bullet)$$

be the $\mathcal{O}_Y$-linear homomorphism that maps $1 \mapsto \sum_i \mathrm{id}_{E^i}$. Clearly,

$$tr_{E^\bullet}(i_{E^\bullet}(1)) = \sum_i (-1)^i \mathrm{rk}(E^i) =: \mathrm{rk}(E^\bullet).$$

If ψ and φ are homogeneous local sections in $\mathcal{H}om^\bullet(E^\bullet, E^\bullet)$, then

$$tr_{E^\bullet}(\varphi \circ \psi) = (-1)^{\deg \varphi \cdot \deg \psi} tr_{E^\bullet}(\psi \circ \varphi). \tag{10.3}$$

This relation can be easily seen as follows: we may assume that $\varphi \in \mathcal{H}om(E^i, E^j)$ and $\psi \in \mathcal{H}om(E^m, E^n)$. Then $tr_{E^\bullet}(\varphi \circ \psi)$ and $tr_{E^\bullet}(\psi \circ \varphi)$ are zero unless $j = m$ and $i = n$. Moreover, $tr_{E^i}(\psi \circ \varphi) = tr_{E^j}(\varphi \circ \psi)$. Hence

$$(-1)^i tr_{E^i}(\psi \circ \varphi) = (-1)^j tr_{E^j}(\varphi \circ \psi) \cdot (-1)^{i-j},$$

and $i - j \equiv \deg(\varphi) \equiv \deg(\psi) \equiv \deg(\varphi) \deg(\psi) \bmod 2$. Let d_E denote the differential in the complex $E^\bullet$. It follows from this and (10.1) that

$$tr_{E^\bullet}(d(\varphi)) = tr_{E^\bullet}(d_E \circ \varphi) - (-1)^{\deg(d_E) \deg(\varphi)} tr_{E^\bullet}(\varphi \circ d_E) = 0.$$

This shows that both $i_{E^\bullet}$ and $tr_{E^\bullet}$ are chain homomorphisms (where $\mathcal{O}_Y$ is a complex concentrated in degree 0) and induce homomorphisms

$$i : H^i(Y, \mathcal{O}_Y) \to \mathrm{Ext}^i(E^\bullet, E^\bullet) \text{ and } tr : \mathrm{Ext}^i(E^\bullet, E^\bullet) \to H^i(Y, \mathcal{O}_Y).$$

Lemma 10.1.3 — *The homomorphisms i and tr have the following properties:*
i) $tr \circ i = \mathrm{rk}(E^\bullet) \cdot \mathrm{id}$.
ii) For any two homogeneous elements $\varphi, \psi \in \mathrm{Ext}^\bullet(F^\bullet, F^\bullet)$ one has

$$tr(\varphi \circ \psi) = (-1)^{\deg(\varphi)\deg(\psi)} \cdot tr(\psi \circ \varphi).$$

Proof. The first assertion clearly follows from the equivalent assertion for $i_{E^\bullet}$ and $tr_{E^\bullet}$. As for the second, suppose $A^\bullet$, $B^\bullet$ are chain complexes and let $T : A^\bullet \otimes B^\bullet \to B^\bullet \otimes A^\bullet$ and $T : \mathbb{H}(A^\bullet) \otimes \mathbb{H}(B^\bullet) \to \mathbb{H}(B^\bullet) \otimes \mathbb{H}(A^\bullet)$ be the twist operator $a \otimes b \mapsto (-1)^{\deg(a) \cdot \deg(b)} b \otimes a$ for any homogeneous elements a and b. Then the diagram

$$
\begin{array}{ccc}
\mathbb{H}(A^\bullet \otimes B^\bullet) & \xrightarrow{\mathbb{H}(T)} & \mathbb{H}(B^\bullet \otimes A^\bullet) \\
\mu \uparrow & & \mu \uparrow \\
\mathbb{H}(A^\bullet) \otimes \mathbb{H}(B^\bullet) & \xrightarrow{T} & \mathbb{H}(B^\bullet) \otimes \mathbb{H}(A^\bullet)
\end{array}
$$

commutes. Specialize to the situation $A^\bullet = B^\bullet = \mathcal{H}om^\bullet(E^\bullet, E^\bullet)$ and let m denote the composition

$$\mathcal{H}om^\bullet(E^\bullet, E^\bullet) \otimes \mathcal{H}om^\bullet(E^\bullet, E^\bullet) \longrightarrow \mathcal{H}om^\bullet(E^\bullet, E^\bullet) \xrightarrow{tr_{E^\bullet}} \mathcal{O}_Y.$$

Then (10.3) can be expressed by saying that $m = m \circ T$. Thus

$$tr = \mathbb{H}(m) \circ \mu = \mathbb{H}(m) \circ \mathbb{H}(T) \circ \mu = \mathbb{H}(m) \circ \mu \circ T = tr \circ T,$$

which is the second assertion of the lemma. $\qquad\square$

An easy modification of the construction leads to homomorphisms

$$i : H^i(Y, \mathcal{N}) \to \mathrm{Ext}^i(E^\bullet, E^\bullet \otimes \mathcal{N}) \text{ and } tr : \mathrm{Ext}^i(E^\bullet, E^\bullet \otimes \mathcal{N}) \to H^i(Y, \mathcal{N})$$

for any coherent sheaf $\mathcal{N}$ on Y which satisfy relations analogous to *i)* and *ii)* in the lemma.

Definition 10.1.4 — *Let F be a coherent sheaf that admits a finite locally free resolution $F^\bullet \to F$. Then $\mathrm{Ext}^i(F^\bullet, F^\bullet \otimes \mathcal{N}) \cong \mathrm{Ext}^i(F, F \otimes \mathcal{N})$ for any locally free sheaf $\mathcal{N}$. Let*

$$\mathrm{Ext}^i(F, F \otimes \mathcal{N})_0 := \ker\left(tr : \mathrm{Ext}^i(F, F \otimes \mathcal{N}) \to H^i(Y, \mathcal{N})\right).$$

10.1.5 The Atiyah class — Let $p_1, p_2 : Y \times Y \to Y$ be the projections to the two factors.

Let $\mathcal{I}$ be the ideal sheaf of the diagonal $\Delta \subset Y \times Y$ and let $\mathcal{O}_{2\Delta} = \mathcal{O}_{Y \times Y}/\mathcal{I}^2$ denote the structure sheaf of the first infinitesimal neighbourhood of Δ. As $\mathcal{O}_\Delta$ is p_2 flat, the sequence

$$0 \to \mathcal{I}/\mathcal{I}^2 \to \mathcal{O}_{2\Delta} \to \mathcal{O}_\Delta \to 0$$

remains exact when tensorized with $p_2^* F$ for any locally free sheaf F on Y. Applying p_{1*}, we get an extension

$$0 \to F \otimes \Omega_Y \to p_{1*}(p_2^* F \otimes \mathcal{O}_{2\Delta}) \to F \to 0,$$

whose extension class $A(F) \in \mathrm{Ext}^1(F, F \otimes \Omega_Y)$ is called the *Atiyah class* of F. Note that

$$p_2^{-1} : \Gamma(F, U) \to \Gamma(p_2^* F \otimes \mathcal{O}_{2\Delta}, (Y \times U) \cap \Delta) = \Gamma(p_{1*}(p_2^* F \otimes \mathcal{O}_{2\Delta}), U)$$

provides a k-linear splitting of the extension. If s is an $\mathcal{O}_Y$-linear splitting, then $\nabla = s - p_2^{-1} : F \to F \otimes \Omega_Y$ is an algebraic connection on F, i.e. ∇ satisfies the Leibniz rule

$$\nabla(\alpha \cdot f) = d\alpha \otimes f + \alpha \cdot \nabla(f)$$

for any local sections $\alpha \in \mathcal{O}_Y$ and $f \in F$. Conversely, if ∇ is a connection, then $s = \nabla + p_2^{-1}$ is $\mathcal{O}_Y$-linear. Thus the Atiyah class $A(F)$ is the obstruction for the existence of an algebraic connection on F.

More generally, if $F^\bullet$ is a finite complex of locally free sheaves, one gets a short exact sequence

$$0 \to F^\bullet \otimes \Omega_Y \to p_{1*}(p_2^* F^\bullet \otimes \mathcal{O}_{2\Delta}) \to F^\bullet \to 0,$$

defining a class $A(F^\bullet) \in \mathrm{Hom}_D(F^\bullet, F^\bullet[1] \otimes \Omega_Y) = \mathrm{Ext}^1(F^\bullet, F^\bullet \otimes \Omega_Y)$.

A quasi-isomorphism $F^\bullet \to G^\bullet$ of finite complexes of locally free sheaves induces an isomorphism $\mathrm{Ext}^1(F^\bullet, F^\bullet \otimes \Omega_Y) \cong \mathrm{Ext}^1(G^\bullet, G^\bullet \otimes \Omega_Y)$ which identifies $A(F^\bullet)$ and $A(G^\bullet)$. In particular, if F is a coherent sheaf that admits a finite locally free resolution $F^\bullet \to F$, then $A(F^\bullet)$ is independent of the resolution and can be considered as the Atiyah class of F.

The class $A(F^\bullet)$ can be expressed in terms of Čech cocyles: choose an open affine covering $\mathcal{U} = \{U_i | i \in I\}$ such that the restriction of the sequence

$$0 \to F^q \otimes \Omega_Y \to p_{1*}(p_2^* F^q \otimes \mathcal{O}_{2\Delta}) \to F^q \to 0,$$

to U_i splits for all q and i. Thus there are local connections $\nabla_i^q : F^q|_{U_i} \to F^q \otimes \Omega_Y|_{U_i}$. (Note that the difference of two (local) connections is an $\mathcal{O}$-linear map.) Define cochains

$$\alpha' \in C^1(\mathcal{Hom}^0(F^\bullet, F^\bullet \otimes \Omega_Y), \mathcal{U}) \text{ and } \alpha'' \in C^0(\mathcal{Hom}^1(F^\bullet, F^\bullet \otimes \Omega_Y), \mathcal{U})$$

as follows:

$$\alpha'^q_{i_0 i_1} = \nabla_{i_0}^q|_{U_{i_0 i_1}} - \nabla_{i_1}^q|_{U_{i_0 i_1}} \text{ and } \alpha''^q_i = d_F \circ \nabla_i^q - \nabla_i^{q+1} \circ d_F,$$

where d_F is the differential of the complex $F^\bullet$. Since

$$d_F(\alpha'_{i_0 i_1}) = d_F \circ \alpha'_{i_0 i_1} - \alpha'_{i_0 i_1} \circ d_F = \alpha''_{i_0}|_{U_{i_0 i_1}} - \alpha''_{i_1}|_{U_{i_0 i_1}} = -(\check{d}\alpha'')_{i_0 i_1},$$

the element $\alpha = \alpha' + \alpha''$ is a cocyle in the total complex associated to the double complex $C^\bullet(\mathcal{H}om^\bullet(F^\bullet, F^\bullet \otimes \Omega_Y), \mathcal{U})$. The cohomology class of α is $A(F^\bullet)$.

This provides an easy way to identify the Atiyah class of the tensor product of two complexes $E^\bullet$ and $F^\bullet$: check that if ∇_E and ∇_F are (local) connections in locally free sheaves E and F, then $\nabla_{E \otimes F} := \nabla_E \otimes \mathrm{id}_F + \mathrm{id}_E \otimes \nabla_F$ is a (local) connection on $E \otimes F$. Whence one deduces that

$$A(E^\bullet \otimes F^\bullet) = A(E^\bullet) \otimes \mathrm{id}_F + \mathrm{id}_E \otimes A(F^\bullet).$$

10.1.6 Newton polynomials — Assume again that $F^\bullet$ is a finite complex of locally free sheaves. Let $A(F^\bullet)^i \in \mathrm{Ext}^i(F^\bullet, F^\bullet \otimes \Omega_Y^i)$ be the image of the i-fold composition $A(F^\bullet) \circ \ldots \circ A(F^\bullet) \in \mathrm{Ext}^i(F^\bullet, F^\bullet \otimes \Omega_Y^{\otimes i})$ under the homomorphism induced by $\Omega_Y^{\otimes i} \to \Omega_Y^i$ and define the i-th *Newton polynomial* of $F^\bullet$ by

$$\gamma^i(F^\bullet) := tr(A(F^\bullet)^i) \in H^i(Y, \Omega_Y^i).$$

These classes differ by a factor $i!$ from the i-th component of the Chern character of $F^\bullet$. As the trace map does not see anything from a Čech cocycle in

$$\prod_{p+q=i} C^p(\mathcal{H}om^q(F^\bullet, F^\bullet \otimes \Omega_Y^i), \mathcal{U})$$

except the components with $p = i$, $q = 0$, it follows that $\gamma^i(F^\bullet)$ depends only on the α'-part of the cocycle $\alpha' + \alpha''$ that gives $A(F^\bullet)$. In particular, $\gamma^n(F^\bullet) = \sum_\ell (-1)^\ell \gamma^n(F^\ell)$. The k-linear differential $d : \Omega_Y^i \to \Omega_Y^{i+1}$ induces k-linear maps $d : H^j(Y, \Omega_Y^i) \to H^j(Y, \Omega_Y^{i+1})$. If F is a locally free sheaf, then $\gamma^i(F)$ is d-closed, i.e. $d(\gamma^i(F)) = 0$ for all i: as γ^i is additive in short exact sequences, we can reduce to the case of line bundles using the splitting principle. If F is a line bundle given by transition functions $f_{ij} \in \mathcal{O}^*(U_{ij})$, then $d \log f_{ij} = f_{ij}^{-1} df_{ij}$ is a Čech cocycle for $A(F)$ that clearly vanishes under d.

10.1.7 Relative versions — Let X be a smooth projective surface, S a base scheme of finite type over k and let $p : S \times X \to S$ and $q : S \times X \to X$ be the projections. Any S-flat family F of coherent sheaves admits a finite locally free resolution $F^\bullet \to F$ so that we can apply the above machinery to F.

Recall that $\mathcal{E}xt_p^j(F, \,.\,)$ are the derived functors of $\mathcal{H}om_p(F, \,.\,) = p_* \circ \mathcal{H}om(F, \,.\,)$. It is easy to see that $\mathcal{E}xt_p^j(F, G)$ is the sheafification of the presheaf

$$U \mapsto \mathrm{Ext}^j(F|_{U \times X}, G|_{U \times X}).$$

If $F^\bullet \to F$ is a finite locally free resolution of F, then $\mathrm{Ext}^j(F^\bullet, F^\bullet) \cong \mathrm{Ext}^j(F, F)$. Thus sheafifying the cup product and the maps i and tr defined for $F^\bullet$, we get maps

$$\mathcal{E}xt^j_p(F, F) \times \mathcal{E}xt^{j'}_p(F, F) \longrightarrow \mathcal{E}xt^{j+j'}_p(F, F),$$

$$tr : \mathcal{E}xt^j_p(F, F) \longrightarrow R^j p_* \mathcal{O}_{S \times X} \cong \mathcal{O}_S \otimes_k H^j(X, \mathcal{O}_X)$$

and

$$i : \mathcal{O}_S \otimes_k H^j(X, \mathcal{O}_X) \longrightarrow \mathcal{E}xt^j_p(F, F),$$

satisfying the relations

$$tr \circ i = \mathrm{rk}(F) \cdot id \quad \text{and} \quad tr(\varphi \circ \psi) = (-1)^{\deg(\varphi) \deg(\psi)} \cdot tr(\psi \circ \varphi).$$

10.1.8 The Kodaira-Spencer map — Let F be an S-flat family on a smooth projective surface X. Choosing a locally free resolution $F^\bullet \to F$ we can define the Atiyah class $A(F) = A(F^\bullet) \in \mathrm{Ext}^1(F^\bullet, F^\bullet \otimes \Omega_{S \times X})$ and consider the induced section under the global-local map

$$\mathrm{Ext}^1(F^\bullet, F^\bullet \otimes \Omega_{S \times X}) \longrightarrow H^0(S, \mathcal{E}xt^1_p(F^\bullet, F^\bullet \otimes \Omega_{S \times X}))$$

coming from the spectral sequence $H^i(S, \mathcal{E}xt^j_p) \Rightarrow \mathrm{Ext}^{i+j}_{S \times X}$. The direct sum decomposition $\Omega_{S \times X} = p^* \Omega_S \oplus q^* \Omega_X$ leads to an analogous decomposition $A(F) = A(F)' + A(F)''$. By definition, the *Kodaira-Spencer map* associated to the family F is the composition

$$KS : \Omega_{\check{S}} \xrightarrow{A(F)'} \Omega_{\check{S}} \otimes \mathcal{E}xt^1_p(F^\bullet, F^\bullet \otimes p^* \Omega_S) \longrightarrow$$

$$\longrightarrow \mathcal{E}xt^1_p(F^\bullet, F^\bullet \otimes p^*(\Omega_{\check{S}} \otimes \Omega_S)) \longrightarrow \mathcal{E}xt^1_p(F^\bullet, F^\bullet).$$

Example 10.1.9 — Let X be a smooth surface as above and $S = \mathrm{Spec}(k[\varepsilon])$. Let $0 \longrightarrow F \xrightarrow{i} \mathcal{F} \xrightarrow{\pi} F \longrightarrow 0$ be a short exact sequence representing an extension class $v \in \mathrm{Ext}^1_X(E, E)$. We can think of $\mathcal{F}$ as an S-flat family by letting ε act on $\mathcal{F}$ as the homomorphism $i \circ \pi$. Decompose the Atiyah class $A(\mathcal{F}) = A(\mathcal{F})' + A(\mathcal{F})''$ according to the splitting

$$\mathrm{Ext}^1_{S \times X}(\mathcal{F}, \mathcal{F} \otimes \Omega_{S \times X}) = \mathrm{Ext}^1_{S \times X}(\mathcal{F}, \mathcal{F} \otimes p^* \Omega_S) \oplus \mathrm{Ext}^1_{S \times X}(\mathcal{F}, \mathcal{F} \otimes q^* \Omega_X).$$

Since $\Omega_S \cong k \cdot d\varepsilon$, and since $\mathcal{F}$ is S-flat, we have

$$\mathrm{Ext}^1_{S \times X}(\mathcal{F}, \mathcal{F} \otimes p^* \Omega_S) \cong \mathrm{Ext}^1_{S \times X}(\mathcal{F}, F) \cong \mathrm{Ext}^1_X(F, F).$$

We want to show that under these isomorphisms $A(\mathcal{F})'$ is mapped to v. According to the definition of the Atiyah class we first consider the short exact sequence of coherent sheaves over $\mathrm{Spec}(k[\varepsilon_1, \varepsilon_2]/(\varepsilon_1, \varepsilon_2)^2) \times X$

$$0 \longrightarrow F \xrightarrow{i'} G \xrightarrow{\pi'} \mathcal{F} \longrightarrow 0, \tag{10.4}$$

where ε_1 and ε_2 act trivially on F and by $i \circ \pi$ on $\mathcal{F}$, and

$$G \cong k[\varepsilon_1] \otimes_k \mathcal{F} \Big/ \varepsilon_1 \varepsilon_2 \mathcal{F} \cong \mathcal{F} \oplus F$$

with actions $\varepsilon_1 = \begin{pmatrix} 0 & \pi \\ 0 & 0 \end{pmatrix}$ and $\varepsilon_2 = \begin{pmatrix} i\pi & 0 \\ 0 & 0 \end{pmatrix}$. Now $A(\mathcal{F})'$ is precisely the extension class of (10.4), considered as a sequence of $k[\varepsilon_1] \otimes \mathcal{O}_X$-modules. But it is easy to see that there is a pull-back diagram

$$
\begin{array}{ccccccccc}
0 & \longrightarrow & F & \xrightarrow{i} & \mathcal{F} & \xrightarrow{\pi} & F & \longrightarrow & 0 \\
& & \| & & \uparrow{\scriptstyle t'} & & \uparrow{\scriptstyle \pi} & & \\
0 & \longrightarrow & F & \xrightarrow{i'} & G & \xrightarrow{\pi'} & \mathcal{F} & \longrightarrow & 0,
\end{array}
$$

which shows that $A(\mathcal{F})' = v$. $\square$

10.2 The Tangent Bundle

Let X be a smooth projective surface and let M^s be the moduli space of stable sheaves on X of rank $r \geq 1$ and Chern classes c_1 and c_2. The open subset $M_0 \subset M^s$ of points $[F]$ such that $\mathrm{Ext}_X^2(F, F)_0$ vanishes is smooth according to Theorem 4.5.4. Suppose there exists a universal family $\mathcal{E}$ on $M_0 \times X$. The Kodaira-Spencer map associated to $\mathcal{E}$ is a sheaf homomorphism

$$KS : \mathcal{T}_{M_0} \longrightarrow \mathcal{E}xt_p^1(\mathcal{E}, \mathcal{E}).$$

It is the goal of this section to show that this map is an isomorphism. In fact one can make sense of the map KS and the $\mathcal{E}xt$-sheaf on the right hand side even if a universal family does not exist. We will prove this first.

There are two ways to deal with the problem that a universal family need not exist: either one uses an étale cover of the moduli space, over which a family exists. Or one works on the Quot-scheme that arises in the construction of M^s and shows that all constructions are equivariant and descend. We will follow this approach.

Let $R^s \subset \mathrm{Quot}(\mathcal{H}, P)$ be the open subset as defined in 4.3 so that $\pi : R^s \to M^s$ is a geometric quotient. If $\mathcal{O}_{R^s} \otimes \mathcal{H} \to \widetilde{F}$ is the universal quotient, we can form the sheaves $\mathcal{E}xt_p^i(\widetilde{F}, \widetilde{F})$. These inherit a natural action of $\mathrm{GL}(V)$ 'by conjugation'. In particular, the center of $\mathrm{GL}(V)$ acts trivially. Moreover, both the cup product and the trace map are equivariant. By descent theory, $\mathcal{E}xt^i_p(\widetilde{F},\widetilde{F})$ and these two maps descend to a coherent sheaf $\widetilde{\mathcal{E}xt}^i_p$ on M^s and homomorphisms

$$\widetilde{\mathcal{E}xt}^i_p \otimes \widetilde{\mathcal{E}xt}^j_p \longrightarrow \widetilde{\mathcal{E}xt}^{i+j}_p \quad \text{and} \quad \widetilde{\mathcal{E}xt}^i_p \xrightarrow{tr} H^i(X,\mathcal{O}_X) \otimes_k \mathcal{O}_{M^s}.$$

Suppose, a universal family $\mathcal{E}$ exists. Then $\pi^*\mathcal{E} \cong p^*A \otimes \widetilde{F}$ for some appropriately linearized line bundle A on R^s. Therefore

$$\pi^*\mathcal{E}xt^i_p(\mathcal{E},\mathcal{E}) \cong \mathcal{E}xt^i_p(\pi^*\mathcal{E},\pi^*\mathcal{E}) \cong \mathcal{E}xt^i_p(\widetilde{F},\widetilde{F}) \otimes \mathcal{E}nd(A) \cong \mathcal{E}xt^i_p(\widetilde{F},\widetilde{F}).$$

Thus in the presence of a universal family $\mathcal{E}$ we have $\widetilde{\mathcal{E}xt}^i_p \cong \mathcal{E}xt^i_p(\mathcal{E},\mathcal{E})$. For this reason we give in to the temptation to use the notation $\mathcal{E}xt^i_p(\mathcal{E},\mathcal{E})$ even if a universal family $\mathcal{E}$ itself does not exist.

Theorem 10.2.1 — *There are natural isomorphisms*

$$\mathcal{E}xt^1_p(\mathcal{E},\mathcal{E})|_{M_0} \cong T_{M_0} \quad \text{and} \quad \mathcal{E}xt^i_p(\mathcal{E},\mathcal{E})|_{M_0} \cong H^i(X,\mathcal{O}_X) \otimes_k \mathcal{O}_{M_0} \text{ for } i=0,2.$$

(The theorem immediately implies Theorem 8.3.2. Note that there we considered the moduli space of sheaves with fixed determinant. However, this has no effect on the canonical bundle, for the difference is the trivial bundle $\det H^*(X,\mathcal{O}_X) \otimes \mathcal{O}_{M_0}$.)

Proof. $R_0 = \pi^{-1}(M_0)$ and M_0 are smooth by Theorem 4.5.4. By the definition of M_0 we have $\text{Ext}^2_X(F,F)_0 = 0$ for all $[F] \in M_0$, and this implies that the homomorphisms

$$i : H^2(X,\mathcal{O}_X) \otimes \mathcal{O}_{R_0} \to \mathcal{E}xt^2_p(\widetilde{F},\widetilde{F}) \text{ and } tr : \mathcal{E}xt^2_p(\widetilde{F},\widetilde{F}) \longrightarrow H^2(X,\mathcal{O}_X) \otimes \mathcal{O}_{R_0}$$

are isomorphisms on the fibres over closed points, and hence are surjective as homomorphisms of sheaves. Since $tr \circ i = r \cdot \text{id}$, both maps are in fact isomorphisms. Similarly, since $\text{Hom}(F,F)_0 = 0$ for stable sheaves, the same argument shows that $\mathcal{E}xt^0_p(\widetilde{F},\widetilde{F}) \cong \mathcal{O}_{R_0}$. From the first isomorphism one deduces that $\mathcal{E}xt^1_p$ commutes with base change, and from the second that $\mathcal{E}xt^1_p(\widetilde{F},\widetilde{F})$ is locally free.

Now consider the Kodaira-Spencer map $KS : T_{R_0} \longrightarrow \mathcal{E}xt^1_p(\widetilde{F},\widetilde{F})$. Let $[\rho : \mathcal{H} \to F] \in R_0$ be a closed point. It follows from Example 10.1.9 and Appendix 2.A that the following diagram commutes:

$$
\begin{array}{ccc}
T_{[\rho]}R_0 & \xrightarrow{\quad T\pi \quad} & T_{[F]}M_0 \\
{\scriptstyle\cong}\downarrow & \searrow{\scriptstyle KS([\rho])} & \downarrow{\scriptstyle\cong} \\
\text{Hom}_X(\ker(\rho),F) & \xrightarrow{\quad\delta\quad} & \text{Ext}^1_X(F,F)
\end{array}
$$

In the diagram the vertical isomorphisms come from deformation theory (cf. 2.A), and δ is the coboundary operator. We conclude that the Kodaira-Spencer map factors through an isomorphism $\pi^* T_{M_0} \to \mathcal{E}xt^1_p(\widetilde{F}, \widetilde{F})$. Since the Atiyah class is invariant, this isomorphism is equivariant and descends to an isomorphism $T_{M_0} \to \mathcal{E}xt^1_p(\mathcal{E}, \mathcal{E})$. $\square$

10.3 Forms on the Moduli Space

We are now going to describe natural one- and two-forms on the moduli space as announced in the introduction.

Let F be an S-flat family of sheaves on a smooth projective surface X. The Newton polynomials $\gamma^i(F) := \gamma^i(F^\bullet) \in H^i(\Omega^i_{S \times X})$ are independent of the choice of a finite locally free resolution $F^\bullet \to F$. Since $\Omega_{S \times X} = p^* \Omega_S \oplus q^* \Omega_X$, and since Ω_X is locally free, there is a Künneth decomposition

$$H^n(S \times X, \Omega^n_{S \times X}) \cong \bigoplus_{i,j} H^i(S, \Omega^j_S) \otimes H^{n-i}(X, \Omega^{n-j}_X).$$

Let $\gamma'(F)$ and $\gamma''(F)$ denote the components of $\gamma^2(F)$ in $H^0(S, \Omega^2_S) \otimes H^2(X, \mathcal{O}_X)$ and $H^0(S, \Omega_S) \otimes H^2(X, \Omega_X)$, respectively.

Definition 10.3.1 — *Let τ_F and δ_F be the homomorphisms given by*

$$\tau_F : H^0(X, K_X) \xrightarrow{\cong} H^2(X, \mathcal{O}_X)^\vee \xrightarrow{\gamma'} H^0(S, \Omega^2_S)$$

and

$$\delta_F : H^0(X, \Omega_X) \xrightarrow{\cong} H^2(X, \Omega_X)^\vee \xrightarrow{\gamma''} H^0(S, \Omega_S).$$

(Here $\cong$ is Serre duality.)

Proposition 10.3.2 — *For any $\alpha \in H^0(X, K_X)$ or $H^0(X, \Omega^1_X)$ the associated two-form $\tau_F(\alpha)$ or one-form $\delta_F(\alpha)$, respectively, on S is closed.*

Proof. The decomposition $d_{S \times X} = d_s \otimes 1 + 1 \otimes d_X$ induces similar splittings for the Künneth components:

$$d_{S \times X} = d_S \otimes 1 + 1 \otimes d_X : H^i(\Omega^j_S) \otimes H^{n-i}(\Omega^{n-j}_X)$$
$$\longrightarrow H^i(\Omega^{j+1}_S) \otimes H^{n-i}(\Omega^{n-j}_X) \oplus H^i(\Omega^j_S) \otimes H^{n-i}(\Omega^{n-j+1}_X).$$

Since X is a smooth projective variety, one has $d_X(\nu) = 0$ for any element $\nu \in H^i(\Omega^j_X)$. Hence, $d_{S \times X}(\gamma(F)) = 0$ implies $d_{S \times X}(\gamma') = 0 = d_{S \times X}(\gamma'')$. Write $\gamma' = \sum_\ell \mu_\ell \otimes \nu_\ell$ for elements $\mu_\ell \in H^0(\Omega^2_S)$ and $\nu_\ell \in H^2(\mathcal{O}_X)$. Then

$$0 = d_{S \times X}(\gamma') = \sum_\ell d_S(\mu_\ell) \otimes \nu_\ell + \sum_\ell \mu_\ell \otimes d_X(\nu_\ell) = \sum_\ell d_S(\mu_\ell) \otimes \nu_\ell.$$

Hence

$$d_S(\tau_F(\alpha)) = d_S\left(\sum_\ell \mu_\ell \cdot \alpha(\nu_\ell)\right) = \sum_\ell d_S(\mu_\ell) \cdot \alpha(\nu_\ell) = 0$$

Similarly one shows $d_S(\delta_F(\alpha)) = 0$. $\qquad\qquad\square$

Lemma 10.3.3 — *Suppose that S is smooth. Then for each $\alpha \in H^0(X, K_X)$ the two-form τ_α on S is the composition of the maps:*

$$T_s S \times T_s S \xrightarrow{\ KS \times KS\ } \mathrm{Ext}^1_X(F_s, F_s) \times \mathrm{Ext}^1_X(F_s, F_s) \xrightarrow{\ \circ\ } \mathrm{Ext}^2_X(F_s, F_s)$$
$$\xrightarrow{\ tr\ } H^2(X, \mathcal{O}_X) \xrightarrow{\ \alpha\ } H^2(X, K_X) \cong k.$$

Proof. This follows readily from the definitions. $\qquad\qquad\square$

In order to define forms on M^s we use quasi-universal families, which always exist by Proposition 4.6.2. The following lemma implies that the construction is independent of the choice of the quasi-universal family:

Lemma 10.3.4 — *Let F be an S-flat family of sheaves on X and let B be a locally free sheaf on S. Then $\gamma'(F \otimes p^*B) = \mathrm{rk}(B) \cdot \gamma'(F)$ and $\gamma''(F \otimes p^*B) = \mathrm{rk}(B) \cdot \gamma''(F)$.*

Proof. We have $A(F \otimes p^*B) = A(F) \otimes \mathrm{id}_B + \mathrm{id}_B \otimes p^*A(B)$. The definition of γ^2 shows, that only $A(F) \otimes \mathrm{id}_B$ contributes to the $H^0(S, \Omega^\bullet_S) \otimes H^2(X, \Omega^\bullet_X)$ component of $\gamma^2(F \otimes p^*B)$, which is relevant for γ' and γ''. Since $tr(A(F)^2 \otimes \mathrm{id}_B) = \mathrm{rk}(B) \cdot tr(A(F)^2)$, the assertion follows. $\qquad\qquad\square$

Definition and Theorem 10.3.5 — *Let $\mathcal{E}$ be a quasi-universal family on $M^s \times X$. Then*

$$\tau := \frac{r}{\mathrm{rk}(\mathcal{E})} \tau_{\mathcal{E}} : H^0(X, K_X) \longrightarrow H^0(M^s, \Omega^2_{M_s})$$

and

$$\delta := \frac{r}{\mathrm{rk}(\mathcal{E})} \tau_{\mathcal{E}} : H^0(X, \Omega_X) \longrightarrow H^0(M^s, \Omega_{M_s})$$

are independent of $\mathcal{E}$.

Proof. If $\mathcal{E}$ and $\mathcal{E}'$ are any two quasi-universal families, then there are locally free sheaves B and B' on M^s such that $\mathcal{E} \otimes p^*B \cong \mathcal{E}' \otimes p^*B'$. The assertion then follows from Lemma 10.3.4. $\qquad\qquad\square$

In order to make use of the differential forms δ and τ in the birational classification of moduli spaces, it is important to extend them from M^s to the compactification $M(r, c_1, c_2)$

(or $M(r, \mathcal{Q}, c_2)$). The case of the one-form is less involved and provides an alternative definition:

Recall that for a smooth projective variety X the Albanese variety is defined as $\mathrm{Alb}(X) = H^0(X, \Omega_X)^{\vee}/H_1(X, \mathbb{Z})$ (cf. Section 5.1). This leads to a canonical isomorphism of the spaces $H^0(\mathrm{Alb}(X), \Omega_{\mathrm{Alb}(X)})$ and $H^0(X, \Omega_X)$. Under this identification the differential $H^0(X, \Omega_X) \to H^0(\mathrm{Alb}(X), \Omega_{\mathrm{Alb}(X)})$ of the Albanese morphism $A : X \to \mathrm{Alb}(X)$ equals the identity map.

Let $M := M(r, \mathcal{Q}, c)$ and fix a point $x \in X$.

Proposition 10.3.6 — *There is a natural morphism $\varphi : M \to \mathrm{Alb}(X)$, which maps $[E]$ to $\tilde{A}(\tilde{c}_2(E))$, where $\tilde{c}_2(E)$ is the second Chern class of E in the Chow-group $CH^2(X)$. If $H^0(\mathrm{Alb}(X), \Omega_{\mathrm{Alb}(X)})$ is identified with $H^0(X, \Omega_X)$, then $\varphi^*(\alpha) = -\delta(\alpha)$ on M^s.*

Proof. Any family F on $S \times X$ defines a morphism $S \to \mathrm{Alb}(X)$ by mapping $t \in S$ to $\tilde{A}(\tilde{c}_2(F_t))$. Since M corepresents the moduli functor, we get a morphism $\varphi : M \to \mathrm{Alb}(X)$. The assertion on $\varphi^*(\alpha)$ is more complicated. For a smooth basis representing locally free sheaves, the proof can be found in [89]. But one can give an algebraic proof for the general case as well, which we omit. $\square$

Note that this provides an alternative definition of δ and immediately shows that $\delta(\alpha)$ is closed and extends to the complete moduli space. In particular, we obtain a one-form on any smooth model of M.

Not much is known about the morphism φ in general. For the rank one case, i.e. $M = \mathrm{Hilb}^c(X)$, there is the following theorem due to M. Huibregtse [110].

Theorem 10.3.7 — *For $c \gg 0$, the morphism $\varphi : \mathrm{Hilb}^c(X) \to \mathrm{Alb}(X)$ is surjective and all the fibres are irreducible of dimension $2c - h^1(X, \mathcal{O}_X)$. If $c \gg 0$, the morphism is smooth if and only if $A : X \to \mathrm{Alb}(X)$ is smooth.* $\square$

Proposition 5.1.5 and Theorem 5.1.6 in Section 5.1 give a first hint for the higher rank case.

Corollary 10.3.8 — *If $c \gg 0$ and $r \geq 1$, then $\varphi : M(r, \mathcal{Q}, c) \to \mathrm{Alb}(X)$ is surjective.* $\square$

This can also be used to show the non-degeneracy of the one-forms $\delta(\alpha)$. In fact, 10.3.8 implies that for any $\alpha \neq 0$ and $c \gg 0$ the one-form $\delta(\alpha)$ is not trivial.

The consequences for the birational geometry of the moduli space will be discussed in Chapter 11.

We certainly cannot expect to have an analogous situation for the two-forms $\tau(\alpha)$. Since the dimension of the moduli space grows with c_2 and, at least in special examples, $\tau(\alpha)$ is generically non-degenerate, $\tau(\alpha)$ cannot be the pull-back of a two-form on a fixed finite dimensional variety Y under a morphism $M^s \to Y$. Work of Mumford on the Chow group of surfaces with $p_g > 0$ suggests that Y should be replaced by $CH^2(X)$, which is neither finite dimensional nor a variety [193]. This 'non-geometric' behaviour of τ makes it more difficult to extend it over a suitable compactification of M^s. For many purposes the following is sufficient.

Corollary 10.3.9 — *There exists a morphism $\psi : \tilde{M} \to M(r, c_1, c_2)$ from a projective variety $\tilde{M}$, which is birational over M^s and such that the pull-back of any two-form $\tau(\alpha)$ on M^s extends to $\tilde{M}$.*

Proof. This is a consequence of 4.B.5. Indeed, if $w(\alpha) := \tau_{\mathcal{E}}(\alpha)$, where $\mathcal{E}$ is the family on $\tilde{M} \times X$, then $w(\alpha)|_{\psi^{-1}(M^s)} = \tau_{\mathcal{E}}(\alpha)|_{\psi^{-1}(M^s)} = \psi^*(\tau(\alpha))$, since the pull-back of a quasi-universal family on $M^s \times X$ to $\psi^{-1}(M^s)$ is equivalent to $\mathcal{E}$ constructed in 4.B.5. $\square$

10.4 Non-Degeneracy of Two-Forms

In the previous section we constructed for each global section $\alpha \in H^0(X, K_X)$ a two-form $\tau(\alpha)$ on the stable part M^s of the moduli space $M(r, c_1, c_2)$. Moreover, if $[E]$ is a closed point in the good part $M_0 \subset M^s$, i.e. if $\mathrm{Ext}^2(E, E)_0 = 0$, then $T_{[E]}M^s \cong \mathrm{Ext}^1_X(E, E)$, and with respect to this identification $\tau(\alpha)([E])$ is given by the map

$$\tilde{\tau} : \mathrm{Ext}^1(E, E) \times \mathrm{Ext}^1(E, E) \xrightarrow{\circ} \mathrm{Ext}^2(E, E)$$
$$\xrightarrow{tr} H^2(X, \mathcal{O}_X) \xrightarrow{\alpha} H^2(X, K_X) \cong k.$$

(cf. 10.2.1 and 10.3.3.) Thus the question whether $\tau(\alpha)$ is non-degenerate in good points $[E] \in M_0$ is answered by the following 'local' proposition:

Proposition 10.4.1 — *The form $\tilde{\tau}$ is non-degenerate if and only if multiplication by α induces an isomorphism $\alpha_* : \mathrm{Ext}^1_X(E, E) \to \mathrm{Ext}^1_X(E, E \otimes K_X)$. Similarly, the restriction of $\tilde{\tau}$ to the subspace $\mathrm{Ext}^1_X(E, E)_0$ is non-degenerate if and only if the homomorphism $\alpha_* : \mathrm{Ext}^1_X(E, E)_0 \to \mathrm{Ext}^1_X(E, E \otimes K_X)_0$ is an isomorphism.*

Proof. In order to prove the proposition, we need to relate the definition of $\tilde{\tau}$ (involving cup product and trace) to Serre duality. Let $E^\bullet \to E$ be a finite locally free resolution. Note that the isomorphism $\mathcal{H}om(E^j, E^i) \cong \mathcal{H}om(E^i, E^j)^\vee$ can be obtained by the pairing

$$\mathcal{H}om(E^i, E^j) \otimes \mathcal{H}om(E^j, E^i) \xrightarrow{\circ} \mathcal{H}om(E^j, E^j) \xrightarrow{tr_{E^j}} \mathcal{O}_X.$$

More generally, if $A^\bullet = \mathcal{H}om^\bullet(E^\bullet, E^\bullet)$, then

$$A^\bullet \otimes A^\bullet \xrightarrow{\;\circ\;} A^\bullet \xrightarrow{\;tr_{E^\bullet}\;} \mathcal{O}_X$$

is a perfect pairing and leads to an isomorphism $A^\bullet \to \mathcal{H}om^\bullet(A^\bullet, \mathcal{O}_X)$. Hence for any section $\alpha : \mathcal{O}_X \to K_X$ there is a commutative diagram

$$
\begin{array}{ccccc}
(A^\bullet \otimes K_X) \otimes A^\bullet & \xrightarrow{\;\cong\;} & \mathcal{H}om^\bullet(A^\bullet, K_X) \otimes A^\bullet & \xrightarrow{\;eval\;} & K_X \\
{\scriptstyle (1\otimes\alpha)\otimes 1}\big\uparrow & & & & \big\uparrow{\scriptstyle \alpha} \\
A^\bullet \otimes A^\bullet & \xrightarrow{\quad\circ\quad} & A^\bullet & \xrightarrow{\quad tr\quad} & \mathcal{O}_X
\end{array}
$$

Passing to cohomology we get

$$
\begin{array}{ccccc}
\mathrm{Ext}^i_X(E, E\otimes K_X) \otimes \mathrm{Ext}^j_X(E, E) & \xrightarrow{\;\cong\;} & \mathbb{E}\mathrm{xt}^i(A^\bullet, K_X) \otimes \mathbb{H}^j(A^\bullet) & \longrightarrow & H^{i+j}(X, K_X) \\
{\scriptstyle \alpha_* \otimes 1}\big\uparrow & & & & \big\uparrow{\scriptstyle \alpha} \\
\mathrm{Ext}^i_X(E, E) \otimes \mathrm{Ext}^j_X(E, E) & \longrightarrow & \mathrm{Ext}^{i+j}_X(E, E) & \xrightarrow{\;tr\;} & H^{i+j}(X, \mathcal{O}_X).
\end{array}
$$

Observe that for $i = j = 1$, $\tilde{\tau}$ is the map from the lower left corner of the diagram to the upper right corner.

Serre duality in its general form says that for a smooth variety X of dimension n and a bounded complex $A^\bullet$ of coherent sheaves the pairing

$$\mathbb{E}\mathrm{xt}^{n-i}(A^\bullet, K_X) \otimes \mathbb{H}^i(X, A^\bullet) \longrightarrow H^n(X, K_X) \xrightarrow{\;\cong\;} k$$

is perfect (cf. [96]). If we apply this to the diagram above with $i = j = 1$ in the case of a surface X, we get: $\tilde{\tau}$ is a non-degenerate if and only if α_* is an isomorphism, thus proving the first part of the proposition.

For the second observe, that for any local section $f \in A^\bullet(U), U \subset X$, and $1 := \sum_i \mathrm{id}_{E^i}$ one certainly has $tr(1 \cdot f) = tr(f)$. Thus the splitting $A^\bullet = \mathcal{O}_X \oplus \ker(tr_{E^\bullet})$ is orthogonal with respect to the bilinear map $A^\bullet \otimes A^\bullet \to A^\bullet \xrightarrow{\;tr\;} \mathcal{O}_X$. This implies that the splitting $\mathrm{Ext}^i_X(E, E) = H^i(X, \mathcal{O}_X) \oplus \mathrm{Ext}^i(E, E)_0$ is orthogonal with respect to $\tilde{\tau}$. It is also respected by Serre duality. Hence one concludes as before. $\qquad\square$

Corollary 10.4.2 — *Let X be a surface with $K_X \cong \mathcal{O}_X$, i.e. X is either abelian or K3. Then the pairing*

$$\tilde{\tau}(1) : \mathrm{Ext}^1_X(E, E) \times \mathrm{Ext}^1_X(E, E) \to k$$

is a non-degenerate alternating form, and the same holds for the restriction of $\tau_E(1)$ to the linear subspace $\mathrm{Ext}^1_X(E, E)_0$. $\qquad\square$

Combining this corollary with the fact that under the same hypotheses the smoothness obstruction group $\text{Ext}^2_X(E, E)_0$ vanishes for any stable sheaf, we get Mukai's celebrated result on the existence of a holomorphic symplectic structure on the moduli space of sheaves on K3 and abelian surfaces:

Theorem 10.4.3 — *If X is a smooth projective surface with $K_X \cong \mathcal{O}_X$, then $M(r, c_1, c_2)^s$ admits a non-degenerate symplectic structure.* □

If $K_X \not\cong \mathcal{O}_X$ one does not expect $\tau(\alpha)$ to be non-degenerate at every point of the moduli space. The best one can hope for is a generic non-degeneracy. Of course, a necessary condition is that the moduli space is of even dimension. Suppose $[F] \in M_0 \subset M^s$ is a good point. According to Proposition 10.4.1, $\tau_F(\alpha)$ is non-degenerate at $[F]$ if and only if $\alpha_* : \text{Ext}^1_X(E, E) \to \text{Ext}^1_X(E, E \otimes K_X)$ is an isomorphism.

Using the exact sequence $0 \to \mathcal{O}_X \to K_X \to K_X|_D \to 0$, where D is the divisor defined by $\alpha \in H^0(X, K_X)$, one sees that a sufficient condition is the vanishing of $\text{Hom}(E, E \otimes K_X|_D)$. If one restricts to the moduli space $M(r, \mathcal{Q}, c_2)$ of sheaves with fixed determinant it suffices to show $\text{Hom}(E, E \otimes K_X|_D)_0 = 0$ in order to have non-degeneracy of $\tau(\alpha)$ at $[E]$.

The following is a crucial result in the theory. It is only known for the rank two case but hopefully true in general. Recently, Brussee pointed out that the assertion is a consequence of the relation between Seiberg-Witten invariants and Donaldson polynomials and the fact that the only Seiberg-Witten class of a minimal surface of general type is $\pm K_X$.

Theorem 10.4.4 — *Let X be a surface of general type and let $D = Z(\alpha) \in |K_X|$ be a reduced connected canonical divisor. If $\chi(\mathcal{O}_X) + c_1^2(\mathcal{Q}) \equiv 0 \bmod 2$, then for $c_2 \gg 0$ the symplectic structure $\tau(\alpha)$ on $M(2, \mathcal{Q}, c_2)$ is generically non-degenerate.*

Proof. Note that the assumption $\chi(\mathcal{O}_X) + c_1^2(\mathcal{Q}) \equiv 0 \bmod 2$ is equivalent to $\dim M \equiv 0 \bmod 2$. Otherwise the symplectic structure could never be non-degenerate.

The proof of the theorem consists of two parts. First, one establishes the existence of a rank two vector bundle F on D such that $\text{Hom}(F, F \otimes K_X|_D)_0 = 0$. Next, one uses this to show that the restriction of the generic bundle $E \in M$ shares the same property. Once the existence of F is known, the proof goes through in the higher rank case as well. The first step is highly non-trivial even when D is smooth. The proof is omitted.

Let us sketch the second part of the proof. Here one makes use of some results in deformation theory. By the same method as in the proof of Theorem 9.3.3 one shows that for $c_2 \gg 0$ the generic sheaf $E \in M(r, \mathcal{Q}, c_2)$ satisfies $\text{Hom}(E, E \otimes K_X^2)_0 = 0$, and hence, $\text{Ext}^2(E, E \otimes (-K_X))_0 = 0$. The infinitesimal deformations of E on X with fixed determinant $\mathcal{Q}$ are given by $\text{Ext}^1(E, E)_0$ and of $E|_D$ by $\text{Ext}^1_D(E_D, E_D)_0$. For a locally free E the cokernel of the natural map $\mathrm{Ext}^1(E, E)_0 \to \mathrm{Ext}^1_D(E|_D, E|_D)_0$ is contained in $\mathrm{Ext}^2(E, E(-K_X))_0$. Thus all infinitesimal deformations of $E|_D$ can be lifted to deformations of E on X. The same procedure works for the deformations of higher order. Consequently, there exists a deformation E' of E which restricts to a generic bundle on D. The assumption on the bundle F implies that the generic bundle $E'|_D$ has vanishing $\mathrm{Hom}(E'|_D, E'|_D \otimes K_X)_0$. $\square$

Comments:

— Mukai was the first to construct algebraically a symplectic structure on the moduli space of simple sheaves on K3 and abelian surfaces [187]. Theorem 10.4.3 is due to him. Later Tyurin [249] generalized his construction for surfaces with $p_g > 0$. He also considered Poisson structures. Trace and pairing were treated by Artamkin in connection with the deformation theory of sheaves [5], though the sign $(-1)^{i \cdot j}$ in 10.1.3 is missing in [5].

— For a very detailed treatment of the Atiyah class of (complexes of) sheaves we refer to the article of Angeniol and Lejeune-Jalabert [2].

— 10.3.2 was proved by O'Grady [207] for smooth S and locally free $\mathcal{E}$. Also compare [33]. A reference for the Albanese mapping is [254].

— The existence of the bundle F in the proof of 10.4.4 is due to Oxbury [214] and O'Grady [207] if D is smooth and to J. Li [149] in general.

Appendix to Chapter 10

10.A Further comments (second edition)

I. Differential forms on moduli spaces.

Symplectic structures. The paper [355] generalizes the construction of regular one- and two-forms on moduli spaces to forms of any degree. Moreover, this is done for moduli spaces of sheaves on higher dimensional varieties as well, which then also covers constructions of forms, e.g. on Fano varieties of lines. The construction involves higher powers of the Atiyah class.

Often moduli spaces of sheaves on K3 surfaces are birational to Hilbert schemes of points on the same surface (see e.g. Theorem 11.3.1). In this case, the regular two-forms constructed in this chapter coincide (up to scaling) under the birational correspondence. This fact is trivial for projective moduli spaces, as then $H^0(M, \Omega_M^2)$ is of dimension one. In other situations, Hilbert scheme and moduli space are not necessarily birationally equivalent, but are related by a functorial construction. E.g. the moduli space of stable bundles of rank two with a section is for sufficiently high degree on the one hand a projective bundle over the moduli space of stable bundles, but also maps to the Hilbert schemes by taking the zero set of each (generic) section. A similar situation has been studied in [281], namely sheaves concentrated on curves in a fixed linear system in a K3 surface together with a section. The main result of the paper is that the pull-back of the two regular two-forms give the same (up to scaling) two-form on the moduli space of such pairs.

Poisson structures. In analogy to the construction of symplectic structures on moduli spaces of sheaves on surfaces with a symplectic structure and on Hilbert schemes on such surfaces, Bottacin has considered Poisson structures on moduli spaces of sheaves on Poisson surfaces. In the series of papers [33, 285, 286, 287] he studied moduli spaces of sheaves, Hilbert schemes of points, moduli spaces of parabolic and framed bundles.

11

Birational properties

Moduli spaces of bundles with fixed determinant on algebraic curves are unirational and very often even rational. For moduli spaces of sheaves on algebraic surfaces the situation differs drastically and, from the point of view of birational geometry, discloses highly interesting features. Once again, the geometry of the surface and of the moduli spaces of sheaves on the surface are intimately related. For example, moduli spaces associated to rational surfaces are expected to be rational and, similarly, moduli spaces associated to minimal surfaces of general type should be of general type. We encountered phenomena of this sort already at various places (cf. Chapter 6).

There are essentially two techniques to obtain information about the birational geometry of moduli spaces. First, one aims for an explicit parametrization of an open subset of the moduli space by means of Serre correspondence, elementary transformation, etc. Second, one may approach the question via the positivity (negativity) of the canonical bundle of the moduli space. The first step was made in Section 8.3. The best result in this direction is due to Li saying that on a minimal surface of general type with a reduced canonical divisor the moduli spaces of rank two sheaves are of general type. This and similar results concerning the Kodaira dimension are presented in Section 11.1. The use of Serre correspondence for a birational description is illustrated by means of two examples in Section 11.3. Both examples treat moduli spaces on K3 surfaces, where this technique can be applied most successfully. In Section 11.2 we survey more results concerning the birational geometry of moduli spaces. For precise statements and proofs we refer to the original articles.

11.1 Kodaira Dimension of Moduli Spaces

For the convenience of the reader we briefly recall some of the main concepts in birational geometry. As a general reference we recommend Ueno's book [254].

Let X be an integral variety of dimension n over an algebraically closed field. X is *rational* if it is birational to $\mathbb{P}^n$. If there exists a dominant rational map $\mathbb{P}^m \to X$, then X is called *unirational*. Note that by replacing $\mathbb{P}^m$ by a general linear subspace we can assume $m = n$.

Definition 11.1.1 — *Let X be a smooth complete variety. Its Kodaira dimension* $\mathrm{kod}(X)$
is defined by:
- *If $h^0(X, \mathcal{O}(mK_X)) = 0$ for all $m > 0$, then $\mathrm{kod}(X) = -\infty$.*
- *If $h^0(X, \mathcal{O}(mK_X)) = 0$ or $= 1$, but not always zero, then $\mathrm{kod}(X) = 0$.*
- *If $h^0(X, \mathcal{O}(mK_X)) \sim m^\kappa$, then $\mathrm{kod}(X) = \kappa \geq 1$.*

It turns out that the Kodaira dimension satisfies $-\infty \leq \mathrm{kod}(X) \leq n$. If X is not smooth or not complete but birational to a smooth complete variety X', then we define $\mathrm{kod}(X) := \mathrm{kod}(X')$. This definition does not depend on X' due to the fact that the Kodaira dimension is a birational invariant. Last but not least, an integral variety X of dimension n is of *general type* if $\mathrm{kod}(X) = n$.

Let us begin with the rank one case. Obviously, $M_H(1, c_1, 0) \cong \mathrm{Pic}^{c_1}(X)$ is either empty or an abelian variety. In particular, in the latter case the Kodaira dimension is zero. The morphism $M_H(r, c_1, c_2) \to \mathrm{Pic}^{c_1}(X)$ defined by the determinant is locally trivial in the étale topology (cf. the proof of 4.5.4). Thus one is inclined to study the geometry of the fibre $M_H(r, \mathcal{Q}, c_2)$ over $\mathcal{Q} \in \mathrm{Pic}^{c_1}(X)$ separately. Since $M_H(1, \mathcal{Q}, c_2) \cong \mathrm{Hilb}^{c_2}(X)$, the following result computes the Kodaira dimension in the rank one case.

Theorem 11.1.2 — *If $n > 0$ then* $\mathrm{kod}(\mathrm{Hilb}^n(X)) = n \cdot \mathrm{kod}(X)$.

Proof. We first introduce some notations: Let $M := \mathrm{Hilb}^n(X)$, $S := S^n(X)$, $X^n := X \times \ldots \times X$, and let $\varphi : X^n \to S$ and $\psi : M \to S$ be the natural morphisms. The tensor product $\bigotimes p_i^* \mathcal{O}(K_X)$ is a line bundle on X^n with a natural linearization for the action of the symmetric group S_n. The isotropy subgroups of all points in X^n act trivially. Therefore, the line bundle descends to a line bundle ω on S. Moreover,

$$H^0(S, \omega^m) = H^0(X^n, \bigotimes p_i^* \mathcal{O}(mK_X))^{S_n} = S^n H^0(X, \mathcal{O}(mK_X)).$$

We use the following facts: i) $\mathcal{O}(K_M) \cong \psi^* \omega$, which follows from a local calculation in points $(x, x, x_3, \ldots, x_n) \in S^n(X)$ with $x_i \neq x$, and ii) $\psi_* \mathcal{O}_M \cong \mathcal{O}_S$, which is an easy consequence of the normality of M and S. Then $H^0(M, \mathcal{O}(mK_M)) \overset{i)}{=} H^0(M, \psi^* \omega^m) \overset{ii)}{=} H^0(S, \omega^m) = S^n H^0(X, \mathcal{O}(mK_X))$. This yields $\mathrm{kod}(M) = n \cdot \mathrm{kod}(X)$. $\square$

Corollary 11.1.3 — *If X is a surface of general type, then $\mathrm{Hilb}^n(X)$ is of general type as well.* $\square$

Let us now come to the higher rank case. Here we first mention a consequence of Theorem 5.1.6 in Chapter 5. Note that a surface with $q(X) = h^1(X, \mathcal{O}_X) \neq 0$ can never be unirational. For such surfaces we have:

Theorem 11.1.4 — *If X is an irregular surface, i.e. $q(X) \neq 0$, then the moduli spaces $M_H(r, \mathcal{Q}, c_2)$ are not unirational for $c_2 \gg 0$.*

Proof. This follows easily from the observation that for $c_2 \gg 0$ the Albanese map defines a surjective morphism $M_H(r, \mathcal{Q}, c_2) \to \mathrm{Alb}(X)$ (cf. 5.1.6). Since $\mathrm{Alb}(X)$ is a torus, any morphism $\mathbb{P}^1 \to \mathrm{Alb}(X)$ must be constant. □

Theorem 11.1.5 — *Let X be a minimal surface of general type. Fix an ample divisor H and a line bundle $\mathcal{Q} \in \mathrm{Pic}(X)$. Assume: i) there exists a reduced canonical divisor $D \in |K_X|$ and ii) $\chi(\mathcal{O}_X) + c_1^2(\mathcal{Q}) \equiv 0(2)$. Then for $c_2 \gg 0$ the moduli space $M_H(2, \mathcal{Q}, c_2)$ is a normal irreducible variety of general type, i.e. $\mathrm{kod}(M_H(2, \mathcal{Q}, c_2)) = \dim(M_H(2, \mathcal{Q}, c_2))$.*

Proof. We first prove the theorem under the additional assumption that X contains no (-2)-curves. This is equivalent to K_X being ample. We will indicate the necessary modifications for the general case at the end of the proof. By the results of Chapter 9 (Theorem 9.3.3 and Theorem 9.4.3) we already know that $M_H(r, \mathcal{Q}, c_2)$ is normal and irreducible for sufficiently large c_2. Thus it remains to verify the assertion on the Kodaira dimension. Theorem 4.C.7 shows that for any two polarizations H and H' the corresponding moduli spaces $M_H(r, \mathcal{Q}, c_2)$ and $M_{H'}(r, \mathcal{Q}, c_2)$ are birational for $c_2 \gg 0$. Therefore, it suffices to prove the theorem in the case $H = K_X$. To simplify notations we write $M = M_{K_X}(2, \mathcal{Q}, c_2)$ and denote by M_0 the open subset of stable sheaves $[E] \in M$ with vanishing $\mathrm{Ext}^2(E, E)_0$. Note that M_0 is smooth (4.5.4).

In order to show that M is of general type we have to control the space of global sections $H^0(\tilde{M}, \mathcal{O}(mK_{\tilde{M}}))$ for some desingularization $\psi : \tilde{M} \to M$. Let W_i denote the irreducible components of codimension one of the exceptional divisor of ψ. Then we claim that $\mathcal{O}(nK_{\tilde{M}}) \cong \psi^* \mathcal{L}_1^n \otimes \mathcal{O}(\sum a_i W_i)$, where n is positive and $\mathcal{L}_1 = \lambda(u_1) \in \mathrm{Pic}(M)$ (for the notation see Section 8.1). Indeed, by Theorem 8.3.3 there exists a positive integer n such that $\mathcal{O}(nK_{M_0}) \cong \mathcal{L}_1^n|_{M_0}$. Moreover, $\mathrm{codim}(M \setminus M_0) \geq 2$, since $M \setminus M_0$ is contained in the subset of sheaves which are not good, i.e. either not μ-stable or $\mathrm{Ext}^2(E, E)_0 \neq 0$, and that this subset has at least codimension two is a consequence of Theorem 9.3.2. This is enough to conclude that $\psi^* \mathcal{L}_1^n$ and $\mathcal{O}(nK_{\tilde{M}})$ only differ by components of the exceptional divisor.

By Corollary 8.2.16 we have $h^0(\tilde{M}, \psi^* \mathcal{L}_1^m) \geq h^0(M, \mathcal{L}_1^m) \geq c \cdot m^d + c'(m)$, where $d = \dim(M)$, the constant c is positive, and $c'(m)$ comprises all terms of lower degree. If $a_i \geq 0$ for all i, then $\psi^* \mathcal{L}_1^n \subset \mathcal{O}(nK_{\tilde{M}})$ and hence $H^0(\tilde{M}, \psi^* \mathcal{L}_1^{mn}) \subset H^0(\tilde{M}, \mathcal{O}(mnK_{\tilde{M}}))$.

Hence $\tilde{M}$ is of general type. The rest of the proof deals with the case that at least one of the coefficients a_i is negative. Here we apply a result of Chapter 10 (see 10.3.9 and also 4.B.5), where we constructed a desingularization $\psi : \tilde{M} \to M$ such that $\tilde{M}$ admits a regular two-form $\omega \in H^0(\tilde{M}, \Omega^2_{\tilde{M}})$ with $\omega|_{\psi^{-1}(M^s)} = \psi^*\tau(\alpha)$. Here $\alpha \in H^0(X, K_X)$ is the section defining D and M^s is the open dense subset of stable sheaves.

From now on let $r = 2$. Then Theorem 10.4.4 applies and shows that $\tau(\alpha)$, and hence ω, is generically non-degenerate. Let β be the Pfaffian of ω. Then $\beta \in H^0(\tilde{M}, \mathcal{O}(K_{\tilde{M}}))$ is a non-vanishing section. We claim that β vanishes on all components W_i: By construction, the desingularization $\psi : \tilde{M} \to M$ has the following properties: There exists a family $\mathcal{E}$ over $\tilde{M} \times X$ of rank $s \cdot r$ such that for all $t \in \tilde{M}$ the sheaf $\mathcal{E}_t$ is isomorphic to $E^{\oplus s}$ for some semistable sheaf E with $[E] = \psi(t)$. Moreover, $\omega = (1/s)\tau_{\mathcal{E}}(\alpha)$. In order to show that β vanishes on a component W_i it suffices to show that ω degenerates at the generic point of W_i. As this is a local problem we may use Luna's Etale Slice Theorem (see Theorem 4.2.12) to assume that $s = 1$, i.e. $\mathcal{E}$ is a family of semistable sheaves of rank r. Fix an integer $m \gg 0$ as in the construction of the moduli space and let $\tilde{\pi} : \tilde{R} \to \tilde{M}$ be the principal $\mathrm{PGL}(P(m))$-bundle associated to $p_*(\mathcal{E} \otimes q^*\mathcal{O}(m))$. With the notations of Chapter 4 and 9 there exists a classifying morphism $\tilde{\Phi} : \tilde{R} \to R \subset \mathrm{Quot}(V \otimes \mathcal{O}(-m), P)$, where $R \to M$ is a good quotient and a principal $\mathrm{PGL}(P(m))$-bundle over M^s. Let W_i be an irreducible component of codimension one of the exceptional divisor of ψ and let $\tilde{W}_i := \tilde{\pi}^{-1}(W_i)$. So we have the following diagram.

$$
\begin{array}{ccc}
\tilde{W}_i \subset & \tilde{R} & \xrightarrow{\tilde{\Phi}} & R \\
\downarrow & \tilde{\pi} \downarrow & & \downarrow \\
W_i \subset & \tilde{M} & \xrightarrow{\psi} & M
\end{array}
$$

Moreover, $\tilde{\Phi}$ is an isomorphism over a dense open subset. The compatibility of the two-forms constructed in Chapter 10 gives $\tilde{\pi}^*\omega = \tilde{\pi}^*\tau_{\mathcal{E}}(\alpha) = \tilde{\Phi}^*\tau_{\tilde{F}}(\alpha)$, where $\tilde{F}$ is the universal quotient sheaf over $R \times X$. If $\psi(W_i) \subset M \setminus M^s$, then $\tilde{\Phi}(\tilde{W}_i) \subset R(0)$, where $R(0)$ is the closed subset of μ-unstable sheaves. By Theorems 9.1.1 and 9.1.2 we have $\mathrm{codim}\, R(0) \geq 2$ for $c_2 \gg 0$. Hence $\tilde{\Phi} : \tilde{W}_i \to R$ has positive fibre dimension. If $\psi(W_i) \cap M^s \neq \emptyset$, then $\psi : W_i \to M^s$ has positive fibre dimension and so has $\tilde{\Phi} : \tilde{W}_i \to R$. Hence, in both cases $\tilde{\Phi}^*\tau_{\tilde{F}}(\alpha)$ degenerates on the component $\tilde{W}_i$. But then the same is true for the two-form ω on W_i.

Having proved that β vanishes along $\sum W_i$ we may consider β as a section of the sheaf $\mathcal{O}(K_{\tilde{M}} - \sum W_i)$. Let $a := \max\{-a_i\}$. Then the multiplication with $\beta^{m \cdot a}$ defines an injection

$$\psi^* \mathcal{L}_1^{mn} \cong \mathcal{O}(m(nK_{\tilde{M}} - \sum a_i W_i))$$
$$\xrightarrow{\beta^{ma}} \mathcal{O}(m((n+a)K_{\tilde{M}} - \sum(a_i + a)W_i))$$
$$\longrightarrow \mathcal{O}(m(n+a)K_{\tilde{M}}).$$

Hence $h^0(\tilde{M}, \mathcal{O}(m(n+a)K_{\tilde{M}})) \geq c \cdot (mn)^d + c'(mn)$ and therefore $\mathrm{kod}(M) = \mathrm{kod}(\tilde{M}) = d = \dim(M)$, i.e. M is of general type.

We now come to the case that X contains (-2)-curves. Then K_X is no longer ample, but still big and nef. If $f : X \to Y$ is the morphism from X to its canonical model Y, then K_X is the pull-back of an ample divisor H_Y on Y. Let H be an arbitrary polarization on X and consider the moduli space $M := M_H(r, \mathcal{Q}, c_2)$. As before, M is normal and irreducible for $c_2 \gg 0$. Moreover, copying the arguments of the proof of Theorem 4.C.7 and using that K_X is in the positive cone we find that for $c_2 \gg 0$ the set of sheaves $[F] \in M$ which are not μ_{K_X}-stable is at least of codimension two.

The Bogomolov Restriction Theorem 7.3.5 (cf. Remark 7.3.7) applied to a smooth curve $C \in |nK_X|$ ($n \gg 0$) yields a rational map $\varphi : M \to M_C$ which is regular on the open subset of μ_{K_X}-stable sheaves with singularities in $X \setminus C$. The complement of this open set has at least codimension two. As before, to conclude the proof it suffices to verify that φ is generically injective. Let E and F be two locally free sheaves and let $G = \mathcal{H}om(E, F)$. The restriction homomorphism $H^0(X, G) \to H^0(C, G|_C)$ is surjective if and only if its Serre dual $H^1(C, G^\vee|_C(K_C)) \to H^2(X, G^\vee(K_X))$ is injective. If C avoids all (-2)-curves, which the generic curve in $|nK_X|$ does, then there is a commutative diagram

$$
\begin{array}{ccc}
H^1(X, G^\vee|_C(K_C)) & \to & H^2(X, G^\vee(K_X)) \\
\downarrow \cong & & \downarrow \cong \\
H^1(C, f_*(G^\vee)|_C(K_C)) & \to & H^2(Y, f_*(G^\vee)(H_Y))
\end{array}
$$

For the second vertical isomorphism use that $R^2 f_*(G^\vee) = 0$ and that $R^1 f_*(G^\vee)$ is zero-dimensional and hence $H^1(Y, R^1 f_*(G^\vee)(H_Y)) = 0$. The kernel of

$$H^1(C, f_*(G^\vee)|_C(K_C)) \to H^2(Y, f_*(G^\vee)(H_Y))$$

is a quotient of $H^1(Y, f_*(G^\vee)((n+1)H_Y))$ which clearly vanishes for $n \gg 0$, since H_Y is ample. Therefore we may assume that for all locally free $[E], [F] \in M$ the restriction $\mathrm{Hom}(E, F) \to \mathrm{Hom}(E|_C, F|_C)$ is surjective. In particular, if $\alpha_C : E|_C \cong F|_C$, then there exists a homomorphism $\alpha : E \to F$ with $\alpha|_C = \alpha_C$. Thus α is generically injective and since $\det(E) \cong \det(F)$, it is in fact bijective. Thus, for $n \gg 0$ the map $\varphi : M \to M_C$ is injective on the locally free part. $\qquad\square$

Remark 11.1.6 — The proof has been presented in a way indicating that the moduli spaces $M_H(r, \mathcal{Q}, c_2)$ for a minimal surface of general type are expected to be of general type without the assumptions *i)*, *ii)* and $r = 2$. In fact, if the singularities of the moduli spaces are canonical, i.e. all coefficients a_i are nonnegative, then the proof goes through: For all three assumptions were only used to ensure the existence of a generically non-degenerate two-form which would not be needed in this case.

Along the same line of arguments, only much simpler, one also proves

Theorem 11.1.7 — *Let X be a surface, H a polarization, $\mathcal{Q} \in \mathrm{Pic}(X)$, and $c_2 \gg 0$.*
i) If X is a Del Pezzo surface, then $M_H(r, \mathcal{Q}, c_2)$ is an irreducible variety of Kodaira dimension $-\infty$ that is smooth in stable points.
ii) If $\mathcal{O}(K_X) \cong \mathcal{O}_X$, i.e. X is abelian or K3, then $M_H(r, \mathcal{Q}, c_2)$ is a normal irreducible variety of Kodaira dimension zero.
iii) If X is minimal and $\mathrm{kod}(X) = 0$, then $\mathrm{kod}(M_H(r, \mathcal{Q}, c_2)) \leq 0$ and equality holds if all $E \in M_H(r, \mathcal{Q}, c_2)$ are stable and $\mathrm{Ext}^2(E, E)_0 = 0$.

Proof. For *i)* and the case $H = -K_X$ we use $\mathcal{O}(-nK_M) \cong \mathcal{L}_1^n$ for some $n > 0$ (cf. Theorem 8.3.3) to conclude $H^0(M, \mathcal{O}(mnK_M)) = 0$ for all $m > 0$. For a polarization different from K_X we again use 4.C.7 *ii)* and *iii)* follow from $\mathcal{O}(nK_{M_0}) \cong \mathcal{O}$ for some $n > 0$ which immediately yields $H^0(\tilde{M}, \mathcal{O}(mnK_{\tilde{M}})) \subset H^0(M_0, \mathcal{O}_{M_0}) = k$ and hence $\mathrm{kod}(M) \leq 0$. If $\mathcal{O}(K_X) \cong \mathcal{O}_X$, then the distinguished desingularization $\tilde{M} \to M$ constructed in Appendix 4.B admits a generically non-degenerate two-form. Hence $H^0(\tilde{M}, \mathcal{O}(K_{\tilde{M}})) \neq 0$. Under the additional assumptions in *iii)* one has $M_0 = M$ and thus $\mathcal{O}(nK_M) \cong \mathcal{O}_M$ which also gives $H^0(M, \mathcal{O}(nK_M)) \neq 0$. Hence $\mathrm{kod}(M) = 0$ in both cases. $\qquad\square$

Remark 11.1.8 — There are explicit numerical conditions on $(H, r, \mathcal{Q}, c_2)$ such that for a minimal surface of Kodaira dimension zero all $[E] \in M_H(r, \mathcal{Q}, c_2)$ are stable (cf. 4.6.8). If the order of K_X does not divide the rank r then $\mathrm{Ext}^2(E, E)_0 = 0$ for any stable sheaf. Thus the conditions in *iii)* are frequently met.

11.2 More Results

Following the Enriques classification of algebraic surfaces we survey known results related to the birational structure of moduli spaces.

11.2.1 Kodaira dimension $-\infty$. Let first X be the projective plane $\mathbb{P}^2$. One certainly expects moduli spaces of sheaves on $\mathbb{P}^2$ to be rational. In general, it is not hard to prove that they are unirational. Already in the seventies moduli spaces of stable rank-two bundles on $\mathbb{P}^2$ were intensively studied. Barth [21] announced that the moduli spaces $N(2, 0, c)$ of stable rank-two bundles with $(c_1, c_2) = (0, c)$ are irreducible and rational. Note that $N(2, 0, c)$ is non-empty if and only if $c \geq 2$. The analogous problem for odd first Chern number, i.e. $c_1 = 1$, was discussed by Hulek [107]: $N(2, 1, c)$ is rational and irreducible. Here $N(2, 1, c)$ is non-empty if and only if $c \geq 1$. Unfortunately, there was a gap in Barth's approach to the rationality. Hulek remarked in [107] that for $c_1 = 1$ this could easily be filled. Ellingsrud and Strømme [59, 60] proved the rationality of $N(2, 0, 2n + 1)$ and $N(2, 1, c)$ with different techniques. They also proved the rationality of an étale $\mathbb{P}^1$-bundle over $N(2, 0, n)$. Maruyama discussed the problem further [169]. In an Appendix to his paper Noruki proved the rationality of $N(2, 0, 3)$. Partial results are known for the higher rank moduli spaces: Le Bruyn [140] proved that $N(r, 0, r)$ is rational for $r \leq 4$. The rationality problem for the rank two case was eventually solved by Katsylo [119]. He proved that $N(r, 0, c)$ is rational if g.c.d.$(r, c) \leq 4$ or $= 6, 12$ with the exception of finitely many cases.

More generally, let X be a rational surface. Ballico [9] showed that there exists a polarization H such that $N_H(r, c_1, c_2)$ is smooth, irreducible and unirational. Combining this with 9.4.2 one finds that the moduli spaces $M_H(r, c_1, c_2)$ are unirational for any polarization if $c_2 \gg 0$.

To complete the case of negative Kodaira dimension, let X be a ruled surface, i.e. X is the projectivization $\pi : \mathbb{P}(E) \to C$ of a rank two vector bundle E on a curve C of genus g. As in the general situation there always exist two canonical morphisms $\det : M_H(r, c_1, c_2) \to \mathrm{Pic}^{c_1}(X) \cong \mathrm{Pic}^0(C)$ and $\varphi : M_H(r, c_1, c_2) \to \mathrm{Alb}(X) \cong \mathrm{Alb}(C) \cong \mathrm{Pic}^0(C)$ (cf. Chapter 10). Loosely speaking, one expects the moduli space $M_H(r, c_1, c_2)$ together with these two morphisms to be birational to a projective bundle over $\mathrm{Pic}^0(C) \times \mathrm{Pic}^0(C)$ or at least to have unirational fibres over $\mathrm{Pic}^0(C) \times \mathrm{Pic}^0(C)$. For rational ruled surfaces, i.e. $g = 0$, this is certainly true ([9]). The rank two case has been studied in detail by many people. Hoppe and Spindler [105] considered the case $E \cong \mathcal{O} \oplus L, r = 2$, and c_1 such that the intersection $c_1.f$ with the fibre class f is odd. They showed that indeed $N_H(2, c_1, c_2)$ is birational to $\mathbb{P}^m \times \mathrm{Pic}^0(C) \times \mathrm{Pic}^0(C)$. Brosius [37], [38] gave a thorough classification of all rank two bundles on ruled surfaces. He distinguishes between bundles of type U and E according to whether $c_1.f$ is odd or even. Bundles of type U are constructed via Serre correspondence as extension $0 \to L \to E \to M \otimes \mathcal{I}_Z \to 0$ and $\varphi(E)$ corresponds to $\mathcal{O}_C(\pi(Z))$. Bundles of type E can be described by means of elementary transformations along fibres of π. In this case $\varphi(E)$ corresponds to the divisor of the fibres where the elementary transformation is performed. Brosius' results allow to generalize the birational description of Hoppe and Spindler. Friedman and Qin combined these results with a detailed investigation of the chamber structure of a ruled surface [218, 219]. One of the ideas is the following: The contribution coming from crossing a wall can be explicitly described. In the case $c_1.f = 1$, when for a polarization near the fibre class the moduli space is always empty (cf. 5.3.4), this is enough to deduce the birational structure of the moduli space with respect to an arbitrary polarization. For rational ruled surfaces and $r > 2$ this was further pursued in [154] and [85]. For related results see also Brinzanescu's article [34]. The explicit example $N(2, 0, 2)$ on the Hirzebruch surfaces $X = \mathbb{P}(\mathcal{O} \oplus \mathcal{O}(n)) \to \mathbb{P}^1$ was treated by Buchdahl in [39]. Vector bundles of rank > 2 on ruled surfaces have also been studied by Gieseker and Li [82]. They use elementary transformations along the fibres to bound the dimension of the 'bad' locus of the moduli space.

11.2.2 Kodaira dimension 0. According to the classification theory of surfaces there are four types of minimal surfaces of Kodaira dimension zero: K3, abelian, Enriques, and hyperelliptic surfaces. The Kodaira dimension of the moduli spaces is by Theorem 11.1.7 known for K3 and abelian surfaces and under additional assumptions also for Enriques and hyperelliptic surfaces. According to a result of Qin [216], with the exception of three special cases, the birational type of the moduli space of μ-stable rank two bundles does not depend on the polarization.

Some aspects of moduli spaces on K3 surfaces were studied in Chapter 5, 6, and 10. The upshot is that sometimes the moduli space is birational to the Hilbert scheme and that it is in general expected to be a deformation of a variety birational to some Hilbert scheme. The birational correspondence to the Hilbert scheme is achieved either by using Serre correspondence or, if the surface is elliptic, by elementary transformations. In the latter case Friedman's result for general elliptic surfaces apply [67, 68]. There are also birational descriptions of some moduli spaces of simple bundles on K3 surfaces available [225, 248]. Two examples of moduli spaces of sheaves on K3 surfaces will be discussed in the next section. Moduli spaces on abelian surfaces behave in many respects similar to moduli spaces on K3 surfaces. In particular, they are sometimes birational to Hilbert schemes or to products of them. For examples see Umemura's paper [256]. The Hilbert scheme itself fibres via the group operation over the surface. Beauville showed that the fibres are irreducible symplectic [25]. The same phenomenon should be expected for the higher rank case. The Fourier-Mukai transformation, which can also be used to study birational properties of the moduli space on abelian surfaces, was introduced by Mukai [186, 189]. It was further studied in [62], [159], [23].

The universal cover $\pi : \tilde{X} \to X$ of an Enriques surface is a K3 surface. The pull-back of sheaves defines a two-to-one map from the moduli space on X to a Lagrangian of maximal dimension in the moduli space on $\tilde{X}$. This and some explicit birational description of moduli spaces using linear systems on the Enriques surface can be found in Kim's thesis [122]. Hyperelliptic surfaces are special elliptic surfaces and, therefore, Friedman's results apply. Special attention to the hyperelliptic structure has been paid in the work of Takemoto and Umemura [257], who studied projectively flat rank two bundles, i.e. bundles with $4c_2 - c_1^2 = 0$.

11.2.3 Kodaira dimension 1. Surfaces in this range all are elliptic. Sheaves of rank two have been studied by Friedman [67, 68]. As for ruled surface, sheaves of even and odd fibre degree are treated differently. [67] deals with bundles with $c_1 = 0$. In particular, they have even intersection with the fibre class. Friedman gives an upper bound (depending on the geometry of the surface) for the Kodaira dimension of the moduli space of rank two bundles with $c_1 = 0$ and $c_2 \gg 0$ which are stable with respect to a suitable polarization (cf. Chapter 5). Moreover, the moduli space is birationally fibred by abelian varieties. If the elliptic surface has no multiple fibres then the base space is rational. Bundles with odd fibre degree are studied in [68] via elementary transformations. Under certain assumptions on the elliptic surface, e.g. if there are at most two multiple fibres, the moduli space is shown to be birational to the Hilbert scheme of an elliptic surface naturally attached to the original one. Hence the Kodaira dimension is known in these cases by 11.1.2. A result in the spirit of 11.1.5 and 11.1.7 is missing. It would be interesting to see which Kodaira dimensions moduli spaces on elliptic surfaces can attain. Do they fill the gap between varieties with Kodaira dimension zero and those of general type?

For the case that rank and fibre degree are coprime O'Grady [211] suggests that the canonical model of the moduli space should be an appropriate symmetric product of the base curve of the elliptic fibration. In particular, he expects that the Kodaira dimension of the moduli space should be half its dimension.

11.2.4 Surfaces of general type. The only known result is Li's Theorem 11.1.5. We do not even know a single example where one can show that the moduli space is of general type without refering to the general theorem. Note that there is a big difference between surfaces of general type with $p_g > 0$ and those with $p_g = 0$. The Chow group of surfaces of the first type is huge [193], but according to a conjecture of Bloch the Chow group is trivial in the latter case. Is this reflected by the birational geometry of the moduli space?

Recently, O'Grady [211] slightly generalized Li's result 11.1.5. He showed that the higher rank moduli spaces $M_H(r, \mathcal{Q}, c_2)$ are also of general type for $c_2 \gg 0$ if one in addition assumes that the minimal surface of general type admits a smooth irreducible canonical curve C such that $h^0(K_X|_C) \equiv \deg(\mathcal{Q}|_C) \bmod r$.

11.3 Examples

We wish to demonstrate how Serre correspondence can be used to give a birational description of moduli spaces. Both examples deal with sheaves on a K3 surface. The techniques can certainly be applied to other surfaces as well, but they almost never work as nicely as in the following two examples.

Let X be a K3 surface and let H be an ample divisor on X. Consider the moduli space $M_H(2, \mathcal{Q}, c_2)$ of semistable sheaves of rank two with determinant $\mathcal{Q}$ and second Chern number c_2. Since twisting with $\mathcal{O}(mH)$ does not change stability with respect to H, the map $E \mapsto E(mH)$ defines an isomorphism $M_H(2, \mathcal{Q}, c_2) \cong M_H(2, \mathcal{Q}(2mH), c_2 + m^2 H^2 + mc_1(\mathcal{Q}).H)$. Thus we may assume that the determinant $\mathcal{Q}$ is ample from the very beginning. We wish to show that in many instances the birational structure of the moduli space $M_H(2, \mathcal{Q}, c_2)$ can be compared with the one of the Hilbert scheme of the same dimension. In order to state the theorem we need to introduce the following quantities: $k(n) := (n^2 + n + 1/2)c_1^2(\mathcal{Q}) + 3$ and $l(n) := (2n^2 + 2n + 1/2)c_1^2(\mathcal{Q}) + 3$.

Theorem 11.3.1 — *If $\mathcal{Q}$ is ample and $n \gg 0$, then the moduli space $M_H(2, \mathcal{Q}, k(n))$ is birational to $\mathrm{Hilb}^{l(n)}(X)$.*

Proof. By Theorem 4.C.7 the two moduli spaces $M_H(2, \mathcal{Q}, k(n))$ and $M_{\mathcal{Q}}(2, \mathcal{Q}, k(n))$ are birational for $n \gg 0$. Thus we may assume $\mathcal{O}(H) \cong \mathcal{Q}$. Theorems 9.4.3 and 9.4.2 say that for $n \gg 0$ the moduli space $M_{\mathcal{Q}}(2, \mathcal{Q}, k(n))$ is irreducible and the generic sheaf $[E] \in M_{\mathcal{Q}}(2, \mathcal{Q}, k(n))$ is μ-stable and locally free. Let $N \subset M_{\mathcal{Q}}(2, \mathcal{Q}, k(n))$ be the open dense subset of all μ-stable locally free sheaves. We will construct a rational map $\mathrm{Hilb}^{l(n)}(X) \to N$ which is generically injective. Since both varieties are smooth and $\dim \mathrm{Hilb}^{l(n)}(X) = 2l(n) = 4k(n) - c_1^2(\mathcal{Q}) - 6 = \dim(N)$, this is enough to conclude that $\mathrm{Hilb}^{l(n)}(X)$ and $M_{\mathcal{Q}}(2, \mathcal{Q}, k(n))$ are birational.

By the Hirzebruch-Riemann-Roch formula

$$h^0(X, \mathcal{O}_X((2n+1)H)) = \frac{(2n+1)^2}{2}H^2 + 2 = l(n) - 1.$$

Hence for the generic $[Z] \in \mathrm{Hilb}^{l(n)}(X)$ we have $H^0(X, \mathcal{I}_Z((2n+1)H)) = 0$. Using the exact sequence

$$0 \to H^0(X, \mathcal{O}_X((2n+1)H)) \to H^0(X, \mathcal{O}_Z) \to H^1(X, \mathcal{I}_Z((2n+1)H)) \to 0,$$

this implies $h^1(X, \mathcal{I}_Z((2n+1)H)) = 1$ for generic Z. In other words, for generic Z there is a unique non-trivial extension

$$0 \to \mathcal{O}_X \to F \to \mathcal{I}_Z((2n+1)H) \to 0.$$

Moreover, such an F is locally free, for $(\mathcal{O}((2n+1)H), Z)$ satisfies the Cayley-Bacharach property (5.1.1). If F is not μ-stable, then there exists a line bundle $\mathcal{L} \subset F$ with $\frac{2n+1}{2} H^2 \leq c_1(\mathcal{L}).H$. Since such a line bundle $\mathcal{L}$ cannot be contained in $\mathcal{O} \subset F$, there exists a curve $C \in |\mathcal{L}^{\vee}((2n+1)H)|$ containing Z. We show that this cannot happen for generic Z. Since X is regular, it suffices to show that $\dim |\mathcal{L}^{\vee}((2n+1)H)|$ can be bounded from above by $l(n) - 1$. Let $C \in |\mathcal{L}^{\vee}((2n+1)H)|$. Then $h^0(\mathcal{O}_C(C)) = h^0(\omega_C) = (2g(C)-2)/2 + 1 = C^2/2 + 1 = \frac{2n+1}{2}((2n+1)H^2 - 2c_1(\mathcal{L}).H) + \frac{c_1^2(\mathcal{L})}{2} \leq \frac{c_1^2(\mathcal{L})}{2}$. Together with the exact sequence

$$0 \to \mathcal{O}_X \to \mathcal{O}_X(C) \to \mathcal{O}_C(C) \to 0$$

this proves

$$h^0(\mathcal{O}_X(C)) \leq c_1^2(\mathcal{L})/2 + 2.$$

By Hodge Index Theorem $c_1^2(\mathcal{L}) \leq (c_1(\mathcal{L}).H)^2/H^2$ and $0 \leq c_1(\mathcal{L}).H \leq (2n+1)H^2$, for $\mathcal{L} \subset \mathcal{O}((2n+1)H)$. Hence $h^0(\mathcal{O}_X(C)) \leq \frac{(2n+1)^2}{2}H^2 + 2 = l(n) - 1$.

Thus, for the generic $[Z] \in \mathrm{Hilb}^{l(n)}(X)$ there exists a unique extension

$$0 \to \mathcal{O} \to F_Z \to \mathcal{I}_Z((2n+1)H) \to 0$$

and F_Z is μ-stable and locally free. Hence, associating the subscheme Z to the sheaf F_Z defines a rational map $\mathrm{Hilb}^{l(n)}(X) \to N$, which is injective, since $h^0(X, F_Z) = 1$. $\quad\square$

Our second example is very much in the spirit of the first one. We use Serre correspondence to prove that certain moduli spaces on a K3 surface of special type are birational to the Hilbert scheme. Specializing to elliptic K3 surfaces enables us to handle a more exhaustive list of moduli spaces; in particular those of higher rank sheaves.

Let $\pi : X \to \mathbb{P}^1$ be an elliptic K3 surface with a section $\sigma \subset X$. We assume that $\mathrm{Pic}(X) = \mathbb{Z} \cdot \mathcal{O}(\sigma) \oplus \mathbb{Z} \cdot \mathcal{O}(f)$, where f is the fibre class. In particular, all fibres are irreducible. Let $v = (v_0, v_1, v_2)$ be a Mukai vector such that $v_1 = \sigma + \ell f$ and consider sheaves E with $r := \mathrm{rk}(E) = v_0$, $c_1(E) = v_1$, and $ch_2(E) + r = v_2$. (For the definition of the Mukai vector see Section 6.1.) If we consider stability with respect to a suitable polarization, then a sheaf is μ-semistable if and only if the restriction to the generic fibre is semistable 5.3.2. Since $(\sigma + \ell f).f = 1$, any semistable sheaf on the fibre is stable. Therefore, with respect to a suitable polarization semistable sheaves on X with $c_1 = \sigma + \ell f$ are μ-stable (5.3.2, 5.3.6). Moreover, since the stability on the fibre is unchanged when the sheaves are twisted with $\mathcal{O}(f)$, a sheaf E is μ-stable with respect to a suitable polarization if and only if $E(f)$ is μ-stable. Thus, by twisting with $\mathcal{O}(f)$ and using $ch_2(E(f)) = ch_2(E) + 1$, we can reduce to the case that the Mukai vector is of the form $(r, \sigma + \ell f, 1 - r)$. The following theorem is Proposition 6.2.6 in Section 6.2, which was stated there without proof.

Theorem 11.3.2 — *Let $v = (r, \sigma + \ell f, 1-r)$ and let $H = \sigma + mf$ be a suitable polarization with respect to v. Then $M_H(v)$ is irreducible and birational to $\mathrm{Hilb}^n(X)$, where $n = \ell + r(r-1)$.*

Proof. Let us begin with a dimension check: $\dim \mathrm{Hilb}^n(X) = 2n = 2\ell + 2r(r-1)$ and $\dim M_H(v) = (v,v) + 2 = 2r(r-1) - 2 + 2\ell + 2 = 2\ell + 2r(r-1)$. Next, since H is suitable, any $[E] \in M_H(v)$ is μ-stable. Hence $M_H(v)$ is smooth. Moreover, if $[E] \in M_H(v)$, then E is μ-stable with respect to $\sigma + m'f$ for all $m' \geq m$. Thus we may assume $m \gg 0$.

The assertion is proved by induction over the rank. Define $v^i := (i, \sigma + (n - i(i-1))f, 1 - i)$ for $i = 1, \ldots, r$. Note that $M_H(v^1) = \mathrm{Hilb}^n(X)$. We will define an open dense subset $U \subset \mathrm{Hilb}^n(X)$ and injective dominant morphisms $\Phi^i : U \to M_H(v^i)$. Since all varieties are smooth of dimension $2n$, this suffices to prove the theorem.

Let $U \subset \mathrm{Hilb}^n(X)$ be the open subset of all $[Z] \in \mathrm{Hilb}^n(X)$ with $H^0(X, \mathcal{I}_Z(\sigma + (n-1)f)) = 0$. We show that U is non-empty: By the Hirzebruch-Riemann-Roch formula $\chi(\mathcal{O}(\sigma + (n-1)f)) = n$. Serre duality gives $h^2(\mathcal{O}(\sigma + (n-1)f)) = h^0(\mathcal{O}(-\sigma - (n-1)f)) = 0$. The vanishing of the first cohomology $H^1(X, \mathcal{O}(\sigma + (n-1)f))$ can be computed as follows: By the exact sequence

$$\to H^1(X, \mathcal{O}(\sigma)) \to H^1(X, \mathcal{O}(\sigma + (n-1)f)) \to \bigoplus_{j=1}^{n-1} H^1(F_j, \mathcal{O}(\sigma)|_{F_j}) \to,$$

where $F_1, \ldots, F_{n-1}$ are distinct generic fibers, one has

$$h^1(X, \mathcal{O}(\sigma + (n-1)f)) \leq h^1(X, \mathcal{O}(\sigma)) + \sum h^1(F_j, \mathcal{O}(\sigma)|_{F_j}) = h^1(X, \mathcal{O}(\sigma)).$$

The exact sequence

$$H^0(X, \mathcal{O}) \twoheadrightarrow H^0(\sigma, \mathcal{O}_\sigma) \to H^1(X, \mathcal{O}(-\sigma)) \to H^1(X, \mathcal{O}) = 0,$$

and Serre duality imply $h^1(X, \mathcal{O}(\sigma + (n-1)f)) \leq h^1(X, \mathcal{O}(\sigma)) = h^1(X, \mathcal{O}(-\sigma)) = 0$. Therefore, $h^0(\mathcal{O}(\sigma + (n-1)f)) = \chi(\mathcal{O}(\sigma + (n-1)f)) = n$. Thus for the generic $[Z] \in \mathrm{Hilb}^n(X)$ the cohomology $H^0(X, \mathcal{I}_Z(\sigma + (n-1)f))$ vanishes, i.e. $U \neq \emptyset$.

Let Φ^1 be the inclusion $U \subset \mathrm{Hilb}^n(X)$ and assume we have already constructed an injective morphism $\Phi^i : U \to M_H(v^i)$ satisfying

(A$_i$) If $[Z] \in U$ and $E^i := \Phi^i(Z)$, then

$$h^0(E^i(-2f)) = h^2(E^i(-2f)) = h^0(E^i(-f)) = 0.$$

(B$_i$) If $i > 1$, then $h^0(E^i) = 1$.

Note that (A$_i$) holds true for $i = 1$ by definition of U. The Hirzebruch-Riemann-Roch formula gives $\chi(E^i(-2f)) = -1$ and by (A$_i$) one knows $h^1(E^i(-2f)) = 1$. Hence there exists a unique non-trivial extension

$$0 \to \mathcal{O} \to E^{i+1} \to E^i(-2f) \to 0.$$

Then $v(E^{i+1}) = v^{i+1}$. Since $c_1(E^{i+1}).f = c_1(E^i).f = 1$ and H is suitable with respect to v^{i+1}, the sheaf E^{i+1} (which is in fact locally free, but we do not need this) is μ-stable if and only if the restriction of the extension to a generic fibre F is non-split. Since the extension space $\mathrm{Ext}^1(E^i(-2f), \mathcal{O}) \cong H^1(X, E^i(-2f))^{\vee}$ is one-dimensional, this is the case if and only if the restriction homomorphism $\mathrm{Ext}^1(E^i(-2f), \mathcal{O}) \to \mathrm{Ext}^1(E^i|_F, \mathcal{O}_F)$ is non-trivial or, dualizing, if and only if $H^0(F, E^i|_F) \to H^1(X, E^i(-2f))$ is non-trivial. The kernel of the latter map is a quotient of $H^0(X, E^i(-f))$ which vanishes by (A$_i$). The space $H^0(F, E^i|_F)$ is non-trivial. Indeed, for $i > 1$ this follows from (B$_i$) and for $i = 1$ from $H^0(F, \mathcal{I}_Z(\sigma + nf)|_F) = H^0(F, \mathcal{O}_F(\sigma)) \neq 0$. Hence E^{i+1} is μ-stable and we define a morphism $\Phi^{i+1} : U \to M_H(v^{i+1})$ by $[Z] \mapsto E^{i+1}$. The map is injective, because $1 \leq h^0(E^{i+1}) \leq h^0(\mathcal{O}) + h^0(E^i(-2f)) = 1$ by (A$_i$). This also shows that E^{i+1} satisfies (B$_{i+1}$) for E^{i+1}. To make the induction work we have to verify (A$_{i+1}$) for E^{i+1}. The vanishing of $H^0(X, E^{i+1}(-2f)) \subset H^0(X, E^{i+1}(-f))$ follows immediately from (A$_i$) and $H^2(X, E^{i+1}(-2f))$ is implied by the stability of E^{i+1} with respect to a suitable polarization.

If we denote by $V_i \subset M_H(v^i)$ the open subset of sheaves E^i satisfying (A$_i$) and (B$_i$), then the arguments show that there exists a morphism $\psi^i : V_i \to M_H(v^{i+1})$ commuting with Φ^i and Φ^{i+1}.

To conclude the proof we have to show that $M_H(v)$ is irreducible or, equivalently, that $\Phi^i : U \to M_H(v^i)$ is dominant for all i. By construction $\Phi^i(U) \subset V_i$ for all i. Let us first assume that the generic $[E] \in M_H(v)$ has exactly one global section, i.e. $h^0(E) = 1$. Then we prove the dominance of Φ^i by induction over i. Assume Φ^i is dominant. Let $[E] \in M_H(v^{i+1})$ with $h^0(E) = 1$, then there exists an exact sequence

$$0 \longrightarrow \mathcal{O} \overset{\alpha}{\longrightarrow} E \longrightarrow E' \longrightarrow 0.$$

Since $E|_F$ is stable for the generic fibre F, α can only vanish along divisors contained in fibres. Since all fibres are irreducible, α could only vanish along complete fibers, which is excluded by $h^0(E) = 1$. Hence, α does not vanish along any divisor. This implies that E' is torsion free and, as one easily checks, also μ-stable. Thus $[E^i := E'(2f)] \in M_H(v^i)$. We claim that $[E^i] \in V_i$. Indeed, the stability of E and $h^0(E) = 1$ imply $h^2(E) = h^1(E) = 0$. Moreover, using the stability of E on the generic fibre, one gets $h^2(E(if)) = h^1(E(if)) = 0$ for $i = 1, 2$. Next, the Hirzebruch-Riemann-Roch formula yields $h^0(E(f)) = 2$ and $h^0(E(2f)) = 1$. Using the long exact cohomology sequence we obtain $h^0(E^i(-2f)) = h^0(E') = h^1(\mathcal{O}) = 0$, $h^0(E^i(-f)) = h^0(E'(f)) = h^1(\mathcal{O}(f)) = 0$, and $h^0(E^i) = h^0(E'(2f)) = h^1(\mathcal{O}(2f)) = 1$. Thus, indeed $[E^i] \in V_i$. Hence, $[E]$ is in the image of $\psi^i : V_i \to M_H(v^{i+1})$. Since by the induction hypothesis $\Phi^i(U)$ is dense in V_i, this is enough to conclude that also $\Phi^{i+1}(U)$ is dense in $M(v^{i+1})$.

We still have to verify that the generic $[E] \in M_H(v)$ has exactly one global section. Consider $[E] \in M_H(v)$ with $h^0(E) = \ell+1$. Since $\pi_* E$ is torsion free of rank $h^0(E|_F) = 1$ (F a generic fibre), it is in fact a line bundle on $\mathbb{P}^1$. Using $\ell + 1 = h^0(E) = h^0(\pi_* E)$, we conclude $\pi_* E \cong \mathcal{O}(\ell)$. Thus there is an exact sequence

$$0 \to \mathcal{O}(\ell f) \to E \to E' \to 0$$

with $h^0(E') = 0$. Moreover, since the restriction of E to the generic fibre is a stable bundle of degree one, an explicit calculation shows that the same is true for E'. The generic deformation of E also has $(\ell + 1)$-dimensional space of global sections if and only if the inclusion $\mathcal{O}(\ell f) \subset E$ deforms in all direction with E. We claim that this implies that the natural map $\text{Ext}^1(E, E) \to \text{Ext}^1(\mathcal{O}(\ell f), E')$ is trivial.

Indeed, consider the relative Quot-scheme Q that parametrizes quotients $E \to E'$. Then $Q \to M_H(v)$ is dominant, and hence for generic E the tangent map is surjective. Using the notations of Proposition 2.2.7 this implies that the obstruction map

$$T_s S = T_{[E]} M_H(v) = \text{Ext}^1(E, E) \to \text{Ext}^1(K, F) = \text{Ext}^1(\mathcal{O}(\ell f), E')$$

vanishes. That the obstruction map is the natural one follows from the arguments in Section 2.A. We show that this leads to a contradiction whenever $\ell > 0$. Note that $\text{Ext}^1(E, E) \to \text{Ext}^1(\mathcal{O}(\ell f), E')$ factorizes through the injection $\text{Ext}^1(\mathcal{O}(\ell f), E) \to \text{Ext}^1(\mathcal{O}(\ell f), E')$. Thus, it suffices to consider the homomorphism $\text{Ext}^1(E, E) \to \text{Ext}^1(\mathcal{O}(\ell f), E)$ which sits in the exact sequence

$$\text{Ext}^1(E, E) \to \text{Ext}^1(\mathcal{O}(\ell f), E) \to \text{Ext}^2(E', E) \to \text{Ext}^2(E, E) \to \text{Ext}^2(\mathcal{O}(\ell f), E).$$

In this sequence $\text{Ext}^2(E, E) \cong k$, and $\text{Ext}^2(E', E) \cong k$ as well, because of $\text{Ext}^2(E', E) \cong \text{Hom}(E, E')^\vee$ and the fact that any deformation of the quotient $E \to E'$ would produce a deformation of $\mathcal{O}(\ell f) \subset E$ and thus more global sections of E (Here we use that E' is stable and, therefore, does not admit any non-scalar automorphisms). Furthermore, since $\text{Ext}^2(\mathcal{O}(\ell f), E) = 0$, the homomorphism $\text{Ext}^1(E, E) \to \text{Ext}^1(\mathcal{O}(\ell f), E)$ is surjective. In the exact sequence

$$\text{Ext}^1(\mathcal{O}(\ell f), \mathcal{O}(\ell f)) \to \text{Ext}^1(\mathcal{O}(\ell f), E) \to \text{Ext}^1(\mathcal{O}(\ell f), E') \to \text{Ext}^2(\mathcal{O}(\ell f), \mathcal{O}(\ell f))$$

the first term vanishes and the last term is isomorphic to k. This yields the lower bound $\text{ext}^1(\mathcal{O}(\ell f), E) \geq h^1(E'(-\ell f)) - 1$. Using the stability of E, one checks that

$$h^0(E'(-\ell f)) = h^2(E'(-\ell f)) = 0,$$

and, hence,

$$h^1(E'(-\ell f)) = -\chi(E'(-\ell f)) = -\chi(E(-\ell f)) + 2 = \ell + 1.$$

Thus $\mathrm{ext}^1(\mathcal{O}(\ell f), E) \geq \ell$ and, therefore, the map $\mathrm{Ext}^1(E, E) \to \mathrm{Ext}^1(\mathcal{O}(\ell f), E)$ does not vanish for $\ell > 0$. $\square$

Comments:
— Theorem 11.1.2 was communicated to us by Göttsche.

— Theorem 11.1.4 in the special case of irrational surfaces of negative Kodaira dimension, which are all irregular, was proved by Ballico and Chiantini in [14].

— Theorem 11.1.5 is due to J. Li [149]. Using the special desingularization $\tilde{M}$ we could simplify some of the arguments. In the original version there is a numerical condition on the intersection of $c_1(\mathcal{Q})$ with the (-2)-curves. Li explained to us how this can be avoided.

— 11.1.7 can be found in [112] and were certainly also known to Li.

— Theorem 11.3.1 was proved by Zuo [264] for $\mathcal{Q} \cong \mathcal{O}_X$ and generalized by Nakashima [198]. The case $\mathrm{Pic}(X) \cong \mathbb{Z}$ was also considered by O'Grady [206]. The result holds in fact without the assumption $n \gg 0$.

— Theorem 11.3.2 is due to O'Grady [210]. Our presentation is slightly different, mostly because we were only interested in the birational description, whereas O'Grady aims for a description of the universal family as well.

Appendix to Chapter 11

11.A Further comments (second edition)

I. Birational properties of moduli spaces.

As explained in the first edition, moduli spaces of sheaves on surfaces are expected to inherit much of the birational properties of the surface. E.g. we have seen that the Hilbert scheme $\text{Hilb}^n(X)$ of zero-dimensional subschemes of length n of a smooth surface X has Kodaira dimension $n \cdot \text{kod}(X)$. In particular, the Hilbert scheme is of general type if X is. This has subsequently been generalized to higher dimensions in [270]. For moduli spaces of sheaves of higher rank one expects a similar behaviour, but to the best of our knowledge Theorem 11.1.5 is still the only general result in this direction.

Moduli spaces on rational surfaces. It seems that we still do not know whether all moduli spaces of stable sheaves on $\mathbb{P}^2$ are rational. In addition to the results mentioned in Section 11.2.1 rationality of $N(r, c_1, c_2)$ has been proved for many choices of (r, c_1, c_2) in [308]. The particular case of rank three bundles has been considered in [369]. Other results of birationality of moduli spaces of stable bundles on $\mathbb{P}^2$ have been obtained by Schofield in [432] by relating these moduli spaces via the Beilinson spectral sequence to moduli spaces of representations of quivers previously studied by the same author in [431].

In [304, 305, 307] rationality of moduli spaces on other rational surfaces has been investigated. E.g. in [307] it is shown that on a rational surface X there always exists an ample line bundle H such that the moduli space $M_H(2, c_1, c_2)$ is rational if its dimension is large enough.

For more existence results for stable rank two bundles on ruled surfaces see [269]

Moduli spaces on surfaces with $\text{kod} = 0$. For K3 surfaces moduli spaces of stable sheaves have been studied intensively. Their birational geometry, at least in the case that M^s is projective, is completely understood. See the comments in Section 6.B.

The birational structure of moduli spaces of sheaves on abelian surfaces has been studied further in [268] and in [453]. In the latter the fibre of the Albanese morphism (cf. Section 10.3.6) has been studied, which is the higher rank analogue of the generalized Kummer variety, i.e. the fibre of $\text{Hilb}^r(A) \longrightarrow \text{Alb}(A) = A$ for an abelian surface A (cf. Theorem 10.3.7). The theory is analogous to the one for K3 surfaces.

Moduli spaces on Enriques surfaces have been studied in [352, 353].

Moduli spaces on elliptic surfaces. In [452] Yoshioka generalizes results of Friedman on the birational structure of moduli spaces of stable sheaves to the higher rank case (cf. Section 11.2.3). The author assumes that all fibres of the elliptic fibration are irreducible and shows that the moduli space of stable sheaves is birationally equivalent to the symmetric product of the relative Jacobian.

II. Numerical invariants of moduli spaces.

The calculation of topological invariants for moduli spaces of semistable sheaves on a surface is, in general, an open problem. There exist partial results for the open subspaces $M^{\mu s} \subset M^s \subset M$ of slope stable respectively stable sheaves.

In [150], Jun Li calculates the first two Betti numbers of the moduli space $M(2, \mathcal{L}, c)^s$ of H-stable sheaves on a smooth projective surface X of rank 2, with fixed determinant $\mathcal{L}$ and $c_2 = c$. More exactly, he shows that $b_1(M(2, \mathcal{L}, c)^s) = b_1(X)$ and $b_2(M(2, \mathcal{L}, c)^s) = 1 + b_1(X) + \binom{b_2(X)}{2}$ for all $c \gg 0$ via a gauge theoretic approach.

Other approaches are based on the Weil conjectures and the counting of points in the moduli space over finite fields. Typically, results are then formulated implicitly in terms of generating functions for series of moduli spaces indexed by, say, the discriminant Δ.

The most complete answer was obtained by Göttsche [83] for moduli of rank 1 sheaves, i.e. Hilbert schemes of points on a surface X. It is also the prototype for later results: Göttsche's formula expresses the generating function for the Betti numbers of $\mathrm{Hilb}^n(X)$, $n \in \mathbb{N}_0$, in terms of the Betti numbers $b_i(X)$ of the underlying surface:

$$\sum_{n=0}^{\infty} \sum_{k=0}^{2n} b_k(\mathrm{Hilb}^n(X)) t^{k-2n} q^n = \prod_{m=1}^{\infty} \prod_{i=0}^{4} (1 - (-1)^i q^m t^{i-2})^{-(-1)^i b_i(X)},$$

where q and t are formal parameters. The calculation uses the Weil conjectures and previous results of Ellingsrud and Strømme [316] for Hilbert schemes of rational surfaces that were obtained by counting fixed points under a torus action on the Hilbert scheme.

Similar formulas exist for generalized Kummer varieties and are due to Göttsche and Soergel [86]. In these two cases the Hodge numbers are known as well. Evaluating at $t = -1$ yields a simpler formula for the topological Euler characteristics in terms of the eta-function $\eta(q) = q^{1/24} \prod_{m>0} (1 - q^m)$:

$$\sum_{n=0}^{\infty} e(\mathrm{Hilb}^n(X)) q^n = \left(\frac{q^{1/24}}{\eta(q)} \right)^{e(X)}.$$

The appearance of the eta-function is not coincidental. Many papers on the subject have been inspired by predictions of Vafa and Witten [435], based on physical reasoning, about relations between appropriately defined Euler characteristics of moduli spaces and modular forms.

For a K3 surface, all moduli spaces $M_H(v)$ of semistable sheaves with primitive Mukai vector and v-generic ample divisor H are smooth and deformation equivalent to Hilbert schemes (see the comments in Appendix 6.B.III). In particular, they have the same Betti numbers as the Hilbert scheme of the same dimension. Similarly, for an abelian surface A and a primitive Mukai vector v the fibres of the morphism $M(v) \to \mathrm{Pic}(A) \times \mathrm{Alb}(A)$ are deformation equivalent to generalized Kummer varieties and have the same Betti numbers.

For other surfaces, the picture is less complete. Yoshioka has obtained results for sheaves of rank 2 on ruled surfaces, formulas relating the generating functions for blow-up's $\widehat{X} \to X$, and for elliptic surfaces [445, 447, 454], see also the survey article [446]. Göttsche's article [323] contains results for rational surfaces including the dependence on the ample divisor and wall crossing formulas. In a recent preprint [354], Kool calculates Euler numbers of moduli spaces of μ-stable sheaves on toric surfaces by counting fixed points for the induced torus action on the moduli space.

References

[1] C. Anghel, *La stabilité de la restriction à une courbe lisse d'un fibré de rang 2 sur une surface algébrique.* Math. Ann. 304 (1996), 53–62.

[2] B. Angeniol, M. Lejeune-Jalabert, *Calcul différentiel et classes caractéristiques en géométrie algébrique.* Hermann, Paris, Travaux en Cours 38 (1989).

[3] A. Altman, S. Kleiman, *Compactifying the Picard Scheme.* Adv. Math. 35 (1980), 50–112.

[4] E. Arbarello, M. Cornalba, P.A. Griffiths, J. Harris, *Geometry of Algebraic Curves.* Volume 1, Springer Grundlehren 267, New York (1985).

[5] I. Artamkin, *On Deformation of sheaves.* Izv. Akad. Nauk SSSR 52, 3 (1988), 660–665. Transl.: Math. USSR Izv. 32, 3 (1989), 663–668.

[6] I. Artamkin, *Stable bundles with $c_1 = 0$ on rational surfaces.* Izv. Akad. Nauk SSSR 54, 2 (1990), 227–241. Transl.: Math. USSR Izv. 36, 2 (1991), 231–246.

[7] I. Artamkin, *Deforming torsion-free sheaves on algebraic surfaces.* Izv. Akad. Nauk SSSR 54, 3 (1990), 435–468. Transl.: Math. USSR Izv. 36, 3 (1991), 449–485.

[8] M. Atiyah, I. Macdonald, *Commutative Algebra.* Addison-Wesley, Reading, Mass. (1969).

[9] E. Ballico, *On moduli of vector bundles on rational surfaces.* Archiv Math. 49 (1987), 267–272.

[10] E. Ballico, *On rank two vector bundles on an algebraic surface.* Forum Math. 4 (1992), 231–241.

[11] E. Ballico, *On moduli of vector bundles on surfaces with negative Kodaira dimension.* Ann. Univ. Ferrara Sez. VII 38 (1992), 33–40.

[12] E. Ballico, *On vector bundles on algebraic surfaces and 0-cycles.* Nagoya Math. J. 130 (1993), 19–23.

[13] E. Ballico, *On the components of moduli of simple sheaves on a surface*. Forum Math. 6 (1994), 35–41.

[14] E. Ballico, L. Chiantini, *Some properties of stable rank 2 bundles on algebraic surfaces*. Forum Math. 4 (1992), 417–424.

[15] E. Ballico, L. Chiantini, *On some moduli spaces of rank two bundles over K3 surfaces*. Boll. Un. Mat. Ital. A (7) (1993), 279–287.

[16] E. Ballico, R. Brussee, *On the unbalance of vector bundles on a blown-up surface*. Preprint (1991).

[17] E. Ballico, P. Newstead, *Uniform bundles on quadric surfaces and some related varieties*. J. London Math. Soc. 31 (1985), 211–223.

[18] C. Banica, J. Le Potier, *Sur l'existence des fibrés vectoriels holomorphes sur les surface non-algébriques*. J. reine angew. Math. 378 (1987), 1–31.

[19] C. Banica, M. Putinar, G. Schuhmacher, *Variation der globalen Ext in Deformationen kompakter komplexer Räume*. Math. Ann. 250, (1980) 135–155.

[20] W. Barth, *Some Properties of Stable Rank-2 Bundles on $\mathbb{P}^n$*. Math. Ann. 226 (1977), 125–150.

[21] W. Barth, *Moduli of vector bundles on the projective plane*. Invent. Math. 42 (1977), 63–91.

[22] W. Barth, C. Peters, A. Van de Ven, *Compact Complex Surfaces*. Erg. Math., 3. Folge, Band 4, Springer Verlag, Berlin (1984).

[23] C. Bartocci, U. Bruzzo, D. Hernández Ruipérez, *A Fourier-Mukai Transform for Stable Bundles on K3 Surfaces*. J. Reine Angew. Math. 486 (1997), 1–16.

[24] S. Bauer, *Some nonreduced moduli of bundles and Donaldson invariants for Dolgachev surfaces*. J. reine angew. Math. 424 (1992), 149–180.

[25] A. Beauville, *Variétés Kähleriennes dont la première classe de Chern est nulle*. J. Diff. Geom. 18 (1983), 755–782.

[26] A. Beauville, *Application aux espaces de modules*. in Géometrie des surfaces K3 ct périodes. Astérisque 126 (1985).

[27] S. Bloch, D. Gieseker, *The positivity of the Chern classes of an ample vector bundle*. Invent. Math. 12 (1971), 45–60.

[28] F. Bogomolov, *Holomorphic tensors and vector bundles*. Izv. Akad. Nauk SSSR 42, 6 (1978), 1227–1287. Transl.: Math. USSR. Izv. 13 (1979), 499–555.

[29] F. Bogomolov, *Stability of vector bundles on surfaces and curves*. Lecture Notes in Pure and Appl. Math. 145 (1993), 35–49.

[30] F. Bogomolov, *Hamiltonian Kähler manifolds*. Soviet Math. Dokl. 19 (1978), 1462–1465.

[31] F. Bogomolov, *Stable vector bundles on projective surfaces*. Mat-Sb. 185 (1994), 3–26. Transl.: Russ. Acad. Sci. Sb. Math. 81 (1995), 397–419.

[32] A. Borel et al., *Intersection cohomology*. Progr. Math. 50 (1984).

[33] F. Bottacin, *Poisson structures on moduli spaces of sheaves over Poisson surfaces*. Invent. Math. 121 (1995), 421–436.

[34] V. Brinzanescu, *Algebraic 2 vector bundles on ruled surfaces*. Ann. Univ. Ferrara 37 (1991), 55–64.

[35] V. Brinzanescu, *Holomorphic Vector Bundles over Compact Complex Surfaces*. Lect. Notes Math. 1624 (1996).

[36] V. Brinzanescu, M. Stoia, *Topologically trivial algebraic 2-vector bundles on ruled surfaces, II*. Lect. Notes Math. 1056 (1984), 34–46.

[37] E. Brosius, *Rank-2 Vector Bundles on a Ruled Surface. I*. Math. Ann. 265 (1983), 155–168.

[38] E. Brosius, *Rank-2 Vector Bundles on a Ruled Surface. II*. Math. Ann. 266 (1983), 199–214.

[39] N. Buchdahl, *Stable 2-Bundles on Hirzebruch Surfaces*. Math. Z. 194 (1987), 143–152.

[40] F. Catanese, M. Schneider, *Bounds for Stable Bundles and Degrees of Weierstrass Schemes*. Math. Ann. 293 (1992), 579–594.

[41] P. Deligne, *Equations différentielles à points singuliers réguliers*. Lect. Notes Math. 163 (1970).

[42] P. Deligne, N. Katz, *Groupes de monodromie en géométrie algébrique*. SGA7 II, Lect. Notes Math. 340, Springer-Verlag, Berlin (1973).

[43] M. Demazure, P. Gabriel, *Groupes Algébriques*. Tome 1. North Holland, Amsterdam (1970).

[44] S. Donaldson, *Anti-self-dual Yang Mills connections over complex algebraic surfaces and stable vector bundles*. Proc. Lond. Math. Soc. 50 (1985), 1–26.

[45] S. Donaldson, *Infinite determinants, stable bundles and curvature*. Duke Math. J. 54 (1987), 231–247.

[46] S. Donaldson, P. B. Kronheimer, *The Geometry of Four-Manifolds*. Clarendon, Oxford (1990).

[47] S. Donaldson, *Polynomial invariants for smooth four manifolds*. Topology 29 (1990), 257–315.

[48] J.-M. Drezet, *Points non factoriels des variétés de modules de faisceaux semistables sur une surface rationelle*. J. Reine Angew. Math. 413 (1991), 99–126.

[49] J.-M. Drezet, *Sur les équations vérifiées par les invariants des fibrés exceptionels*. Forum Math. 8 (1996), 237–265.

[50] J.-M. Drezet, *Groupe de Picard des variétés de modules de faisceaux semi-stables sur* $\mathbb{P}_2(\mathbb{C})$. Ann. Inst. Fourier 38 (1988), 105–168.

[51] J.-M. Drezet, J. Le Potier, *Fibrés stables et fibrés exceptionels sur* $\mathbb{P}_2$. Ann. Scient. Ec. Norm. Sup. 18 (1985), 193–244.

[52] J.-M. Drezet, M. S. Narasimhan, *Groupe de Picard des variétés de modules de fibrés semi-stables sur les courbes algébriques*. Invent. Math. 97 (1989), 53–94.

[53] L. Ein, *Stable vector bundles in char* $p > 0$. Math. Ann. 254 (1980), 52–72.

[54] G. Elencwajg, O. Forster, *Bounding cohomology groups of vector bundles on* $\mathbb{P}_n$. Math. Ann. 246 (1979/80), 251–270.

[55] G. Elencwajg, O. Forster, *Vector bundles on manifolds without divisors and a theorem on deformations*. Ann. Inst. Fourier 32 (1982).

[56] G. Ellingsrud, *Sur l'irreducibilité du module des fibrés stables sur* $\mathbb{P}^2$. Math. Z. 182 (1983), 189–192.

[57] G. Ellingsrud, L. Göttsche, *Variation of moduli spaces and Donaldson invariants under change of polarization*. J. Reine Angew. Math. 467 (1995), 1–49.

[58] G. Ellingsrud, M. Lehn, *The Irreducibility of the Punctual Quot-Scheme*. Ark. Mat. 37 (1999), 245–254.

[59] G. Ellingsrud, S. Strømme, *On the rationality of the moduli space for stable rank-2 vector bundles on* $\mathbb{P}_2$. Lect. Notes Math. 1273 (1987), 363–371.

[60] G. Ellingsrud, S. Strømme, *The Picard group of the moduli space for stable rank-2 vector bundles on* $\mathbb{P}^2$ *with odd first Chern class*. Preprint (1979).

[61] G. Faltings, *Stable G-bundles and projective connections*. J. Alg. Geom. 2 (1993), 507–568.

[62] R. Fahlaoui, Y. Laszlo, *Transformée de Fourier et stabilité sur les surfaces abéliennes*. Comp. math. 79 (1991), 271–278.

[63] H. Flenner, *Restrictions of semistable bundles on projective varieties*. Comment. Math. Helv. 59 (1984), 635–650.

[64] H. Flenner, *Eine Bemerkung über relative Ext-Garben*. Math. Ann. 258 (1981), 175–182.

[65] J. Fogarty, *Algebraic Families on an Algebraic Surface*. Am. J. Math. 90 (1968), 511–521.

[66] O. Forster, A. Hirschowitz, M. Schneider, *Type de scindage généralisée pour les fibrés stables*. in Progr. Math. 7, Vector bundles and differential equations. Birkhäuser 1979, 65–81.

[67] R. Friedman, *Rank two vector bundles over regular elliptic surfaces*. Invent. Math. 96 (1989), 283–332.

[68] R. Friedman, *Vector Bundles and* $SO(3)$*-Invariants for Elliptic Surfaces*. J. AMS 8, 1 (1995), 29–139.

[69] R. Friedman, *Algebraic Surfaces and Holomorphic Vector Bundles*, Springer Verlag, New York (1998).

[70] R. Friedman, J. Morgan, *On the diffeomorphism types of certain algebraic surfaces I*. J. Diff. Geom. 27 (1988), 297–369; II. J. Diff. Geom. 27 (1988), 371–398.

[71] R. Friedman, J. Morgan, *Smooth four-manifolds and complex surfaces*. Erg. Math. (3. Folge) Band 27, Springer Verlag (1994).

[72] R. Friedman, Z. Qin, *Flips of moduli spaces and transition formulas for Donaldson invariants of rational surfaces*. Comm. Anal. Geom. 3 (1995), 11–83.

[73] A. Fujiki, *On the de Rham cohomology group of compact Kähler symplectic manifolds*. Alg. Geom., Sendai 1985, Adv. Stud. Pure Math. 10 (1987), 105–167.

[74] W. Fulton, *Intersection Theory*. Erg. Math. (3. Folge) Band 2, Springer Verlag (1984).

[75] D. Gieseker, *p-ample vector bundles and their Chern classes*. Nagoya Math. J. 43 (1971), 91–116.

[76] D. Gieseker, *Stable vector bundles and the Frobenius morphism*. Ann. Scient. Ec. Norm. Sup. 6 (1973), 95–101.

[77] D. Gieseker, *On the moduli of vector bundles on an algebraic surface*. Ann. Math. 106 (1977), 45–60.

[78] D. Gieseker, *On a theorem of Bogomolov on Chern classes of stable bundles*. Am. J. Math. 101 (1979), 77–85.

[79] D. Gieseker, *A construction of stable bundles on an algebraic surface*. J. Diff. Geom. 27 (1988) 137–154.

[80] D. Gieseker, *Geometric Invariant Theory and applications to moduli problems*. Lect. Notes Math. 996 (1982), 45–73.

[81] D. Gieseker, J. Li, *Irreducibility of moduli of rank 2 bundles on algebraic surfaces*. J. Diff. Geom. 40 (1994), 23–104.

[82] D. Gieseker, J. Li, *Moduli of high rank vector bundles over surfaces*. J. AMS 9 (1996), 107–151.

[83] L. Göttsche, *Hilbert schemes of zero dimensional subschemes of smooth varieties*. Lect. Notes Math. 1572, Springer Verlag, Berlin (1994).

[84] L. Göttsche, *Change of polarization and Hodge numbers of moduli spaces of torsion free sheaves on surfaces*. Math. Z. 223 (1996), 247–260.

[85] L. Göttsche, *Rationality of Moduli Spaces of Torsion Free Sheaves over Rational Surfaces*. Manuscripta math. 89 (1996), 193–201.

[86] L. Göttsche, W. Soergel, *Perverse sheaves and the cohomology of Hilbert schemes of smooth algebraic surfaces*. Math. Ann. 296 (1993), 235–245.

[87] L. Göttsche, D. Huybrechts, *Hodge numbers of moduli spaces of stable bundles on K3 surfaces*. Int. J. Math. 7 (1996), 359–372.

[88] H. Grauert, G. Mülich, *Vektorbündel vom Rang 2 über dem n-dimensionalen komplex projektiven Raum*. Manuscripta math. 16 (1975), 25–100.

[89] P. Griffiths, *Some results on algebraic cycles on algebraic manifolds*. nt. Coll. Tata Inst. Bombay (1968), 93–191.

[90] P. Griffiths, J. Harris, *On the Noether-Lefschetz Theorem and Some Remarks on Codimension-two Cycles*. Math. Ann. 271 (1985), 31–51.

[91] P. Griffiths, J. Harris, *Principles of Algebraic Geometry*. Wiley-Interscience (1978).

[92] A. Grothendieck, *Sur la classification des fibrés holomorphes sur la sphère de Riemann*. Am. J. Math. 79 (1957), 121–138.

[93] A. Grothendieck, *Techniques de construction et théorèmes d'existence en géométrie algébrique IV: Les schémas de Hilbert*. Séminaire Bourbaki, 1960/61, no. 221.

[94] A. Grothendieck, J. Dieudonné, *Eléments de Géométrie Algébrique*. Publ. Math. IHES, 28 (1966).

[95] G. Harder, M. Narasimhan, *On the cohomology groups of moduli spaces of vector bundles on curves*. Math. Ann. 212 (1975), 215–248.

[96] R. Hartshorne, *Residues and Duality*. Lect. Notes Math. 20, Springer Verlag, Heidelberg (1966).

[97] R. Hartshorne, *Ample vector bundles*. Publ. Math. IHES, 29 (1966), 319–350.

[98] R. Hartshorne, *Algebraic Geometry*. GTM 52, Springer Verlag, New York (1977).

[99] R. Hartshorne, *Ample vector bundles on curves*. Nagoya Math. J. 43 (1971), 73–90.

[100] R. Hartshorne, *Ample Subvarieties of Algebraic Varieties*. Lect. Notes Math. 156, Springer Verlag, Berlin (1970).

[101] G. Hein, *Shatz-Schichtungen stabiler Vektorbündel*. Thesis (1996).

[102] A. Hirschowitz, *Sur la restriction des faisceau semi-stables*. Ann. Scient. Ec. Norm. Sup. 14 (1980), 199–207.

[103] A. Hirschowitz, K. Hulek, *Complete families of stable families of stable vector bundles over* $\mathbb{P}^2$. Complex Analysis and Algebraic Geometry. Lect. Notes Math. 1194 (1986), 19–33.

[104] A. Hirschowitz, Y. Laszlo, *A propos de l'existence de fibrés stables sur les surfaces*. J. reine angew. Math. 460 (1995), 55–68.

[105] H. Hoppe, H. Spindler, *Modulräume stabiler Vektorraumbündel vom Rang 2 auf Regelflächen*. Ann. Math. 249 (1980), 127–140.

[106] H. Hoppe, *Modulräume stabiler Vektorraumbündel vom Rang 2 auf rationalen Regelflächen*. Math. Ann. 264 (1983), 227–239.

[107] K. Hulek, *Stable Rank-2 Vector Bundles on* $\mathbb{P}^2$ *with* c_1 *odd*. Math. Ann. 242 (1979), 241–266.

[108] K. Hulek, J. Le Potier, *Sur l'espace de modules des faisceaux semistables de rang 2 de classes de Chern* $(0, 3)$ *sur* $\mathbb{P}^2$. Ann. Inst. Fourier 39 (1989), 251–292.

[109] M. Huibregtse, *The Albanese mapping for a punctual Hilbert scheme I: Irreducibility of the fibres.* Trans. AMS 251 (1979), 267–285.

[110] M. Huibregtse, *The Albanese mapping for a punctual Hilbert scheme II: Symmetrized differentials and singularities.* Trans. AMS 274 (1982), 109–140.

[111] D. Huybrechts, *Stabile Vektorbündel auf algebraischen Flächen – Tyurins Methode zum Studium der Geometrie der Modulräume.* Thesis (1992).

[112] D. Huybrechts, *Complete curves in moduli spaces of stable bundles on surfaces.* Math. Ann. 298 (1994), 67–78.

[113] D. Huybrechts, *Birational symplectic manifolds and their deformations.* J. Diff. Geom. 45 (1997), 488–513.

[114] D. Huybrechts, M. Lehn, *Stable pairs on curves and surfaces.* J. Alg. Geom. 4 (1995), 67–104.

[115] D. Huybrechts, M. Lehn, *Framed modules and their moduli.* Int. J. Math. 6 (1995), 297–324.

[116] A. Ishii, *On the moduli of reflexive sheaves on a surface with rational double points.* Math. Ann. 294 (1992), 125–150.

[117] B. Iversen, *Linear determinants with applications to the Picard scheme of a family of algebraic curves.* Lect. Notes Math. 174, Springer Verlag, Berlin (1970).

[118] Y. Kametani, Y. Sato, *0-Dimensional Moduli Spaces of Stable Rank 2 Bundles and Differentiable Structures on Regular Elliptic Surfaces.* Tokyo J. Math. 17(1) (1994), 253–267.

[119] P. Katsylo, *Birational Geometry of Moduli Varieties of Bundles over* $\mathbb{P}^2$. Izv. Akad. Nauk SSSR Ser. Mat. 55 (1991), 429–438. Transl.: Math. USSR Izv. 38 (1992), 419–428.

[120] Y. Kawamata, *Unobstructed Deformations — A Remark on a Paper of Z. Ran.* J. Alg. Geom. 1 (1992), 183–190. Erratum in J. Alg. Geom. 6 (1997), 803–804.

[121] Y. Kawamata, *Unobstructed Deformations. II.* J. Alg. Geom. 4 (1995), 277–279.

[122] H. Kim, *Stable vector bundles on Enriques surfaces.* Thesis Michigan (1990).

[123] A. King, *Moduli of Representations of Finite Dimensional Algebras.* Quart. J. Math. Oxford 45 (1994), 515–530.

[124] S. Kleiman, *Les théorèmes de finitude pour le foncteur de Picard.* SGA de Bois Marie, 1966/67, exp. XIII.

[125] S. Kleiman, *Towards a numerical theory of ampleness.* Ann. Math. 84 (1966), 293–344.

[126] F. Knudson, D. Mumford, *The projectivity of the moduli space of stable curves I: Preliminaries on "det" and "Div".* Math. Scand. 39 (1976), 19–55.

[127] S. Kobayashi, *Differential Geometry of Complex Vector Bundles.* Iwanami Shoten, Princeton University Press (1987).

[128] J. Kollár, *Rational curves on algebraic varieties*. Erg. Math. 32 (3. Folge), Springer Verlag, Berlin (1996).

[129] D. Kotschick, *Moduli of vector bundles with odd c_1 on surfaces with $q = p_g = 0$.* Am. J. Math. 114 (1992). 297–313.

[130] S. Kosarew, C. Okonek, *Global moduli spaces and simple holomorphic bundles.* Publ. RIMS Kyoto Univ. 25 (1989), 1–19.

[131] H. Kraft, P. Slodowy, T. Springer, *Algebraische Transformationsgruppen und Invariantentheorie, Algebraic Transformation Groups and Invariant Theory.* DMV Seminar Band 13, Birkhäuser Basel, Boston, Berlin (1989).

[132] M. Kreuzer, *On 0-dimensional complete intersections.* Math. Ann. 292 (1992), 43–58.

[133] S. Kuleshov, *An existence theorem for exceptional bundles on K3 surfaces.* Izv. Akad. Nauk SSSR 53, 2 (1989), 363–378. Transl.: Math. USSR Izv. 34, 2 (1990), 373–388.

[134] S. Kuleshov, *Stable bundles on K3 surfaces.* Izv. Akad. Nauk SSSR 54, 1 (1990), 213–220. Transl.: Math. USSR Izv. 36, 1 (1991), 223–230.

[135] S. Langton, *Valuative criteria for families of vector bundles on algebraic varieties.* Ann. Math. 101 (1975), 88–110.

[136] O. Laudal, *Formal Moduli of Algebraic Structures.* Lect. Notes Math. 754, Springer Verlag, Berlin (1979).

[137] O. Laudal, *Matric Massey Products and Formal Moduli I.* Algebra, Algebraic Topology and their interactions, edited by Jan Erik Roos. Lect. Notes Math. 1183, Springer Verlag (1986), 218–240.

[138] R. Lazarsfeld, *Brill-Noether theory without degenerations.* J. Diff. Geom. 23 (1986), 299–307.

[139] R. Lazarsfeld, *A sampling of Vector Bundle Techniques in the Study of Linear Systems.* Lectures on Riemann surfaces, M. Cornalba et. al (eds.), World Scientific Press (1989), 500–599.

[140] L. Le Bruyn, *Some Remarks on Rational Matrix Invariants.* J. Alg. 118 (1988), 487–493.

[141] J. Le Potier, *Fibrés stables de rang 2 sur $\mathbb{P}^2(\mathbb{C})$.* Math. Ann. 241 (1979), 217–250.

[142] J. Le Potier, *Sur le group de Picard de l'espace de modules de fibrés semistables sur $\mathbb{P}^2$.* Ann. Scient. Ec. Norm. Sup. 14 (1981), 141–155.

[143] J. Le Potier, *Fibrés vectoriels sur les surfaces K3.* Séminaire Lelong-Dolbeault-Skoda. Lect. Notes Math. 1028 (1983), 225–238.

[144] J. Le Potier, *Fibré déterminant et courbes de saut sur les surfaces algébriques.* Complex Projective Geometry, London Mathematical Society: Bergen (1989), 213–240.

[145] J. Le Potier, *L'espace de modules de Simpson*. Séminaire de géométrie algébrique, Jussieu, fév. 1992.

[146] J. Le Potier, *Module des fibrés semi-stables et fonctions thêta*. in: Moduli of vector bundles, ed. M. Maruyama. Lect. Notes in Pure and Appl. Math. 179 (1996), 83–101.

[147] J. Le Potier, *Systèmes cohérents et structures de niveau*. Astérisque 214 (1993).

[148] J. Li, *Algebraic geometric interpretation of Donaldson's polynomial invariants of algebraic surfaces*. J. Diff. Geom. 37 (1993), 416–466.

[149] J. Li, *Kodaira dimension of moduli space of vector bundles on surfaces*. Invent. Math. 115 (1994), 1–46.

[150] J. Li, *The first two Betti numbers of the moduli spaces of vector bundles on surfaces*. Comm. Anal. Geom. 5 (1997), 625–684.

[151] J. Li, *Picard groups of the moduli spaces of vector bundles over algebraic surfaces*. in: Moduli of vector bundles, ed. M. Maruyama. Lect. Notes in Pure and Appl. Math. 179 (1996), 129–146.

[152] J. Li, *The Geometry of Moduli Spaces of Vector Bundles over Algebraic Surfaces*. Proc. Int. Congress Math. 1994. Birkhäuser (1995).

[153] W.-P. Li, Z. Qin, *Lower degree Donaldson polynomial invariants of rational surfaces*. J. Alg. Geom. 2 (1993), 413–442.

[154] W. P. Li, Z. Qin, *Stable vector bundles on algebraic surfaces*. Trans. AMS 345 (1994), 833–852.

[155] M. Lübke, C. Okonek, *Stable bundles on regular elliptic surfaces*. J. Reine Angew. Math. 378 (1987), 32–45.

[156] M. Lübke, *The analytic moduli space of framed bundles*. J. Reine Angew. Math. 441 (1993), 45–59.

[157] M. Lübke, A. Teleman, *The Kobayashi-Hitchin Correspondence*. World Scientific (1995).

[158] D. Luna, *Slices étales*. Bull. Soc. Math. France 33 (1973), 81–105.

[159] A. Macioccia, *Gieseker stability and the Fourier-Mukai transform for abelian surfaces*. Oxford Quart. J. Math. 47 (1996), 87–100.

[160] M. Maruyama, *Stable vector bundles on an algebraic surface*. Nagoya Math. J. 58 (1975), 25–68.

[161] M. Maruyama, *Openness of a family of torsion free sheaves*. J. Math. Kyoto Univ. 16 (1976), 627–637.

[162] M. Maruyama, *Moduli of stable sheaves I*. J. Math. Kyoto Univ. 17 (1977), 91–126.

[163] M. Maruyama, *Moduli of stable sheaves II*. J. Math. Kyoto Univ. 18 (1978), 557–614.

[164] M. Maruyama, *Boundedness of semistable sheaves of small ranks.* Nagoya Math. J. 78 (1980), 65–94.

[165] M. Maruyama, *On boundedness of families of torsion free sheaves.* J. Math. Kyoto Univ. 21 (1981), 673–701.

[166] M. Maruyama, *The theorem of Grauert-Mülich-Spindler.* Math. Ann. 255 (1981), 317–333.

[167] M. Maruyama, *Elementary transformations in the theory of algebraic vector bundles.* Algebraic Geometry (La Rábida). Lect. Notes Math. 961 (1982), 241–266.

[168] M. Maruyama, *Stable rationality of some moduli spaces of vector bundles on* $\mathbb{P}^2$. Lect. Notes Math. 1194 (1986), 80–89.

[169] M. Maruyama, *The rationality of the moduli spaces of vector bundles of rank two on* $\mathbb{P}^2$. Adv. Stud. Pure Math. 10 (1987), 394–414.

[170] M. Maruyama, *On a compactification of a moduli space of stable vector bundles on a rational surface.* in: Alg. geom. and com. alg. in honor of Masayashi Nagata, Vol. 1 (1988), 233–260.

[171] K. Matsuki, R. Wentworth, *Mumford-Thaddeus Principle on the Moduli Space of Vector Bundles on an Algebraic Surface.* Int. J. Math. 8 (1997), 97–148.

[172] H. Matsumura, *Commutative Algebra*, Benjamin, London (1980).

[173] Y. Matsushima, *Fibrés holomorphes sur un tore complexe.* Nagoya Math. J. 14 (1959), 1–14.

[174] V. Mehta, *On Some Restriction Theorems for Semistable Bundles.* Lect. Notes Math. 996 (1983), 145–153.

[175] V. Mehta, A. Ramanathan, *Semistable sheaves on projective varieties and the restriction to curves.* Math. Ann. 258 (1982), 213–226.

[176] V. Mehta, A. Ramanathan, *Restriction of stable sheaves and representations of the fundamental group.* Invent. Math. 77 (1984), 163–172.

[177] N. Mestrano, *Poincaré bundles for projective surfaces.* Ann. Inst. Fourier 35 (1985), 217–249.

[178] J. Milne, *Etale Cohomology.* Princeton University Press, Princeton (1980).

[179] J. Milnor, *Morse Theory.* Ann. Math. Studies 51, Princeton Univ. Press, Princeton (1963).

[180] Y. Miyaoka, *The Chern classes and Kodaira dimension of a minimal variety.* Advanced Studies in Pure Math. 10 (1987), 449–476.

[181] J. Morgan, *Comparison of the Donaldson polynomial invariants with their algebro-geometric analogues.* Top. 32 (1993), 449–488.

[182] S. Mori, *Projective manifolds with ample tangent bundles.* Ann. Math. 110 (1979), 593–606.

[183] H. Morikawa, *A note on holomorphic vector bundles over complex tori.* Nagoya Math. J. 41 (1971), 101–106.

[184] D. Morrison, *On K3 surfaces with large Picard number.* Invent. Math. 75 (1984), 105–121.

[185] S. Mukai, *Semi-homogenous vector bundles on an abelian variety.* J. Math. Kyoto Univ. 18 (1978), 239–272.

[186] S. Mukai, *Duality between $D(X)$ and $D(\hat{X})$ with its application to Picard sheaves.* Nagoya Math. J. 81 (1981), 153–175.

[187] S. Mukai, *Symplectic structure of the moduli space of sheaves on an abelian or K3 surface.* Invent. Math. 77 (1984), 101–116.

[188] S. Mukai, *On the moduli space of bundles on K3-surfaces I.* in: Vector Bundles on Algebraic Varieties. Bombay (1984).

[189] S. Mukai, *Fourier Functor and its Application to the Moduli of bundles on an Abelian Variety.* Adv. in Pure Math. 10 (1987), 515–550.

[190] S. Mukai, *Moduli of vector bundles on K3-surfaces and symplectic manifolds.* Sugaku Expositions 1,2 (1988), 138–174.

[191] D. Mumford, *Projective invariants of projective structures and applications.* in Proc. Int. Cong. Math. Stockholm (1962), 526–530.

[192] D. Mumford, *Lectures on curves on an algebraic surface.* Annals Math. Studies 59, Univ. Press, Princeton (1966).

[193] D. Mumford, *Rational equivalence of 0-cycles on surfaces.* J. Math. Kyoto Univ. 9 (1968), 195–204.

[194] D. Mumford, *Abelian Varieties.* Oxford Univ. Press. Oxford (1970).

[195] D. Mumford, J. Fogarty, *Geometric Invariant Theory.* Erg. Math. 34, 2nd ed., Springer Verlag, Berlin, Heidelberg (1982).

[196] J. Murre, *Lectures on an Introduction to Grothendieck's Theory of the Fundamental Group.* Lecture Notes, Tata Insitute of Fundamental Research, Bombay (1967).

[197] T. Nakashima, *Moduli of stable bundles on blown-up surfaces.* J. Math. Kyoto Univ. 33 (1993), 571–581.

[198] T. Nakashima, *Moduli of stable rank 2 bundles with ample c_1 on K3-surfaces.* Arch. Math. 61 (1993), 100–104.

[199] T. Nakashima, *On the moduli of stable vector bundles on a Hirzebruch surface.* Math. Z. 212 (1993), 211–221.

[200] T. Nakashima, *A note on Donaldson polynomials for K3 surfaces.* Forum Math. 6 (1994), 385–390.

[201] T. Nakashima, *Some moduli spaces of stable bundles on elliptic K3 surfaces.* Math. Nachr. 170 (1994), 201–209.

[202] M. Narasimhan, C. S. Seshadri, *Stable and unitary bundles on a compact Riemann surface.* Ann. Math. 82 (1965), 540–567.

[203] P. Newstead, *Lectures on Introduction to Moduli Problems and Orbit Spaces.* Tata Institute of Fundamental Research, Bombay. Springer Verlag, Berlin (1978).

[204] A. Norton, *Analytic moduli of complex vector bundles.* Indiana Univ. Math. J. 28 (1979), 365–387.

[205] T. Oda, *Vector bundles on abelian surfaces.* Invent. Math. 13 (1971), 247–260.

[206] K. O'Grady, *Donaldson's polynomials for K3 surfaces.* J. Diff. Geom. 35 (1992), 415–427.

[207] K. O'Grady, *Algebro-geometric analogues of Donaldson polynomials.* Invent. Math. 107 (1992), 351–395.

[208] K. O'Grady, *The irreducible components of moduli spaces of vector bundles on surfaces.* Invent. Math. 112 (1993), 585–613.

[209] K. O'Grady, *Moduli of vector bundles on projective surfaces: Some basic results.* Invent. Math. 123 (1996), 141–207.

[210] K. O'Grady, *The weight-two Hodge structure of moduli spaces of sheaves on a K3 surface.* J. Alg. Geom. 6 (1997), 599–644.

[211] K. O'Grady, *Moduli of Vector Bundles on Surfaces.* Algebrai Geometry – Santa Cruz 1995. Proc. Sympos. Pure Math. 62,2. AMS (1997), 101–126.

[212] C. Okonek, M. Schneider, H. Spindler, *Vector bundles on complex projective spaces.* Progress in Math. 3, Birkhäuser (1980).

[213] T. Ono, *Simple vector bundles of rank 2 on a Del Pezzo surface.* TRU Math. 19 (1983), 125–131.

[214] W. Oxbury, *Theta-characteristics and stable vector bundles.* Proc. London Math. Soc. 71 (1995), 481–500.

[215] Z. Qin, *Birational properties of moduli spaces of stable locally free rank-2 sheaves on algebraic surfaces.* Manuscripta math. 72 (1991), 163–180.

[216] Z. Qin, *Chamber structures of algebraic surfaces with Kodaira dimension zero and moduli spaces of stable rank two bundles.* Math. Z. 207 (1991), 121–136.

[217] Z. Qin, *Symmetric polynomials constructed from moduli of stable sheaves with odd c_1 on ruled surfaces.* Manuscripta math. 73 (1991), 373–383.

[218] Z. Qin, *Moduli spaces of stable rank 2 bundles on ruled surfaces.* Invent. Math. 110 (1992), 615–626.

[219] Z. Qin, *Moduli of stable sheaves on ruled surfaces and their Picard groups.* J. Reine Angew. Math. 433 (1992), 201–219.

[220] Z. Qin, *Stable rank 2 bundles on simply connected elliptic surfaces.* Duke Math. J. 67 (1992), 557–569.

[221] Z. Qin, *Simple sheaves versus stable sheaves on algebraic surfaces.* Math. Z. 209 (1992), 559–579.

[222] Z. Qin, *On smooth structures of potential surfaces of general type homeomorphic to rational surfaces.* Invent. Math. 113 (1993), 163–175.

[223] Z. Qin, *Equivalence classes of polarizations and moduli spaces of sheaves.* J. Diff. Geom. 37 (1993), 397–415.

[224] Z. Qin, *On the existence of stable rank 2 sheaves on algebraic surfaces.* J. Reine Angew. Math. 439 (1993), 213–219.

[225] Z. Qin, *Moduli of simple rank 2 sheaves on $K3$ surfaces.* Manuscripta math. 79 (1993), 253–265.

[226] S. Ramanan, A. Ramanathan, *Some remarks on the instability flag.* Tôhoku Math. J. 36 (1984), 269–291.

[227] Z. Ran, *Deformations of Manifolds with Torsion or Negative Canonical Bundle.* J. Alg. Geom. 1 (1992), 279–291.

[228] M. Reid, *Bogomolov's theorem $c_1^2 \leq 4c_2$.* in Proc. Int. Symp. Alg. Geom. Kyoto (1977), 633–642.

[229] I. Reider, *Vector bundles of rank 2 and linear systems on algebraic surfaces.* Ann. Math. 127 (1988), 309–316.

[230] A. Rudakov et al., *Helices and Vector Bundles.* Cambridge Univ. Press, London Math. Soc. Lecture Note Series 148 (1990).

[231] M. Schlessinger, *Functors of Artin Rings.* Trans. AMS 130 (1968), 208–222.

[232] R. Schwarzenberger, *Vector bundles on algebraic surfaces.* Proc. London Math. Soc. 11 (1961), 601–622.

[233] R. Schwarzenberger, *Vector bundles on the projective plane.* Proc. London Math. Soc. 11 (1961), 623–640.

[234] J.-P. Serre, *Espaces fibrés algébriques.* Séminaire C. Chevalley (1958): Anneaux de Chow et applications.

[235] C. Seshadri, *Space of unitary vector bundles on a compact Riemann surface.* Ann. Math. 85 (1967), 303–336.

[236] C. Seshadri, *Fibrés vectoriels sur les courbes algébriques.* Astérisque 96 (1982).

[237] C. Seshadri, *Vector Bundles on Curves.* in: Linear algebraic groups and their representations. Contemp. Math. 153 (1993), 163–200.

[238] S. Shatz, *The decomposition and specializations of algebraic families of vector bundles.* Comp. math. 35 (1977), 163–187.

[239] S. Shatz, *Fibre equivalence of vector bundles on ruled surfaces.* J. Pure Appl. Alg. 12 (1978), 201–205.

[240] C. Simpson, *Moduli of representations of the fundamental group of a smooth projective variety I.* Publ. Math. IHES, 79 (1994), 47–129.

[241] C. Sorger, *Sur l'existence de fibrés semi-stables sur une surface algébrique*. Prépublication (1993).

[242] H. Spindler, *Der Satz von Grauert-Mülich für beliebige semistabile holomorphe Geradenbündel über dem n-dimensionalen komplex-projektiven Raum*. Math. Ann. 243 (1979), 131–141.

[243] S. Strømme, *Ample divisors on fine moduli spaces on the projective plane*. Math. Z. 187 (1984), 405–423.

[244] F. Takemoto, *Stable vector bundles on algebraic surfaces I*. Nagoya Math. J. 47 (1972), 29–43.

[245] F. Takemoto, *Stable vector bundles on algebraic surfaces II*. Nagoya Math. J. 52 (1973), 173–195.

[246] M. Thaddeus, *Stable pairs, linear systems and the Verlinde formula*. Invent. Math. 117 (1994), 317–353.

[247] A. Tyurin, *Cycles, curves and vector bundles on an algebraic surface*. Duke Math. J. 54 (1987), 1–26.

[248] A. Tyurin, *Special 0-cycles on a polarized K3 surface*. Izv. Akad. Nauk SSSR 51,1 (1987), 131–151.

[249] A. Tyurin, *Symplectic structures on the varieties of moduli of vector bundles on algebraic surfaces with $p_g > 1$*. Izv. Akad. Nauk SSSR 52,4 (1988), 813–852. Transl.: Math. USSR Izv. 33 (1989), 139–177.

[250] A. Tyurin, *The Weil-Peterson metric on the moduli space of stable bundles and sheaves on an algebraic surface*. Izv. Akad. Nauk 55, 3 (1991), 608–690. Transl.: Math. USSR Izvestiya 38 (1992), 599–620.

[251] A. Tyurin, *The moduli space of vector bundles on threefolds, surfaces and curves*. Preprint Erlangen (1990).

[252] A. Tyurin, *A slight generalisation of Mehta-Ramanathan's theorem*. Alg. Geom. Chicago 1989, ed. Fulton, Bloch, Lect. Notes Math. 1479 (1991), 258–271.

[253] A. Tyurin, *The geometry of the special components of moduli of vector bundles over algebraic surfaces of general type*. Complex Algebraic Varieties, Bayreuth 1990, Lect. Notes Math. 1507 (1992), 166–175.

[254] K. Ueno, *Classification theory of algebraic varieties and compact complex spaces*. Lect. Notes Math. 439, Springer Verlag, Berlin (1975).

[255] K. Uhlenbeck, S.-T. Yau, *On the existence of Hermitian Yang-Mills connections on stable vector bundles*. Commun. Pure Appl. Math. 39 (1986), 257–293; 42 (1989), 703–707.

[256] H. Umemura, *Moduli spaces of the stable vector bundles over abelian surfaces*. Nagoya Math. J. 77 (1980), 47–60.

[257] H. Umemura, *Stable vector bundles with numerically trivial Chern classes over a hyperelliptic surface*. Nagoya Math. J. 59 (1975), 107–134.

[258] J. Varouchas, *Kähler Spaces and Proper Open Morphisms*. Math. Ann. 283 (1989), 13–52.

[259] A. Van de Ven, *On Uniform Vector Bundles*. Math. Ann. 195 (1972), 245–248.

[260] E. Viehweg, *Quasi-projective moduli for polarized manifolds*. Erg. Math. 30, (3. Folge), Springer Verlag, Berlin (1995).

[261] J. Wehler, *Moduli Space and Versal Deformation of Stable Vector Bundles*. Rev. Roumaine Math. Pures-Appl. 30 (1985), 69–78.

[262] Yan-Gang Ye, *Lagrangian subvarieties of the moduli space of stable bundles on a regular algebraic surface*. Math. Ann. 295 (1993), 411–425.

[263] K. Zuo, *Generic smoothness of the moduli spaces of rank two stable vector bundles over algebraic surfaces*. Math. Z. 207 (1991), 629–643.

[264] K. Zuo, *The moduli spaces of some rank 2 stable vector bundles over algebraic K3 surfaces*. Duke Math. J. 64 (1991), 403–408.

[265] K. Zuo, *Regular 2 forms on the moduli space of rank 2 stable bundles on an algebraic surface*. Duke Math. J. 65 (1992), 45–84.

New References (second edition)

[266] T. Abe, *A remark on the 2-dimensional moduli spaces of vector bundles on* $K3$ *surfaces*. Math. Res. Lett. 7 (2000), 463–470.

[267] L. Álvarez-Cónsul, A. King, *A functorial construction of moduli of sheaves*. Invent. Math. 168 (2007), 613–666.

[268] C. Anghel, *Fibrés vectoriels stables avec* $\chi = 0$ *sur une surface abélienne simple*. Math. Ann. 315 (1999), 497–501.

[269] M. Aprodu, V. Brinzanescu, *Moduli spaces of vector bundles over ruled surfaces*. Nagoya Math. J. 154 (1999), 111–122.

[270] D. Arapura, S. Archava, *Kodaira dimension of symmetric powers*. Proc. AMS 131 (2003), 1369–1372.

[271] D. Arcara, A. Bertram, M. Lieblich, *Bridgeland-Stable Moduli Spaces for K-Trivial Surfaces*. arXiv:0708.2247v1.

[272] V. Balaji, J. Kollár, *Holonomy groups of stable vector bundles*. Publ. Res. Inst. Math. Sci. 44 (2008), 183–211.

[273] V. Balaji, A. Dey, R. Parthasarathi, *Parabolic bundles on algebraic surfaces. I. The Donaldson-Uhlenbeck compactification*. Proc. Indian Acad. Sci. Math. Sci. 118 (2008), 43–79.

[274] V. Baranovsky, *Moduli of sheaves on surfaces and action of the oscillator algebra*. J. Diff. Geom. 55 (2000), 193–227.

[275] A. Beauville, *Sur la cohomologie de certains espaces de modules de fibrés vectoriels*. Geometry and analysis. Bombay 1992. Oxford Univ. Press. Stud. Math. Tata Inst. Fundam. Res. 13 (1995), 37–40.

[276] A. Beauville, *Counting rational curves on* $K3$ *surfaces*. Duke Math. J. 97 (1999), 99–108.

[277] A. Beauville, *Vector bundles on the cubic threefold*. Contemp. Math. 312 (2002), 71–86.

[278] K. Behrend, B. Fantechi, *The intrinsic normal cone*. Invent. Math. 128 (1997), 45–88.

[279] P. Belkale, *The strange duality conjecture for generic curves*. J. AMS 21 (2008), 235–258.

[280] U. Bhosle, *Parabolic sheaves on higher-dimensional varieties*. Math. Ann. 293 (1992), 177–192.

[281] I. Biswas, A. Mukherjee, *On the symplectic structures on moduli space of stable sheaves over a* $K3$ *or abelian surface and on Hilbert scheme of points*. Arch. Math. 80 (2003), 507–515.

[282] I. Biswas, T. Gómez, *Restriction theorems for principal bundles*. Math. Ann. 327 (2003), 773–792.

[283] I. Biswas, J. Biswas, G. Ravindra, *On some moduli spaces of stable vector bundles on cubic and quartic threefolds*. J. Pure Appl. Alg. 212 (2008), 2298–2306.

[284] S. Boissière, M. Nieper-Wißkirchen, *Generating series in the cohomology of Hilbert schemes of points on surfaces*. LMS J. Comput. Math. 10 (2007), 254–270.

[285] F. Bottacin, *Poisson structures on Hilbert schemes of points of a surface and integrable systems*. Manuscripta math. 97 (1998), 517–527.

[286] F. Bottacin, *Poisson structures on moduli spaces of parabolic bundles on surfaces*. Manuscripta math. 103 (2000), 31–46.

[287] F. Bottacin, *Poisson structures on moduli spaces of framed vector bundles on surfaces*. Math. Nachr. 220 (2000), 33–44.

[288] F. Bottacin, *Closed differential forms on moduli spaces of sheaves*. Rend. Mat. Appl. 28 (2008), 139–162.

[289] S. Bradlow, O. Garca-Prada, P. Gothen, *Moduli spaces of holomorphic triples over compact Riemann surfaces*. Math. Ann. 328 (2004), 299–351.

[290] L. Brambila-Paz, *Non-emptiness of moduli spaces of coherent systems*. Int. J. Math. 19 (2008), 779–799.

[291] H. Brenner, *There is no Bogomolov type restriction theorem for strong semistability in positive characteristic*. Proc. AMS 133 (2005), 1941–1947.

[292] T. Bridgeland, A. Maciocia, *Complex surfaces with equivalent derived categories*. Math. Z. 236 (2001), 677–697.

[293] T. Bridgeland, A. Maciocia, *Fourier-Mukai transforms for $K3$ and elliptic fibrations*. J. Alg. Geom. 11 (2002), 629–657.

[294] T. Bridgeland, *Stability conditions on triangulated categories*. Ann. Math. 166 (2007), 317–345.

[295] M. Britze, *On the cohomology of generalized Kummer varieties*. PhD thesis, Köln 2002.

[296] I. Ciocan-Fontanine, M. Kapranov, *Derived Quot schemes*. Ann. Scient. Ec. Norm. Sup. 34 (2001), 403–440.

[297] I. Ciocan-Fontanine, M. Kapranov, *Derived Hilbert schemes*. J. AMS 15 (2002), 787–815.

[298] A. Căldăraru, *Derived categories of twisted sheaves on Calabi-Yau manifolds*. PhD-thesis Cornell (2000).

[299] A. Căldăraru, *Non-fine moduli spaces of sheaves on $K3$ surfaces*. Int. Math. Res. Not. 20 (2002), 1027–1056.

[300] J. Choy, Y.-H. Kiem, *On the existence of a crepant resolution of some moduli spaces of sheaves on an abelian surface*. Math. Z. 252 (2006), 557–575.

[301] J. Choy, Y.-H. Kiem, *Nonexistence of a crepant resolution of some moduli spaces of sheaves on a* $K3$ *surface*. J. Korean Math. Soc. 44 (2007), 35–54.

[302] J.-L. Colliot-Thélène, *Algèbres simples centrales sur les corps de fonctions de deux variables (d'après A. J. de Jong)*. Séminaire Bourbaki. Astérisque 307 (2006), 379–413.

[303] K. Corlette, C. Simpson, *On the classification of rank-two representations of quasiprojective fundamental groups*. Compos. Math. 144 (2008), 1271–1331.

[304] L. Costa, R. Miró-Roig, *On the rationality of moduli spaces of vector bundles on Fano surfaces*. J. Pure Appl. Algebra 137 (1999), 199–220.

[305] L. Costa, R. Miró-Roig, *Rationality of moduli spaces of vector bundles on Hirzebruch surfaces*. J. Reine Angew. Math. 509 (1999), 151–166.

[306] L. Costa, R. Miró-Roig, *Moduli spaces of vector bundles on higher-dimensional varieties*. Michigan Math. J. 49 (2001), 605–620.

[307] L. Costa, R. Miró-Roig, *Rationality of moduli spaces of vector bundles on rational surfaces*. Nagoya Math. J. 165 (2002), 43–69.

[308] L. Costa, R. Miró-Roig, *Elementary transformations and the rationality of the moduli spaces of vector bundles on* $\mathbb{P}^2$. Manuscripta math. 113 (2004), 69–84.

[309] G. Danila, *Résultats sur la dualité étrange sur le plan projectif*. Bull. SMF 130 (2002), 1–33.

[310] G. Danila, *Sur la cohomologie de la puissance symétrique du fibré tautologique sur le schéma de Hilbert ponctuel d'une surface*. J. Alg. Geom. 13 (2004), 81–113.

[311] O. Debarre, *On the Euler characteristic of generalized Kummer varieties*. Am. J. Math. 121 (1999), 577–586.

[312] I. Dolgachev, *Lectures on invariant theory*. London Mathematical Society Lecture Note Series, 296. Cambridge University Press, Cambridge (2003).

[313] J.-M. Drézet, *Faisceaux cohérents sur les courbes multiples*. Collect. Math. 57 (2006), 121–171.

[314] J.-M. Drézet, *Moduli spaces of coherent sheaves on multiples curves*. in: Algebraic cycles, sheaves, shtukas, and moduli, Trends in Math., Birkhäuser, Basel, (2008), 33–43.

[315] S. Druel, *Espace des modules des faisceaux de rang 2 semi-stables de classes de Chern* $c_1 = 0, c_2 = 2$ *et* $c_3 = 0$ *sur la cubique de* $\mathbb{P}^4$. Int. Math. Res. Not. 19 (2000) 985–1004.

[316] G. Ellingsrud, S. A. Strømme, *On the homology of the Hilbert scheme of points in the plane*. Invent. Math. 87 (1987), 343–352.

[317] G. Ellingsrud, S. A. Strømme, *Towards the Chow ring of the Hilbert scheme of* $\mathbb{P}^2$. J. Reine Angew. Math. 441(1993), 33–44.

[318] G. Ellingsrud, L. Göttsche, M. Lehn, *On the cobordism class of the Hilbert scheme of a surface.* J. Alg. Geom. 10 (2001), 81–100.

[319] G. Ellingsrud, L. Göttsche, *Hilbert schemes of points and Heisenberg algebras.* School on Algebraic Geometry (Trieste, 1999), 59–100. ICTP Lect. Notes, 1, Abdus Salam ICTP Trieste, 2000.

[320] G. Faltings, *Thetafunktionen auf Modulräumen von Vektorbündeln.* Jahresber. DMV 110 (2008), 3–17.

[321] B. Fantechi, L. Göttsche, L. Illusie, S. Kleiman, N. Nitsure, A. Vistoli, Angelo, *Fundamental algebraic geometry. Grothendieck's FGA explained.* Mathematical Surveys and Monographs, 123. AMS (2005).

[322] R. Friedman, *Algebraic surfaces and holomorphic vector bundles.* Universitext. Springer Verlag, Berlin, Heidelberg (1998).

[323] L. Göttsche, *Theta Functions and Hodge Numbers of Moduli Spaces of Sheaves on Rational Functions.* Comm. Math. Phys. 206 (1999), 105–136.

[324] T. Gómez, T. Ramadas, *Parabolic bundles and representations of the fundamental group.* Manuscripta math. 103 (2000), 299–311.

[325] T. Gómez, *Brill-Noether theory on singular curves and torsion-free sheaves on surfaces.* Comm. Anal. Geom. 9 (2001), 725–756.

[326] T. Gómez, I. Sols, *Moduli space of principal sheaves over projective varieties.* Ann. Math. 161 (2005), 1037–1092.

[327] T. Gómez, A. Langer, A. Schmitt, I. Sols, *Moduli spaces for principal bundles in arbitrary characteristic.* Adv. Math. 219 (2008), 1177–1245.

[328] T. Gómez, *Lectures on principal bundles over projective varieties.* In: Algebraic cycles, sheaves, shtukas, and moduli. Trends Math. (2008), 45–68.

[329] I. Grojnowski, *Instantons and affine algebras. I. The Hilbert scheme and vertex operators.* Math. Res. Lett. 3 (1996), 275–291.

[330] M. Gross, D. Huybrechts, D. Joyce, *Calabi-Yau manifolds and related geometries.* Universitext. Springer Verlag, Berlin (2003).

[331] M. Gulbrandsen, *Computing the Euler characteristic of generalized Kummer varieties.* Ark. Mat. 45 (2007), 49–60.

[332] M. Gulbrandsen, *Lagrangian fibrations on generalized Kummer varieties.* Bull. SMF 135 (2007), 283–298.

[333] G. Hein, *Duality construction of moduli spaces.* Geom. Dedicata 75 (1999), 101–113.

[334] G. Hein, *Generalized Albanese morphisms.* Comp. math. 142 (2006), 719–733.

[335] G. Hein, *Restriction of stable rank two vector bundles in arbitrary characteristic.* Comm. Alg. 34 (2006), 2319–2335.

[336] G. Hein, D. Ploog, *Postnikov-stability versus semistability of sheaves*. arXiv:0901. 1554v1.

[337] N. Hoffmann, U. Stuhler, *Moduli schemes of generically simple Azumaya modules*. Doc. Math. 10 (2005), 369–389.

[338] D. Huybrechts, *Fourier-Mukai transforms in algebraic geometry*. Oxford Math. Mon. Clarendon Press, Oxford University Press (2006).

[339] D. Huybrechts, *Derived and abelian equivalence of K3 surfaces*. J. Alg. Geom. 17 (2008), 375–400.

[340] D. Huybrechts, P. Stellari, *Equivalences of twisted K3 surfaces*. Math. Ann. 332 (2005), 901–936.

[341] D. Huybrechts, R. Thomas, *Deformation-obstruction theory for complexes via Atiyah and Kodaira-Spencer classes*. arXiv:0805.3527v1.

[342] D. Hyeon, *Principal bundles over a projective scheme*. Trans. AMS 354 (2002), 1899–1908.

[343] A. Iliev, D. Markushevich, *The Abel-Jacobi map for a cubic threefold and periods of Fano threefolds of degree 14*. Doc. Math. 5 (2000), 23–47.

[344] M. Inaba, *Moduli of parabolic stable sheaves on a projective scheme*. J. Math. Kyoto Univ. 40 (2000), 119–136.

[345] M. Inaba, *Toward a definition of moduli of complexes of coherent sheaves on a projective scheme*. J. Math. Kyoto Univ. 42 (2002), 317–329.

[346] M. Inaba, *On the moduli of stable sheaves on a reducible projective scheme and examples on a reducible quadric surface*. Nagoya Math. J. 166 (2002), 135–181.

[347] M. Inaba, *On the moduli of stable sheaves on some nonreduced projective schemes*. J. Alg. Geom. 13 (2004), 1–27.

[348] M. Inaba, *Moduli of stable objects in a triangulated category*. arXiv:0612078v2.

[349] J. Iyer, C. Simpson, *A relation between the parabolic Chern characters of the de Rham bundles*. Math. Ann. 338 (2007), 347–383.

[350] D. Kaledin, M. Lehn, *Local structure of hyperkähler singularities in O'Grady's examples*. Moscow Math. J. 7, 4 (2007), 653 - 672.

[351] D. Kaledin, M. Lehn, C. Sorger, *Singular symplectic moduli spaces*. Invent. Math. 164 (2006), 591–614.

[352] H. Kim, *Moduli spaces of stable vector bundles on Enriques surfaces*. Nagoya Math. J. 150 (1998), 85–94.

[353] H. Kim, *Moduli spaces of bundles mod Picard groups on some elliptic surfaces*. Bull. Korean Math. Soc. 35 (1998), 119–125.

[354] M. Kool, *Euler characteristics of moduli spaces of torsion free sheaves on toric surfaces*. arXiv:0906.3393.

[355] A. Kuznetsov, D. Markushevich, *Symplectic structures on moduli spaces of sheaves via the Atiyah class*. arXiv:0703264v3.

[356] A. Langer, *Semistable sheaves in positive characteristic*. Ann. Math. 159 (2004), 251–276. Addendum: Ann. Math. 160 (2004), 1211–1213.

[357] A. Langer, *Moduli spaces of sheaves in mixed characteristic*. Duke Math. J. 124 (2004), 571–586.

[358] A. Langer, *Semistable principal G-bundles in positive characteristic*. Duke Math. J. 128 (2005), 511–540.

[359] A. Langer, *Moduli spaces and Castelnuovo-Mumford regularity of sheaves on surfaces*. Am. J. Math. 128 (2006), 373–417.

[360] A. Langer, *Moduli spaces of sheaves and principal G-bundles*. Algebraic geometry – Seattle 2005. Proc. Symp. Pure Math. 80,1. AMS (2009), 273–308.

[361] M. Lehn, *On the cotangent sheaf of Quot-schemes*. Int. J. Math. 9 (1998), 513–522.

[362] M. Lehn, *Lectures on Hilbert schemes*. in: Algebraic structures and moduli spaces, CRM Proc. Lect. Notes, 38, AMS (2004), 1–30.

[363] M. Lehn, *Chern classes of tautological sheaves on Hilbert schemes of points on surfaces*. Invent. Math. 136 (1999), 157–207.

[364] M. Lehn, C. Sorger, *Symmetric Groups and the Cup Product on the Cohomology of Hilbert Schemes*. Duke Math. Journal, 110,2 (2001), 345–357.

[365] M. Lehn, C. Sorger, *The cup product of Hilbert schemes for $K3$ surfaces*. Invent. Math. 152 (2003), 305–329.

[366] M. Lehn, C. Sorger, *La singularité de O'Grady*. J. Alg. Geom. 15 (2006), 753–770.

[367] J. Le Potier, *Lectures on vector bundles*. Cambridge Studies in Advanced Mathematics, 54. Cambridge University Press, Cambridge (1997).

[368] J. Li, G. Tian, *Virtual moduli cycles and Gromov-Witten invariants of algebraic varieties*. J. AMS 11 (1998), 119–174.

[369] W.-P. Li, *Relations between moduli spaces of stable bundles over P_2 and rationality*. J. Reine Angew. Math. 484 (1997), 201–217.

[370] W.-P. Li, Z. Qin, W. Wang, *Vertex algebras and the cohomology ring structure of Hilbert schemes of points on surfaces*. Math. Ann. 324 (2002), 105–133.

[371] W.-P. Li, Z. Qin, *Stable rank-2 bundles on Calabi-Yau manifolds*. Int. J. Math. 14 (2003), 1097–1120.

[372] M. Lieblich, *Moduli of complexes on a proper morphism*. J. Alg. Geom. 15 (2006), 175–206.

[373] M. Lieblich, *Moduli of twisted sheaves*. Duke Math. J. 138 (2007), 23–118.

[374] M. Lieblich, *Twisted sheaves and the period-index problem*. Comp. math. 144 (2008), 1–31.

[375] M. Lieblich, *Compactified moduli of projective bundles.* arXiv:0706.1311v3.

[376] W. Lowen, *Obstruction theory for objects in abelian and derived categories.* Comm. Algebra 33 (2005), 3195–3223.

[377] C. Lundkvist, R. Skjelnes, *Non-effective deformations of Grothendieck's Hilbert functor.* Math. Z. 258 (2008), 513–519.

[378] C. Madonna, V. Nikulin, *On a classical correspondence between K3 surfaces. II.* in: Strings and geometry, Clay Math. Proc., 3, AMS (2004), 285–300.

[379] C. Madonna, *On some moduli spaces of bundles on K3 surfaces.* Monatsh. Math. 146 (2005), 333–339.

[380] A. Marian, D. Oprea, *A tour of theta dualities on moduli spaces of sheaves.* in: Curves and abelian varieties, Contemp. Math. 465, (2008), 175–202.

[381] A. Marian, D. Oprea, *Sheaves on abelian surfaces and Strange Duality.* Math. Ann. 343 (2009), 1–33.

[382] A. Marian, D. Oprea, *On the strange duality conjecture for elliptic K3 surfaces.* arXiv:0902.3052.

[383] E. Markman, *Brill-Noether duality for moduli spaces of sheaves on K3 surfaces.* J. Alg. Geom. 10 (2001), 623–694.

[384] E. Markman, *Generators of the cohomology ring of moduli spaces of sheaves on symplectic surfaces.* J. Reine Angew. Math. 544 (2002), 61–82.

[385] E. Markman, *On the monodromy of moduli spaces of sheaves on K3 surfaces.* J. Alg. Geom. 17 (2008), 29–99.

[386] D. Markushevich, *Lagrangian fibrations on punctual Hilbert schemes of K3 surfaces.* Manuscripta math. 120 (2006), 131–150.

[387] M. Maruyama, K. Yokogawa, *Moduli of parabolic stable sheaves.* Math. Ann. 293 (1992), 77–99.

[388] M. Maruyama, *Construction of moduli spaces of stable sheaves via Simpson's idea.* Moduli of vector bundles, ed. M. Maruyama. Lect. Notes in Pure and Appl. Math. 179 (1996), 147–187.

[389] D. Maulik, N. Nekrasov, A. Okounkov, R. Pandharipande, *Gromov-Witten theory and Donaldson-Thomas theory. I.* Comp. math. 142 (2006), 1263–1285.

[390] V. Mehta, V. Trivedi, *Restriction theorems for homogeneous bundles.* Bull. Sci. Math. 131 (2007), 397–404.

[391] V. Mehta, *Representation of algebraic groups and principal bundles on algebraic varieties.* Proc. Int. Cong. Math. Beijing (2002), 629–635.

[392] E. Mistretta, *Stable vector bundles as generators of the Chow ring.* Geom. Dedicata 117 (2006), 203–213.

[393] S. Mozgovyy, *The Euler number of O'Grady's 10-dimensional symplectic manifold.* PhD-thesis Mainz (2006). (Variant spelling of author's name: Mozgovoy)

[394] S. Mukai, *An introduction to invariants and moduli*. Cambridge Studies in Advanced Mathematics, 81. Cambridge University Press, Cambridge, (2003).

[395] H. Nakajima, *Lectures on Hilbert schemes of points on surfaces*. University Lecture Series, 18. AMS (1999).

[396] H. Nakajima, K. Yoshioka, *Instanton counting on blowup. I. 4-dimensional pure gauge theory*. Invent. Math. 162 (2005), 313–355.

[397] T. Nakashima, *Restriction of stable bundles in characteristic p*. Trans. AMS 349 (1997), 4775–4786.

[398] T. Nakashima, W. Watanabe, *Vector bundles on del-Pezzo fibrations*. Geom. Dedicata 100 (2003), 1–9.

[399] T. Nakashima, *Stable rank two bundles on Del-Pezzo fibrations of degree 1 or 2*. Arch. Math. 81 (2003), 258–265.

[400] T. Nakashima, *Moduli spaces of stable bundles on $K3$ fibered Calabi-Yau threefolds*. Commun. Contemp. Math. 5 (2003), 119–126.

[401] T. Nakashima, *A construction of stable vector bundles on Calabi-Yau manifolds*. J. Geom. Phys. 49 (2004), 224–230.

[402] Y. Namikawa, *Extension of 2-forms and symplectic varieties*. J. Reine Angew. Math. 539 (2001), 123–147.

[403] T. Nevins, *Representability for some moduli stacks of framed sheaves*. Manuscripta Math. 109 (2002), 85–91.

[404] T. Nevins, *Moduli spaces of framed sheaves on certain ruled surfaces over elliptic curves*. Int. J. Math. 13 (2002), 1117–1151.

[405] M. Nieper-Wißkirchen, *On the Chern numbers of generalised Kummer varieties*. Math. Res. Lett. 9 (2002), 597–606.

[406] K. O'Grady, *Desingularized moduli spaces of sheaves on a $K3$*. J. Reine Angew. Math. 512 (1999), 49–117.

[407] K. O'Grady, *A new six-dimensional irreducible symplectic variety*. J. Alg. Geom. 12 (2003), 435–505.

[408] M. Olsson, J. Starr, *Quot functors for Deligne-Mumford stacks*. Comm. Algebra 31 (2003), 4069–4096.

[409] M. Olsson, *On proper coverings of Artin stacks*. Adv. Math. 198 (2005), 93–106.

[410] N. Onishi, K. Yoshioka, *Singularities on the 2-dimensional moduli spaces of stable sheaves on $K3$ surfaces*. Int. J. Math. 14 (2003), 837–864.

[411] D. Orlov, *On equivalences of derived categories and K3 surfaces*. J. Math. Sci. (New York) 84 (1997), 1361–1381.

[412] D. Orlov, *Derived categories of coherent sheaves on abelian varieties and equivalences between them*. Izv. Math. 66 (2002), 569–594.

[413] R. Pandharipande, R. Thomas, *Curve counting via stable pairs in the derived category*. Invent. Math. 178, (2009).

[414] A. Polishchuk, *Symplectic biextensions and a generalization of the Fourier-Mukai transform*. Math. Res. Lett. 3 (1996), 813–828.

[415] A. Polishchuk, *Abelian varieties, theta functions and the Fourier transform*. Cambridge Tracts in Math. 153. Cambridge University Press (2003).

[416] M. Popa, M. Roth, *Stable maps and Quot schemes*. Invent. Math. 152 (2003), 625–663.

[417] Z. Qin, Q. Zhang, *On the crepancy of the Gieseker-Uhlenbeck morphism*. Asian J. Math. 12 (2008), 213–223.

[418] A. Ramanathan, *Moduli for principal bundles over algebraic curves: I,II*. Proc. Indian Acad. Math. Sci. 106 (1996), 301–328 and 421–449.

[419] A. Rapagnetta, *Topological invariants of O'Grady's six dimensional irreducible symplectic variety*. Math. Z. 256 (2007), 1–34.

[420] A. Rapagnetta, *On the Beauville form of the known irreducible symplectic varieties*. Math. Ann. 340 (2008), 77–95.

[421] J. Santos, *Framed holomorphic bundles on rational surfaces*. J. Reine Angew. Math. 589 (2005), 129–158.

[422] J. Sawon, *Lagrangian fibrations on Hilbert schemes of points on $K3$ surfaces*. J. Alg. Geom. 16 (2007), 477–497.

[423] L. Scala, *Dualité étrange de Le Potier et cohomologie du schéma de Hilbert ponctuel d'une surface*. Gaz. Math. 112 (2007), 53–65.

[424] L. Scala, *Cohomology of the Hilbert scheme of points on a surface with values in representations of tautological bundles*. arXiv:0710.3072.

[425] A. Schmitt, *Singular principal bundles over higher-dimensional manifolds and their moduli spaces*. Int. Math. Res. Not. 23 (2002), 1183–1209.

[426] A. Schmitt, *A closer look at semistability for singular principal bundles*. Int. Math. Res. Not. 62 (2004), 3327–3366.

[427] A. Schmitt, *Singular principal G-bundles on nodal curves*. J. Eur. Math. Soc. 7 (2005), 215–251.

[428] A. Schmitt, *Moduli for decorated tuples of sheaves and representation spaces for quivers*. Proc. Indian Acad. Sci. Math. Sci. 115 (2005), 15–49.

[429] A. Schmitt, *Geometric invariant theory and decorated principal bundles*. Zürich Lectures in Advanced Mathematics. (EMS) Zürich, (2008).

[430] U. Schlickewei, *Hodge classes on self-products*. PhD thesis, Bonn (2009).

[431] A. Schofield, *Birational classification of moduli spaces of representations of quivers*. Indag. Math. 12 (2001), 407–432.

[432] A. Schofield, *Birational classification of moduli spaces of vector bundles over* $\mathbb{P}^k$. Indag. Math. 12 (2001), 433–448.

[433] R. Thomas, *A holomorphic Casson invariant for Calabi-Yau 3-folds, and bundles on K3 fibrations.* J. Diff. Geom. 54 (2000), 367–438.

[434] Y. Toda, *Moduli stacks and invariants of semistable objects on K3 surfaces.* Adv. Math. 217 (2008), 2736–2781.

[435] C. Vafa, E. Witten, *A strong coupling test for S-duality* Nuclear Phys. B. 431 (1994), 3–77.

[436] E. Vasserot, *Sur l'anneau de cohomologie du schéma de Hilbert de* $\mathbf{C}^2$. C. R. Acad. Sci. Paris Sér. I Math. 332 (2001), 7–12.

[437] P. Vermeire, *Moduli of reflexive sheaves on smooth projective 3-folds.* J. Pure Appl. Algebra 211 (2007), 622–632.

[438] E. Viehweg, K. Zuo, *Arakelov inequalities and the uniformization of certain rigid Shimura varieties.* J. Diff. Geom. 77 (2007), 291–352.

[439] C. Voisin, *On the Chow ring of certain algebraic hyper-Kähler manifolds.* Pure Appl. Math. Q. 4 (2008), 613–649.

[440] X. Wang, *Balance point and stability of vector bundles over a projective manifold.* Math. Res. Lett. 9 (2002), 393–411.

[441] X. Wang, *Canonical metrics on stable vector bundles.* Comm. Anal. Geom. 13 (2005), 253–285.

[442] J.-G. Yang, *Coherent sheaves on a fat curve.* Japan J. Math. (N.S.) 29 (2003), 315–333.

[443] K. Yokogawa, *Compactification of moduli of parabolic sheaves and moduli of parabolic Higgs sheaves.* J. Math. Kyoto Univ. 33 (1993), 451–504.

[444] K. Yoshioka, *A note on moduli of vector bundles on rational surfaces.* J. Math. Kyoto Univ. 43 (2003), 139–163.

[445] K. Yoshioka, *The Betti numbers of the moduli space of stable sheaves of rank 2 on a ruled surface.* Math. Ann. 302 (1995), 519–540.

[446] K. Yoshioka, *Betti numbers of moduli of stable sheaves on some surfaces.* Nuclear Phys. B. (Proc. Suppl.) 46 (1996), 263–268.

[447] K. Yoshioka, *Numbers of F_q-rational points of the moduli of stable sheaves on elliptic surfaces.* Moduli of vector bundles, ed. M. Maruyama. Lect. Notes in Pure and Appl. Math. 179 (1996), 297–305.

[448] K. Yoshioka, *The Picard group of the moduli space of stable sheaves on a ruled surface.* J. Math. Kyoto Univ. 36 (1996), 279–309.

[449] K. Yoshioka, *Chamber structure of polarizations and the moduli of stable sheaves on a ruled surface.* Int. J. Math. 7 (1996), 411–431.

[450] K. Yoshioka, *A note on the universal family of moduli of stable sheaves.* J. Reine Angew. Math. 496 (1998), 149–161.

[451] K. Yoshioka, *Some examples of Mukai's reflections on K3 surfaces.* J. Reine Angew. Math. 515 (1999), 97–123.

[452] K. Yoshioka, *Some notes on the moduli of stable sheaves on elliptic surfaces.* Nagoya Math. J. 154 (1999), 73–102.

[453] K. Yoshioka, *Moduli spaces of stable sheaves on abelian surfaces.* Math. Ann. 321 (2001), 817–884.

[454] K. Yoshioka, *A note on moduli of vector bundles on rational surfaces.* J. Math. Kyoto Univ. 43 (2003), 139–163.

[455] K. Yoshioka, *Moduli spaces of twisted sheaves on a projective variety.* Adv. Stud. Pure Math., 45 (2006), 1–30.

[456] K. Yoshioka, *Moduli of vector bundles on algebraic surfaces.* Sugaku Expositions 20 (2007), 111–135.

Glossary of Notations

General notations

$\lfloor x \rfloor$	round down of a real number x.
$\lceil x \rceil$	round up of a real number x.
$[x]_+$	$= \max\{x, 0\}$ for a real number x.
$(\leq), (<)$	*convention used in the definition of semistability, see p. 11.*
$\mathbb{Q}[T]_d$	vector space of polynomials of degree $\leq d$, p. 26.
$\mathbb{Q}[T]_{d,d'}$	quotient vector space $\mathbb{Q}[T]_d/\mathbb{Q}[T]_{d'-1}$, p. 26.
$\mathfrak{m}$	maximal ideal in a local ring.
∇	connection, p. 67.
$V^{\vee}$	$= \mathrm{Hom}_k(V, k)$ dual of a k-vector space V.

Schemes, varieties, morphisms

k	field, most of the time algebraically closed, in the second part of the book in general of characteristic zero.
X	in general scheme of finite type over k, in the second part of the book a surface, which always means an irreducible smooth projective surface.
C	mostly a smooth projective curve.
Δ	Diagonal in a product $X \times X$, *but see also:* $\Delta(E)$ *for a sheaf E.*
$\mathcal{I}_Z$	ideal sheaf of a subscheme $Z \subset X$.
$\ell(Z)$	$= \ell(\mathcal{O}_Z)$, length of a zero-dimensional scheme Z.
H	often an ample or very ample divisor on X.
$\deg(X)$	degree of X with respect to some fixed ample divisor H.
$\dim_x(X)$	dimension of X at $x = \dim(\mathcal{O}_{X,x})$.
$\mathrm{kod}(X)$	Kodaira dimension of a variety X, p. 273.
ω_X, K_X	canonical sheaf of a smooth variety.
$\mathrm{Pic}(X)$	Picard group of X.
$\mathrm{Pic}^G(X)$	equivariant Picard group of G-linearized line bundles on X, p. 98.
$\mathrm{Alb}(X)$	Albanese variety of X, p. 149.
$CH(X)$	Chow group of X, p. 149.
$K_0(X)$	Grothendieck group of coherent sheaves on X, p. 39.

316

$K^0(X)$	Grothendieck group of locally free sheaves on X, p. *39*.
$K(X)$	$= K^0(X) = K_0(X)$, if X is smooth, p. *213*.
K_c	$= c^\perp \subset K$, p. *215*.
$K_{c,H}$	$= K_c \cap \{1, h, h^2, \dots\}^{\perp\perp}$, p. *215*.
$\mathrm{Num}(X)$	$= \mathrm{Pic}(X)/\equiv$, $\equiv$ numerical equivalence, p. *206*.
K^+	open cone in Num, p. *206*.
$\mathrm{Coh}(X)$	category of coherent sheaves on X, p. *3*.
$S^\ell(X)$	symmetric product of X, p. *102*.
$\mathrm{Hilb}^\ell(X)$	Hilbert scheme of subschemes of X of length ℓ, p. *44*.

Categories

(Sch/k)	schemes of finite type over a field k.
(Sch/S)	schemes of finite type over a scheme S.
$(Sets)$	category of sets.
$\mathcal{C}^o$	category opposite to $\mathcal{C}$, i.e. with all arrows reversed.
$\mathrm{Ob}(\mathcal{C})$	objects of the category $\mathcal{C}$.
$\mathrm{Mor}(\mathcal{C})$	morphisms of the category $\mathcal{C}$.
$\mathrm{Coh}(X)$	coherent sheaves on X, p. *3*.
$\mathrm{Coh}(X)_d$	coherent sheaves on X of dimension $\leq d$, p. *26*.
$\mathrm{Coh}(X)_{d,d'}$	quotient category $\mathrm{Coh}(X)_d/\mathrm{Coh}(X)_{d'-1}$, p. *26*.
$\mathcal{C}(p)$	category of semistable sheaves with reduced Hilbert polynomial p, p. *25*.

Coherent sheaves

$E^\vee$	$= \mathcal{H}om(E, \mathcal{O}_X)$ dual sheaf of E, *but compare E^D*.
$E\vert_Y$	$= i^*E$ restriction of E to a subscheme $i : Y \to X$.
E_x	stalk of E in $x \in X$.
$E(x)$	$= E_x/\mathrm{m}_x E_x$ fibre of E in $x \in X$.
$\mathrm{Supp}(E)$	support of a coherent sheaf E, p. *3*.
$\dim(E)$	dimension of a coherent sheaf E, p. *3*.
$\mathrm{dh}(E)$	homological dimension of a coherent sheaf E, p. *4*.
$T_i(E)$	maximal subsheaf of E of dimension $\leq i$, p. *4*.
$T(E)$	$= T_{\dim(E)-1}(E)$ torsion subsheaf of E, p. *4*.
$\ell(T)$	$= \mathrm{length}(T)$, length of a zero-dimensional sheaf.
E^D	$= \mathcal{E}xt^c_X(E, \omega_X)$, the dual of E, p. *6*.
E^{DD}	$= ((E^\vee)^\vee$, reflexive hull of E.
E^{**}	reflexive hull of the graded object associated to a μ-Jordan Hölder filtration of E, p. *227*.
$h^i(E)$	$= \dim H^i(X, E)$ for a coherent sheaf on a scheme X.
$\chi(E)$	$= \sum(-1)^i h^i(E)$, Euler characteristic of a coherent sheaf, p. *10*.
$P(E)$	Hilbert polynomial of E, $P(E, m) = \chi(E(m))$, p. *10*.
$\alpha_i(E)$	coefficients of the Hilbert polynomial in the expansion $P(E, m) = \sum_{i=0}^{\dim(E)} \alpha_i(E)\frac{m^i}{i!}$, p. *10*.
$\mathrm{rk}(E)$	rank of a sheaf E, p. *11*.
$p(E)$	$= \frac{P(E)}{\alpha_{\dim(E)}(E)}$, reduced Hilbert polynomial of E, p. *11*.

$p_{\max}(E), p_{\min}(E)$	reduced Hilbert polynomial of the first (last) factor in the Harder-Narasimhan filtration of E, p. *16*.
$p_{d,d'}(E)$	class of $p(E)$ in $\mathbb{Q}[T]_{d,d'}$, p. *26*.
$\deg(E)$	$= c_1(E).H^{d-1}$, degree of a sheaf E with respect to an ample divisor on a d-dimensional variety, p. *14*.
$\mu(E)$	$= \frac{\deg(E)}{\mathrm{rk}(E)}$, slope of a non-torsion sheaf E on a projective variety, p. *14*.
$\hat{\mu}(E)$	$= \frac{\alpha_{d-1}(E)}{\alpha_d(E)}$, generalized slope of a d-dimensional sheaf E, p. *28*.
$\mu_{\max}, \mu_{\min}$	minimal and maximal slope of the first (last) factor in a Harder-Narasimhan filtration, p. *28*.
$\delta\mu$	maximal distance of the slopes in a Harder-Narasimhan filtration, p. *65*.
$\hom(E, F)$	$= \dim \mathrm{Hom}(E, F)$ for two coherent sheaves on a projective scheme X.
$\mathrm{ext}^i(E, F)$	$= \dim \mathrm{Ext}^i_X(E, F)$.
$\chi(E, F)$	$= \sum(-1)^i \mathrm{ext}^i(E, F)$, Euler characteristic of the pair (E, F).
$\xi_{E,F}$	$= c_1(E)/\mathrm{rk}(E) - c_1(F)/\mathrm{rk}(F)$, p. *206*.
$tr : \mathrm{Ext}^i(E, E) \to H^i(\mathcal{O}_X)$	trace map, p. *258*.
$\mathrm{End}(F)_0$	$= \ker(tr : \mathrm{End}(F) \to H^i(\mathcal{O}_X))$, traceless endomorphisms, p. *258*.
$\mathrm{Ext}^i(F, F)_0$	$= \ker(tr : \mathrm{Ext}^i(F, F) \to H^i(\mathcal{O}_X))$, traceless extensions, p. *258*.
$\mathrm{Ext}^i_\pm(E, F)$	*defined for filtered sheaves E and F, see Appendix 2.A.*
$\det(E)$	determinant line bundle of a sheaf E that admits a finite locally free resolution, p. *9, 39*.
$A(E)$	Atiyah class of E, p. *259*.
$\mathcal{F}_s$	$= \mathcal{F} \otimes k(s)$ restriction of an S-flat coherent sheaf on $S \times X$ to the fibre $k(s) \times X$ over a point $s \in S$, p. *34*.
$\mathrm{HN}_\bullet(E)$	Harder-Narasimhan filtration of E, p. *16*.
$gr^F(E)$	$= \bigoplus_i F_iE/F_{i-1}E$ graded object associated to a filtration $F_\bullet$ of E.
$\mathrm{reg}(E)$	Mumford-Castelnuovo regularity of E, p. *30*.
$c_i(E)$	i-th Chern class of E, p. *79*.
$\Delta(E)$	discriminant of E, p. *79*.
$ch(E)$	Chern character of E. p. *79*
$v(E)$	Mukai vector of E, p. *167*.
$\beta(E)$	$= \mathrm{ext}^2(E, E)_0$, p. *242*.
β_∞	uniform bound for $\beta(E)$, p. *115*.
$S_{k,c}$	Serre's condition, p. *5*.

Group actions and invariant theory

G	an algebraic group, p. *91*.
$\mu : G \times G \to G$	group multiplication, p. *91*.
$\sigma : X \times G \to X$	action of G on X, p. *92*.
G_x	stabilizer subgroup of a point $x \in X$, p. *92*.
V^G	subspace of invariant elements in a G-representation V, p. *92*.
$\underline{X/G}$	quotient functor, p. *92*.
$X/\!/G$	GIT quotient of X by G, p. *93*.

$\mathbb{G}_m$	multiplicative group scheme $R \mapsto R^\times, = \mathbb{A}^1 \setminus \{0\}$.
$\lambda : \mathbb{G}_m \to G$	1-parameter subgroup, p. *96*.
$\lim_{g \to 0} \sigma(x, \lambda(g))$	limit point of the orbit of x under λ, p. *96*.
$\mu(x, \lambda)$	weight of the action of lambda at the limit point of x, p. *96*.
$\Phi : \sigma^* F \to p_1^* F$	G-linearization of a sheaf F, p. *94*.
$X^{ss}(L)$	set of semistable points in X with respect to the linearization of L, p. *96*.
$X^s(L)$	set of stable points in X with respect to the linearization of L, p. *96*.
$\mathrm{Pic}^G(X)$	Picard group of G-linearized line bundles on X, p. *98*.

Constructions related to sheaves or families of sheaves

$p : S \times X \to S$	projection to the 'base'.
$q : S \times X \to X$	projection to the 'fibre'.
$\mathbb{P}(E)$	Proj $S^* E$ projectivization of a coherent sheaf E.
$\mathrm{Grass}(E, r)$	Grassmann scheme of locally free quotients of E of rank r, p. *41*.
$\mathrm{Quot}(E, P)$	Quot-scheme of flat quotients of E with Hilbert polynomial P, p. *42*.
$\mathrm{Hilb}^\ell(X)$	$= Quot(\mathcal{O}_X, \ell)$, Hilbert scheme, p. *44*.
$\mathrm{Drap}(E, P_\bullet)$	flag-scheme of flags in E with flat factors with Hilbert polynomials P_i, p. *51*.
$\det : S \to \mathrm{Pic}(X)$	morphism associated to a flat family of sheaves on a smooth variety X parametrized by S.
$\Phi_F : S \to M$	classifying morphism associated to an S-flat family F of semistable sheaves.
$\mathcal{E}xt_p^i(F, G)$	relative $\mathcal{E}xt$-sheaf for a morphism p, right derived functors of the composite functor $p_* \circ \mathcal{E}xt$, p. *260*.
$KS : \Omega_S^\vee \to \mathcal{E}xt_p^1(F^\bullet, F^\bullet)$	Kodaira-Spencer map, p. *261*.

Construction of the moduli space, objects on the moduli space

$\mathcal{M}', (\mathcal{M}')^s$	moduli functor for semistable and stable sheaves, resp., p. *90*.
$\mathcal{M} = \mathcal{M}'/\sim$	quotient functor of $\mathcal{M}'$ for the equivalence $F \sim F' \Leftrightarrow F \cong F' \otimes p^* L$ for some $L \in Pic(S)$, p. *90*.
$\mathcal{M}^s = (\mathcal{M}')^s/\sim$	as above for families of stable sheaves, p. *90*.
M	$= M(P)$ moduli space of semistable sheaves with Hilbert polynomial P, p. *90*.
M^s	$\subset M$ open subspace of points corresponding to geometrically stable sheaves.
M_H	moduli space of semistable sheaves with respect to a polarization H, in a context where the polarization varies.
$M(v)$	moduli space of semistable sheaves with Mukai vector v, p. *168*.
$\tilde{M}$	scheme birational to M, constructed in Appendix 4.B.
$M^{\mu ss}$	moduli space of μ-semistable sheaves, p. *226*.
M_C	moduli space of semistable sheaves on a curve C.
M_0	moduli space of stable sheaves F with $\mathrm{Ext}^2(F, F)_0 = 0$, p. *262*.
m	integer which is sufficiently large so that the conditions of Thm. 4.4.1 are satisfied.

V	k-vector space of dimension $= P(m)$.
V_n	direct summand of the weight space decomposition of V for a one-parameter subgroup λ, p. *107*.
$\mathcal{H}$	$= V \otimes_k \mathcal{O}_X(-m)$, p. *99*.
R	open subset in $\mathrm{Quot}(\mathcal{H}, P)$ of all quotients $[\rho : \mathcal{H} \to F]$ with F semistable and $V \to H^0(F(m))$ an isomorphism, p. *99*.
$R^s \subset R$	subset of points $[\rho : \mathcal{H} \to F]$ with F stable, p. *99*.
$\pi : R \to M$	quotient morphism constructed with GIT, p. *102*.
$[\rho]$	$= [\rho : \mathcal{H} \to F]$, a point in R or $\mathrm{Quot}(\mathcal{H}, P)$, p. *99*.
$\tilde{\rho} : q^*\mathcal{H} \to \widetilde{F}$	the universal quotient family on $\mathrm{Quot}(\mathcal{H}, P) \times X$, p. *101*.
L_ℓ	$= \det(p_*(\widetilde{F} \otimes q^*\mathcal{O}_X(\ell))$, determinant line bundle on $\mathrm{Quot}(\mathcal{H}, P)$, p. *101*.
$R(\mathcal{F})$	frame bundle associated to a family $\mathcal{F}$, pp. *100, 93*.
$\widetilde{\Phi}_{\mathcal{F}} : R(\mathcal{F}) \to \mathrm{Quot}(\mathcal{H}, P)$	classifying morphism for the frame bundle associated to a family $\mathcal{F}$.
$\mathcal{A}$	line bundle on R arising in Gieseker's construction App. 4.A.
$\lambda : K_{c,H} \to \mathrm{Pic}(M)$	group homomorphism, p. *215*.
$u_i = u_i(c)$	classes in $K_{c,H}$, p. *219*.
$\mathcal{L}_i$	line bundles on M constructed in Chapter 8, p. *219*.
$\tau(\alpha)$	two-form on M^s, p. *265*.
$\delta(\alpha)$	one-form on M^s, p. *265*.

Index

321

Printed in the United States
By Bookmasters